THE PARALYSIS IN ENERGY DECISION MAKING

THE PARALYSIS IN ENERGY DECISION MAKING

EUROPEAN ENERGY POLICY IN CRISIS

MALCOLM GRIMSTON

Whittles Publishing

To my parents, Arnold and Ena Grimston

Published by
Whittles Publishing,
Dunbeath,
Caithness KW6 6EG,
Scotland, UK

www.whittlespublishing.com

© 2016 Malcolm Grimston

978-184995-167-8

CONTENTS

1 | INTRODUCTION – THE FATAL SCLEROSIS IN ENERGY DECISION-MAKING

1.1 AN EXTRAORDINARY COMBINATION OF PRONOUNCEMENTS

In July 2015, Ofgem, the UK energy regulator, published its annual assessment of future electricity generating capacity. Traditionally it has been regarded as prudent to keep a 'capacity margin' – a gap between the amount of dispatchable (coal, gas, nuclear or hydro) generating capacity available and the expected maximum electricity demand – of about 20% in case demand was higher than expected and a number of major plants should break down or be otherwise unavailable. Ofgem's projection of capacity margin for the following winter was as low as -1% - in other words in the worst scenario it was possible that the system could not meet projected demand even without major unexpected events. Although major power outages remained unlikely, this figure represented the biggest risk to secure supplies for 40 years.

The following week the Department of Energy and Climate Change stated that the total amount of declared power capacity held by major power producers in the UK, when intermittency was taken into account, fell by over 2GW to 76.5 GW, as older plant closed but in effect no replacement capacity came on line: this figure had fallen by over

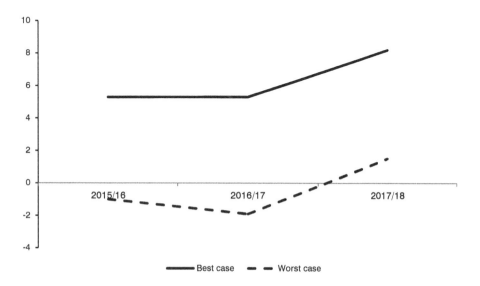

Figure 1.1 Projected UK capacity margins (%), 2014/15–2018/19 (FES: Future Energy Scenarios)[1]

7.5% in just five years. Progressive closure of coal-fuelled power stations owing to the European Large Plant Directive and nuclear plants through age over the course of the following decade would further diminish the amount of existing capacity.

In December 2014 the government announced the results of the first round of auctions for electricity generating capacity. The measure was introduced in the Energy Act of 2013 as a response to growing concerns that the liberalised power market was not building sufficient new reliable (or 'dispatchable') capacity to ensure power supplies in the medium to long term. Companies which guaranteed to make capacity available in the winter of 2018/19 would be paid £19.40 per kW, in addition to any income they might receive for selling electricity at the time. 49.3 GW of capacity was contracted through this process, at a cost to the electricity consumer of over £950 million. Only 2.6 GW of capacity yet to be built won support (of which, the biggest project, a 1.9 GW CGT at Trafford in Greater Manchester, failed to win financial backing and was effectively abandoned before the end of 2015). Another 12.9 GW went to plants which required refurbishment. The vast bulk of the total (33.6 GW) went to existing power plants, something which could clearly only be justified if there was a realistic prospect of this plant being closed before 2018/19.[2] A similar outcome was to follow in December 2015.

Taken together these three announcements painted a clear picture. The UK was rapidly reaching a point where a huge amount of new investment was required in power generation capacity. Estimates suggested a total of more than £350 billion would need to be found by 2030 to keep the power system in a fit state, not just in terms of low-carbon generating capacity required to meet environmental goals, but also with regard to the effectiveness and reliability of transmission and distribution systems, gas storage and so on.

In normal commodity markets involving heavily capital-intensive industries, impending shortages result in increasing prices, which in turn encourage investment in new capacity that ultimately brings prices down again. In the UK electricity market, this was not happening. Investors were clearly unconvinced that, despite the predicted shortage of capacity, prices would rise far enough and remain high for long enough to underpin the levels of investment required. Far from investors putting up the capital to build sufficient new plant to preserve a workable capacity margin, the power consumer was being required to find nearly £1 billion just to prevent companies closing down their existing plant 'early'.

This was obviously not a market which was working. There was a patent need for very rapid decisions on investment, given the amount of time it takes to design and finance a project, receive necessary permissions, construct it and connect it to the grid; yet no major new power plants began operation in 2014. Nor was it a market where companies were making an inflated return on capital. UK end user power prices remained below the European Union average and well below those in western EU nations, such as Denmark, Germany, Italy or Spain.[3]

Power stocks, not just in the UK but across Europe, had been among the worst-performing of all the industrial sectors.[4]

Yet, the political rhetoric, as the 2015 election approached, seemed based on a very different analysis. At the Labour Party Conference in September 2013, leader Ed Miliband enthused delegates as follows:

Take the gas and electricity companies. We need successful energy companies in Britain. We need them to invest for the future. But you need to get a fair deal and frankly, there will never be public consent for that investment unless you do get a fair deal. And the system is broken and we are going to fix it.

If we win the election 2015 the next Labour government will freeze gas and electricity prices until the start of 2017 … Now the companies aren't going to like this because it will cost them more but they have been overcharging people for too long because of a market that doesn't work. It's time to reset the market. So we will pass legislation in our first year in office to do that … .[5]

In the event, Mr Miliband was unsuccessful in that election and the more business-friendly Conservatives won an overall majority. But although the price freeze was dead and buried, the chief whip in the previous Conservative-led coalition government, Michael Gove, had said that Miliband was 'absolutely right' to warn about energy price rises and said of the energy companies: 'I do take what they say with a pinch of salt actually. The way in which the major energy companies have behaved in the past does not give me confidence in everything they say.'[6]

Of course, if companies are not making a reasonable rate of return on their existing assets and are facing hostile comments from government and opposition alike, the appetite among potential investors to make the decisions necessary to deliver a power system needed by a modern economy is likely to be severely blunted. If the power market had been 'working' in 2003, for example, then there is a real likelihood that not just 40 per cent of the generating capacity would have gone bust, as actually happened, but the whole system would have followed it.

The aim of this book is to look at how the UK and other countries had found themselves in such an impasse, the potential consequences, and possible ways out of the sclerosis before it leads to an inevitable thrombosis in the arteries of a modern economy for which electricity has become the lifeblood.

1.2 The lost art of decision-making

Countries like the United States and those of western Europe were not always so inept at taking decisions necessary to protect their long-term interests and create a better tomorrow for generations yet to come.

The first electricity from a wind turbine was generated in 1887 – by 1900 there was around 30 MW of wind generating capacity in Denmark, while by 1908 there were 72 wind-driven power stations operating in the US.

In 1926 the UK prime minister, Stanley Baldwin, introduced the Electricity (Supply) Act, creating the Central Electricity Board. Within five years 4,000 miles of transmission cable connecting 122 power stations had been laid despite protests by 'impractical aesthetes', as they were dubbed by *The Times*, including Rudyard Kipling, John Maynard Keynes, Hilaire Belloc and John Galsworthy.

In 1936 the US Congress endorsed Roosevelt's Rural Electrification Act, part of the New Deal, which was designed to assist the American people during the Great Depression – for example, before the Act only 10 per cent of people living in south-eastern Ohio, had access to electricity in their homes and businesses while by 1939 it had reached 50 per cent.[7] These measures rendered variable generators such as wind turbines redundant and heralded the new age of electricity.

Between 1967 and 1977 Britain converted its entire gas industry from town gas to natural gas, an enormous project.

The first electricity generated by nuclear fission lit four 50 W light bulbs in December 1951: in August 1956 the world's first commercial-scale nuclear power station began operating at Calder Hall in the UK.

By contrast, in January 2008 the government of the day said that it had 'concluded that nuclear should have a role to play in the generation of electricity'. Dictionaries offer 'decide' as a synonym for 'conclude'; but if by conclude the government meant 'determined that something would happen' then they would be sorely disappointed: it took almost eight years even for a realistic funding arrangement to emerge, involving major investment from China.

The distrust between the political and industrial communities which underlays the potentially impending capacity crisis of the mid-2010s is not the only contribution to the decision paralysis that has beset electricity generation and indeed many other activities, such as climate change mitigation, in recent years. Other factors include:

- the arrogance and overplaying of their hand by practitioners of science and technology, almost behaving like a new priesthood, leading to disillusionment;
- an obsession with technological possibilities, notably, for example, in the field of nuclear safety, almost irrespective of whether considerations of proportionality or economics, which has both created an unwarranted fear of radiation and added unsustainability to costs, in effect pricing the technology out of many markets;
- the degradation of the role of evidence-based scientific findings in favour of the apparently more 'compassionate' anti-industrial movement and the anecdotal testimony of individuals;
- a shift in the broad societal ethic from a utilitarian one to a more rights-based one, following a long period of relative prosperity and calm;
- a clever campaign by Big Green, the multinational association of large 'environmental' corporations, of portraying ancient, failed approaches, such as variable renewables and energy efficiency, as 'new' and easy to exploit, only having failed to flourish because of an oil-funded conspiracy and lack of funding;

- the muddying of what a 'decision' actually is in a liberalised power market, as opposed to a centralist approach dictated by the state, and who is responsible for taking such decisions.

1.3 ENERGY RETURNS TO THE POLITICAL AGENDA

Around the turn of the millennium 'energy' as an issue hardly made an appearance on the national agenda. The Labour Party manifesto for the 2001 general election, *Ambitions for Britain*, a document of over 20,000 words, devoted just 211 of those words to energy. The thrust was hardly radical, focussing mainly on environmental and social concerns with little mention, for example, of security of supply:

> Labour is committed to a secure, diverse and sustainable supply of energy at competitive prices. We have brought full competition to the gas and electricity markets. Coal and nuclear energy currently play important roles in ensuring diversity in our sources of electricity generation. We are putting an obligation on electricity companies to deliver 10 per cent of the UK's electricity from renewable sources by 2010 … We will consider setting further targets for renewables supported by a £100 million fund. It will back up the Climate Change Levy, which includes agreements to improve efficiency in energy-intensive sectors, and the new Carbon Trust … We will support research into clean coal technology and investigate its commercial possibilities. We will double the expenditure on energy efficiency. Fuel poverty blights lives: our aim is that by 2010 no vulnerable household in the UK need risk ill-health due to a cold home.

The more modest Conservative manifesto in that year, *Time for Common Sense*, a mere 13,000 words plus, boasted one reference to a review of nuclear power and 49 words about energy's role in environmental protection:

> The biggest global environmental challenge is to prevent climate change causing long-term damage through extreme weather conditions. We will meet the commitments made by successive British governments by a comprehensive package of emission permit trading, energy conservation measures, tax incentives, greater encouragement of renewable energy and cleaner energy generation.

The environment had perhaps been a deeper theme in Liberal Democratic Party pronouncements than in the two largest parties. Its 2001 manifesto – *Freedom, Justice, Honesty* – the longest at over 21,000 words, does refer to energy about ten times, but again in reference to tariff and tax reform, energy efficiency, boosting renewables and phasing out nuclear power as ways of reducing fuel poverty and greenhouse gas emissions. A flavour can be gained by a few quotes:

We will improve health by requiring an energy efficiency audit to be completed on all homes before sale to promote high energy efficiency standards throughout the housing market ... Our homes insulation programme, to provide decent levels of home insulation within 15 years by speeding up the current 30year programme funded by the energy utilities, will also cut fuel bills ... Our top priorities are to reduce energy consumption overall, improving the efficiency with which it is used, and to switch from polluting forms to clean energy sources ... We will require a minimum of 10 per cent of the UK's energy to be generated from UK-based renewable energy sources by 2010, increasing by 1 per cent a year thereafter. As well as benefiting the environment, this will create thousands of new jobs in the green energy sector. We will ensure that local authority structure plans incorporate targets for CO2 emission reductions to encourage the development of renewable energy facilities and account for the climate change consequences of their policies, including transport.

In effect energy, rather than being treated as the multifaceted, complex and fascinating issue that it is, became relegated to a supporting role in other debates.

1.4 THE ENERGY TETRALEMMA

Energy, and perhaps electricity in particular, has often been characterised as presenting a 'trilemma'. Our energy supply systems need to provide energy which is:

- secure – energy outages, especially power cuts, have potentially severe detrimental effects on a modern economy, in all kinds of social, financial, industrial, health and other ways;
- economic – high energy costs are damaging to an economy's international competitiveness, while high energy prices can seriously undermine quality of life, especially for those on low or fixed incomes;
- environmentally acceptable – climate change is the biggest, most salient and urgent issue in the environmental arena if current scientific thinking is to be believed; but there is a wide range of other potential environmental consequences of energy use, at a local (air quality, visual intrusion or noise from power plants and power cables or pipelines, managing waste heat, and so on), regional (acid rain, radioactive emissions from a major nuclear accident) and global (climate change) level.

In reality energy might be better described as presenting a tetralemma. As well as addressing the above requirements, energy policy must also be socially and politically acceptable. Many policies which have made sense with regard to one or more of the requirements have been rejected, either at point of origin or later, because of public protests or changes in political fashion.

One reason why energy may have dropped off the political radar during the late 1990s was because the UK found itself in the rare position that a single policy – the dash for gas – was delivering positive outcomes against all four horns of the tetralemma. The UK had its own North Sea gas reserves – indeed, it was a net gas exporter for a decade from 1994 – and the structure of the new, progressively liberalised electricity market was encouraging the construction of large amounts of gas-fired generating capacity, thereby finally breaking the power of the National Union of Mineworkers (NUM) which had so threatened the security of coal and electricity supplies for many decades. Not only did gas continue to penetrate the heating market, it also came from nowhere in 1990 to generate some 40 per cent of the nation's electricity in 2000, while coal's share fell from 75 to 30 per cent. As far as economics was concerned, gas was cheap and a new, highly efficient electricity generation technology, Combined Cycle Gas Turbine (CCGT), was emerging from its military origins. The introduction of competition into electricity generation and supply, at a time when there was, if anything, a surplus of generating capacity, drove wholesale electricity prices down below levels at which many companies could cover their operating costs, let alone consider investment in new capacity – a problem not recognised, as the political manifestoes of that time demonstrate. Environmentally, the dash for gas delivered a welcome side benefit of reducing greenhouse gas emissions from electricity production, CCGT emitting less than half as much carbon dioxide per unit of electricity delivered as the coal-fired plants it was replacing. And gas, a familiar feature of everyday life since the country converted its gas system from 'town gas', a rather noxious mixture including hydrogen, carbon monoxide and methane, to much cleaner natural gas in the late 1960s and early 1970s, was not subject to the societal protests that have accompanied, for example, nuclear power stations or wind farms. (Although the former proved particularly popular with the Conservative administration of the 1980s and 1990s, many of whose key characters, especially the prime minister, Margaret Thatcher, remembered the role that the NUM had played in bringing down the Conservative administration of 1970–74, this verbal support did not translate into a major new construction programme.)

This brief moment was neither typical nor sustainable. In the 1970s, for example, security of supply and soaring costs in the wake of the oil shocks of 1973 and 1979 and the outfall of miners' strikes in the UK during the winters of 1971/2 and 1973/4 kept energy in a much more prominent place on the political agenda. Fears over security of supply led to several countries following policies that increased costs and levels of greenhouse gas and acidic emissions, notably by switching to coal from the marginally less polluting and previously considerably cheaper oil for power generation. More generally, maintaining sufficient reliable generating capacity to fulfil demand at peak times requires servicing plant for long periods when it will not be needed for generation because electricity cannot be stored in significant quantities and demand varies very considerably during the day and throughout the year. By the 2010s energy was edging back onto the political agenda to become a major issue for the 2015 general election and beyond as prices once again rose.

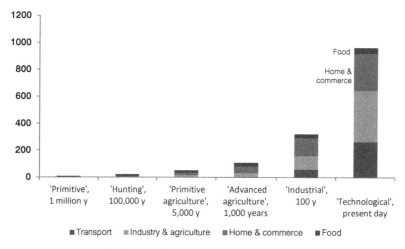

Figure 1.2 Energy use per capita (kJ) at various stages of history[8]

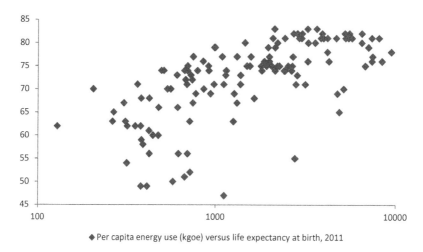

Figure 1.3 Per capita energy use (toe) versus life expectancy at birth, 151 countries, 2011[10]

1.5 THE IMPORTANCE OF BEING ENERGETIC

Perhaps inevitably when energy, for a considerable period of time, seemed no longer to be a major challenge in developed countries including the UK, its vital importance to a modern economy somewhat slipped from sight.

Yet a global perspective shows the true picture. The rise of modern society has been based on the growing availability of energy. A widely-cited study from the early 1970s estimated energy consumption for humans at various stages of development, as shown in figure 1.2.

Figures from the World Bank show that in 2011 the unweighted average life expectancy among those countries with a per capita energy use of less than 0.5 tonnes of oil equivalent (toe) per year was 62.5 years, while among those using more than 5 toe per capita per year life expectancy was 78.8 years.[9] (See also figure 1.3.)

In countries with below global average life expectancy at birth the average is 66 years and per capita energy use is 74 per cent of the global average; in the countries with above average life expectancy the average is 78 years and per capita energy use is 212 per cent.

The difference in infant mortality rates is even more striking. Among those countries with an average annual energy use of below 0.5 toe per capita, unweighted average infant mortality rate in 2011 was 44 per 1,000 infants dying before their first birthday, while the unweighted average in countries using more than 5 toe per capita per year was six per 1,000.

Interruptions to electricity supplies are especially damaging. In 2014 the UK prime minister's Council for Science and Technology published a report on the consequences of falling short of fulfilling electricity demand.[11] Its main message was that the UK was now so dependent on its electricity networks and electronic communications systems that any significant interruption to the electricity supply would have even more severe economic consequences than had been the case during the previous widespread power outages 40 years earlier, potentially running into billions of pounds. The pace of change was so great that understanding of the potential magnitude of social and economic impacts was constrained by limited knowledge about the knock-on consequences that could occur across different sectors of the economy and society.

A nationwide blackout lasting for longer than 48 hours could have a severe impact on society. There are obvious political impacts as well – our historical high levels of supply in the UK mean we are accustomed to continuous power supply. Current estimates show that we need to spend upwards of £200 billion over the next decade to update and decarbonise our electricity supply infrastructure. We have not invested on that scale since the 1960s and we do not have a good yardstick to compare the potential cost of infrastructure investment with the cost to society of major power outages.

The failure to take timely decisions ultimately risks a national disaster of enormous consequences. However, experience from Japan in the wake of the Great East Japan earthquake and tsunami in 2011 and consequent Fukushima nuclear accident, following which the nation closed its nuclear power plants that had been generating nearly 30 per cent of the country's electricity in the previous year, suggests that power outages can be prevented for several years by a combination of reduced demand for energy from industry and households and a major increase in the use of fossil fuels. The balance of payments suffered enormously, going into deficit for the first time in over 30 years. The fuel cost of generating electricity doubled until the fall in the oil price came to the rescue

– prices for small and medium businesses were almost 30 per cent higher four years after the earthquake than they had been at the time, while the power companies were losing money at an unsustainable level. Greenhouse gas emissions targets were simply abandoned. Something similar was seen in Germany which also took a decision to phase out its nuclear plants – power prices were the second highest in Europe, greenhouse gas emissions rose as brown coal use reached its highest levels of use since reunification in 1990 and Germany was regularly dumping excess production onto its neighbours, often at negative price. But despite a few near misses, the system did not collapse.

So the tetralemma is not a fellowship of equals. When trouble strikes it seems that environmental measures are the first to be abandoned – or to put it another way, unless secure supplies can be provided at manageable cost, emissions may well continue their inexorable global rise which has seen carbon dioxide emissions grow by well over 50 per cent since the Rio Convention called for serious action against climate change in 1992. High prices can be sustained for some time if they are regarded as the necessary price for keeping supplies secure. Maintaining reliability clearly trumps both of these. But, perhaps, ultimately it is the political and social dimension which takes first place. Both Japan and Germany could have lowered costs, improved security of supply and reduced greenhouse gas emissions by continuing to operate their nuclear power stations at least until the end of their economic and technical lifetimes – it was a purely political decision not to do so.

Furthermore, the default position for energy production is the use of fossil fuels. Storable, transportable, reliable, relatively cheap and easy to use at small or large scale, their short-term advantages are considerable while some of their disadvantages lie sometime in the future. They are eventually limited. Although the periodic doomsaying by a variety of vested interests who would like governments to believe that 'peak oil' is just round the corner has rather discredited the idea, one day they will become sufficiently limited or expensive that alternatives will need to be found purely on resource grounds. They are also the main contributors to climate change.

This book will survey the history of our use of major approaches to fulfilling energy requirements – fossil fuels, nuclear power, renewables, and energy efficiency – to determine what the past tells us about some of the more fanciful claims made today which are acting as a barrier to effective decision-making. It will look at some key issues that are relevant to energy policymaking as a whole – the organisation of power markets; the interaction among the political, scientific/technical and public spheres; the challenge of climate change; the German Energiewende; and the situation in Japan after the Fukushima nuclear accident.

It will look at the pernicious influence of Big Green, which seeks to pervert the scientific process to its own interests in the same way that Big Business has done and still does; how it stands in relation to the large number of honest local environmental groups as some multinational business corporations do to the vast number of honest local businesses; and at the technocratic 'product focused' establishment that has dominated decision-making, notably in the nuclear field, and has ended up creating mistrust and unnecessary costs, exemplified in the changing public relationship with radiation.

Finally it will suggest that it may be possible to create a society which is neither in thrall to science and technology, thereby allowing necessary decisions to be taken but not subjecting those decisions to a level of scrutiny and resulting in systematic biases in that decision-making; nor so paralysed by attacks from the anti-industrial movement, including Big Green, that appropriate decisions do not get taken and inevitably fossil fuels win out. To get there, however, will require a level of leadership and open-mindedness on all sides, which is hard to foresee.

Endnotes

1 https://www.ofgem.gov.uk/ofgem-publications/88523/electricitycapacityassessment2014-fullreportfinalforpublication.pdf, Ofgem, 'Electricity Capacity Assessment Report 2014', 30 June 2014.

2 https://www.gov.uk/government/uploads/system/uploads/attachment_data/file/389832/Provisional_Results_Report-Ammendment.pdf, National Grid, 'Provisional auction results', December 2014.

3 http://ec.europa.eu/eurostat/tgm/table.do?tab=table&init=1&language=en&pcode=ten00117&plugin=1, Eurostat (2015), 'Electricity prices by type of user'.

4 http://www.rense.com/general96/euopres.html, McKillop A (2014), 'Europe's energy transition paradox', rense.com.

5 http://www.newstatesman.com/politics/2013/09/ed-milibands-speech-labour-conference-full-text, 'Ed Miliband's speech to the Labour conference: full text', New Statesman, 24 September 2013

6 http://www.independent.co.uk/news/uk/politics/michael-gove-says-ed-miliband-right-to-attack-energy-companies-8844864.html, Morris N, 'Michael Gove says Ed Miliband right to attack energy companies', *The Independent*, 27 September 2013.

7 http://www.ohiohistorycentral.org/w/Rural_Electrification_Act, 'Rural Electrification Act', *Ohio History Connection*.

8 http://www.scientificamerican.com/article/the-flow-of-energy-in-an-industrial/, Cook E (1971), 'The flow of energy in an industrial society', *Scientific American*, 225 (3).

9 http://data.worldbank.org/, World Bank (2015), 'Open Data'.

10 http://rogerpielkejr.blogspot.co.uk/2013/02/graph-of-day-life-expectancy-vs-energy.html, Pielke R, 'Graph of the day: life expectancy vs. energy use', *Roger Pielke Junior's blog*, 14 February 2013.

11 http://www.raeng.org.uk/publications/reports/counting-the-cost, Walker A, et al., 'Counting the cost: the economic and social costs of electricity shortfalls in the UK', Royal Academy of Engineering, November 2014.

2 | Fossil fuels

In a way the fossil fuels represent an obvious place to start a book on energy. Notwithstanding the argument, the enthusiasm, the antagonism, the column inches, the subsidies, the agonising over alternative forms of energy in the shape of nuclear power and renewables, the fossil fuels – oil, coal and gas – continue to reign supreme. In 2014, for example, global traded primary energy was sourced as shown in figure 2.1.

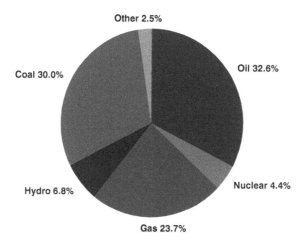

Figure 2.1 Global traded primary energy by fuel, 2014

In the two decades since the Rio Convention saw the world committing itself to a low-carbon future, the fuel which had shown the greatest growth, by some way, was coal, increasing its consumption by over 75 per cent and its market share from 26.2 to 30.0 per cent. Taken together, the share of primary energy provided by oil, coal and gas fell from 94.0 per cent in 1973 to 87.1 per cent in 1993 – while the global nuclear programme was taking hold – and then effectively remained constant, standing at 87.4 per cent in 2003 and 86.3 per cent in 2014, as Big Green and others decided that the best use of growing renewables was to reduce the use of nuclear power rather than to tackle climate change – see figure 2.2.

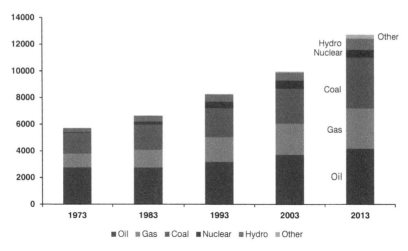

Figure 2.2 Primary energy consumption by fuel (mtoe), 1973–2013

2.1 A HISTORY OF FOSSIL FUEL USE – COAL

For most of mankind's development humans were dependent on what we would now call renewables – of low power density, resource intensive to collect and sometimes unreliable, but not unsuited for a lifestyle in which most of the relatively tiny world population, with the exception of the very rich, were used – devoting much of their short lives to foraging to collect enough food and of course avoiding predators, a situation which sadly persists in some areas today, though the predators may be different.

To break out of this cycle required new sources of energy which offered the prospect of reliability and a much higher power density than can be found in wind, water, sunlight or hot springs.

The first to step into this role was coal. Wood has an energy density of around 16 MJ per kg; coal's is typically around 24 MJ per kg. As well as yielding 50 per cent more energy mass for mass than wood, coal can be collected in greater quantities from a suitable source – trees require a certain distance between them to flourish – and with less damage to the local environment.

In fact there are several types of coal. As the peat from which coal is formed is compressed it is first converted into lignite or 'brown coal', which is relatively soft with a colour that can range from black to various shades of brown. Lignite is generally regarded as the dirtiest coal, having the lowest calorific value per unit of carbon dioxide produced. Over millions of years the continuing effect of temperature and pressure produces further change in the lignite, progressively transforming it into the range known as 'sub-bituminous' coals with higher calorific values. Further chemical and physical changes occur until these coals became harder and blacker, forming 'bituminous' or 'hard' coal. The process can continue until anthracite, the highest grade of coal, is formed.[1]

In addition to carbon, coals contain hydrogen, oxygen, nitrogen and varying amounts of sulphur. High-rank coals are high in carbon and therefore heat value, but low in hydrogen and oxygen. Low-rank coals are low in carbon, but high in hydrogen and oxygen content.

In China evidence of surface mining and use of coal goes back to 4000 BC. There was an underground coal mining industry by the time of the Han Dynasty (third century BC to third century AD) and coal briquettes and ashes have been found in the ruins of houses in Fushun in the north-east of the country. The ancient compilation of myth and geography known as the Shan Hai Jing described the Bingjingtai Coal Mine, built by statesman Cao Cao in AD 210 in what is now Henan Province. The mine was 50 metres underground, with an annual output of several thousand tonnes. Several large coal mines have been found dating back to the Song Dynasty (tenth to thirteenth century AD): a dynasty writer recorded that several million people were using coal as fuel instead of firewood.

In Europe the Greek philosopher Theophrastus (371–287 BC) refers to metalworking using coal and the Romans exploited many coalfields in the UK between the second and fifth centuries AD. By this time coal was being used across Europe and the Middle East. However, the use of coal in Britain seems practically to have died out when the Romans left in AD 410 and the country sank into what has become known as the Dark Ages, only re-emerging in the late eleventh century. Coal, often known as 'sea coal' as it was won from carboniferous seams that were uncovered by marine erosion in coastal areas, was traded in Scotland and north-east England from the time of the Magna Carta in 1215. The first charter to mine coal was granted to Newcastle-upon-Tyne by Henry III in 1239. By the fourteenth century coal was being used for domestic heating and as early as 1306 the burning of coal was causing such pollution to London air that a Royal proclamation was issued prohibiting the use of coal for industry and requiring artisans to return to traditional fuels such as wood and charcoal. By the sixteenth century coal was increasingly replacing wood as supplies of the latter ran short. More sophisticated deep mining techniques were developed through the sixteenth to eighteenth centuries.

It was the coming of the Industrial Revolution that saw the use of coal expand enormously, as shown in figure 2.3.

The stimulus to this rapid growth in coal use was the development of the steam engine. The use of boiling water to create motive power goes back over 2,000 years but it was in the seventeenth century that steam power began to become a reality when major improvements in energy efficiency made increased energy use a practical proposition. The first steam engine was patented in 1606 by the Spanish soldier, painter, musician and inventor Jerónimo de Ayanz y Beaumont and was designed to pump water out of mines. This created a powerful synergy – the steam engine both created a greater demand for coal and made coal mining to fulfil that demand more efficient. In 1698 engineer Thomas Savery, whose inventions included a machine for polishing glass or marble and another

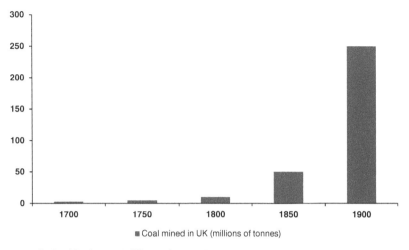

Figure 2.3 Coal mined in the UK (millions of tonnes), 1700–1900[2]

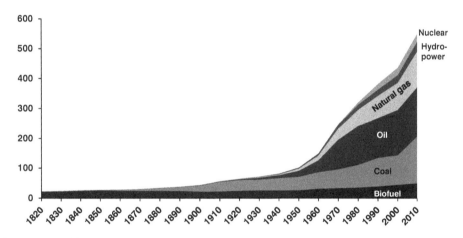

Figure 2.4 World energy consumption (Exajoules per year), 1820–2010[3]

for 'rowing of ships with greater ease and expedicion than hitherto beene done by any other', patented a steam pump that used condensing steam to create a partial vacuum and draw water into a chamber where pressurised steam would pump the water out.

Perhaps the breakthrough came around 1710 when Thomas Newcomen, sometimes called the 'father of the Industrial Revolution', introduced a piston into the workings of his 'atmospheric engine'. By 1733 some 125 of Newcomen's engines were operating in the UK.

In the Newcomen engine a high proportion of the energy released simply went on reheating the piston between strokes. James Watt overcame this and other problems around 1775, developing a separate steam condensation chamber and surrounding the piston with a steam jacket which would maintain its temperature. Watt also developed an engine that produced a rotary motion that could be used for a much wider range of purposes than Newcomen's rising and falling motion. As discussed in the chapter 5, this significantly increased the thermal efficiency of steam engines and acted as a further incentive for economic growth and increased energy demand.

Coal dominated global energy use, at least in the developed world, throughout the nineteenth century – see figure 2.4. By 1864, the first year for which annual figures seem to be available, global coal production had reached around 180 million tonnes per year. From this date until 1914 production grew at an average 4 per cent per year to reach 1,300 million tonnes, doubling every 17 years.[4] By the end of 1947 the cumulative production of coal during human history had amounted to approximately 80 billion tonnes. Of this around 40 billion tonnes had been mined and consumed since 1920, and over 60 billion tonnes since 1900. The growth of ready energy was accompanied by increased life expectancy, from about 40 years old in 1800 to over 50 by 1900, and unparalleled economic expansion. Mining in the UK peaked at almost 300 million tonnes in 1913, employing well over a million miners.

The advantages that coal could bring in terms of storability and energy density gave it an insuperable edge over the renewable technologies it supplanted with the exception of hydropower, as discussed in chapter 4. The coming of large-scale electricity production in the mid twentieth century created a new market. In 2014 global coal production stood at a near-record 8,165 million tonnes, nearly double its level of 30 years previously. Almost half was being mined in China alone, whose use had quintupled over the 30-year period.

But coal was under challenge from a 'newcomer' that offered not only a power density almost twice as high (44 MJ per kg for petrol) but had another great advantage – it is a liquid. Unlike coal, it does not require hacking and hewing, or the modern equivalent, and it can be pumped both over long distances and through the pipework of engines and machinery. It has a high calorific value per unit volume when stored under normal conditions of temperature and pressure, unlike gas which requires a high pressure and/ or low temperature to store it as a liquid, and is easier to collect from the oil well, being less volatile.

2.2 THE EMERGENCE OF OIL

The use of crude oil, as a source for building materials, such as asphalt, and for lighting, can be traced back more than 4,000 years to Babylon, Persia and ancient Rome. China had oil wells by AD 350, the oil being used largely to evaporate seawater to produce salt. Petroleum was known as 'burning water' in seventh century Japan. In 1088 Shen Kuo of the Song Dynasty coined the term 'rock oil', or in Latin 'petra-oleum'. Distillation seems to have reached western Europe from the Middle East in the twelfth century, and extraction and use of oil continued in a modest way during the following centuries. For example, oil sands were mined from 1745 in Alsace.[5]

It was in the mid nineteenth century that oil began its century-long journey to dominance in the world's energy constellation. In 1847 Scottish chemist James Young, the 'father of the modern oil industry', noticed natural petroleum seepage in the Ridding Colliery in Derbyshire. He distilled a light liquid for use in lamps and a heavier material for lubricating machinery, setting the scene for Benjamin Stillwell to invent oil refining at Yale University in 1854 and the building of the first modern refinery in Alsace in 1857. Refining involves separating the complex mixture that represents the raw material into several 'fractions' containing different numbers of carbon atoms with different densities and properties and which are suitable for different purposes, from combustion for light and heat to liquids for lubrication and solids for construction, waterproofing etc. (Hydrocarbon fuels, though containing many impurities such as sulphur compounds, mainly conform to the general chemical formula C_nH_{2n+2} – the higher the value of 'n' the denser, the less volatile and the less flammable the substance. Methane gas, the simplest, has the formula CH_4 while a bitumen molecule may contain between 400 and 600 carbon atoms.) The products so produced have a much wider range of uses than the original unrefined material; for example, the lighter fractions burn much more cleanly and efficiently than crude oil itself and represent a cleaner and cheaper alternative to traditional fuels such as whale oil used for lighting. At first the main source of oil was seepage from the ground or from coal mines, but the sinking of drilled oil wells from the mid-1850s increased the available resource hugely, as shown in figure 2.5.

Impressive though these increases appear, in 2014 the US was producing over 11.5 million barrels of oil per day (from a world total of 89 million barrels per day) and consuming almost two-thirds as much again. At the turn of the nineteenth and twentieth

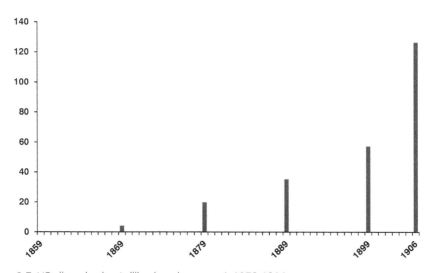

Figure 2.5 US oil production (million barrels per year), 1859–1906

centuries coal still remained supreme for transportation, through the steam locomotive, space heating, industry and, alongside hydropower, the beginnings of the production of electricity. Again to give a perspective, cumulative global oil production by the end of 1947 was 58 billion barrels, half having been produced and consumed since 1937 and 97 per cent since 1900.

In the early years of the twentieth century oil found a major new niche within the realm of transportation. As early as 1882 Britain's Captain Jack Fisher, later to be Admiral Lord Fisher, argued that Britain must convert its naval fleet from bulky coal-fired propulsion to the new oil fuel. He argued that a battleship powered by petroleum issued no tell-tale smoke, while a coal ship's plume was visible up to 10 km away. It required between four and nine hours for a ship's coal-fired engines to reach full power while an oil motor required a mere 30 minutes. To provide oil fuel for a battleship required the work of 12 men for 12 hours: the same equivalent of energy for a coal ship required the work of 500 men for five days. For equal horsepower propulsion the oil-fired ship required one-third the engine weight and little more than a quarter of the daily tonnage of fuel. The radius of action of an oil-powered fleet was therefore up to four times as great as that of the comparable coal ship.[6] One of Winston Churchill's earliest decisive acts on becoming First Lord of the Admiralty in 1911 was to ensure that this transition was carried out, despite the risk associated with Britain's lack of oil reserves.

Subsequently, the development of more energy-efficient internal combustion engines brought motoring within the financial reach of growing numbers of people – as is so often the case improvements in energy efficiency led to a huge increase in the use of energy. Oil replaced coal as the world's largest source of energy in the 1950s, even making inroads into electricity generation.

The geopolitics of oil was never far from the horizon. Britain's dependence on imports before the discovery of its own North Sea reserves in the mid-1960s led to international humiliation during the Suez Crisis of 1956. The Suez Canal had been opened in 1869 and assumed enormous strategic importance as the only direct access for shipping between Africa and Asia. By the 1880s it was effectively under British ownership and control, and the fading Ottoman Empire agreed that it should become a neutral zone under British protection and that international shipping should be allowed to pass freely through it. Initially vital as a link between Britain and its colonies in the Far East in particular, this role became less important with the dissolution of the British Empire after the Second World War. However, the canal was to gain a new role as a conduit for oil. By 1955 two-thirds of Europe's oil imports from the Middle East came through the canal, representing half of the canal's traffic.[7] When the revolutionary government of President Gamal Abdel Nasser nationalised the canal in July 1956 Israel, France and Britain launched military action, only to have to back down under pressure from the US and the Soviet Union and loss of political support at home and internationally. The crisis cost the prime minister, Anthony Eden, his long-awaited job after just two years and besmirched his reputation forever.

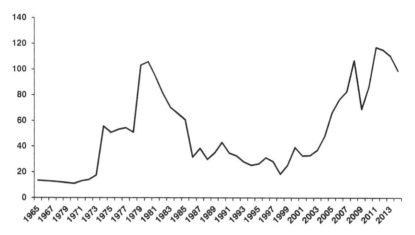

Figure 2.6 Oil price in US$ (2014 values – 1965–83 Arabian Light, 1984–2014 Brent dated)

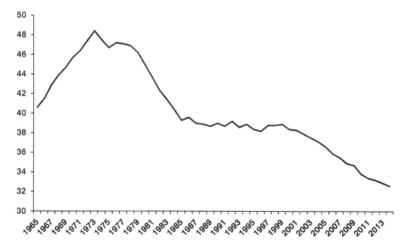

Figure 2.7 Oil consumption as a percentage of global traded primary energy use, 1965–2014

In constant 2013 dollar values the price of crude had typically been in the $10 to $25 per barrel range for most of the twentieth century. The emergence of OPEC (the Organisation of the Petroleum Exporting Countries) in 1960, which was dominated by countries of the Middle East, Africa and South America, would change this decisively. In October 1973 OPEC declared an oil embargo in response to the United States' and western Europe's support for Israel in the Yom Kippur War. The result was a rise in oil prices from $3 to $12 per barrel (in money of the day). Outfall included the rationing of petrol in the US, consumer panic reaction and a decline in the value of the US dollar. The Iranian Revolution of 1979 led to another bout of oil price increases, as shown in figure 2.6.

The 1970s oil shocks had a profound influence on the global economy, causing an overall slowdown of economic activity which in itself acted to reduce demand growth and create something of a downward pressure on prices, as well as prompting many governments to diversify away from oil as an energy source, by promoting nuclear power, renewables, coal and, increasingly, natural gas as alternatives. Oil's share of global primary energy fell from its peak of 48.4 per cent in 1973 to 32.6 per cent in 2014 – see figure 2.7.

However, in absolute terms this still represented a 50 per cent increase, as shown in figure 2.8.

In due course relations within OPEC became strained; Saudi Arabia increased its output considerably as did non-OPEC producers increased their production. From the mid-1980s onwards the oil price gradually collapsed back to its pre-crisis levels.

Oil has remained a source of international tension, most notably in the first Gulf War of 1990/1 following the Iraqi invasion of Kuwait. Although unconventional oil reserves were being discovered and exploited, by the end of 2013 just eight countries – Venezuela, Saudi Arabia, Canada, Iran, Iraq, the Russian Federation, Kuwait and the United Arab Emirates – held over 80 per cent of the world's proven oil reserves of 240 billion tonnes, as shown in figure 2.9.

The mismatch between production and consumption demonstrates the key position which the Middle East continued to hold, as shown in figure 2.10.

In the early twenty-first century oil prices were once again to increase, actually exceeding the peaks they had reached in 1979 and crossing $147 per barrel in mid-2008. The price remained at or above the $100 per barrel mark until mid-2014 when it began another dramatic decline, halving within six months.

2.3 NATURAL GAS

Natural gas consists primarily of methane, along with other gaseous hydrocarbons and various impurities including mainly carbon dioxide – it is in effect a gaseous version of oil.

The ancient Greeks were among those awed by mystical flames coming from the ground, often ignited by lightning, one of the most famous being found on Mount Parnassus around 1000 BC. The temple built at the site of this sacred flame housed the priestess known as the Oracle of Delphi who pronounced prophesies vouchsafed to her by the flame itself.

The Chinese were using natural gas alongside oil to produce salt by 500 BC[8], and by AD 300 were using natural gas for lighting and heating, the gas being transported in bamboo pipelines similar to the ones which had been developed for the use of oil.

It was to be many centuries before the use of gas became routine. As cities grew through the Middle Ages, street lighting became an increasingly important issue, not least in an attempt to reduce crime levels. In 1417 Henry Barton, Lord Mayor of London, required lanterns to be lit on evenings throughout November, December and January, while the first ordinance to light Paris was issued in 1524. However, most lighting was provided by using oils extracted from animals or plants. Meanwhile the growth of coal

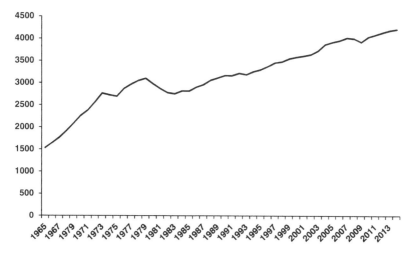

Figure 2.8 Global oil consumption (million tonnes), 1965–2014

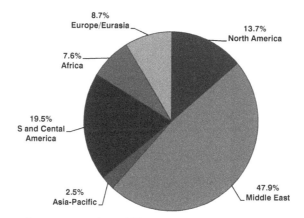

Figure 2.9 Proven oil reserves (total 240 billion tonnes), 2014

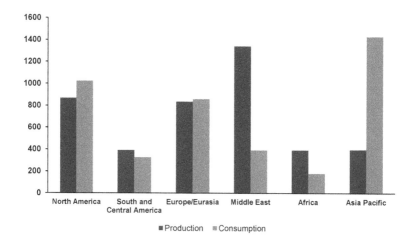

Figure 2.10 Oil production and consumption (million tonnes per year), 2014

mining was accompanied by several explosions involving gas emitted from the mines, notably 'firedamp', a mixture containing methane and a number of impurities. At this stage such gas was regarded as a nuisance rather than a potential fuel.

Interest in using gas for lighting began to grow in the 1790s, when William Murdoch, an employee of James Watt, started to experiment with various forms of flammable gas, eventually settling on 'coal gas', a noxious mixture containing hydrogen, methane, carbon monoxide and non-combustibles such as nitrogen. He lit his own house in Cornwall in 1792 and the Soho Foundry in 1798, installing the first outdoor gas lighting in 1802. Public street lighting was demonstrated in Pall Mall in 1807, and the London and Westminster Gas Light and Coke Company became the world's first chartered gas business. Gas lighting allowed for not only safer streets but also longer winter working hours in factories, even allowing for 24-hour working in some cases. Westminster Bridge was being lit by gas in 1813. Paris adopted street lighting using gas in 1820. Baltimore became the first US city with gas lighting in the 1810s.

Using gas for lighting costs only about a quarter as much as oil lamps or candles. By 1859 there were over a thousand gasworks in the UK making 'town gas' – as manufactured coal gas became known. By the time town gas was finally abandoned in the 1980s, between 12,000 and 20,000 production sites had been created[9] – a fine example of distributed energy, but one which was to collapse through the inherent inefficiencies of that approach.

The first emergence of gas proved to be something of a false dawn. The gas was poisonous, principally because of the carbon monoxide, but also because of hydrogen sulphide, a foul-smelling 'bad eggs' impurity. The lack of an efficient central production and distribution network worked against the economics and when electricity emerged on the scene in the late nineteenth century the use of gas for lighting declined rapidly, although gas was still commonly used for cooking.

Methane had first been isolated by Alessandro Volta in 1776, after he read a paper by Benjamin Franklin concerning 'flammable air'. The first dedicated natural gas well was sunk in 1821 in New York by William Hart, 'the father of natural gas'. Through the nineteenth century natural gas was used for street lighting close to where it was produced; Robert Bunsen invented his famous burner in 1885 which greatly expanded the useful purposes to which natural gas could be put. However, the difficulty of transporting natural gas for long distances limited its potential. Much of the natural gas which was produced from oilfields was simply burned off. A major pipeline, 120 miles long and carrying natural gas from wells in central Indiana to the city of Chicago, was constructed in 1891 but it was not very efficient. It was only in the 1920s that significant effort was put into building a pipeline infrastructure in the US. After the Second World War new welding techniques, along with advances in pipe rolling and metallurgy, further improved pipeline reliability. This post-war pipeline construction boom lasted well into the 1960s and allowed for the construction of thousands of miles of pipeline in America.

Europe had to wait even longer. In 1959 the British Gas Council showed that liquefied natural gas (LNG) could be transported effectively over long distances by sea and a 500 km high pressure pipeline was constructed from London to Leeds. When North Sea gas was discovered in 1965 the impetus for converting the UK from town gas to natural gas became irresistible. Coal, already suffering from the effects of oil, lost further territory as natural gas not only replaced the need for the manufacture of town gas but also started to be used for fulfilling peak electricity demand. Growing output from the North Sea made the development of a 5,000 km national distribution grid viable. Most impressively, all gas equipment in Great Britain was converted from burning town gas to natural gas over the period from 1967 to 1977. Northern Ireland followed and town gas was manufactured for the last time in 1987.

It is worth taking a moment to remember the sheer scale of this enterprise and the efficiency with which it was carried out. A single body (the British Gas Corporation, successor to the Gas Board) took the decision and implemented it at a cost of £100 million (perhaps £1.5 billion today) which included the costs of writing off redundant town gas facilities. All the gas-using equipment of around 13 million domestic, 400,000 commercial and 60,000 industrial customers was converted.

Natural gas requires little processing before use, obviating the need for the thousands of local manufacturing facilities referred to above. It can be brought ashore in large quantities from single sources and distributed through a national pipeline network. Methane is not toxic, unlike carbon monoxide which was a central component of town gas, though of course it can explode and if not properly ventilated its partial combustion can create carbon monoxide. Methane is also odourless so a small amount of an unpleasant-smelling additive, mercaptan, is introduced so leaks can be detected.

As well as being produced by decaying plant matter there are very large methane reserves to be found as separate deposits or associated with oilfields and coalfields. The first use of natural gas for electricity generation by a utility was in 1939 for an emergency power station in Neuchatel, Switzerland. Nonetheless, until the late years of the twentieth century it was regarded as a limited resource which should be reserved for premium purposes, principally space heating and cooking. The lifting of a European Commission Directive banning the use of electricity for electricity generation, coupled with the emergence of the highly efficient Combined Cycle Gas Turbine, the first of which in Britain began generation in 1991 at Roosecote in Cumbria, and, in the UK, the liberalisation of power markets created a major new market for gas and consumption grew steadily through the 1990s.

2.4 Pros and cons

Why should the fossil fuels have developed such a resilient stranglehold on the world's energy supplies, having so far proved resistant to every attempt to unseat them with the minor and short-lived exception of nuclear power in the 1970s to 1990s? Was it simply a matter of the power of the international oil conspiracy or was there more to it than that? And what might challenge this hegemony?

The advantages of these three brothers – or perhaps ugly sisters from the point of view of the atmosphere – can be outlined as follows:

- availability;
- storability – with the associated benefits for reliability;
- flexibility;
- transportability;
- power density;
- familiarity – and public acceptability;
- affordability.

However, the fossil fuels have a dark side as well. Critics, with considerable justification, claim that they are limited, dangerous and dirty.

2.5 SUBSIDIES

It should be noted that fossil fuels do routinely and, relatively speaking, uncontroversially attract subsidies. According to the OECD International Energy Agency (IEA), by the mid-2010s these were running at some $550 billion a year. Although this in total represented some four times the $120 billion subsidy going to non-hydro renewables it was of course a much smaller figure per unit of energy produced – around $50 per mtoe for fossil fuels, and $430 per mtoe for 'new' renewables despite their being broad contemporaries in terms of their initial development and deployment. Subsidies were of particular importance in the Middle East, largely to keep energy prices low: the IEA made the rather peculiar observation that some two million barrels per day of oil were being burned to generate power there that could otherwise come from renewables, which would be competitive with unsubsidised oil.[10] On the face of it this seem unlikely at a time when several countries in the region, notably Iran and the United Arab Emirates, were developing nuclear power programmes specifically to allow them to export more of their fossil fuels. Presumably the IEA was saying that heavily subsidised renewables would be competitive with unsubsidised oil, something which of course must become true if the level of subsidy is high enough. Nonetheless, while low energy prices seriously reduce incentives to use energy efficiently, reforming the subsidy system is proving difficult in countries where the population is still generally quite poor and political instability makes governments nervous of radical action.[11] In 2012, for example, Nigeria's newly re-elected president, Goodluck Jonathan, lifted fuel subsidies on petrol in an effort to cut $8 billion a year from the national budget. Petrol, which had been selling for 45 cents per litre, doubled in price overnight. A nationwide strike was launched, accompanied by clashes that left 16 people dead and 205 injured.[12] Within a week the government had backed down.

That being said, oil extraction is also a major tax contributor in many areas. In the UK companies operating North Sea oil and gasfields discovered before 1992 would have their profits taxed at anything up to 80 per cent, with more recently developed fields paying 60 per cent. The fall in the oil price from mid-2014 was met with proposed tax concessions, the industry calling for a cut of 30 per cent.[13]

2.6 THE END IS NIGH?

Ever since the Industrial Revolution began to take hold there have been periodic panics about the imminent exhaustion of the main fossil fuel of the day. In *The Coal Question* (1865)[14], W. Stanley Jevons, 'father of neoclassical economics', and great man though he was, being the first to recognise that improvements in energy efficiency often tend to boost energy use rather than curtailing it, was predicting that the UK would run out of coal, and therefore of energy, within perhaps 50 years. Speaking about 'peak coal' in terms remarkably portentous of the 'peak oil' brigade a century and a half later, he said:

> Suppose our progress [in expanding coal production] to be checked within half a century, yet by that time our consumption will probably be three or four times what it now is; there is nothing impossible or improbable in this; it is a moderate supposition, considering that our consumption has increased eight-fold in the last sixty years. But how shortened and darkened will the prospects of the country appear, with mines already deep, fuel dear and yet a high rate of consumption to keep up if we are not to retrograde.

A fairly typical forecast from 1900 came from one John Elfreth Watkins Jnr, who was asked by the *Ladies' Home Journal* in December 1900 to come up with a series of predictions for the year 2000.[15] Watkins was no dummy – among his premonitions were television, the mobile phone, high speed trains moving at more than 150 mph, cars replacing horses, ready-cooked meals, the streaming of high fidelity music into the home through telephone wires, digital colour photography, slowing population growth and an increase in the average height of people in rich countries. His view of the need for 'alternative energy' was widely held.

> Coal will not be used for heating or cooking. It will be scarce but not entirely exhausted. The earth's hard coal will last until the year 2050 or 2100; its soft-coal mines until 2200 or 2300. Meanwhile both kinds of coal will have become more and more expensive. Man will have found electricity manufactured by waterpower to be much cheaper. Every river or creek with any suitable fall will be equipped with water-motors, turning dynamos, making electricity.

No such future awaited mankind. Techniques for mining coal improved, both in their economics and in their efficiency in winning coal from the seams. More importantly perhaps, other fossil fuels entered the scene. So fear of 'peak coal' gradually gave way to the 'peak oil' industry.

Eventually peak oil must come. Oil is after all a limited natural resource, the product of millions of years of natural biology and chemistry. Mankind is depleting it at a rate thousands of times faster than it was created. The same inevitability may be ascribed, for example, to Malthus's predictions on global population. Writing in 1798, he said:

> The power of population is so superior to the power of the earth to produce subsistence for man that premature death must in some shape or other visit the human race. The vices of mankind are active and able ministers of depopulation. They are the precursors in the great army of destruction, and often finish the dreadful work themselves. But should they fail in this war of extermination, sickly seasons, epidemics, pestilence and plague advance in terrific array, and sweep off their thousands and tens of thousands. Should success be still incomplete, gigantic inevitable famine stalks in the rear and with one mighty blow levels the population with the food of the world.

While it is the case that the earth could not support an unlimited population, the growth in the number of humans from around one billion in 1800 to seven billion in the mid-2010s was not accompanied by a Malthusian catastrophe but actually by a growing life expectancy as the agricultural revolution led to enormous increases in the availability of food. Some of the more comical predictions of forthcoming calamity from Big Green gurus such as Ehrlich and Lovins are discussed in chapter 10.

When it comes to making practical use of the world's fossil reserves, a realistic sense of how quickly they might be expected to become depleted is clearly of considerable importance.

It has not been difficult to find individuals prepared to warn of the imminent exhaustion of reserves of one kind or another. In 1885 the United States Geological Survey announced that there was little or no chance of oil being discovered in California, a few years later saying the same about Kansas and Texas. In 1919 David White, chief geologist of the US Geological Survey, was predicting: 'The peak of production will soon be passed, possibly within three years. There are many well-informed geologists and engineers who believe that the peak in the production of natural petroleum in this country will be reached by 1921 and who present impressive evidence that it may come even before 1920.'[16]

In 1939 the US Department of the Interior said American oil supplies would last only another 13 years; in 1949 the Secretary of the Interior said the end of US oil supplies was in sight.

The American geologist M. King Hubbert suggested in 1949 that the period of the fossil fuels would prove to be a very short interlude in the history of mankind, let alone of the planet. He suggested that the extraction of oil from a particular region would follow a bell-shaped curve, growing to a maximum and then declining.at a similar rate once peak production has been passed. He predicted the peak production of oil in the US would occur between 1965 and 1971, with the global point being reached in 1995.[17] Hubbert predicted that by 1998, presuming nuclear power grew only at 10 per cent per year (the actual rate might be twice this amount), that nuclear would have overtaken oil by 1998.[18]

In the event, the recession caused by the oil shocks of the 1970s reduced the growth in oil consumption to well below previous predictions. Nevertheless, alarm calls continued to be heard. In 1998, for example, geologists Colin Campbell and Jean Laherrère published a paper entitled *The End of Cheap Oil* which has attracted much attention over the years. In it they said:

> Our analysis of the discovery and production of oil fields around the world suggests that within the next decade the supply of conventional oil will be unable to keep up with demand. This conclusion contradicts the picture one gets from oil industry reports, which boasted of 1,020 billion barrels of oil (Gbo) in 'proved' reserves at the start of 1998. Dividing that figure by the current production rate of about 23.6 Gbo a year might suggest that crude oil could remain plentiful and cheap for 43 more years – probably longer, because official charts show reserves growing. Unfortunately, this appraisal makes three critical errors. First, it relies on distorted estimates of reserves. A second mistake is to pretend that production will remain constant. Third and most important, conventional wisdom erroneously assumes that the last bucket of oil can be pumped from the ground just as quickly as the barrels of oil gushing from wells today. In fact, the rate at which any well – or any country – can produce oil always rises to a maximum and then, when about half the oil is gone, begins falling gradually back to zero. We calculate that the oil industry will be able to recover only about another 1,000 billion barrels of conventional oil.

A few more examples:

- 2001: Professor Kenneth Deffeyes, Princeton University – peak oil would be in 2005 (an idea he clung to at least until 2009);
- 2006: Matthew Simmons, founder of the Ocean Energy Institute – peak oil had been reached in 2005 (in 2005 Simmons had bet a journalist and the widow of an economist $2,500 each that the price of oil would average over $200 per barrel in 2010, though sadly, at least from his punters' point of view, he died before he could settle up, the price having remained stubbornly below $100 throughout that year);
- 2007: Energy Watch Group, Germany – production had peaked in 2006 and would decline by several percentage points annually;

- 2007: Sadad Al Husseini, former head of production and exploration at Saudi Aramco – peak reached in 2006;

- 2008: US Association for the Study of Peak Oil and Gas – peak in all oil, including unconventional sources, would occur in 2010.

- 2005: T. Boone Pickens, alternative energy activist – oil was close to peak;

- 2008: UK Industry Taskforce on Peak Oil and Energy Security (ITPOES), convened by solar power industry lobbyist Jeremy Leggett – peak oil was most likely to occur by 2013;

- 2009: UKERC (UK Energy Research Council) – a peak in conventional oil production before 2030 appeared likely and there was a significant risk of a peak before 2020;

- 2011: Jeremy Leggett – 'perhaps as early as 2012 global production will have hit the highest level it ever will and given that demand won't be abating the oil price will soar'.

Nor were these dire warnings confined to oil: in 1974 the US Geological Survey, building on its long record of getting oil wrong, advised that the US had only a ten-year supply of natural gas.[19]

Sixteen years after Campbell and Laherrère's paper things did not look quite as they had predicted. The world consumed some 500 billion barrels of oil between 1998 and 2014. Yet in 2014 stated reserves stood at 1,700 billion barrels (see figure 2.11), some 56 per cent higher than they had been in 1998 (and 14 per cent higher than in 2008). Campbell and Laherrère predicted that reserves should have reduced by half in that period – in fact they increased by half. In just 16 years Campbell and Laherrère's

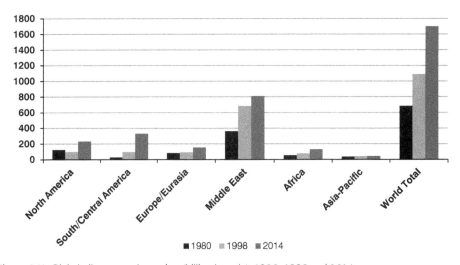

Figure 2.11 Global oil reserves by region (billion barrels), 1980, 1998 and 2014

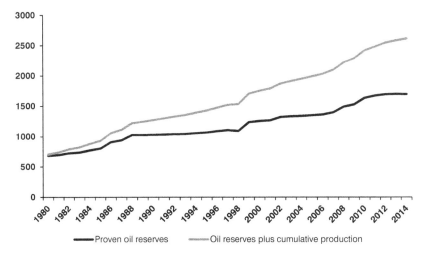

Figure 2.12 Global oil reserves (billion barrels), 1980–2014

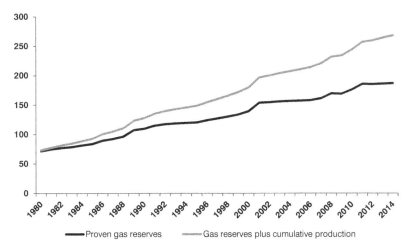

Figure 2.13 Global gas reserves (trillion m³), 1980–2014

figures had proved to be out by the equivalent of the entire known oil reserve at the time.

At the end of 2014 the 'reserve-to-production' ratio for oil, i.e. the size of the proven reserve divided by the current rate of extraction, stood at 52.5 years – as opposed to the 43 years which Campbell and Laherrère had criticised so robustly 16 years previously. Those of gas and coal were 54.1 years and 110 years respectively. For some decades the proven reserves of oil and gas had been increasing year-on-year, despite the consumption of another year's supplies in the interim. When cumulative consumption is added to stated reserves, the world knew of approximately 3.7 times as much gas and oil in 2014 as it had in 1980 – see figures 2.12 and 2.13.

Figure 2.14 US field production of crude oil (thousand barrels per day monthly average), 1990–2015[20]

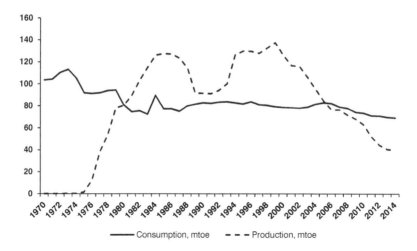

Figure 2.15 UK oil production and consumption (million tonnes per year), 1970–2014

Perhaps most strikingly, US oil production, which had more or less followed the Hubbert curve from its peak of around 10,000 barrels per day in 1970 to half that figure in the late 2000s, saw a dramatic upturn, as shown in figure 2.14, which took it back towards its previous peak as unconventional oil reserves came into play.

To be fair, not all commentators took the same line. The IEA's World Energy Outlook 2013 took the view that estimates of ultimately recoverable resources of oil would continue to increase as technologies unlocked types of resources such as 'light tight oil' that were not considered recoverable only a few years earlier.[21]

As is generally the case within commodity markets, the very fact of sustained high oil prices carries within it the seeds of its own downfall. A period of high oil price may have several consequences, including:

- a relative reduction in energy demand below expectations owing both to a slowdown in the economy and the greater implementation of energy efficiency measures;
- more vigorous exploration for new 'conventional' oilfields and investment in techniques to increase the amount of oil extracted from existing wells;
- the emergence of new oil-producing countries and areas;
- the exploitation of less attractive oil reserves, including unconventional reserves, to increase supply
- development of alternative energy supplies to displace oil;
- growing political differences of opinion among oil producers leading to a decline in the cohesiveness of their position and a potential oil glut.

All of these were factors in the decline of the oil price between the mid-1980s and the late 1990s. The UK is an especially pronounced example of how the oil shock led both to a reduction in the use of oil, notably for electricity production, and a greater incentive to develop North Sea reserves, as shown in figure 2.15.

It is unlikely that the UK would have developed its North Sea oil (and gas) reserves as rapidly had the oil price in 1980 been at the same level as it had been in 1970.

As the cycle proceeds, so the success of these measures to reduce oil demand, and hence price, is likely to promote economic growth and make oil more attractive again, leading to increasing prices.

The unpredicted fall in the oil price that started in mid-2014 apparently shared a similar genesis. The sluggish world economy was accompanied by the significant increase in production from oil in shale formations in Texas and North Dakota, which as noted previously brought US production levels up towards those of Saudi Arabia. The costs of deploying the relatively new technology of fracking – hydraulic fracturing of shale formations to release trapped oil – were falling significantly, from $70 per barrel to $57 a barrel in one year according to the research firm IHS.[22] In addition Saudi Arabia, with very large foreign currency reserves, seems to have been prepared to allow the oil price to fall to create disproportionate problems for its regional rivals, mainly Iran, and to put pressure on the reborn US oil production industry with its relatively high costs. In such a way periods of low oil prices make both alternative energy sources and unconventional oil reserves less attractive and so tend to work in oil's benefit in the long term.

Ultimately the oil will not 'run out' and perhaps the whole debate on peak oil is a rather esoteric one. As Sheikh Yamani, Saudi oil minister, said at the height of the first modern oil shock in 1973: 'The Stone Age did not end for lack of stone and the Oil Age will end long before the world runs out of oil.'[23] If economics were the only driver of policy one would expect that when most oil reserves had been identified and exploited, the price signals would be quite sufficient to divert energy use towards more competitive alternatives.

One might speculate, then, as to why, despite the long historical record of predictions of forthcoming doom generally being followed by major new discoveries and falling prices, so many commentators are still prepared to risk and indeed lose their reputations by repeating the century-old mantra of oil being about to run out. However, whether it is sincere incompetence, a reflection of vested interests hoping to divert more subsidies into their own businesses or research coffers, or some other reason, such a strategy ultimately and simply discredits the serious argument about resources and developing alternatives.

And serious argument there is, even on resource grounds. In 2000 the European Commission published a Green Paper on Europe's energy security.[24] In it the Commission said:

> The European Union is consuming more and more energy and importing more and more energy products. [European] Community production is insufficient for the Union's energy requirements. As a result, external dependence for energy is constantly increasing. The dramatic rise in oil prices which could undermine the recovery of the European economy, caused by the fact that the price of crude oil has tripled since March 1999, once again reveals the European Union's structural weaknesses regarding energy supply, namely Europe's growing dependence on energy, the role of oil as the governing factor in the price of energy and the disappointing results of policies to control consumption. Without an active energy policy, the European Union will not be able to free itself from its increasing energy dependence. If no measures are taken, in the next 20 to 30 years 70 per cent of the Union's energy requirements, as opposed to the current 50 per cent, will be covered by imported products.

In the following decade and a half the situation worsened. The EU's dependence on imported energy continued to grow, from 46.2 per cent in 2000 to 54.7 per cent in 2008 before reaching something of a plateau, as shown in figure 2.16.

Heightened tensions in Ukraine between the European Union and the Russian Federation in 2014, for example, brought into particular focus the EU's dependence on imports from the Russian Federation, the single biggest source of imports of hard coal (25.9 per cent of total EU-28 imports), crude oil (33.6 per cent) and natural gas (32.0 per cent).

The situation in the UK with regard to energy import dependency changed more

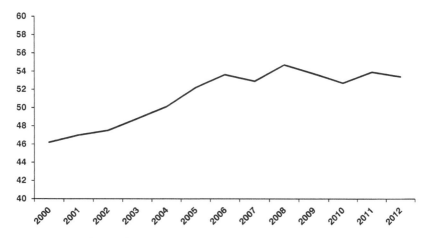

Figure 2.16 EU energy import dependency (%), 2000–12[25]

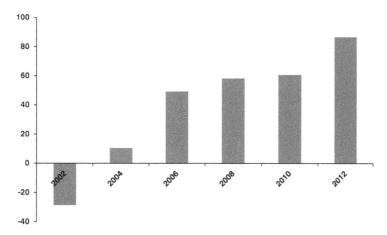

Figure 2.17 UK net energy imports (mtoe), 2002–12

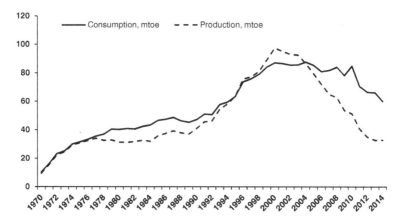

Figure 2.18 UK gas production and consumption (mtoe), 1970–2014

rapidly than in any other EU member. While the EU-28's energy production fell by 15.7 per cent between 2002 and 2012 (from 942 mtoe to 794 mtoe), that of the UK fell by 54.4 per cent (from 256 mtoe to 116 mtoe). Although in 2010 energy imports per head of the population were still below the EU average, the UK had moved from being a net energy exporter in 2002 to a net energy importer, as shown in figure 2.17.

The key factor was the decline in production of gas from the North Sea – indeed, UK oil and gas production broadly adhered to a Hubbertian analysis, at least over this period – see figure 2.18.

Of course countries with fossil fuel reserves generally wish to sell them. Being a net energy importer per se need not be a barrier to a successful economy. However, the UK finds itself at the end of some long pipelines across Europe. Furthermore, when the UK had its own North Sea reserves of gas and oil, storage was less of an issue – in effect, the oilfields and gasfields themselves were the storage capacity. As those reserves became depleted storage became a concern. In 2011 the House of Commons heard that the UK was consuming around 100 billion cubic metres (m^3) of gas per year but only had storage capacity equivalent to a little over 4 per cent of this, much less than other European countries. The UK's storage capacity then was equivalent to around 14 days' worth of supply, compared to 69 in Germany, 59 in Italy, 87 in France and 66 days in the US.[26]

To summarise, it is difficult to see any reason for believing that the recent rash of claims that the world is reaching peak oil are more likely to be accurate than similar claims which have been made for a century or more, any more than we should foresee a sudden reverse in the growth of global energy demand simply because some commentators are saying we should use energy more efficiently. The geopolitics of fossil energy remains a challenge – it will be fascinating to see how US foreign policy changes as it becomes markedly less dependent on Middle Eastern oil – but those who hope to use scare stories about security of supply to drive governments away from a reliance on fossil fuels, for commercial, ideological or environmental reasons, may be seriously disappointed while damaging the credibility of the wider case.

The climate change implications of fossil fuels are profound and considered in detail in chapter 9. It should be noted once again that at the power station the use of gas for electricity generation emits less than half of the carbon dioxide per unit of electricity generated than does the use of coal. It is important to ensure that the gas extraction and delivery system does not result in significant atmospheric leakage of methane, which is almost 30 times as powerful a greenhouse gas as is carbon dioxide. But if this can be achieved then it is unarguable that by far the fastest way of reducing carbon dioxide emissions globally would be to encourage power producers to switch from coal to gas.

However, climate change is not by any means the only environmental challenge created by fossil fuels. Among the adverse environmental effects of coal mining and oil/gas extraction are the following:

- alteration to the landscape, notably in rural areas, as discussed in chapter 6 – this can involve not only visual impact and potential destruction of archaeological sites but also the displacement of entire towns to make way for the new coalfield etc.;

- damage to vegetation, wildlife and soil quality, both in sinking and operation of the mine or oil well and in the transportation of the product;

- damage to local air quality;

- creation of large amounts of waste and spoil;

- contamination of waterways – acidification, toxic trace elements, sediment;

- damage caused by mine or wells collapsing – in the Saar region of Germany in 2008 a mine collapse created an earthquake measuring 4.0 on the Richter magnitude scale;

- potential damage caused by fires – a coal fire has burned under the borough of Centralia in Pennsylvania since 1962, resulting in its population dropping from over 1,000 to 10;

- requirements for large amounts of water, especially in opencast mining for dust suppression.

Such issues are of particular concern in developing countries such as Nigeria, where it is often felt that multinational fossil fuel companies do not pay the same attention to environmental protection or the safety of workers and other local people as they might 'at home', sometimes with the support of the local government. In 1993, for example, the Ogoni people in Nigeria expelled Shell from its land, ending oil exploration and extraction in the area. However, pollution persisted from the many pipelines which cross the area. The issue became a focus for political unrest, with an estimated 2,000 Ogonis dying and 80,000 being displaced by the Nigerian security forces. In 1995 Ogoni leader Ken Saro-Wiwa and the rest of the 'Ogoni 9' were executed by the Nigerian authorities, then a military dictatorship, on what many commentators believed were invented charges. At least two witnesses who testified that Saro-Wiwa was involved in the murders of the Ogoni elders later recanted, stating that they had been bribed with money and offers of jobs with Shell to give false testimony.[27] In August 2011 a United Nations report confirmed the levels of pollution and declared that a clean-up should commence without delay, but four years later there had reportedly been no meaningful action from the Nigerian government or the oil companies.[28]

Fossil-fired power stations also have local and regional environmental implications:

- significant volumes of waste – in 2008 the Kingston Fossil Plant in Tennessee suffered a release of some 4 million m^3 of coal fly ash slurry which spread across the Emory River, covering 1.2 km^2 of the surrounding land, damaging homes and causing devastation to local wildlife;

- air pollution – in addition to greenhouse gases, fossil fuel plants, especially coal, release large quantities of sulphur and nitrogen oxides which contribute to acid rain, alongside a cocktail of toxins such as particulates, hydrogen cyanide, arsenic, lead, mercury, nickel, vanadium, cadmium, barium, chromium, copper, radium and uranium – a traditional coal-fired power station in normal operation also releases more radioactive material than does a nuclear plant of the same output;

- water pollution – in addition to the chemical pollution risk, the need for cooling water for inland power stations requires the removal of large amounts of water from local rivers, to be returned at a considerably higher temperature and thereby affecting local wildlife.

2.7 UNCONVENTIONAL RESERVES AND FRACKING

'Conventional' gas is defined by the US Geological Survey, some of whose pronouncements we have already met, as 'gas sourced from discrete pools or pools localised in structural traps by the boundary of gas and water'. Unconventional gas is sourced from 'accumulations with large spatial dimensions and indistinctly defined boundaries existing more or less independently of the water column'.

One unconventional source is 'shale gas'. Shale is a sedimentary rock made up of a muddy mixture of flakes of clay and fragments of other minerals such as quartz (a form of silica) and calcite (calcium carbonate). It is quite 'fissile', not in the nuclear sense but in a geological sense, i.e. it breaks easily along the boundaries between its thin layers – slate is a form of shale. As it was laid down in oceans the bodies of small animals became caught up in its structure, gradually turning to oil and/or gas which became trapped in its structure.

A particular controversy which emerged as Britain tried to find a replacement for its dwindling North Sea gas reserves has involved 'fracking' of shale. The term is logical enough – it refers to 'hydraulic fracturing' of rocks to release trapped oil or gas – though it has a sound that lends itself to all sorts of mildly rude-sounding jibes. Fracking involves injecting a high-pressure mixture of water, sand and chemicals into underground shale rocks to fracture (or 'fission') them to release the trapped gas which then flows out of the wellhead. The process can be carried out vertically or, more commonly, by drilling horizontally into the rock layer. The process can create new pathways to release gas or can be used to extend existing channels.

Shale gas was first extracted in New York State in 1821, but it was another century before drilling began in the 1930s. The first use of fracking was recorded in the US in 1947.[29]

In the 1970s the US federal government introduced price controls on natural gas which led to shortages – and therefore ultimately to higher prices, a good example of how such measures tend to be counterproductive. One response was the Eastern Gas Shales Project, lasting from 1976 to 1992. Subsequently, the US Department of Energy

(DOE) entered into partnership with private gas companies to develop the first air-drilled multifracture horizontal well in shale in 1986. New diagnostic techniques, notably microseismic imaging which had been developed in the coal industry, allowed for promising reserves to be detected with much greater accuracy.

Despite such progress, in 2000 shale gas still only accounted for less than 1 per cent of US gas production. In 1998, George P. Mitchell, 'the father of the shale gas industry', developed a new technique which brought the costs of fracking down to the level of many conventional gasfields. Shale gas subsequently became the fastest growing contributor to primary energy in the US, by 2010 accounting for over 20 per cent of US gas production, a figure expected to reach almost 50 per cent by 2035. The IEA estimates that shale gas could increase technically recoverable natural gas resources by as much as 50 per cent.[30] The effect of the shale gas revolution, as it was rightly termed, on US natural gas production since 2000 is shown clearly in figure 2.19.

The effects of the growth of shale gas were significant. Gas prices fell as the US reduced its imports, creating something of a glut of liquefied natural gas (LNG). As the US switched from coal to gas for its power production it began to increase its coal exports, leading to a depressed coal price elsewhere. Paradoxically, then, from Europe's point of view, and especially the UK's, the most immediate effect was a reduction in the use of gas for power production in favour of coal. Between 2011 and 2012 the proportion of UK electricity generated using coal grew from 30 to 39 per cent while that generated using gas fell from 40 to 28 per cent, with significantly unhelpful implications for greenhouse gas emissions. Big Green has used this as an argument against fracking – the US may be reducing its greenhouse gas emissions significantly by shifting to gas but this is pointless if the upshot is that it exports coal for use elsewhere. This seems a particularly odd proposition as it could be applied equally to argue against the US or any other country with coal switching to nuclear or renewables. In the event, when exports

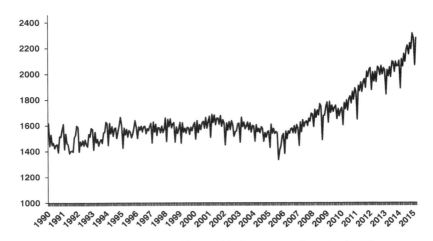

Figure 2.19 US dry natural gas production (billion cubic feet per month), 1990–2015[31]

of natural gas 'caught up' and gas prices started to fall the shift from gas to coal in the UK was soon reversed.

In the UK, drilling companies estimated that trillions of cubic feet of shale gas might be recoverable from underneath parts of northern England. However, especially since 2010, growing environmental concerns led in effect to a moratorium on new development. Several potential problems have been identified by Big Green and others.

- Fracking inevitably uses huge amounts of water that must be transported to the fracking site, at significant environmental cost.
- Potentially toxic chemicals may escape and contaminate groundwater around the fracking site.
- The fracking process can cause small earth tremors: two small 'earthquakes' (of 1.5 and 2.2 magnitude, roughly equivalent to a heavy lorry going past the front door) hit the Blackpool area in 2011 following fracking.
- The UK is a more densely populated country than the US, with fewer areas where fracking can be carried out with little implication for local settlements.
- Fracking may be distracting energy firms and governments from investing in renewable sources of energy and encouraging continued reliance on fossil fuels.

The counterargument is that developing a fracking industry in the UK would go some way towards compensating for falling North Sea gas production, though the potential is unclear at this point. Whether this would reduce gas prices or improve the UK's balance of payments position, it would still be of considerable economic benefit. Some two million fracking wells have been sunk globally – of course some of these have led to some local environmental damage, but there is now a huge body of best practice into which the UK could tap to prevent such problems recurring. Gas does not displace renewables or vice versa – indeed, as explored in chapter 4, the variability of output from wind and solar in particular requires the system to burn large amounts of gas as the most flexible, dispatchable technology.

From the point of view of a decision waiting to be taken it seemed clear by the mid-2010s that crucial factors in whether fracking would emerge in the UK were the relationship that developed between the potential developers and local communities and whether government would once again fall into the trap of encouraging fears by over-regulating the industry, thereby creating the impression that the activity was inherently more unpleasant than it actually is. Initial signs were mixed – the hiatus in development between 2010 and 2015 suggests that ministers effectively might be susceptible to choking off fracking before it has a chance to prove itself. However, government did seem prepared to defend the entirely sensible (and essential) approach of allowing companies to frack underneath homes without the owner's permission.[32] Such operations would be taking place at a depth of several hundred metres; one can imagine how difficult things would have become if every single homeowner on the route of a new underground line in London had to give permission before the line could be developed. In the terms proposed in chapter 10 this may be an encouraging sign that decision-making is beginning to take a wider utilitarian stance once more, though the Brent Spar debacle considered in that chapter suggests that such common sense cannot always be relied upon when political waters become choppy.

2.8 SAFETY

As discussed in chapter 11, there is something of a paradox in public perceptions of the dangers associated with various power sources, the most dangerous seeming to cause the least concern. Even excluding the health effects of climate change, the fossil fuels, particularly oil and coal, are by far the most dangerous ways of making energy (see figure 2.20 and table 2.1), though having said this the biggest accident in energy history, in terms of directly attributable human deaths, was at the Banqiao hydro dams in China in 1975, with an estimated 170,000 deaths including 26,000 from flooding and 145,000 from disease and starvation.[33]

For example, the independent consensus concerning the health effects of the Chernobyl accident (e.g. World Health Organisation, United Nations Scientific Committee on the Effects of Atomic Radiation) is about 50 deaths among those who were onsite at the time (31 immediate), 6,000 thyroid cancers with about ten deaths, and possibly 4,000 to 7,000 people facing a shortened lifespan over the 70 years following the accident (undetectable against natural level of cancer). In April 2013 a NASA paper estimated that the use of nuclear power rather than fossil fuels had saved some 1.84 million air pollution-related deaths, while saving the emission of 64 billion tonnes of carbon dioxide, with potentially up to a further seven million lives to be saved over the following four decades should a major new nuclear programme be initiated globally (and depending on which fuel it displaced).[36] One recalls Stalin: 'One death is a tragedy, a million deaths a statistic'.

The dangers of fossil fuels are manifold. Perhaps coal mining accidents spring to mind most readily. The first large-scale mining disaster for which reliable evidence seems to be available was at Gateshead, County Durham, in October 1705, where a fire took over 30 lives. The worst disasters in UK mines happened in 1866, when two explosions at the Oaks pit near Barnsley in South Yorkshire caused the deaths of 388 workers, and in 1913 when 440 miners died at the Senghenydd Colliery near Caerphilly in Wales. The worst in Europe was at Courrières, France, in 1906 when 1,099 miners died; and even recently in 2014 over 300 miners died at Soma in Turkey. The worst recorded mining disaster in history happened at the Benhixu mine in Liaoning, China, in 1942, with 1,549 fatalities. Over 50,000 miners were killed in China between 2000 and 2010 (see figure 2.21), although the reported rate per tonne of coal extracted fell sharply as older and more dangerous mines were gradually closed.

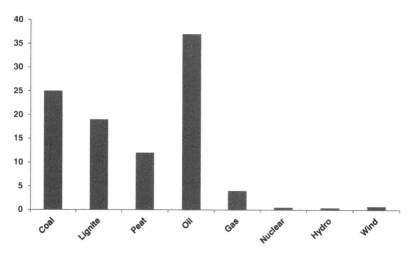

Figure 2.20 Average deaths per MWh of generating electricity in the EU (excluding climate change deaths)[34]

Table 2.1 Accidents involving five or more direct fatalities in fossil, hydro and nuclear chains, 1969–2000[35]

Energy chain	OECD			Non-OECD		
	Accidents	Fatalities	Fatalities per GW-year	Accidents	Fatalities	Fatalities per GW-year
Coal	75	2,259	0.157	1,044	18,017	
Coal (China, 1994–1999)				819	11,334	6.17
Coal (without China)				102	4,831	0.60
Oil	165	3,713	0.13	232	16,505	0.90
Natural gas	90	1,043	0.09	45	1,000	0.11
LPG	59	1,905	1.96	46	2,016	14.9
Hydro	1	14	0.003	10	29,924	10.3
Nuclear	0	0	0	1	31	0.05
TOTAL	390	8,934		1,480	72,324	

Collapsing spoil tips at collieries or power stations also represent a considerable threat if not properly managed. One of the worst occurred in the Welsh village of Aberfan, near Merthyr Tydfil, in 1966, when some 40,000 m³ of spoil slurry slid downhill into Pantglas Junior School, killing 116 children and 28 adults. And inevitably the major transportation operations necessary to move fossil fuels around in quantity lead to road, rail and shipping accidents.

Deaths are also directly associated with the oil and gas industry. There were 545 reported fatalities in US oilfields from 2008 to 2012, Texas being responsible for 216.

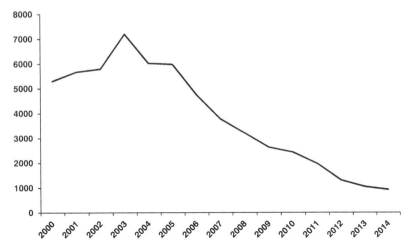

Figure 2.21 Annual mining deaths in China, 2000–14[37]

Fatalities in the US offshore oil industry were running at an average of 16 per year from 2003 to 2010. The worst single oil rig disaster was Occidental's Piper Alpha explosion in 1988 with 167 fatalities. A natural gas explosion which destroyed the New London School in Texas in 1937 killed 300 students and teachers.

The wider health effects of burning fossil fuels, even ignoring the consequences of climate change, appear to be devastating. Outdoor air pollution, largely from particulates and other emissions associated with fossil fuels, is estimated to cause 3.7 million premature deaths per year, 88 per cent in developing countries.[38] The last of the Great Smogs in London is thought to have killed over 12,000 people in a fortnight in December 1952.[39]

2.9 CARBON CAPTURE AND STORAGE – THE GREAT TECHNICAL FIX?

The direct death toll from fossil fuels may be severe but it pales into insignificance against the potential consequence of climate change. Disruption of crops, flooding and salination of coastal land, interference in the food chain, threats to biodiversity, water shortages, extreme weather events … the list is apparently endless.

If switching to alternative low-carbon fuels is expensive or politically unpopular, might it be possible simply to scrub the carbon dioxide from the flue gases of fossil-fuelled power stations? After all, Flue Gas Desulphurisation has been very successful in addressing acid rain.

Carbon capture and storage (CCS) is a three-stage process. Carbon is captured from the source where it arises; transported to the storage or disposal facility; and 'stored' or disposed of, ideally for the very long term.

Simple it may sound, but technically and economically it is anything but. Capturing (or 'sequestering') carbon dioxide can be achieved in a number of ways – cryogenically, by freezing the carbon dioxide out of the flue gases; by adsorption onto graphite; or by chemical means, dissolving it in an alkaline solution – perhaps an amine. In each case the carbon dioxide must then be reconstituted for transport to the disposal field. The transport itself is relatively straightforward, the key challenge perhaps being to avoid leakage in the pipes or containers being used. Disposal requires identification of geological structures which will contain the carbon dioxide in such a way as to offer high certainty that it will not escape into the atmosphere.

CCS is not a new concept. It has been used for some decades to increase the amount of oil or gas which can be extracted from a particular gasfield or oil well – carbon dioxide is pumped into the resource, forcing the oil or gas out. Since the field itself tends to be quite secure geologically, usually lying under a 'shale cap' which is impervious to carbon dioxide – that is why the oil or gas has accumulated and remained there – the carbon dioxide tends to remain trapped. Interest in its potential in combating climate change began to grow around the turn of the twenty-first century. A 2001 paper stated:

Although it has received relatively little attention as a potential method of combating climate change in comparison to energy reduction measures and development of carbon-free energy technologies, sequestration of carbon dioxide in geologic or biospheric sinks has enormous potential. Considerable quantities of carbon dioxide separated from natural gas deposits and from hydrogen production from steam reforming of methane are already used in enhanced oil recovery and in extraction of coalbed methane, the carbon dioxide remaining sequestered at the end of the process. A number of barriers lie in the way of its implementation on a large scale. There are concerns about possible environmental effects of large-scale injection of carbon dioxide especially into the oceans. Available technologies, especially of separating and capturing the carbon dioxide from waste streams, have high costs at present, perhaps representing an additional 40 to 100 per cent onto the costs of generating electricity. In most of the world there are no mechanisms to encourage firms to consider sequestration. Considerable R & D (research and development) is required to bring down the costs of the process, to elucidate the environmental effects of storage and to ensure that carbon dioxide will not escape from stores in unacceptably short timescales. However, the potential of sequestration should not be underestimated as a contribution to global climate change mitigation measures.[40]

A decade later and journalists were saying:

If there was a prize for false starts, dashed hopes and failed promises, the carbon capture and storage industry would be a strong contender. For more than a decade it has offered a tantalising solution to a central dilemma of climate change: the need to curb carbon dioxide emissions from the fossil fuel burning energy systems on which most economies depend. In reality the idea has proved more troublesome.[41]

One of the main research projects was located at the Sleipner West gasfield some 250 km west of Stavanger, operated by Norway's Statoil. Natural gas from the field has a relatively high carbon dioxide concentration – almost 10 per cent – which reduces the calorific value of the fuel and has to be stripped out. Carbon dioxide is removed using amine and injected into an undersea aquifer containing saltwater, where it dissolves. Between its inception in 1996 and early 2013 the project had captured and stored some 14 million tonnes of carbon dioxide, equivalent to the emissions of Norway's cars for two years.

In 2003 President Bush announced his commitment to FutureGen, a project with the US DOE which would build the first coal power plant with near zero carbon emissions. Ten years later it was hoped the project might go live in 2017. In 2009, the first CCS facility in West Virginia began operating; after a year or so the project ended and American Electric Power decided to halt its plans to build a carbon capture plant for a 235 MW generation unit at its 1.3 GW Mountaineer power plant in New Haven, WV. In 2014 the 110 MW SaskPower CCS plant at Boundary Dam in Canada became the world's first operational large-scale CCS project. The total cost was initially expected to be around $1.24 billion, though by 2013 it was $115 million over budget.[42]

The main cost is in the sequestering of the carbon dioxide. One attractive approach is to do the separation upstream by running the plant on pure oxygen, producing a flue stream which is highly concentrated carbon dioxide, as acid gases and other impurities will be filtered out before reaching the flue gas. Another would involve 'steam stripping' of methane to create carbon dioxide and hydrogen, separating the carbon dioxide out and using the hydrogen as a clean fuel.

However, the disposal phase is not without its hazards. In 1986 naturally sequestered carbon dioxide deep in Lake Nyos in Cameroon was released explosively, resulting in some 1,700 deaths and wiping out animals and vegetation over a 15 mile radius. Some 100 to 300 tonnes of carbon dioxide are believed to have been released – a relatively small amount compared to the quantities which are proposed for CCS.[43] The granting of environmental licenses for disposal is a contentious issue – in the UK in the mid-2010s storage of carbon dioxide underground, as opposed to offshore, was banned in the UK.

Financially CCS faces another challenge. One can imagine a world in which nuclear power, or even renewables, appears the lowest cost option for producing power even ignoring carbon costs. This can never be the case for CCS – it will always be an extra cost both in terms of initial capital investment and in operational terms because CCS requires energy and so reduces the efficiency of the plant. As such it depends, even more than renewables and nuclear do, on a very high degree of certainty that the regulatory and subsidy framework will provide sufficiently large sums for a sufficiently long time to compensate for the initial investment. The first decade and a half of the twenty-first century gave little confidence that this would be the case.

2.10 THE SWING FUELS

It is a perfectly defensible position to take to say that our use of fossil fuels is destroying the planet, at least from the point of view of the sustainability of a human population inevitably heading towards 10 or 11 billion unless a Malthusian disaster intervenes. Compared to nuclear power or renewables their direct consequences are dire, their indirect ones worse. Yet since the Rio and Kyoto processes were instigated fossil fuels have doggedly defended their relative position in world energy, growing enormously in absolute terms as they do so. As noted earlier, during only one brief period – the growth of nuclear power from the mid-1970s to the mid-1990s – has the proportion of global primary energy from fossil fuels fallen, and then only from about 94 to 87 per cent.

To put it another way, the paralysis in decision-making has affected fossil fuels less than any other part of the energy picture. Fossil fuels remain the 'swing fuels', the ones which inevitably step up to fill the gap if attempts to persuade investors, public or private, to fund something else should fail.

The real limitation on fossil fuels may not be 'peak oil' or the size of the resource, nor indeed the changing economics as the reserve runs short or becomes ever more concentrated in a small number of hands – at present perceptions on both of these are moving in the other direction as new unconventional reserves are opened up in the US and perhaps Europe. It is more likely to be environmental. As noted in chapter 9, the atmosphere simply cannot survive the burning of all our fossil reserves, especially coal.

It must be recognised that although many players would quite like to reduce fossil fuel use it is nobody's top priority. Governments of countries which import large volumes of fossil fuels regularly bemoan the fact, especially at times of high prices, but are often unwilling to take the action necessary to force unwilling investors to fund alternatives. After all, as has been seen time and time again, most recently in the mid-1980s and the mid-2010s, the oil price might go up but it is likely to come down again. As many producers found in 2014 and 2015, the money they had poured into unconventional and more difficult-to-reach conventional hydrocarbon supplies looked as though it might be wasted, or at least not make sustainable returns, when oil prices came down to below the cost of extraction of some of these reserves. Meanwhile Big Green has shown in its post-Fukushima attitude to Japan and Germany that it regards it as perfectly acceptable to increase use of fossil fuels, including 'brown coal', the worst of them all from an environmental point of view, if the output from doctrinally unacceptable nuclear power can be reduced by so doing.

And of course for others – countries and companies – whose economies and political stability depends heavily on exploiting and exporting their fossil reserves, the ongoing bickering between advocates of renewables, nuclear power, carbon capture and storage – an expensive technology which would presumably increase the cost of and so reduce demand for fossil fuels – and energy efficiency suits them very well.

Grounds for optimism could still be found. The Paris climate change talks at the end of 2015 were hailed as presaging the 'end of the fossil fuel era'. $100 billion would be committed in annual climate financing by 2020 from public and private sources and 'by some point after 2050' man-made emissions should be reduced to a level that forests and oceans can absorb.

But, as discussed in chapter 9, experiences of a quarter of a century following the Rio Convention may lead one to conclude that projections for the future which assume that the use of fossil fuels are about to decline dramatically might go seriously awry unless all sides come together to put climate change at the very top of the agenda. That appears highly unlikely to happen.

ENDNOTES

1 http://www.worldcoal.org/coal/what-is-coal/, World Coal Association, 'What is coal?'.

2 http://www.historylearningsite.co.uk/coal_mines_industrial_revolution.htm, Trueman C N, 'Coal Mines in the Industrial Revolution', *The History Learning Site*.

3 http://ourfiniteworld.com/2012/03/12/world-energy-consumption-since-1820-in-charts/, Tverberg G, 'World Energy Consumption Since 1820 in Charts', *Our Finite World*, March 2012.

4 Roush G (ed.), (1942), '*The mineral industry: its statistics, technology and trade during 1941*', McGraw-Hill.

5 http://oilandgasinvestmentcompanies.com/what-is-the-history-of-petroleum/, Oil and Gas Investment Companies, 'History of petroleum'.

6 http://www.engdahl.oilgeopolitics.net/History/Oil_and_the_Origins_of_World_W/oil_and_the_origins_of_world_w.HTM, Engdahl F (2007), 'Oil and the origins of the War to make the world safe for Democracy', *Geopolitics-Geoeconomics*.

7 Yergin D (1991), *The Prize*, Simon & Schuster.

8 http://naturalgas.org/overview/history/, 'History', *naturalgas.org*, 2013.

9 http://iaeg2006.geolsoc.org.uk/cd/PAPERS/IAEG_564.PDF, Hatheway A and Doyle B (2006), 'Technical history of the town gas plants of the British Isles,' IAEG 564.

10 http://www.worldenergyoutlook.org/publications/weo-2014/, IEA (2014), 'World energy outlook 2014', OECD: Paris.

11 http://www.imf.org/external/pubs/ft/dp/2014/1403mcdsum.pdf, Sdralevich C *et al.* (2014), 'Subsidy reform in the Middle East and North Africa – a summary of recent progress and challenges ahead', International Monetary Fund.

12 http://edition.cnn.com/2012/01/11/business/nigeria-oil-economics/index.html, Defterios J, 'Nigeria's oil economics fuel deadly protests', *CNN*, 11 January 2012.

13 http://www.telegraph.co.uk/finance/newsbysector/energy/oilandgas/11338769/UK-oil-firms-warn-George-Osborne-Without-big-tax-cuts-we-are-doomed.html, Gribben R, 'UK oil firms warn Osborne: Without big tax cuts we are doomed', *Telegraph*, 11 January 2015.

14 Jevons W S (1865), *The Coal Question* (2nd ed.), London: Macmillan.

15 http://yorktownhistory.org/wp-content/archives/homepages/1900_predictions.htm, Yorktown History (2000), 'Predictions of the year 2000 from *The Ladies' Home Journal* of December 1900'.

16 White D (1919), 'The unmined supply of petroleum in the United States', *Transactions of the Society of Automotive Engineers*, 14 (1) 227.

17 http://www.hubbertpeak.com/hubbert/natgeog.htm, Grove N (1974), 'Oil, the dwindling treasure', *National Geographic*.

18 Hubbert M K (1956), 'Nuclear energy and the fossil fuels', *Drilling and Production Practice*.

19 http://www.creators.com/opinion/walter-williams/environmentalists-wild-predictions.html, Williams W (2008), 'Environmentalists' wild predictions', *creators.com*.

20 http://www.eia.gov/dnav/pet/hist/LeafHandler.ashx?n=PET&s=MCRFPUS2&f=M, US Energy Information Administration (2015), 'US field production of crude oil'.

21 http://www.worldenergyoutlook.org/media/weowebsite/factsheets/WEO2013_Factsheets.pdf, International Energy Agency (2013), 'World Energy Outlook 2013 factsheet'.

22 http://www.economist.com/node/21635472/print, 'The new economics of oil', *The Economist*, 6 December 2014.

23 http://www.economist.com/node/2155717, 'Ways to break the tyranny of oil are coming into view: governments need to promote them', *The Economist*, 23 October 2003.

24 http://iet.jrc.ec.europa.eu/remea/sites/remea/files/files/documents/com_2000_769_enegy_security.pdf, European Commission (2000), Green Paper, 'Towards a European strategy for the security of energy supplies', COM/2000/0769 final.

25 http://ec.europa.eu/eurostat/statistics-explained/index.php/Energy_production_and_imports#Further_Eurostat_information, Eurostat (2014), 'Energy production and imports'.

26 http://www.publications.parliament.uk/pa/cm201012/cmselect/cmenergy/1065/106508.htm#note132, Energy and Climate Change Select Committee (2011), 'UK energy supply: security or independence? – infrastructure resilience', parliament.uk.

27 http://www.aei.org/publication/seeds-of-ngo-activism-shell-capitulates-in-saro-wiwa-case/, Entine J (2009), 'Seeds of NGO activism: Shell capitulates in the Saro-Wiwa case', NGO Watch.

28 http://www.aljazeera.com/indepth/features/2015/04/drenched-oil-nigerians-demand-shell-spill-clean-150427101051609.html, Whitehead E (2015), 'Drenched in oil: Nigerians demand Shell spill clean-up', *Al-Jazeera*.

29 http://www.chathamhouse.org/publications/papers/view/185311, Stevens P (2012), 'The shale gas revolution – developments and changes', The Royal Institute of International Affairs, Chatham House, London.

30 http://www.worldenergyoutlook.org/media/weowebsite/2012/goldenrules/WEO2012_GoldenRulesReport.pdf, IEA (2012), 'Golden rules for a golden age of gas', OECD, Paris.

31 http://www.eia.gov/dnav/ng/hist/n9070us1m.htm, US Energy Information Administration (2014), 'US dry natural gas production'.

32 http://www.telegraph.co.uk/news/earth/energy/fracking/11124212/Fracking-under-homes-99-per-cent-opposed-to-law-change.html, Gosden E, 'Fracking under homes: 99% opposed to law change,' *Telegraph*, 26 September 2014.

33 http://engineeringfailures.org/?p=723, 'The Banqiao reservoir dam failure', *Engineering Failures*, (2012).

34 http://ec.europa.eu/research/energy/pdf/kina_en.pdf , European Commission (2005), 'ExternE: externalities of energy – methodology 2005 update', EUR 21951, Brussels.

35 http://www.oecd-nea.org/ndd/reports/2010/nea6862-comparing-risks.pdf, OECD (2010), 'Comparing nuclear accident risks with those from other energy sources'.

36 http://pubs.giss.nasa.gov/abs/kh05000e.html, Kharecha P. and Hansen J. (2013), 'Prevented mortality and greenhouse gas emissions from historical and projected nuclear power', *Environmental Science and Technology*, **47**, 4889–95.

37 http://www.rfa.org/english/commentaries/energy_watch/china-coal-deaths-03162015103452.html, Lelyveld M (2015), 'China cuts coal mine deaths, but count in doubt,' Radio Free Asia.

38 http://www.who.int/mediacentre/factsheets/fs313/en/, World Health Organisation (2014), 'Ambient (outdoor) air quality and health – factsheet 313'.

39 http://www.ncbi.nlm.nih.gov/pmc/articles/PMC1241789/pdf/ehp0112-000006.pdf, Bell M, Davis D and Fletcher T (2004), 'A retrospective assessment of mortality from the London smog episode of 1952: the role of influenza and pollution', *Environmental Health Perspectives*.

40 Grimston M et al. (2001), 'The European and global potential of carbon dioxide sequestration in tackling climate change', *Climate Policy*, **1**, 155–77.

41 http://www.ft.com/cms/s/0/95c83fa8-71ec-11e1-90b5-00144feab49a.html, Clark P, 'New technologies: carbon capture dogged by volatile history', *Financial Times*, 28 March 2012.

42 https://sequestration.mit.edu/tools/projects/boundary_dam.html, MIT, January 2015, 'Boundary Dam fact sheet: carbon dioxide capture and storage project'.

43 http://www.nature.com/scientificamerican/journal/v293/n1/full/scientificamerican0705-49.html, Socolow R (2005), 'Can we bury global warming?', *Scientific American*, **293**, 49–55.

3 | Nuclear energy

Notwithstanding the misleading description 'new renewables' applied to a number of ancient technological concepts by their advocates, nuclear power (from nuclear fission) is by some way the newest of the large-scale energy technologies currently available. The power of atomic technology made itself known to the public in stark and horrifying circumstances with the dropping of the atomic bombs on Hiroshima and Nagasaki in 1945, coming less than three years after the first demonstration of 'large-scale' nuclear fission. The first nuclear-generated electricity, enough to light four 50W light bulbs, was created in Idaho in December 1951. Within five years the UK was opening the world's first commercial-scale nuclear power station, at Calder Hall in Cumberland.

The energy field is characterised by a range of entrenched disputes. Large-scale hydropower, fracking and onshore wind power, for example, have their staunch advocates and impassioned opponents. But it is probably fair to say that the debate over nuclear power has been particularly acrimonious for some decades, even predating the first of the major accidents in a nuclear power reactor at Three Mile Island in 1979.

Advocates of nuclear power point to the relative abundance of uranium as a fuel source, especially if used in 'breeder' reactors; to the greater geographical diversity of its mineral deposits, in contrast with the concentration of hydrocarbon reserves in the Middle East and the former Soviet Union; to its very low emissions of carbon dioxide and other pollutants such as acidic gases and particulates; to its resistance to fuel price inflation; to its impressive safety record. Opponents raise concerns over nuclear weapons proliferation; the very heavy and growing capital investment costs of nuclear power plants; the very poor record of the industry on delivering construction programmes on time and within budget; the environmental damage associated with uranium mining; the failure to resolve the issue of radioactive waste management and disposal; the huge costs and psychological damage that arise from major nuclear power accidents; alleged uncertainties concerning the health effects of low-level radiation exposure; the centralist and secretive society necessary to sustain such a technology.

The debate is sometimes couched in terms of the 'need' for nuclear power. This is on the face of it a rather odd idea. To say that we 'need' nuclear power would seem to be tantamount to saying that if nuclear fission was not a feature of nature, or if we had not discovered it, then mankind could have no future. That is patently absurd. Yet the question which one would like to address – on balance does nuclear power contribute to a better, more sustainable future or not? – is strangely difficult to evaluate. There is a theological absolutism on both sides of the debate, or rather among the extreme apostles on both sides, which can be dismissive of any alternative view and perhaps represents a clash between the crypto-religions of science and anti-industrialism. This dispute is one that includes not just differences of opinion over facts, for example about the effects of low-level radiation, but also of values. Indeed, though one suspects the hard line advocates on the edge of either side of the argument would recoil at the thought that they have anything in common, their style and rhetoric are often strikingly similar, as shown in table 3.1.

The advocates	The opponents
Belief that major elements of the future are predictable; certainty about general projections for various energy sources. (For example, renewables demonstrably do not have the practical potential to be other than relatively minor players in world energy supply.)	Belief that major elements of the future are predictable; certainty about general projections for various energy sources. (For example, renewables demonstrably have the practical potential to predominate in world energy supply.)
Certainty about the future role of nuclear power (a major one) and about issues such as nuclear waste (not a difficult technical problem).	Certainty about the future role of nuclear power (no role at all) and about issues such as nuclear waste (a technically insoluble problem).
Arrogance born of belief in infallibility of own analysis.	Arrogance born of belief in infallibility of own analysis.
Belief that the public is irrationally frightened of nuclear power. If only people could be properly educated they would become more pro-nuclear and support the nuclear industry.	Belief that the public is irrationally complacent about nuclear power. If only people could be properly educated they would become more anti-nuclear and support anti-nuclear campaigns.
Characterisation of opponents as either foolish or ill-intentioned.	Characterisation of opponents as either foolish or ill-intentioned.
Belief that government is not to be trusted to take wise decisions as it is too much influenced by the anti-nuclear media and pressure groups.	Belief that government is not to be trusted to take wise decisions as it is too much influenced by the nuclear industry and its supporters.

Table 3.1 Attitudes towards nuclear power: advocates and proponents[1]

3.1 RAPID RISE, SLOW DECLINE

In July 2014 the leading light of an anti-nuclear pressure group calling itself the World Information Service on Energy (WISE – most organisations that include 'Information' in their title turn out to be impassioned propagandists for one side or the other in some dispute, especially if they describe themselves as 'independent') reported that nuclear power's percentage share of global power generation had fallen to 10.8 per cent in 2013, the lowest since the 1980s and well behind the 17.6 per cent in its peak year of 1996.[2]

In absolute terms nuclear still generated more power in 2013 than it had in 1996 (2,489 TWh against 2,408 TWh), some 25 per cent more than it had in 1989. But it was unarguable that this figure, albeit skewed somewhat by the withdrawal of the Japanese fleet after Fukushima, several of whose reactors had at least a prospect of reopening at some point, was considerably lower than nuclear's peak year of 2006 in terms of output (2,806 TWh). The source which had been the fastest growing of all the major energy sources in the 1970s, 1980s and 1990s had ground to something of a halt. The WISE report drew a considered and certainly not dismissive response from nuclear supporters.[3]

The decline could be ascribed to a range of factors including worsening relative economics, growing confidence that security of supply was not as challenging as had been thought and a set of public perception problems.

3.2 THE ORIGINS OF NUCLEAR TECHNOLOGY

The basics of nuclear science are introduced in chapter 11. The rise of civil nuclear power technology was as rapid and impressive as had been that of electricity generated from wind or hydropower half a century earlier.

In the UK in July 1941 the MAUD Committee (codename for the research programme into fission and its potential applications) produced two summary reports: *Use of uranium for a bomb* and *Use of uranium as a source of power*. The latter concluded that the controlled fission of uranium could be used to provide energy in the form of heat for use in machines. It referred to the use of heavy water and possibly graphite as 'moderators' for the fast neutrons, as discussed later, suggesting that even ordinary ('light') water could be used if the uranium was enriched in the uranium-235 isotope. It concluded that the 'uranium boiler' had considerable promise for future peaceful uses but that it was not worth considering during wartime conditions.

After the US entered the Second World War in late 1941, collaboration between the UK and the US on developing atomic weapons deepened through the Manhattan Project, which had been established by Leo Szilard, 'the reluctant father of the atomic bomb'. On 2 December 1942, in a squash court in Chicago, the first controlled nuclear chain reaction was achieved by a team headed by Enrico Fermi, 'father of the atomic bomb', also called 'the father of atomic power'. Full-scale plutonium production reactors were established at Argonne in Illinois, Oak Ridge in Tennessee, and at Hanford in Washington which also had a reprocessing plant to extract the plutonium. The first explosive atomic device, based on plutonium, was tested successfully at Alamogordo in New Mexico on 16 July 1945, followed by the first and so far only use of atomic bombs in conflict in Japan at Hiroshima (uranium) on 6 August and at Nagasaki (plutonium) on 9 August 1945.

3.3 REVIVAL OF THE 'NUCLEAR BOILER' AND A RANGE OF CONCEPTS

After the Second World War, weapons development continued on both sides of the Iron Curtain but with a new focus on harnessing atomic power, now dramatically if tragically demonstrated, for making steam and electricity. In principle anything that can explode can also be harnessed to provide useful energy for everyday purposes. In an atom bomb one wants to release as much energy as quickly as possible by allowing as many as possible of the neutrons produced by each fission, generally two or three, to cause new fissions. The number of fissions per microsecond increases very rapidly when two 'sub-critical masses' are brought together. (If the mass of fissile material is too small, too many neutrons escape from the surface and sustainable fission never starts.)

For a source of useful energy, however, one wants each fission to produce on average exactly one neutron which can cause a further fission, so the process neither fizzles out – if there were too few neutrons available, perhaps because they are being lost at the edges of the uranium block or being absorbed by something else apart from the fuel – nor runs away with itself as in a bomb.

A variety of reactor concepts emerged in the 1950s. In all cases the ultimate purpose of the nuclear core was to produce heat which would be transferred to a circuit containing water. The steam produced in that circuit was used to drive turbines in much the same way as it was harnessed in power stations fuelled by coal, oil or more recently gas.

A uranium-based nuclear power reactor requires something to transfer the heat from the 'core' where the fission occurs to the circuit that boils the water to make the electricity (a 'coolant'); something to slow the neutrons down so they will be absorbed by the next uranium-235 atom rather than just bouncing off (a 'moderator'); and something to absorb any excess neutrons, so ensuring the process proceeds at a steady rate ('control rods').

The coolant obviously has to be a fluid which can be circulated into the core and out again to boil the water. Candidates include light water; 'heavy water' (a compound of oxygen with the second isotope of hydrogen, known as deuterium and consisting of one proton and one neutron); a gas such as helium or carbon dioxide; or a liquid metal like the very light reactive metal sodium or a mixture of the heavy metals lead and bismuth.

The moderator has to consist of a small atom which can absorb some of the neutron's momentum and so slow it down. Neutrons tend simply to bounce off big atoms at more or less the same speed they were going when they hit them. (In snooker the cue ball slows down significantly when it hits another ball but comes off a cushion at much the same speed as when it strikes it, a consequence of the Law of Conservation of Momentum.) Good candidates include graphite (made up of carbon atoms, mass number 12, i.e. whose nuclei consist of six protons and six neutrons), light water (one proton) and heavy water (a proton and a neutron). If the nuclear fuel is plutonium there is no need to moderate the neutrons because plutonium undergoes fission with 'fast' as well as 'slow' neutrons. Reactors using plutonium fuel in this way are therefore called 'fast reactors', as opposed to 'thermal reactors' which employ a moderator.

All reactor types make use of external neutron absorbers, such as silver, boron, indium or cadmium, usually in the form of rods that can be inserted or removed from the core to slow down or speed up the fission reaction. Emergency measures may include the option of flooding the core with water containing dissolved boron. However, the coolant and/or the moderator may also play a significant role in absorbing neutrons. Liquids such as light or heavy water are generally speaking quite good neutron absorbers, while gases like steam or carbon dioxide are much less effective. Indeed, light water is such a good absorber of neutrons that it can create a problem of keeping the reactor from shutting itself down, leading the Canadians to develop a pressurised water reactor using much more expensive heavy water instead.

The main reactor types can be characterised as shown in table 3.2 – ignoring the role played by control rods in absorbing neutrons.

Table 3.2 A brief description of major reactor concepts

Reactor Type	Coolant	Moderator	Absorption
Pressurised Water Reactor (PWR, also VVER)	Light water	Light water	Light water
Boiling Water Reactor (BWR)	Light water	Light water	Light water
RBMK	Light water	Graphite (and water)	Light water (at low power)
Gas-cooled (e.g. Magnox; Advanced Gas-cooled Reactor, AGR)	Carbon dioxide	Graphite	-
(Very) High Temperature Reactor ((V)HTR)	Helium	Graphite	-
CANDU (Canadian Deuterium-Uranium)	Heavy water	Heavy water (a bit less effective than light water)	Heavy water (much less effective than light water)
Fast Reactor (FR)	Sodium or lead/bismuth	-	-

One difference among reactor designs is what happens to the fission processes in a 'loss of coolant accident' (LOCA), whereby a breach in the pipework carrying the coolant to, through or from the core allows coolant to escape. In light water- or heavy water-cooled reactors, such as the American PWR or BWR, the Soviet VVER or the Canadian CANDU, escape of the coolant might be expected to result in a tendency for the fission processes to increase, as the water plays a significant role in absorbing neutrons. However, the water also provides moderation. So the fission process shuts down at once because, although for an instant fewer neutrons are being absorbed by the coolant, the neutrons that are being produced are going too fast to support a sustainable chain reaction. In a gas-cooled reactor using carbon dioxide or helium loss of the coolant does not result in a significant loss in absorption as the coolant does not play much of a role in that respect, so there is no tendency for the fission processes to increase and control rods can be introduced in an orderly fashion to shut down the fission processes. In a fast reactor the coolant is a metal which would not boil away if the circuit was breached so a LOCA is less plausible.

The RBMK, only ever built at sites in the former Soviet Union including at Chernobyl in Ukraine, mixed the two concepts. In normal operation water boils in the tubes going through the core, so does not play an important part in neutron absorption – or moderation, thereby requiring the presence of a fixed moderator, graphite. However, if for any reason operators want to run the reactor at very low power it would be necessary to remove practically all of the control rods, as the tubes would be largely filled with liquid water which would absorb neutrons efficiently. If anything were then to happen to cause that liquid water to boil – either a breach in the tubes or some slight increase in temperature – the situation would be very serious indeed. As the water boiled there would be an increase in the number of neutrons. Unlike a PWR or BWR, these neutrons would still be moderated by the fixed graphite structures. This could lead therefore to an increase in the amount of fission occurring, in turn leading to an increase in temperature, causing more water to boil and a further increase in neutrons and therefore of fission … . This situation is called a 'positive void coefficient' – if the 'void' (i.e. the degree to which steam replaces liquid water) increases, a positive feedback loop develops causing more heat production, more voiding and yet more heat production. This could only happen at very low power output, which is why operators were unequivocally required never to operate the station below 25 per cent of normal power output or to remove more than an equivalent of 181-196 of the 211 control rods from the core. Unfortunately, there was no insuperable physical barrier to operators breaching these conditions.

The rate of development and deployment of new reactor types in the 1940s and particularly the 1950s shows that there is nothing inherent in nuclear technology to prevent rapid growth if decision-making processes allow. The first nuclear reactor to produce electricity was a fast reactor, the Experimental Breeder Reactor (EBR-1), designed and operated by Argonne National Laboratory and situated at Idaho in the US. This reactor started up in August 1951 and in December of that year generated the world's first nuclear electricity, some 200 kW which powered four light bulbs. The first nuclear station to export electricity into a local grid (around 5 MW) was at Obninsk in 1954, some 100 km from Moscow, in a reactor close in concept to the RBMK. The first nuclear power station to approach what we might call commercial scale was at Calder Hall in the UK in 1956: when its four reactors were commissioned it had a rated capacity of 240 MW.

Electricity production was not regarded as the only possible use of fission power. Admiral Hyman Rickover, 'father of the nuclear navy', led a US project to develop a light-water cooled reactor concept that could, inter alia, be used in nuclear submarines. The prototype naval PWR started up in March 1953 in Idaho and the first nuclear-powered submarine, USS Nautilus, was launched in 1954. In 1959 both the US and USSR would launch their first nuclear-powered surface vessels. But it was in electricity generation that civil nuclear technology was to make its mark.

3.4 THE UK NUCLEAR PROGRAMME – POST-WAR ORIGINS

The British nuclear programme after the Second World War was established with remarkable speed. In 1946 research and production facilities were established at Harwell in Oxfordshire, Risley in Cheshire, Springfields in Lancashire, and Windscale in Cumbria. Motivations included the desire to create an independent UK nuclear deterrent and a wish to develop a new way of making electricity which would not suffer from some of the problems associated with coal, such as its effect on air quality and the power of the mining unions. The first, non-electricity producing nuclear reactors in the UK were built at what was then the Atomic Energy Research Establishment (AERE) at Harwell. Europe's first reactor, GLEEP – the 3 kW Graphite Low Energy Experimental Pile – began operating in 1946 and ran until 1990. The 6 MW British Experimental Pile 0 (BEPO) followed in 1948. In 1947 an ex-Royal Ordnance factory at Sellafield in what was then Cumberland, now Cumbria, was chosen for construction of Britain's military reactors. The site was renamed 'Windscale' to avoid confusion with the uranium manufacturing establishment at Springfields near Preston, though it remained known as Sellafield locally. Windscale Pile No. 1 became operational in October 1950 and Windscale Pile No. 2 in June 1951.[4] These reactors did not generate electricity and heat was simply vented to atmosphere by blowing air over the graphite cores. Their role was to create plutonium for the weapons programme which had been revealed to the House of Commons by the Atlee government in 1948. An 'enrichment plant' was opened at Capenhurst in Cheshire in 1952 to create uranium with a higher proportion of the fissile uranium-235 isotope than found in the naturally occurring metal. In the mid-to-late 1950s two more sites were established: at Dounreay, nine miles from Thurso in Caithness on Scotland's north coast, for the UK's fast reactor research programme; and at Winfrith Heath in Dorset for researching other novel reactor concepts.

Soon after the Windscale Piles began operating attention turned to the next generation of plutonium-producing reactors. At Prime Minister Winston Churchill's request, Christopher Hinton, 'the father of the UK civil nuclear industry', designed a reactor capable of producing both plutonium – its main purpose – and electricity and supervised the construction of the first examples of a PIPPA (Pressurised Pile Producing Power and Plutonium) at Calder Hall, next to the Windscale site. Construction of the planned single 60 MW unit began in 1953; and soon afterwards it was decided to add a second unit, followed in 1955 by orders for two more, eventually creating a power station with an output of 240 MW. (As a comparison, Battersea Power Station, the country's third-largest when completed in 1955, had a capacity of just over 500 MW.) Calder Hall involved gas-cooled, graphite-moderated reactors in which the fuel was kept in a non-oxidising magnesium alloy (Magnox) which was to lend its name to the design.

Calder Hall would be connected to the grid in August 1956 but while it was still under construction in 1955 the UK government published a White Paper, *A Programme of Nuclear Power.*[5] Power demand had been increasing rapidly and peak metered demand in England and Wales would more than treble in the fifteen years from that of 1948 – see figure 3.1.

The White Paper set out a ten-year programme for the construction of a fleet of Magnox civil nuclear power stations intended to supply between 1,500 and 2,000 MW of electricity to the national grid. The Magnox programme was to provide 25 per cent of the UK's electricity needs at a total cost of £300 million (£6.5 billion in 2015 prices). The case for nuclear power received a further boost the following year: the Suez Crisis of 1956 can be regarded as one of the first wars for oil. The unsuccessful military operation in Egypt left Britain's access to oil from the Persian Gulf looking very precarious, leading to problems such as petrol rationing. Although oil was a minor fuel for electricity generation in the UK at the time, its use was growing.[7] This further strengthened the determination to develop a source of energy that was less vulnerable to foreign disruption and led to plans to treble the Magnox programme to 6,000 MW.[8] Although the programme was subsequently scaled back – a theme which would be repeated more than once with respect to UK nuclear ambitions over the years – some 26 Magnox reactors were eventually being built at 11 nuclear sites across the UK, including the eight military reactors at Calder Hall and Chapelcross, with a total capacity 4,200 MW. Two Magnox stations were exported: to Italy and Japan. The North Koreans based their 5 MW experimental military reactor

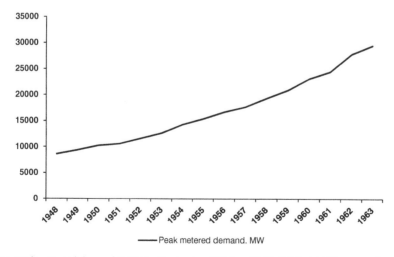

Figure 3.1 Peak metered demand (MW) in England and Wales, 1948–63 (total UK system demand is typically some 10–15% higher)[6]

at Yongbyon on the Magnox design; this plant was closed down in 1994, reactivated in 2003, closed again in 2007, reactivated again in 2013. North Korea also had plans to build two larger Magnox-style reactors but these were abandoned as a result of international agreements reached in 1994. The first UK Magnox station at Berkeley to cease generating electricity closed in March 1989; the last at Wylfa operated until December 2015.

The UK as trailblazer – pros and cons

The first UK nuclear programme was also the world's. By the end of 1964 the UK had 1,245MW of nuclear capacity operating, including prototype power reactors, followed by the US with 599MW, Italy with 413MW and the USSR with 304MW. A number of themes with relevance beyond the UK nuclear industry emerged from this first programme and the way it led into its successor, of which perhaps the most important are:

- the dangers in making sweeping economic and technical claims about unproven technology – especially based on scaling up relatively small prototypes by a factor of 10 or 20;
- the folly of building a large number of what were effectively prototypes, each to a different design, rather than identifying a single design and batch-producing them;
- the importance of maintaining a strong link with international developments in the technological field.

During this period there were five industrial consortia competing to build Magnox stations. All of the 11 Magnox power stations built in the UK, with the exception of the military reactors at Calder Hall and Chapelcross, were built to different designs. One was refuelled from below; one had heat exchangers external to the reactor building; two were built with concrete 'pressure vessels' – which protect the reactor core – while the rest used steel; and so on. Each of the purely commercial reactors took a different size and design of fuel element. This practice of building a series of prototypes did not serve the UK nuclear industry well, severely limiting both learning curve benefits and economies of scale during manufacture and servicing.

From the start of the Magnox programme it was recognised that electricity generated from Magnox stations would be more expensive than that generated by burning coal. Nuclear energy was nonetheless to be pursued, for a number of reasons. It represented an added degree of fuel diversity, and therefore a potentially improved security of supply, in a world in which coal and coal mining unions remained dominant, global oil supplies were questionable and the UK was as yet unaware of the hydrocarbon resource lying under its territorial waters. The environmental effects of burning coal remained starkly apparent.[9] It was believed that in time, as the cost of other fuels rose and nuclear technology became cheaper, nuclear power would become more competitive than other options, with the exception of hydropower. And, one suspects, membership, indeed leadership of the hi-tech 'nuclear club' would demonstrate to the world that the great imperial power Great Britain was still to be taken seriously on the world stage despite the economic burden of the Second World War and the subsequent dissolution of the Empire.

In engineering terms this first British programme was a remarkable achievement. Calder Hall generated electricity from 1956 to 2003, a period of 47 years. All of the Magnox reactors exceeded their 25-year 'design life' – in effect a guess as to their longevity to give a basis for accounting policy as nobody really knew for how long they could operate safely when they were opened – with the exception of the Latina plant which was exported to Italy and whose operating life was cut short at the tender age of 24 owing to a post-Chernobyl accident referendum on nuclear power in the country. However, the Magnox design was relatively inefficient, both thermally, i.e. a relatively low proportion of the heat generated was turned into electricity, and in terms of use of uranium, with the result that costs were high. In addition, relatively large quantities of operational and, eventually, decommissioning waste was created, alongside waste arising from the research projects of the 1940s, 1950s and 1960s. Some nominal financial provision was made to finally discharge these liabilities but, inadequate as it undoubtedly was, this provision was never made available, leaving a large liability to future taxpayers in the form of what have become known as the 'legacy wastes', a problem which seems to grow with every announcement from the responsible agency, the Nuclear Decommissioning Authority, which assumed responsibility for the issue in 2005.

In 1957 a fire broke out at Windscale Pile No. 1 which in terms of releases of radioactive materials still ranks as the world's third most serious incident in a nuclear reactor, albeit not a power reactor. The accident's technical cause stemmed from 'Wigner energy', named after Eugene Wigner, 'father of the quantum theory of chaos'. When the first pile had been shut down in 1952 the temperature in its core of graphite bricks continued to rise. Graphite's stable structure involves sheets of hexagonally arranged carbon atoms with relatively long weak bonds between the layers – this accounts for graphite's slippery feel to the touch. Wigner suggested that as neutrons bumped into the carbon atoms the energy that was transferred pushed some carbon atoms out of formation and into the gaps between the layers. This potential energy became stored up over time until there was enough to be given out quite rapidly. To prevent this, the piles were to be 'annealed' every year to prevent Wigner energy building up. Annealing involved switching off the fans that blew cooling air over the graphite pile, allowing the temperature to rise until the Wigner energy was safely released at a temperature of around 250°C. In the 1957 annealing process, however, it seems that a badly-placed thermistor (for measuring temperature) suggested that in one part of the core the annealing had not begun. A second burst of heating was applied but in fact the area in question had been heating up as planned. The outcome was that the graphite and the uranium fuel caught fire. Switching the cooling circulation on again simply and literally 'fanned the flames' and the fire burned for more than a day. Eventually the decision was taken to douse the fire with water – a risk because at the temperatures being measured of over 1,000°C the water might have pyrolysed into hydrogen and oxygen, resulting in a hydrogen explosion like those seen in the outer buildings of the Fukushima reactors more than half a century later. Fortunately, the fire was brought under control. Restrictions were placed on local milk production for three months until the released iodine-131 had decayed away and it is not thought that the releases of radioactivity will have led to any health problems.

Reports commissioned after the fire made several criticisms of the management structure of the United Kingdom Atomic Energy Authority (UKAEA), the government quasi-department responsible for nuclear research and policy advice which had been created in 1954. In 1959 what was to become the Nuclear Installations Inspectorate (NII) was set up. It is now part of the Office for Nuclear Regulation (ONR), a statutory public corporation established in 2014 which acts as the government's watchdog on the industry.

Though the Windscale fire did relatively little real damage, it did perhaps mark the end of a honeymoon, the early acceptance or even enthusiasm which had accompanied developments in the nuclear field thus far. At the same time, opposition to the nuclear arms race and the atmospheric testing of atomic weapons was growing. In 1957 there were protests by the likes of Albert Schweitzer, J.B. Priestley, the Japanese government and an influential group of West German physicists. The forerunner of the Campaign for Nuclear Disarmament (CND) was formed in the UK and in the following year the first march on the nuclear weapons establishment at Aldermaston took place.

The second UK nuclear power programme

By the late 1950s, with the Magnox programme under construction, the UKAEA turned its attention to finding a successor. This heralded a time of many different research programmes. In 1957 it was decided to build an improved version of the Magnox design at Windscale. A 33 MW prototype Advanced Gas-cooled Reactor (AGR) started operating in 1962. In 1959 a small fast reactor at Dounreay went critical, i.e. started up – nuclear language is often careless as to the impression it gives to lay people – and an international project to develop a gas-cooled High Temperature 'Dragon' Reactor began at Winfrith. Construction of another reactor concept, with some similarities to the Canadian CANDU programme and called the Steam Generating Heavy Water Reactor (SGHWR), began in 1963, also at Winfrith. There was also enthusiastic research into nuclear fusion.

The impression that in the UK, as in other countries, decisions on policy as well as how to implement that policy had been delegated to the nuclear community is inescapable. Politicians seemed to feel that nuclear energy was both essential, for concrete reasons and also perhaps as proof of the technological credentials of the nation, but at the same time too complicated for anyone but the 'experts' to provide governance. A vast amount of research resource was made available without proper democratic or commercial oversight, resulting in the development of a wide range of dead-end technologies at the end of which arguably the UK chose the wrong one – at least with the benefit of hindsight. 'Technocratic capture' is a potentially serious challenge to the theoretical benefit of a state-controlled approach to electricity supply and is perhaps unavoidable in a 'phase 1' society as discussed in chapter 10.

Following the success of the Windscale prototype the chosen successor to Magnox was the AGR, a decision which in effect cut the UK off from the global nuclear mainstream for 25 years, most of the rest of the world pursuing light water technology, mainly PWR or BWR. In 1964 the Minister of Power, Fred Lee, announced the decision by telling Parliament, 'we have hit the jackpot … we have the greatest breakthrough of all time'.[10] It was apparently assumed that scaling up the prototype by a factor of 20 would be a relatively straightforward technical matter.

This decision and the decision to give the work for the first three AGR stations to the three surviving construction consortia in the nuclear field were not without their critics. There were those who preferred the essentially American PWR. Others felt that the wrong version of AGR was chosen as the first to be constructed.

Whatever one's view on such matters the AGR programme was a disappointment. The first station, Dungeness B in Kent, ordered in 1965, only began working reasonably well in 1993 and ran heavily over budget. The second design, built at Hinkley Point B in Somerset and Hunterston B in Strathclyde, was considerably more successful; but the third design, Heysham I in Lancashire and Hartlepool, County Durham, also took some time to start performing adequately. A fourth design, based on the Hinkley Point and Hunterston design, was built at Torness in Lothian and at Heysham II in the late 1970s as something of a stopgap as the UK prepared to adopt PWR technology. Reminiscent of the Magnox programme, there were eventually just seven AGR stations but built to four different designs. By the mid-2010s closure of the AGR programme was expected to span the period 2023–33.

A third programme?

The inconsistent performance of the AGR programme led to considerable debate over the third stage of Britain's nuclear power programme. Despite a decision in 1966 to build a one-quarter scale (250 MW) prototype fast reactor at Dounreay, which operated from 1973 to 1994, and the relative technical success of the Dragon programme, which was nonetheless abandoned in 1974, the choice lay between continuing with the AGR, developing the SGHWR or adopting the PWR. SGHWR was favoured in an announcement by the Secretary of State for Energy in 1974, Sizewell in Suffolk and Torness being chosen for the first stations, but within two years rumours were circulating that the decision was to be reversed. Demand for electricity was not growing as rapidly as expected, obviating the need for as much new plant of any description as had been assumed, and the task of scaling up the 100 MW prototype SGHWR at

Winfrith to a 600 MW commercial design was proving problematic and expensive. A government direction to the Central Electricity Generating Board (CEGB) to build a new coal-fired plant, Drax B, further damaged the case for new nuclear investment at that stage.

In 1978 the SGHWR was abandoned following a critical report by the National Nuclear Corporation (NNC). The last two AGRs were ordered but the PWR belatedly became the favoured technology for development of a third nuclear programme in Britain. On the face of it this seemed a strange compromise but was one forced by the bitter rivalry between the AGR and PWR camps within the industry. This decision, like those which had led to the Windscale fire 20 years earlier, served to demonstrate the downside of a phase 1 relationship between science and society in which scientists and technologists are more or less left to 'get on with it': without proper scrutiny plenty of decisions can be taken quickly but some of those rushed decisions turn out to be less than optimal. Figure 3.2 shows the locations of the nuclear power sites in the UK.

A new UK government

When Margaret Thatcher was elected as prime minister in 1979 she took power as a firm advocate of nuclear energy, not least because she had seen the National Union of Coalminers (NUM) destroy the previous Conservative administration (1970–74) and was determined to break its powers. To do that, in a country where coal was responsible for some 80 per cent of electricity generation, would require a number of measures, of which having a large-scale alternative way of generating electricity was one. The final two AGRs were under construction, the UK had decided to rejoin the global nuclear mainstream by buying in PWR technology from the US and oil prices were in the throes of the second major hike within a decade. In December 1979 the energy secretary, David Howell, made a statement implying support for building 15,000 MW of PWR capacity, a family of ten reactors.

Contrary to popular misrepresentation this was never a commitment, the actual wording being: 'Looking ahead, the electricity supply industry has advised that even on cautious assumptions it would need to order at least one new nuclear power station a year in the decade from 1982, or a programme of the order of 15,000 MW over ten years. The precise level of future ordering will depend on the development of electricity demand and the performance of the industry but we consider this a reasonable prospect against which the nuclear and power plant industries can plan. Decisions about the choice of reactor for later orders will be taken in due course.'

Nonetheless, the implication looked clear enough and Sizewell was chosen as the site for the lead plant of such a programme. A public inquiry was held between 1983 and 1985 and the decision to build followed in 1987. By now profound decision paralysis was setting in but at least a decision did eventually emerge. Hinkley Point was identified as the site for the second reactor, intended to be built to a design as close to Sizewell B's as possible, to be followed by Wylfa in north Wales and then by Sizewell again. Howell's statement about 'later orders' hints perhaps that few lessons had been learned about the dangers of changing the design for each project but in the event that point turned out to be academic.

The global nuclear industry was coming to terms with the accident at Three Mile Island (TMI) in Pennsylvania in March 1979. In the UK the direct implications of TMI were limited as Britain still did not use light water technology, but the incident was influential in persuading the NII that if the UK were indeed to pursue the PWR for its next programme of reactors a 'belt-and-braces' approach to safety licensing would have to be followed, further adding to costs. And then came the liberalisation of the electricity power supply and nuclear was withdrawn from privatisation as being unfinanceable. The rest of the UK nuclear industry meanwhile did little to strengthen the case for new build: in 1999 British Nuclear Fuels plc (BNFL), the state-owned nuclear fuel cycle company, was at the centre of a scandal involving the falsification of quality assurance data concerning the export of mixed oxide fuel pellets to Japan. Nuclear power had lost what shine it had left and the bankruptcy of the by then privatised nuclear generator British Energy in 2003 seemed simply to be the final nail in the coffin.

3.5 GLOBAL DEVELOPMENTS

The UK's early lead in civil nuclear technology did not last long. In 1953 President Eisenhower proposed his 'Atoms for Peace' programme, which redirected significant research effort towards electricity generation and set the course for civil nuclear energy development in the US. The following year businessman and philanthropist Lewis Strauss, then chairman of the US Atomic Energy Commission (AEC), made what has become an infamous reference to electricity as 'too cheap to meter',[11] which Big Green has long claimed to be evidence of how unrealistic all advocates of nuclear power are. Strauss was actually referring to nuclear fusion and in any case was making a wider, if wildly overblown, claim for the role that science and technology could play in improving lives:

> It is not too much to expect that our children will enjoy in their homes electrical energy too cheap to meter. It is not too much to expect that our children will know of great periodic regional famines in the world only as matters of history and will experience a lifespan far longer than ours, as disease yields and man comes to understand what causes him to age.[12]

One can argue that using this quote to show that 'the nuclear industry' was making outrageous claims is no more justified than using it to show that 'the medical profession' was promising to eradicate disease and eliminate ageing, or 'the agriculture industry' to eliminate food shortages.

As a counterweight, in the same year Strauss's predecessor as chairman of the AEC, Gordon Dean, offered a more mainstream assessment of the thinking of the day.

> Among all the questions hanging over the future of atomic power, perhaps the most fundamental is this: is it all really worth the effort? I have heard many people express shock and surprise when they learned that about all they can expect from atomic power, at least at first, is a new source of electricity that will only take a few pennies a month, if that, off their monthly light bill. This is, however, the case, and here is why: to produce electricity an atomic power plant needs all of the electrical generating and distribution equipment that a coal-burning plant needs. The only difference is that in the atomic plant the coal hopper and steam boiler would be replaced by a nuclear reactor and a different kind of steam boiler. There is no chance, therefore, of reducing the cost of the plant by going to the atom for fuel. As a matter of fact, it seems quite possible that atomic power plants will always cost more to build than coal plants – they certainly do now – because a nuclear reactor is, by its very nature, vastly more expensive than a coal furnace. The place where you can save money by going over to atomic power is in the cost of the fuel. And here you do save money, because the atom packs so much energy into such a small space. This means that your fuel, per unit of heat, not only comes more cheaply in the first place; it also means that you save money all along the line on transportation, handling and storage charges. So great is this saving that some economists, when calculating the cost of atomic power, put the cost of the nuclear fuel down as virtually zero. But it is important to remember that, even if coal were mined and distributed free to electric generating plants today, the reduction in your monthly electricity bill would amount to but 20 per cent, so great is the cost of the plant itself and the distribution system. To express it in the simplest terms: you can save a lot of money on fuel if you have an atomic power plant but it will cost a great deal more to build than a coal-burning plant.[13]

It is perhaps illuminating to note that Big Green still prefers to use Strauss's ancient quote than the more reasonable and equally authoritative, contemporary one from Dean, let alone Jürgen Trittin's, Germany's former environment minister, much more recent (2004) implication that Energiewende would cost each household the equivalent of 'a scoop of ice cream a month'.

In the two decades following the end of the Second World War nuclear energy occupied a privileged position in many countries. For example, the preamble to the

European EURATOM Treaty of 1957 states: 'Nuclear energy represents an essential resource for the development and invigoration of industry and will permit the advancement of the cause of peace'.[14] However, lest it be thought that nuclear power was always welcomed with open arms in its early days, in 1958 Pacific Gas & Electric proposed to build a commercial nuclear power plant at Bodega Bay in California, 80 km from San Francisco, at a site on the San Andreas fault and close to the centre of the area's dairy and fishing industries; thermal pollution from the cooling water was one particular concern. The Northern California Association to Preserve Bodega Head (NCAPBH) was formed, instigating a press campaign and submitting appeals to various state and federal bodies. Plans to build the plant were abandoned in 1964, by which time a pit had been dug for the foundations of the reactor near the tip of Bodega Head. Since the abandonment of the site the pit has partially filled with water and become a pond, informally called the 'Hole in the Head'.[15]

At the same time Westinghouse was designing the first fully commercial PWR power reactor, with a 250 MW output, at Yankee Rowe in Massachusetts, which started up in 1960. Meanwhile the Boiling Water Reactor (BWR) was developed by the Argonne National Laboratory. This involved water boiling as it passed through the core containing the fissile material and the steam so produced being used to drive the turbines directly. A prototype BWR, Vallecitos, ran from 1957 to 1963 and the first commercial unit, Dresden 1 (250 MW), started up in 1960. By the end of the 1960s orders were being placed for PWR and BWR reactor units of more than 1,000 MW electrical output.

Canadian reactor development set off on a quite different, but no less rapid, journey using heavy water as both moderator and coolant and natural uranium for fuel. The first unit started up in 1962. The CANDU design continued to be refined throughout the next half-century. France started out with a gas-graphite design similar to the UK Magnox used at Calder Hall; its first prototype reactor started up in 1956, with commercial models operating from 1959.

In 1964 the first two commercial-scale Soviet nuclear power plants were commissioned, a decade after the 5 MW Obninsk reactor had become the world's first to feed power into a grid. A 100 MW boiling water graphite channel RBMK reactor began operating in Beloyarsk in the Urals; the first large RBMK started up at Sosnovy Bor near Leningrad in 1973. A different design, a 210 MW pressurised water reactor known as VVER, began operating at Novovoronezh on the Volga, to be followed in 1969 by a 365 MW prototype of the same design which became the basis of a programme of reactors of 440 MW output. In turn this was superseded in the 1980s by a 1,000 MW version which became a standard design. In Kazakhstan the world's first commercial prototype fast reactor, the BN-350, started up in 1972 on the Caspian Sea, producing 120 MW of electricity and also energy to desalinate seawater. A larger version, BN-600, was opened in Beloyarsk in 1980.

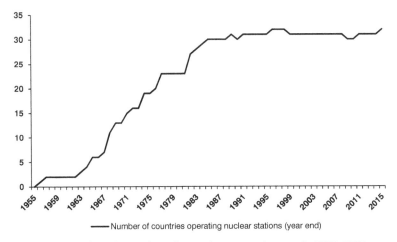

Figure 3.3 Rise in the number of countries using nuclear power (year-end), 1955–2015

By the end of 1974 there were 19 countries operating nuclear power reactors; by the end of 1984 the figure was 29 – see figure 3.3.

Installed capacity and amount of power generated expanded rapidly between 1975 and 1995, as shown in figures 3.4 and 3.5 respectively.

Around the world, with few exceptions countries chose light water designs for their nuclear power programmes, so that by the mid-2010s 68 per cent of the world capacity was PWR and 20 per cent BWR. France built a fleet of 58 reactors based on just three related PWR designs initially imported from the US.

In addition to the 31 countries which went on to develop commercial civil nuclear energy, the International Atomic Energy Agency (IAEA) lists a further 38 which have operated research reactors, mostly commissioned between the 1950s and 1980s. These include nations as diverse as Algeria (first operation in 1989), Australia (1958), Bangladesh (1986), Colombia (1965), Congo (1959), Egypt (1961), Ghana (1994), Jamaica (1984), Norway (1959), Peru (1978), Poland (1958), Thailand (1977), Turkey (1962), Venezuela (1960) and Vietnam (1963).[16]

So, clearly a nuclear golden age? Well, perhaps not. One of the consequences of the relatively long construction period involved in nuclear power is that a change in policy or sentiment can often take several years, if not decades, to become reflected in the raw statistics.

The TMI accident was undoubtedly a significant turning point in the global development of nuclear power.[17] Between 1963 and 1979 the number of reactors under construction had increased globally in every year except 1971 and 1978. However, following the event the number of reactors under construction declined every year from 1980 to 1998.[18]

It would be an exaggeration to ascribe plant cancellations during this period purely to the accidents at TMI and subsequently at Chernobyl in 1986. By 1979 the world was

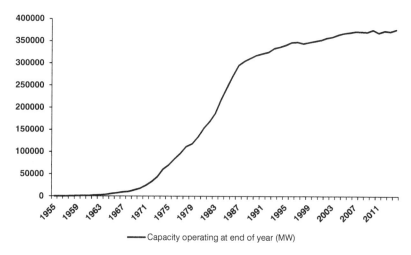

Figure 3.4 Global nuclear power capacity operating at end of year (MW), 1955–2014

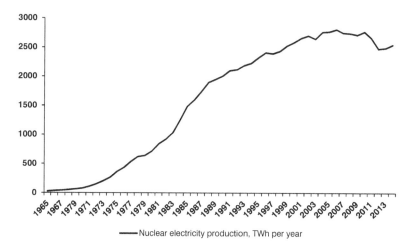

Figure 3.5 Global nuclear electricity production (TWh per year), 1965–2014

experiencing economic recession caused by the hikes in oil prices through the mid-to-late 1970s. The resulting overcapacity in power production had already led to some 40 planned nuclear power plants being cancelled between 1973 and 1979. Nor were cost or time overruns afflicting nuclear projects confined to the post-TMI period.

As shown in figure 3.6, the average cost overrun in the decade from 1966 onwards was 207 per cent. In the words of one commentator critical of the nuclear industry: 'It was not the Three Mile Island accident that caused the nuclear industry's collapse. By the time TMI happened a wave of cancellations of new nuclear plant orders was already underway. If anything, the accident simply capped off a trend which was already occurring.'[20]

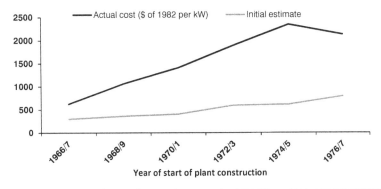

Figure 3.6 Projected and actual cost of US nuclear capacity (1982$ per kW), 1966/7–1976/7[19]

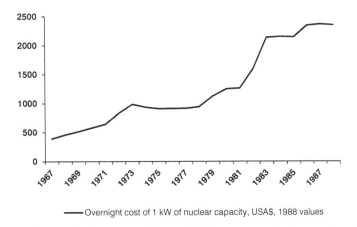

Figure 3.7 Increase in overnight costs of US nuclear installation (corrected for inflation), 1967–88[23]

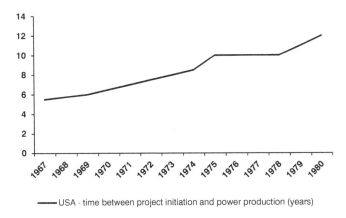

Figure 3.8 US – time between project initiation and power production (years), 1967–80

The escalation in costs for nuclear new builds in the 1970s and 1980s was caused by at least four factors.[21] Between 1976 and 1988 the labour costs associated with nuclear construction increased enormously – by an average of 18.7 per cent per year. This was largely accounted for by an increased use of professionals, such as design, construction and quality control engineers, owing to 'regulatory ratcheting', especially following TMI. Regulators constantly raised the 'safety standards' required – though it is questionable as to whether this actually led to any improvement in operational safety – an expensive enough approach if introduced at the design stage and much more so if, as happened post-TMI, plants already under construction had to be redesigned and back-fitted with extra safety features. At the same time, materials costs increased by an average of 7.7 per cent per year. Between the early and late 1970s regulatory requirements individually increased the quantity of steel needed in a power plant of equivalent electrical output by 41 per cent, the amount of concrete by 27 per cent, the length of piping by 50 per cent and the length of electrical cable by 36 per cent, again a trend which increased after TMI.[22] The cumulative effect of these increases can be seen in figure 3.7.

Secondly, the high inflation and therefore interest rates of the 1970s and early 1980s had a particularly detrimental effect on the investment costs of nuclear power, whose projects have a long construction phase compared to the coal-fired and oil-fired plants that were prevalent in the early 1970s and so incur heavier costs in servicing the capital before there is a compensating cash flow from electricity sales.

Thirdly, the effect of regulatory ratcheting on construction schedules had the effect of tying up capital for longer periods of time and hence significantly increasing the cost of the project. Between 1971 and 1982 the period of time from project initiation to first power production rose from 7 to 12 years, in effect further doubling the costs of the power produced, as shown in figure 3.8.

Again this trend accelerated after TMI. Construction of the Watts Bar 1 plant in Tennessee began in 1973 but the plant was not connected to the grid until 1996. Watts Barr 2, where work actually started in 1972, was to come online in 2016 following a 32-year 'suspension' since 1985.

Crudely extrapolating the cost experience from 1967 to 1979 to the later years of the following decade is shown in figure 3.9.

This might suggest that the increase in cost as a result of TMI was considerable, amounting to some $1,200 per kW by 1988 – approximately half of the total cost in that year.

Finally, growing public concerns led to an increasingly negative atmosphere for nuclear investment. As noted in chapter 11, the prospect that a new plant might never receive an operating license, as happened at the Shoreham plant on Long Island, New York State, introduced a major new element of economic risk around nuclear investment.

At the time of the TMI accident 129 nuclear power plants had been approved in the US – of these only 53 were completed. Ultimately, of the 253 nuclear power

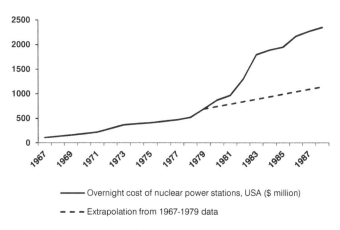

Figure 3.9 Extrapolation of 1967–79 overnight costs

reactors ordered between 1953 and 2008 in the US 48 per cent were cancelled, 11 per cent prematurely shut down and 14 per cent suffered at least one outage lasting a year or more.[24] Several large nuclear power plants were completed in the early 1970s at a typical cost of $170 per kW, whereas plants of the same size completed in 1983 cost a non-inflation-corrected average of $1,700 per kW, a tenfold increase. Some plants completed in the late 1980s cost as much as $5,000 per kW, 30 times what they would have cost 15 years earlier. Even taking inflation into account costs increased fourfold in the decade up to 1983 and, in the most extreme cases, more than doubled again by the end of that decade. So, for example, Commonwealth Edison, the utility serving the Chicago area, completed its Dresden nuclear plants in 1970/1 for $146 per kW, its Quad Cities plants in 1973 for $164 per kW and its Zion plants in 1973/4 for $280 per kW. Its LaSalle reactors completed in 1982–84 cost $1,160 per kW and its Byron and Braidwood plants completed in 1985–87 cost $1,880 per kW – a 13-fold increase over a 17-year period. When the Marble Hill plant in Indiana was abandoned in 2004 with construction only 20 per cent complete some $2.8 billion had been spent on it by Public Service Indiana against an initial estimated total cost of $700 million.[25] The Washington Public Power System's commitment to build five large nuclear plants in the 1970s was blamed for its financial collapse.[26] As one contemporary commentator put it: 'The failure of the US nuclear power programme ranks as the largest managerial disaster in business history, a disaster on a monumental scale. It is a defeat for the US consumer and for the competitiveness of US industry, for the utilities that undertook the programme and for the private enterprise system that made it possible.'[27]

In addition there were the direct costs of the accidents themselves. The clean-up phase for TMI, which lasted until 1993, cost in the order of $1 billion in prices of the day in addition to the cost of the plant itself which had to be written off. In the case of Chernobyl, with far greater off-site releases of radioactive materials, estimates of the total cost reach several hundred billion dollars[28], quite apart from the human costs in terms of illness (notably thyroid cancer), the social effects of evacuation for many thousands of citizens and the anxiety caused. The risk of incurring such costs severely reduces the attractiveness of nuclear investment and therefore increases the demanded rate of return. The cost which would fall to the operating utility in the case of a major accident is capped under legislation such as the Price-Anderson Act in the US, which limits the industry's liability to $10 billion for any accident, and the Paris and Vienna Conventions. However, this is a quid pro quo for nuclear operators accepting 'strict liability' – in other words, claimants would not have to prove any negligence on the part of the operator in a major accident scenario and simply show that radioactive materials had been released, which is a severe financial imposition when compared to other industries. Nuclear plant operators must carry considerable insurance against eventualities that may not be their fault.

The growth in nuclear capacity during the 1980s in particular, was actually more of a reflection of growing delays in completing nuclear projects rather than the perseverance of the enthusiasm of the 1960s and early 1970s. No nuclear plant ordered after TMI has so far been completed in the US.

Shifting focus and a much-trumpeted revival of sorts

While new nuclear investment practically died out in western Europe and North America from the 1990s onwards, this was not the case in Asia–Pacific. Japan, South Korea and, increasingly, China continued to expand their numbers of reactors (figure 3.10).

Interest in nuclear power within central and eastern Europe was also maintained, although the economic crisis following the collapse of communism in the late 1980s undermined the case for investment in any new power capacity for some years. So although by mid-2015 the US remained the largest user of nuclear power with 99 of

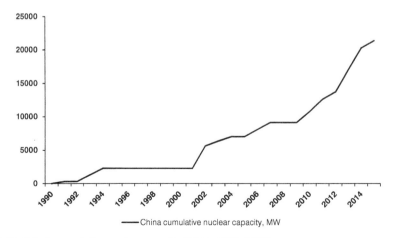

Figure 3.10 China's cumulative nuclear capacity (MW), 1990–2015

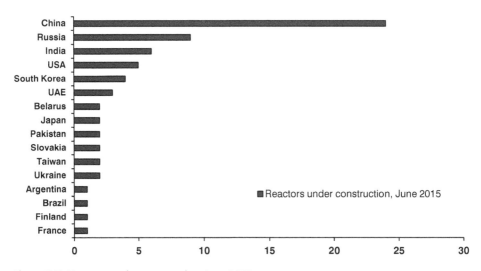

Figure 3.11 Reactors under construction, June 2015

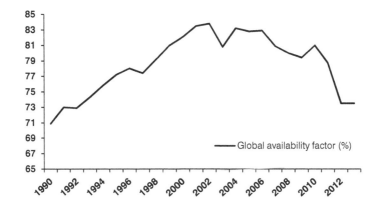

Figure 3.12 Average world nuclear energy 'availability factor' (%)

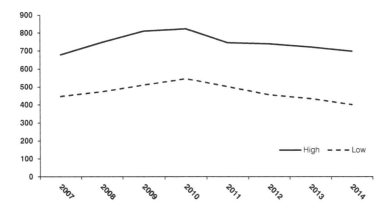

Figure 3.14 IAEA low and high projections for nuclear capacity in 2030 by year of estimate

the world's 438 'operational' reactors, followed by France with 58 and Japan with 48, though the status of several of the Japanese ones was unclear post-Fukushima, 24 of the 67 reactors described as 'under construction' were in China with nine in Russia, while almost all were in countries which had not liberalised their power industries along the Western model – see figure 3.11.

The early years of the twenty-first century saw a significant revival in countries talking about building new nuclear power stations, boosted mainly by high fossil fuel prices but also by concerns over security of supply given the ongoing tension in the Middle East, especially after the 9/11 atrocity. In addition, the global performance of nuclear plants had been improving significantly through the 1990s as the technology matured and a body of expertise developed – see figure 3.12. (The headline figure dipped after the Fukushima accident in 2011 owing to most of Japan's fleet being technically 'operational' but not generating any electricity.)

In 2003 three US utilities – Entergy, Exelon and Dominion Resources – announced their intention to apply for licenses for new build and it was reported that officials in the George W. Bush administration believed that the first new reactors would be finished in 2010. In the UK, September 2008 saw French energy giant EDF purchase British Energy, at that point partly nationalised following the company's financial crisis of 2003, for a handsome £12.5 billion. The prime minister, Gordon Brown, said: 'New nuclear is becoming a reality. This deal is good value for the taxpayer and a significant step towards the construction of a new generation of nuclear stations to power the country. Nuclear is clean, secure and affordable: its expansion is crucial for Britain's long term energy security, as we reduce our oil dependence and move towards a low-carbon future.'[29] For neither the first nor the last time the nuclear renaissance looked well and truly under way.

However, as Alan Nogee of the Union of Concerned Scientists noted at the time: 'There's talk of a nuclear second coming every few years and so far, obviously, without success on their part.'[30] By 2009 the US Nuclear Regulatory Commission had received applications for Combined Construction and Operating Licenses (COLs) for at least 27 new plants, as shown in figure 3.13; five years later only four were under construction alongside the reactivated Watt's Bar 2 reactor, a hangover from the pre-TMI days;

As an indicator of sentiment, the projections for nuclear capacity to be operating in 2030 made by the IAEA grew every year from 2003 to 2010 and then started to fall – see figure 3.14.

In a remarkable parallel with the 1970s, the 2000s saw a steady growth in estimates of the cost of nuclear power capacity even before a major accident (in this case at Fukushima in 2011). As a 'reality check', experience with the five units completed in Japan and Korea between 2004 and 2006 suggested overnight costs of between $2,700 and $3,830 per kW with an average of just over $3,400 per kW (expressed in $2015).[31]

In 2004 the US Senate Committee on Energy and Natural Resources, referring to the Westinghouse AP1000 reactor, had reported: 'The industry estimates the capital cost of the first few nuclear plants built would be in the range of $1,400 per kilowatt. After these plants are built and the first-of-a-kind design and engineering costs have been recovered, subsequent plants of the series will have capital costs in the $1,000 to $1,100 per kilowatt range, which is fully competitive with other sources of baseload electricity.'[32]

There was considerable general inflation in power plant construction costs for all fuels in the US in the decade 2000–10.[33] An average power plant costing $1 billion in 2000 would have cost $2.31 billion in constant 2000 money values in May 2008, representing an average increase of some 130 per cent – see figure 3.15. However, while increases in the cost of non-nuclear capacity grew by an average of 82 per cent, that of nuclear was considerably higher.

Among the factors behind these increases was a high ongoing demand for new power generation facilities worldwide, leading to cost increases, supply issues and longer delivery times as manufacturers struggled to keep up with demand. Steep increases in commodity prices, exacerbated by a skilled labour shortage, led to significant increases in the overall cost estimates for major construction projects around the world. The materials and the worldwide supply network associated with new nuclear projects was also being called upon to build other generation facilities better suited to liberalised markets. Nuclear operators were competing with major oil, petrochemical and steel companies for access to these resources.[35] For instance, there were only two companies that have the heavy forging capacity to create the largest components for new nuclear plants, namely the Japan Steel Works and Creusot Forge in France. Indeed concerns emerged in mid-2015 as to production and quality control standards at the latter with regard to the reactor pressure vessel for the Flamanville plant in France. The demand for heavy forgings is significant because the nuclear industry would be competing with the petrochemical industry and new refineries, as well as other electricity generation projects.

The estimated costs for new nuclear power plants in the US begin to increase significantly in the second half of the 2000s.[36] A June 2007 report by the Keystone Center estimated an overnight cost of $2,950 per kW for a new nuclear plant – between $3,600 per kW and $4,000 per kW when interest was included.[37] In October 2007, Moody's Investor Services estimated a range of between $5,000 per kW and $6,000 per kW for the total cost of new nuclear units including escalation and financing costs, but expressed the opinion that this cost estimate was 'only marginally better than a guess'.[38] Also in October 2007 Florida Power and Light (FPL) announced a range of overnight costs between $3,108 per kW and $4,540 per kW for its two proposed nuclear power plants with a total output 2,200 MW. FPL estimated the total cost of the project including escalation and financing costs as being between $5,492 per kW and $8,081 per kW, giving a projected total cost of $12.1 billion to $17.8 billion for two 1,100 MW plants.[39] Progress Energy, which filed an application for new build at Levy in Florida, projected a cost of around $10.5 billion for two new nuclear units, with financing costs bringing the total up to around $13 billion to

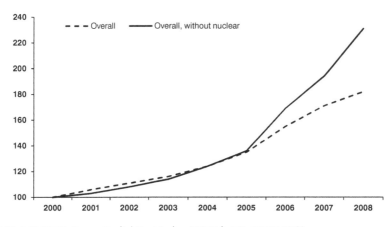

Figure 3.15 IHS–CERA Power Capital Cost Index (PCCI) in US, 2000–08[34]

Forecast	Overnight cost ($ per kW)	Total plant cost ($ per kW)	Total plant cost – two 1100MW units ($ billions).
US Department of Energy (2002)	1,200 1,500		
MIT (2003)	2,000		
Keystone Center (2007)	2,950 2,950	3,600 4,000	
Moody's Investor Services (2007)		4,000 6,000	
Florida Power and Light (2007)	3,108 4,540	5,492 8,081	12.1 17.8
Progress Energy (2008)		6,360	14.0
Georgia Power (2008)		6,500	6.4 for 45% stake

Table 3.3 The escalation in nuclear construction cost estimates in the US

$14 billion.[40] Georgia Power estimated that the cost of its 45 per cent share in two proposed nuclear plants at Vogtle would be $6.4 billion, consistent with Progress Energy's estimates.[41]

The escalation in nuclear construction cost estimates in the US is illustrated in table 3.3.

Olkiluoto-3 in Finland, a 1,600 MW Areva EPR design located alongside two 1970s BWRs, was expected to cost some €3.2 billion and to be available in May 2009 when it was ordered in 2005. By early 2014 Teollisuuden Voima (TVO) had effectively given up trying to estimate when it might be available amid rumours that the timetable had slipped to 2018 at the earliest.[42] Projected costs had reached at least €8.5 billion ($11.4 billion or $7,200 per kW). The costs of the Flamanville-3 EPR in France had also reached €8.5 billion by December 2012 from an original €3.4 billion just five years earlier, even before concerns arose concerning the integrity of the reactor pressure vessel in 2015. In Taiwan two Advanced Boiling Water Reactors (ABWRs) at Lungmen were ordered in 1999, with operation expected in 2005 and at an expected cost of $3.7 billion. By the time pre-operational tests were carried out in 2014 nearly $10 billion had been spent equating to $3,700 per kW. The plants were then mothballed for an expected further three years until a referendum on their operation could be held, a delay which alone was expected to cost a further $2 billion.[43] By mid-2015 the construction programme at Vogtle was already some $1 billion over its initial $14 billion budget ($6,800 per kW) and over three years late.

Cost and schedule overruns are not unique to nuclear power, of course. In 1994 the Channel Tunnel came in some 80 per cent over its 1988 budget at £4.65 billion.[44] One analysis of 401 power plant and transmission projects in 57 countries suggests that costs were underestimated in three out of every four projects. Hydroelectric dams, nuclear power plants, wind farms and solar facilities each have their own unique set of construction risks.[45] For example, as noted in chapter 4 initial capital costs of the Greater Gabbard wind farm off the east coast of England stood at $1.8 billion in 2008. Original owner Fluor then added a total of an extra $819 million between 2010 and 2012 to cover cost overruns.[46] The cost of new transmission connections to wind farms in Texas grew from an initial $4.9 billion in 2005 to $6.8 billion in 2013 (corrected for inflation). But the idea of building a 'merchant' nuclear plant without some very long term, generous guarantees concerning the price it could command for its output was a non-starter.

One practical aspect of liberalisation was potentially more beneficial. In the days of state-run or state-controlled power systems each country tended to pursue its own technical route – even the French, once they had bought in American PWR technology altered it somewhat for their own use. Furthermore, in a relatively small market with a new plant being ordered perhaps every two of three years, the temptation to play around with the design to 'stretch' its output – bringing with it inevitable (in a general sense) but unpredictable (in the specifics) technical challenges – affected both the economies of scale and the speed with which experience of operating the design could be gathered.

The growth of international companies operating in several liberalised markets – and indeed increasingly in non-liberalised markets as well, at least as contractors to the national government in question or its institutions – created a growing international market for a handful of different plant designs. In principle this would allow for the development of series economies of scale across national boundaries instead of the potential market for a particular design being largely restricted to the country that invented it (the Russian VVER being an exception since it was built in a number of communist bloc countries).

Two theoretical ways of funding nuclear plants emerged. One was the Finnish model. TVO, which ordered the new station at Olkiluoto, was a consortium of major electricity users and suppliers who didn't want to become too dependent on imports of Russian gas. Members contracted to take electricity at a cost in proportion to their initial investment, thereby in effect creating lifetime contracts for the output within a mature liberalised market. The other was that which served as the basis of the new French plant at Flamanville – a consortium of big European electricity generators with deep enough pockets to invest in both long-term and short-term power sources. However, even the biggest utilities in Europe found investment in new nuclear plant an economic risk too great for them to take on board on their own.

Pointless progress

Following a UK nuclear White Paper of 2008 there was, in a way, a considerable amount of progress around the edges of a positive decision to build new reactors – at least from the point of view of fans of bureaucracy and procedure. The following description is designed simply to provide a flavour of the monumental barriers involved in getting to a point where anyone can actually decide whether to build a nuclear reactor or not.

National Policy Statements were issued on energy, setting out the national need for nuclear power, and on-site selection. Eight sites were declared appropriate for nuclear new build to be operating by 2025. Planning permission and environmental permits for a new nuclear power station, consisting of two EPR reactors (each of capacity 1,650 MW), at Hinkley Point were granted in March 2013.

In 2009 three commercial consortia were formed to build new nuclear power stations in the UK.

- EDF Energy, owner of the existing UK nuclear plants as well as those in France and others in a variety of countries, teamed up with Centrica, owner of another of the UK 'Big Six' energy companies, British Gas, to create NNC GenCo, with an initial view of building EPRs at Hinkley Point and at Sizewell.
- German giants RWE and E.ON, two more of the Big Six with pan-European interests in a variety of electricity projects including nuclear, formed Horizon Energy which was looking to build new stations (possibly AP1000s) at Wylfa and at Oldbury in Gloucestershire.
- GDF Suez, the French-based energy giant which, inter alia, owned the Belgian nuclear company Electrabel, teamed up with Iberdrola of Spain and SSE, another of the UK Big Six, to establish NuGen, focussing on a site in Cumbria near Sellafield.

Over the following five years the consortia were to undergo several changes in ownership, reflecting differing levels of confidence in the viability of a nuclear new build programme despite the apparent support of both Labour and the Coalition government before and after the election of 2010. Centrica dropped out of NNC GenCo in 2013. Following Fukushima and the German government's decision to close all of its country's nuclear plants RWE and E.ON were left with much diminished confidence and cash flow; they sold Horizon to GE-Hitachi in 2012. SSE and Iberdrola sold their stakes in NuGen to Toshiba-Westinghouse in 2011 and 2013 respectively.

In effect each consortium ended up championing its own technology, EDF having had a decades-long, close working relationship with its compatriot Areva. Through the early years of the twenty-first century a number of competing large plant designs emerged. They included:

- Areva: European Pressurised Water Reactor, EPR, 1,650 MW;
- Toshiba-Westinghouse: Advanced Passive Pressurised Water Reactor, AP1000, 1,100 MW;
- GE-Hitachi: Economic Simplified Boiling Water Reactor, ESBWR, 1,600 MW;
- Hitachi-GE: Advanced Boiling Water Reactor, ABWR, 1,380 MW;
- Atomic Energy of Canada Ltd (AECL): Advanced CANDU reactor, ACR, 1,200 MW;
- KEPCO (Korea Electric Power Corporation): APR, 1,400 MW;
- Rosatom (Russia): VVER, 1,200 MW.

A new Generic Design Assessment (GDA) process had been introduced after an energy review in 2006. Under this mechanism, once a reactor design has been licensed nationally it then does not need to be justified at each and every public inquiry for a new plant. Initially four designs were submitted for possible consideration under the GDA, i.e. EPR, AP1000, ESBWR and ACR. Subsequently, AECL withdrew the ACR and GE-Hitachi 'suspended' its interests. Decisions on the Areva and Toshiba-Westinghouse designs were expected in June 2011 but were delayed by the ONR's request for specific information on possible modifications to the designs following the Fukushima accident in March 2011. In December 2011 the ONR and the Environment Agency issued interim approvals but several outstanding issues requiring resolution were identified. Toshiba-Westinghouse also suspended its involvement as there was no licensee expressing an interest in constructing the AP1000 at that point. In December 2012 full approval was granted to the EPR; in early 2013 an application was entered for the ABWR; and in 2014 NuGen said that it wished to reactivate the GDA for the AP1000.

A 'strike price', which is discussed in chapter 8, for the electricity generated by Hinkley Point C was agreed in 2013. A preliminary funding agreement was also reached with the soon-to-be-merged state-owned Chinese nuclear companies China General Nuclear Power Corporation (CGN) and China National Nuclear Corporation (CNNC), but by now the expected cost of the project had risen to £16 billion. The European Competition Commission immediately called in this deal for an investigation into whether the agreement had breached European rules on 'state aid' – a process which lasted almost a year Although the Commission approved the deal, in 2015 the Austrian government announced that it was going to launch a legal challenge to the decision, introducing further potential delay.[47]

Meanwhile a nuclear tipping point was approaching. The AGRs received successive lifetime extensions – but this could not go on forever. Eventually, for example, physical deterioration, of the graphite blocks, which act as a moderator, would force their closure. Assuming AGR lifetime extensions of between 45 and 50 years, which is well above the 'design lifetime' when they were conceived, a programme of new build bringing 1.65GW online every two years from 2023 onwards would preserve UK nuclear capacity at around the 10GW mark. As older plants closed they could be replaced with new ones in a more or less orderly fashion.

It was only late in 2015 that a credible funding mechanism emerged for the proposed Hinkley Point C, depending controversially on a major investment from the Chinese state. By this point the timing was precarious. Although 2023 may just have looked feasible – EDF set its sights on 2025 – the construction record of PWRs elsewhere gave little cause for comfort. A five-year delay to starting such a programme would see nuclear output drop considerably in the mid-2020s before recovering. Should the delay be longer, or if lifetime extension should prove impossible for any reason, clearly that drop would be more pronounced. In that case some alternative capacity – almost certainly CCGT – would be built to fill the gap, capacity that would then be in competition with the plants later on in the proposed nuclear programme. (This analysis, shown in figure 3.16, is of course somewhat simplistic as nuclear is not operating within a bubble. For example, considerable amounts of coal-fuelled capacity would close in the same timeframe.)

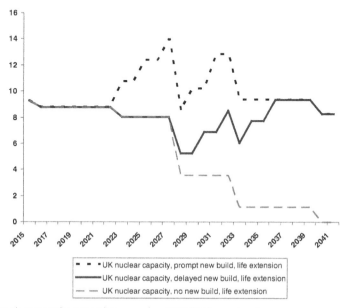

Figure 3.16 Nuclear capacity scenarios assuming AGR lifetime extension

Delay is also harmful in industrial policy terms. In order for supply chain firms to feel confident to invest the considerable sums necessary to prepare for a programme of new nuclear plant, they need a strong reason to believe that such a programme would commence imminently. A rolling programme of six twin PWR stations would employ an average of 10,000 workers over the period of construction.[48] Potentially as much as 70 to 80 per cent of the value of a nuclear new build programme could be provided by British firms. This would represent a much higher proportion of the levelised cost of nuclear generation, with benefits for the UK's balance of payments, than would be the case with a CCGT where much of the cost would be accounted for by imported gas. However, for the UK to take advantage of this opportunity, rather than having to source a larger proportion of the project from overseas, UK firms would need to be gearing up some time before the construction began. Ongoing delays weaken the confidence of firms to do this. This is not only the case with smaller firms – in mid-2013 and again in mid-2015 EDF began to lay off staff at Hinkley Point as negotiations over price guarantees and the commitment of funding apparently stalled, saying: 'We cannot afford to burn money every day, every week, every month without a clear understanding of where it's leading us.'[49] Big Green understands this point well and often uses legal processes simply to slow down the whole process. Winning or losing on any particular point of appeal is almost irrelevant – the delay itself both pushes up costs and diminishes investor confidence.

3.6 THE AVAILABILITY OF URANIUM

Uranium is not a renewable resource. Like fossil fuels it is a finite reserve with a limited lifetime. Unlike hydrocarbons, however, uranium is, as shown in table 3.4, a relatively widespread resource without some of the geopolitical complications which surround the geographical concentration of oil and gas reserves and supplies.

At global rates of usage of 66,000 tonnes per year, these proven uranium reserves at $130 per kg would be exhausted in about 90 years. To put it another way – assuming a reactor life of 60 years, this resource could sustain a nuclear programme of about half as much again compared to current levels for one 60-year life cycle.

If nuclear power is to make a significant contribution to reducing greenhouse gas emissions it will need to replace a considerable proportion of the global fossil fuel use over the next few decades. In order to do so, nuclear power capacity will need to grow at a faster rate than electricity demand as a whole. It follows that, even if other obstacles to nuclear development can be overcome, for nuclear power to play a major role in meeting demand and mitigating climate change then one or both of two things must happen: considerably more uranium must be found and/or new nuclear technologies must be deployed.

Major efforts to locate uranium reserves are effectively a post-war phenomenon. Prior to the discovery of fission in 1938 uranium had few industrial uses and was assumed to be an extremely rare element. Between the mid-1980s and the mid-

	Thousand tonnes uranium	% of global reserves
Australia	1,706	29
Kazakhstan	679	12
Russia	506	9
Canada	494	8
Niger	405	7
Namibia	383	6
South Africa	338	6
Brazil	276	5
USA	207	4
China	199	4
Mongolia	142	2
Ukraine	118	2
Uzbekistan	91	2
Botswana	69	1
Tanzania	58	1
Jordan	34	1
Other	191	3
WORLD TOTAL	5,902	

Table 3.4 Known recoverable resources of uranium at $130 per kg, 2013[50]

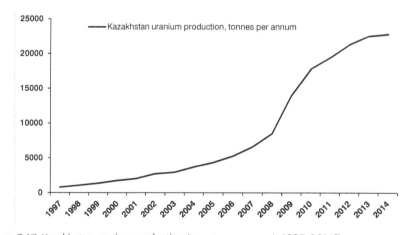

Figure 3.17 Kazakhstan uranium production (tonnes per annum), 1997–2014[51]

2010s, however, proven uranium reserves grew at about the same rate as did oil and gas reserves, despite the global glut of uranium that followed the decline of the first wave of nuclear investment coupled with the availability of uranium from the destruction of nuclear weapons from the 1980s onwards, suppressing the search for new reserves. Figure 3.17 shows the example of Kazakhstan, which rose from a minor player to second in the world reserves ranks in less than two decades. Assuming that this rate of discovery continues, and using the experiences gained from oil and gas as a benchmark, one might expect the proven uranium reserves in 35 years' time to be perhaps fourfold what is currently stated at the $130 per kg price. When the increasing efficiency of nuclear plants is taken into account – both in the terms of the 'burn up' of uranium, i.e. the amount of useful energy extracted from a tonne of uranium before it has to be removed from the reactor core, and the efficiency of converting the heat produced into electricity – it is likely that there is sufficient uranium to power a programme of six or seven times more than the current size for a full 60-year reactor cycle, even if the uranium ore mined towards the end of such a programme would be of a lower grade than the reserves currently being exploited, with potential implications for both the cost of extraction and the energy input required.

As a testament to the eternal optimism of the scientist, speculation is raised on occasion as to the possibility of extracting uranium from seawater. The uranium concentration of seawater is about three parts per billion or 3mg of uranium per cubic metre of water. The total volume of the world's oceans is around 1.37 billion km^3 – a useful fact to know – so there is a total of approximately four billion tonnes of uranium in seawater. This could fuel current demand for over 60,000 years. However, extraction may prove a challenge. It has been suggested that the cost of extraction for the foreseeable future might be around $250 per kg of uranium[52] but other estimates are rather higher.

In any case, by the mid-2010s such speculation seemed even more fanciful. The spot price of uranium fell from $153 per kg in 2011 to $63 per kg in 2014 owing to persistent oversupply.[53] The mining industry had geared up for a nuclear renaissance that had still not materialised; there could be no certainty that it ever would.

3.7 RADIOACTIVE WASTE MANAGEMENT

When asked to state the problems associated with nuclear power, nuclear waste disposal is consistently one of – if not the – most-mentioned public concerns.

Uranium is an extremely concentrated form of energy: one tonne of uranium in a modern reactor producing about as much electricity as 20,000 tonnes of coal. The amounts of waste being produced are therefore very small by industrial standards. Less than 15,000m^3 of radioactive waste is produced every year by the UK nuclear industry's power stations, reprocessing plants, enrichment, fuel fabrication and other facilities, compared to 40 million m^3 of industrial waste and 40 million m^3 of domestic rubbish. Within the industrial waste are some 5 million m^3 of potentially deadly toxic wastes.

Radioactive waste can vary enormously in the amount of radiation it emits so it is categorised into low-level, intermediate-level and high-level waste. A nuclear power station produces around 100 m³ of solid radioactive waste each year – about the volume of a big lorry – more than 90 per cent of which is categorised as 'low-level' waste.

- Low-level waste (LLW). This includes solids and liquids, e.g. used protective clothing or air filters, which might be contaminated with traces of radioactive materials. It also arises because of use of radioactive materials in industry, medicine, and so on. Granite, Brazil nuts, fertilisers and coffee beans would be categorised as low-level waste if they were man-made. Approximately 10,000 m³ of low-level solid waste is produced each year in the UK, the size of a four-bedroomed family home. Most low-level waste is solidified, squashed into a small volume and disposed of in metal containers at a site outside Drigg, near Sellafield in Cumbria. This has been operating since the 1960s without attracting much public interest.

- Intermediate-level waste (ILW). This consists of solid and liquid materials from nuclear power stations, notably the metal cans within which the fuel was contained in the reactor, and fuel reprocessing and defence establishments. Less than 4,000 m³ of intermediate-level waste is produced every year in the UK, much of it from the reprocessing operation at Sellafield rather than from the operating power stations.

- High-level waste (HLW). This is made up of the 'fission products' and heavy metals contained in spent nuclear fuel. Fission products are the smaller atoms created when the atoms of uranium or plutonium fuel are split and are often extremely radioactive, also producing considerable amounts of heat. Roughly speaking, the amounts of radiation and heat are proportional – if radioactivity drops by half over a particular period of time so does the heat produced. A few minutes after a reactor is shut down, e.g. to allow for maintenance or in response to some fault or external event such as an earthquake, the fission products will still be producing heat equivalent to about 7 per cent of the heat output in normal operation. Although the task of managing this waste energy diminishes over time as production output falls off rapidly, the heat must be sustainably and reliably removed for several years. Technically HLW arises after spent fuel has been reprocessed (see later), but management of spent fuel that has not been reprocessed raises similar challenges.

Since the advent of UK nuclear power in the 1950s a total of around 1,500 m³ of high-level waste has been produced, about enough to fill half an Olympic-sized swimming pool. Each year the industry produces an amount equivalent in volume to a taxi. HLW has been stored as a thick liquid but it is being progressively 'vitrified', although the performance of the vitrification plant at Sellafield has been woeful and many targets have been missed. The process involves turning the waste into a stable glass form, reducing its volume by two-thirds and aiding its safe storage.

The most serious accident in a radioactive waste facility occurred in September 1957 at a processing and storage facility at Kyshtym in Chelyabinsk, Russia, on the Europe–Asia boundary. Liquid high-level waste was stored in underground stainless-steel tanks in concrete trenches fitted with heat exchangers to remove heat build-up in the tanks as the fission products decayed away. The heat exchanger on one tank was shut down in 1956 due to a fault and apparently it was decided that it was safe to leave the tank uncooled. Water evaporation resulted in a concentration of flammable nitrates and ethanoates on the surface of the tank's contents in contact with air. A spark from monitoring equipment detonated the contents, creating a chemical explosion equivalent to that of 5 to 10 tonnes of TNT and releasing around 70 to 80 tonnes of wastes. Ninety per cent of the activity was deposited nearby with 10 per cent dispersed over a distance of 100 km or more downwind. Eighteen months after the accident over 10,000 people had been evacuated and over 1,000 km^2 of land had been removed from agricultural use, most of which had been returned to use by 1978. The accident was first discussed by Soviet academics in 1976 and only acknowledged by the Soviet government in 1989, just before the USSR collapsed.[54] It seems that around 50 people subsequently died of radiation-related cancer.[55]

However, this major accident notwithstanding, in normal circumstances radioactive waste does not seem to have been associated with any discernible health consequence. All waste can have an effect on the environment if it is released and much industrial waste is unpleasant or dangerous. Domestic waste tips give off methane gas which adds to climate change and may be explosive in sufficient concentrations. Industrial wastes contain chemicals and asbestos, some of which will remain toxic forever.

The commercial reactors available by the mid-2010s were much more efficient in their use of fuel than the pioneering designs of half a century earlier, giving rise to correspondingly lower volumes of waste. As Friends of the Earth noted: 'A new build nuclear programme would not add significantly to the quantity of waste but could increase the overall radioactivity of the waste inventory by around 265 per cent.'[56] In one of the internecine disputes that characterises most of the world's major religions, hard line anti-nuclear group No2NuclearPower criticised this statement. Fearing that 'when we start to use phrases like "would not add significantly" we start to lose that sense of moral outrage that we need to oppose a 265 per cent increase in this dangerous waste that we have almost no idea what to do with', No2NuclearPower argued that the statement should read along the lines of: 'the volume of the waste is irrelevant.'[57] This is a constant Big Green line – unless, of course, the volume sounds very big, in which case Big Green uses phrases like: 'The UK now has enough radioactive waste to fill the Royal Albert Hall five times over' (we are rarely told that the UK makes enough domestic rubbish to fill the Albert Hall every two hours) or '400 tonnes of radioactive water is entering the sea every day at Fukushima.' (Of course 'the volume is irrelevant' since all water is radioactive – but not if it makes a good scare story.)

The approach taken to nuclear waste has always been rather different from that associated with fossil fuel waste, which is generally released into the atmosphere and ground, whatever its long-term effects may be. Laudable as this may be, it has inevitably created something of an impression that radioactive waste is uniquely dangerous.

In an effort to be seen that they are ensuring that radioactive waste does not cause problems for future generations, most governments of countries which have nuclear power programmes have expressed a preference to dispose of radioactive waste underground, rather than continued storage on or near the surface. The main arguments for 'deep geological disposal' include:

- dealing with the waste issue in this generation rather than passing it on to our children;
- requirement for a single site (or a very small number) rather than several surface sites, allowing redundant nuclear sites to be returned to other uses;
- lower worker doses;
- greater protection against accidental or deliberate damage.

Radioactive waste management is a matter of a race between two forces of nature. On the one hand, the radioactive material itself becomes less radioactive every day that it can be kept away from living systems. It takes about 800 years for the radioactivity of high-level waste to fall to that of the uranium ore from which the fuel was mined. Over rather longer periods its activity will eventually fall to levels comparable to or lower than 'background'. On the other hand there is a tendency for nature to spread material around the environment unless it is properly contained. The job of a repository designer is to make sure that material does not start to escape from the facility until its radioactivity has died away to 'background' levels – a standard not applied to any other category of waste and certainly not to emissions of greenhouse gases.

By far the most likely way that material will escape from a repository too early is via movement of groundwater. To counter this, a 'multi-barrier' approach is proposed.

The first barriers would be physical. Much intermediate-level waste, for example, is made up of the metal cans that held the fuel in the reactor core. Once the waste has been separated from the fuel it is intended that it would be put into stainless-steel barrels which would then be filled with cement or bitumen before being placed in the concrete vaults of the repository. When the vaults were full they too would be backfilled with cement. So any groundwater which seeped towards the waste would have to eat through the concrete and cement of the vaults, the stainless steel of the barrels and the cement which directly surrounded the waste. As described above, high-level waste is slowly turned into glass blocks which would be stored in thick-walled stainless-steel flasks and be even more resistant to being dissolved.

The next level of barrier would be chemical. Water would become alkaline as it dissolved the cement in the vaults and barrels. The materials in the waste tend to be less soluble in alkaline conditions than they are in acidic or neutral water. The facility would also be placed in an area where the rocks were of such a type as to adsorb the waste rather than let it move with the water when it finally did escape from the repository.

The final level of barrier would be geological. A site would be chosen where there was very little groundwater and what there was moved slowly and predictably. There should be little evidence of past earthquakes or volcanic activity. And lastly there would be several hundred metres of rock between the waste and the outside world.

Of course, all this takes us into the realm of philosophy. We know of uranium deposits which have not moved for millions of years, but that is a very different thing from 'proving' that a particular site would remain intact for the thousands of years under discussion. In effect the science is being asked to prove a negative – that the material cannot escape – something beyond logical possibility. By setting such standards the regulators are not only creating an image of unique hazard, they are also saying the even in principle the waste cannot be managed safely against the entirely meaningless and largely unnecessary criteria which have been set for it. No wonder Big Green has such fun and such success with the waste message.

Spent fuel and reprocessing

When nuclear fuel has been in the reactor for typically three to five years it needs to be removed and replaced with fresh fuel. This 'spent fuel' is mainly unused uranium (96 per cent) but also contains about 1 per cent of plutonium, which can also be used as a nuclear fuel, and 3 per cent of fission products, some of which are effective neutron absorbers. A point comes where these fission products are absorbing so many neutrons that the fuel has to be replaced despite still containing a significant amount of uranium.

In some countries such as the UK, France and China, spent fuel is regarded not as waste but as a potential source of future nuclear fuel once it has been 'reprocessed', or recycled. UK policy, set in 1997, has been to allow spent fuel from the AGRs and Sizewell B to be either reprocessed or held in long-term storage, although Magnox and most of the AGR spent fuel has been or will be reprocessed. The assumption on which the proposed new build programme of the 2010s was predicated was that spent fuel from new reactors would not be reprocessed. It would therefore be regarded as waste at some point in the future, as it is in countries such as the US and Sweden which do not reprocess.

In practice neither reprocessing nor disposal happens immediately. Spent fuel is both vastly more radioactive and produces much more heat than does unirradiated fuel before it enters the reactor. There is a considerable period of time during which it must be kept cool before being reprocessed or disposed of. Removing and disposing of the heat from spent fuel is therefore a particularly important issue in the first few months after it is removed from the reactor. At Fukushima, for example, there were fears that ponds containing spent fuel, which unlike in most reactor designs were part of the outer building of the reactor and quite high from the ground, might leak or boil dry owing to structural damage sustained during the earthquake or the loss of power supplies. In such circumstances the spent fuel could potentially melt its way out of its metal can and become exposed to the environment.

In the short term most spent fuel is kept under water. It can be stored for the medium to long term either in longer-term storage ponds or in 'dry stores' which use natural convection of air to remove the heat; several countries have carried out research into this approach. Dry storage options include:

- cask fuel storage – fuel is placed in sealed metal containers within concrete or metal flasks housed in a weather-proof building;
- 'Modular Vault Dry Storage' – fuel is placed in vertical storage tubes within a concrete vault that can be extended for as more storage space as needed.

The alternative approach, which also in principle offers a way of stretching out the world's uranium reserves, is by reprocessing the fuel some time after it comes out of the reactor. Reprocessing was introduced in the 1940s to extract plutonium from irradiated uranium for use in nuclear weapons. In the UK it was also recognised that spent Magnox fuel, made mainly of natural uranium metal, could not be stored under water for long periods of time – as was then the practice – as it would corrode. In principle, reprocessing reduces the amount of fresh uranium which has to be mined, so extending the lifetime of uranium reserves and reducing local environmental impact. It liberates plutonium for use either in fast reactors or as a component of mixed oxide (MOx) fuel – a mixture of uranium and plutonium oxides used in a number of reactors worldwide – for thermal reactors. Reprocessing also reduces the volume, though not the activity, of highly radioactive material sent for eventual disposal.

In the early days it was assumed, at least in countries such as the UK, that because nuclear power was going to grow extremely rapidily and uranium was a very rare metal, reprocessing and the fast reactor would be necessary within a very few years. But, events proved different, in both respects. As time went by the downside of reprocessing came more into focus. Some countries decided to pursue a reprocessing route and some did not. The US, from the time of President Carter's administration in 1977–81, exerted considerable pressure on countries not to reprocess spent fuel. The UK was one of a small group of countries, France being the main competitor, which could offer the business of reprocessing to other countries through the facility at Sellafield in Cumbria.

In summary, reprocessing involves the use of chemicals and plant which become contaminated with radioactivity and causes emission of radioactive materials into the environment. These emissions have reduced considerably in recent years but there is a historic legacy both of reprocessing wastes and of environmental contamination round the principal UK reprocessing plant at Sellafield in Cumbria – decommissioning will be a very long costly business. Separated plutonium could represent a potential risk regarding the proliferation of nuclear weapons. Although the concepts behind reprocessing involve very simple chemistry, building plant that operates efficiently has proved elusive. Reprocessing is expensive but for much of the 1990s and early 2000s uranium was not, by historical standards. The increase in uranium prices in the middle and early years of the 2000s did briefly cause some countries to look again at reprocessing as an option but such muted enthusiasm as there was soon petered out, with the exception of China and Japan.

Radioactive waste in society

It is one thing having a policy to dispose of radioactive waste and/or spent fuel, but quite another to carry it out. Whether or not disposal of radioactive waste is an insuperable technical challenge or a relatively routine clean-up operation is in a sense irrelevant if social and political factors prevent implementation of agreed policy. Radioactive waste is already in a difficult position – it is unlikely ever to be in the interests of the elected government of the day to take a decision to proceed with constructing a repository, as it will face all of the political outcry and initial cost of doing so while not remaining in office, or perhaps even remaining alive, long enough to see the benefits. There is a permanent temptation, then, to launch another in-depth study of the issue, as if anything very substantial could come from it.

In terms of the model explained in chapter 10, in the 1970s, 1980s and 1990s the UK authorities attempted a classic 'phase 1' utilitarian approach of picking the best site based on national criteria and trying to impose the facility on the local community. This failed at Billingham in 1985, at four sites around the country in 1987 and at Sellafield in 1997 where it was proposed to build an underground rock laboratory. Within a growing rights-based ethic the communities involved ran highly effective public relations and political campaigns for the policy to be abandoned.

Other countries have done rather better. In Finland a site at the nuclear power station at Olkiluoto has been identified for final waste disposal, with full involvement of the local communities. Considerable progress has also been made in Sweden and France which have also identified their preferred sites. There is no fundamental reason why waste should not be disposed of safely but first lessons have to be learnt about managing the tension between national and local needs and interests.

In 2003 the British government, in the form of veteran anti-nuclear campaigner and environment minister, the late Michael Meacher, set up the Committee on Radioactive Waste Management (CoRWM) to 'provide independent advice on the long-term management of the UK's solid higher activity radioactive waste'. CoRWM once again went right back to the start of the argument and in 2006 published its main recommendations – that the waste should be managed by means of deep geological disposal. In other words, it had effectively taken three more years to restate pre-existing policy, though with a greater emphasis on monitorability and retrievability. CoRWM did suggest that, rather than the scheme being imposed on an area, local communities should be invited to express an interest in hosting a waste repository and be regarded as full partners in any proposals that emerge. This appeared similar to the successful approach taken in Finland and Sweden which was the outcome of two years of discussions with the public and key stakeholders – but without the level of trust which had been built up between industry, regulator and public over many years.

A White Paper of 2008 set out the framework for delivering a deep repository within the voluntarist framework.[58] It stated:

> An Expression of Interest will enable without commitment discussion between local communities and Government to begin. The scope of initial discussions will be for mutual agreement between the local community/ies and government. It could include discussion of what support might be available to assist continuing community engagement up until the next stage and of the point at which the Nuclear Decommissioning Authority (NDA) (and others) might become involved in discussions. At the same time the British Geological Survey (BGS) will be asked to apply sub-surface screening criteria in order to eliminate from the process any area that is obviously geologically unsuitable.

Various comments were made about the need for evidence of local support at each stage, based on good local consultation and communication.

However, the White Paper was somewhat vague about what is meant by the 'local community' – indeed, it made a (dubious) virtue of 'not wishing to be over-prescriptive about the way that the voluntarism and partnership arrangements should work at the outset as individual local circumstances differ and, to a degree, a tailored approach to any discussions will need to be taken.' It identified three types of community.

- Host community – the community in which any facility will be built. This will be a small geographically defined area and include the population of that area and the owners of the land. For example, it could be a town or village.
- Decision making body – the local government decision-making authority for the host community.
- Wider local interests – other communities that have an interest in whether or not a facility should be built in the host community, such as the next village, a neighbouring district or a community on the local transport routes to the host community.

All three levels of community would need to liaise closely with one another as the process was taken forward. Both the government and the NDA would need to engage with all three 'communities'.

Unfortunately, the White Paper did not identify what the procedure should be for communities where there was two-tier local government, with both the County and the Borough(s) or District(s) involved in having a say on the planning of major projects. In the absence of clarity over this issue, in 2011 Cumbria County Council and the Borough Councils of Copeland and Allerdale agreed a memorandum of understanding which in effect stipulated that there would have to be agreement at both county and borough level, i.e. the County Council and at least one of the two affected Borough Councils. The government endorsed this approach in letters from the then energy minister to the three council leaders in November 2011. The letters said unequivocally:

> The government accepts that if either a Borough Council (in respect of its area), or the County Council in a Cabinet decision, or the government, after considering the issues, continues to have genuine concerns and no longer wishes to participate, then the principles of partnership to which we have all been committed cannot be met. Accordingly, we would not proceed with the Managing Radioactive Waste Safely process in west Cumbria.[59]

Big Green saw its chance. For many years it had argued that it was inequitable simply to find an area with suitable geology and impose the scheme on the local community – the community should be asked first whether it would be interested. Logically, of course, there would then be no point in carrying out detailed geological surveys until an area had expressed an interest. Now the argument was effortlessly turned on its head. It was outrageous that local communities should be asked to make a decision before the geology had been properly characterised.

In 2013 the cabinets of Copeland and Allerdale Borough Councils voted in favour of moving to stage 4 of the process but the Cumbria County Council cabinet did not.[60] Local MP Jamie Reed, an advocate of the area remaining in the process, expressed his intention to try and persuade the government to allow Copeland to go ahead with its own preliminary surveys of the area's geology. He introduced a private members' bill aimed at getting Parliament's backing to recognise Copeland's wishes for desktop studies without any commitment to eventually hosting an underground repository, but the firm nature of Minister of State Charles Hendry's, undertaking was a formidable obstacle to any change in government's stance on the matter. Baroness Verma, the parliamentary Under-Secretary of State, gave an assurance that the right of withdrawal from the process right up to the start of repository construction would be made legally binding – a lack of clarity over this issue had been cited by the County Council as the main reason for it withdrawing from the process.[61]

Kent County Council and Shepway District Council, the only others considering participation, withdrew their involvement with respect to a site near Dungeness in Kent in 2012, leaving the process with no active expressions of interest. Having swung from an over-centralist approach in the 1980s to an over-localist one in the 2010s, the government seemed to have left itself without recourse except to try again with other communities. On the face of it this would seem to be rather less likely than with the people of Cumbria with their long history of living with nuclear waste. Evidence from Sweden and Finland suggested that patient relationship-building over several years could deliver a solution but in both of those cases it was the communities round the countries' largest nuclear establishments which took part.

For Big Green, one suspects, the essential outcome was to ensure that no route for long-term waste management was implemented, thereby keeping the 'they don't know what to do with the waste' mantra firmly in play. The significant amounts of waste being produced from say medical and industrial uses were rarely, if ever, mentioned, but the (identical) materials coming from nuclear power stations were deadly and shouldn't be produced until the authorities could prove the negative that material would never escape.

Ultimately, society has a perfect right to decide to apply higher 'standards' to radioactive waste than to other potentially harmful wastes, though the outcome would of course be suboptimal in terms of human well-being. It is quite hard to think of any policy which would deliver worse value for money in terms of saved lives, improved health and better standards of living than burying intermediate-level waste, and arguably even vitrified high-level waste, deep underground. Nonetheless, such decisions should be taken openly and honestly. An extraordinary alliance of the technocrats of the industry, 'technically we can treat this stuff far more carefully than it merits so we should and you should pay us to do so'; the paymasters of the political world, 'you seem to think this is really dangerous so we'd better pay you to get rid of it or at least make it look like we are doing something about it'; and Big Green, 'we can't believe our luck!', has led not only to vast sums being 'wasted' on waste but also to the whole issue becoming a severe barrier to timely and effective decision-making in the nuclear field.

3.8 NUCLEAR PROLIFERATION

It is outside the scope of this book to delve deeply into the complex world of safeguards against nuclear weapons proliferation. It is, however, worth a moment's reflection on the scale of achievement that led to the first atomic weapons in the 1940s, perhaps something which could only have arisen during wartime. From the first demonstration that nuclear fission could be made to work outside of laboratory conditions to the production of the first practical explosive device took barely two and a half years. One can disapprove of the application while still noting again the stark difference between the ability to deliver on hard decisions in the 1940s as compared to today's paralysis. It also shows that there is no technical fix to nuclear weapons proliferation. Developing a nuclear weapon, even with 1940s technology, was a much easier task than developing a working nuclear power station. The nuclear safeguards regime, administered through the IAEA and which in effect allows for help with developing civil nuclear power for nations which guarantee not to divert the material for military purposes, has been largely successful. One suspects that when China became the fifth nuclear weapons state in 1964, people then may have been surprised to learn that over the subsequent 50 years only two more nations, India and Pakistan, would explode nuclear devices, with Israel also having a weapons capability, and North Korea suspected to have one. None of these signed the Nuclear Non-Proliferation Treaty (NNPT) except North Korea which had something of an in-out history. However, at various times concerns had been raised about South Africa, Libya, Iraq (mistakenly) and Iran, demonstrating that eternal vigilance was essential.

3.9 THE ACCIDENTS – CHERNOBYL, FUKUSHIMA AND THREE MILE ISLAND

Accidents, of course, are a normal part of industrial operations. Generally speaking industries learn a lesson each time and move on. The paradox that nuclear power, perhaps the safest of the major energy technologies and certainly much more so than oil, coal and gas, even if the health effects of climate change are ignored, is regarded as being the most dangerous by considerable numbers of people is explored in chapter 11. Here we look at the three major accidents in the history of civil nuclear power – at Chernobyl, Fukushima and Three Mile Island – in order of decreasing severity and see what lessons they offer.

Chernobyl

Remarkably, Chernobyl remains the only accident in the history of civil nuclear power with clearly discernible off-site health consequences caused by radiation. There were some thousands of thyroid cancers among the affected populations, a figure which was boosted both by the relatively iodine-poor diet in much of the region and the failure of the authorities to distribute iodine tablets which flood the thyroid with stable iodine and thereby prevent the key radioactive isotope iodine131 from being taken up from the plume that was produced by the accident. Application of the Linear No-Threshold model of radiation damage suggests around 4,000 to 7,000 extra deaths from cancer over 70 years, assuming of course there is no progress in curing or managing cancer in the interim. Whether this is in any sense a meaningful calculation is examined in chapter 11.

The Chernobyl nuclear power plant, now shut down, was in northern Ukraine. Apart from the company-built town of Pripyat, with a population of 45,000, located 3km from the power station, there were no major centres of population near the facility. Chernobyl was the third site in the USSR to use the RBMK design, the first reactor of 1,000 MW rating entering full service in 1978. A second unit began operating in 1979, a third in 1982 and unit 4, the one which exploded, in 1986.

Ironically, one of the potentially more serious events that can occur at a nuclear power station is a power cut. Even if the nuclear fission processes are themselves stopped immediately the same cannot be achieved for the residual heating caused by radioactive decay of fission products in the fuel. At the moment of shutdown this heating would represent some 7 per cent of normal operating power. If emergency cooling systems were unable to operate because of interruption in their power supplies it is feasible that considerable damage could be done to the reactor. For this reason back-up emergency systems running on different energy sources such as diesel are fitted. It was the failure of these back-up plants as a result of the Great East Japan earthquake and tsunami which would result in the Fukushima accident some 25 years later.

Chernobyl unit 4 was scheduled to come out of service for routine maintenance on 25 April 1986. As the station came offline it was proposed to carry out an experiment to find out whether, even in the event of a power cut, there would be enough energy in the turbines as they 'coasted down' to operate the plant's emergency core cooling system (ECCS) for the 30 or 40 seconds required to get the diesel back-up power supply working. The ECCS could not be used in this experiment as it would shut down the reactor before the experiment could start. It was therefore decided to use the main circulating pumps that keep the coolant in the core circulating to simulate the load on the turbines which the ECCS would represent in a real-life situation.

The experiment was designed by the establishment's electrical engineers. It seems they simply assumed that the test would not involve the nuclear side of the plant, which would presumably have been shut off completely before the process began. However, the plant operators decided to keep the reactor operating at low power so that the experiment could be repeated if it didn't work first time. Under such conditions, the effect of using the circulating pumps to simulate the load of the ECCS was highly relevant to the nuclear side of the plant, as they would change the rate of flow and hence the temperature of water passing through the core and this would affect how much neutron absorption the coolant would carry out – the colder the water the more it would absorb neutrons, so the less neutron absorption from the control system could be allowed if the reactor was not shut off.

From 0100 hours on 25 April the reactor's power was gradually reduced from its normal thermal output of 3,200MWth so that by 1200 it was operating at half power. The experiment was scheduled to be carried out at 700–1,000MWth – the thermal output of a nuclear power station is typically about three times as great as its electrical output as the efficiency of conversion of heat into electricity is only about one-third.

Running the station on half power exacerbated one of the classic problems of nuclear power generation – 'xenon poisoning'. The isotope xenon135, which is created as a result of nuclear fission and, more importantly, from another fission product, iodine135, is a very good absorber of neutrons. If too much of it builds up in the fuel it can therefore prevent nuclear fission occurring. In normal reactor operation the amount of xenon remains constant, as the rate at which it is created is balanced by the rate at which it absorbs neutrons and turns into other elements. However, if the reactor is brought down from full power to lower output levels there will be a considerable period during which the rate at which xenon135 is created is greater than that at which it is destroyed. This is because iodine135 is still decaying and producing xenon135, but xenon135 is not being destroyed so quickly. The concentration of xenon135 therefore increases.

At 1400 the reactor's ECCS was disconnected from the primary circuit. This system would have shut down or 'tripped' the reactor during the experiment and so prevented a repetition being attempted. Now the first hitch in the plan appeared. The coal-fuelled power station which was supposed to take over from Chernobyl-4 didn't start up properly and the call came from the Ukrainian electricity authority, Kievenergo, for Chernobyl-4 to keep operating for a further nine hours, until 2310. Such demands were apparently not uncommon and operators were expected to comply by any means necessary. The station was therefore run at half power for this period with the ECCS disengaged.

At 2310 the operators could finally start to reduce power in accordance with the experiment by inserting control rods into the core to absorb neutrons. However, at 0028 on 26 April an operator failed to reset the power regulation system to the desired 700–1,000 MWth – the one genuine 'human error' in the whole train of events – and as a result the power kept falling, slumping to below 30 MWth. This led to greatly increased levels of xenon135; it was still being produced from iodine135 at the 1,600 MWth output rate, but was hardly being burned up at all by the very low neutron levels in the fuel at this low power output.

The increase in xenon poisoning made it much more difficult for the operators to reverse the mistake and bring the power back up to what was required. The vast majority of the total 211 control rods had to be removed and even then the power could only be raised back up to 200 MWth. At this relatively low power output most of the water in the pipes passing through the core was liquid. This created an unstable and completely forbidden regime. Several alarm signals were sent from various parts of the station ordering a reactor trip but they were overridden by the operators who had disabled the automatic control system.

In essence what happened next was as follows. The circulating pumps were shut off to mimic a power cut. This cut the flow of water so the temperature rose and the liquid water in the tubes began to boil. However, the fixed graphite moderator was not affected so the nuclear fission started to accelerate. At precisely 0123.40 the shift manager tried to insert more control rods but these would take around eighteen seconds to be fully introduced. Within three seconds the power output had reached 530 MWth, causing more boiling, less absorption, more fission, and a higher temperature. Two explosions were heard, the first from an interaction between the fuel and the coolant once some of the fuel rods had been destroyed; the second involving hydrogen, formed when the very hot water came into contact with the graphite moderator, and air which had entered the reactor space. The second explosion represented something like 480 times the normal operating power of the reactor which had been running at 6 per cent of its normal power just five seconds earlier. No structure could withstand such stress.

At least 3 per cent of the 190 tonnes of fuel in the core were thrown from the building and over 30 fires were started. Highly radioactive material was strewn around the wreckage of the reactor building. Emissions to the atmosphere continued for ten days until 5 May.

Human error, component failure – or something else?

It has been common practice, at least until recently, to regard accidents as being caused either by 'component failure' or by 'human error'. In fact, of course, these are not mutually exclusive categories. Simple component or system failure, for example, can often be traced back to human mistakes in the original design of the plant or during maintenance and monitoring. Similarly, it can be argued that a properly designed plant should be capable of 'forgiving' considerable deviation from instructions on the part of the operators.

Public accounts of Chernobyl tend to stress that it was caused by 'human error'. The Soviets' own report found six crucial examples:

- the number of reserve control rods in the reactor core was allowed to drop below permissible levels, making it more difficult to control the reactor;
- the automatic controls for the reactor's power level were shut off, allowing the power to drop below safe levels and rendering the nuclear reaction more difficult to manage;
- both the main water-circulation pumps and the back-up pumps were turned on at the same time, forcing the coolant to flow too quickly; this created air bubbles that hampered the cooling process;
- in attempting to prevent a reactor shutdown the staff cut off automatic blocking devices that would have shut off the reactor when steam failed to reach the generator;
- staff also switched off systems that controlled water level and steam pressure;
- staff turned off the emergency safety cooling system for the reactor. [62]

However, closer analysis suggests that there is an intermediate class of errors between the two, which can be referred to as 'organisational'. Organisational inadequacies might involve excess pressure on operators to ignore safety procedures, or indeed inadequate safety procedures in the first place; poor management or other factors leading to low morale; poor supervision of the operation; poor flow of information throughout the operating utility etc. Such organisational factors can result in operators quite deliberately acting contrary to safety and other operating codes or being unaware of them. Arguably only the first of the six cited 'causes' involved the operator doing something he did not mean to do. In the other cases the term 'human error' would not seem appropriate. Chernobyl exhibited many examples of this phenomenon, as a 'HOT' (Human, Operational and Technological) analysis shows.

Human factors

The Chernobyl accident is remarkable in at least two respects. The first is that there was no component failure of any description: all 'technological' factors apply purely to the design of the plant. The second is that, with the exception of the failure of the operator to reset the power regulation system to stabilise the reactor power at 700–1,000 MWth, none of the human actions can be described as 'errors'. All of the other precipitating actions were deliberate violations of operating rules, presumably made within the context of a determination to carry out the experiment under all circumstances.

The first set of human factors relevant to the accident concern the conception of the experiment itself. This experiment was designed by a consulting electrical engineer who did not appear to consider nuclear implications.

- The members of the state inspectorate, the Gosatomenergoadzor, had all gone to the local clinic for medical inspections on 25 April, so nobody was on site to prevent breaches of the operating code.[63]
- The drawing up of the test programme, without agreement with the station physicists, the reactor's designers or representatives of the inspectorate, was inadequate.
- The operators deviated very significantly from this programme.
- When the Kievenergo control centre demanded that Chernobyl-4 be run for a further nine hours the plant was run at half power with the ECCS switched off, in contravention to strict operating instructions.
- The original test had been scheduled for the afternoon of 25 April but was carried out in the early hours of the next morning, when most of the site's professional scientists and engineers had left and perhaps also when operators were not at a peak of alertness.
- Out of 211 control rods, the plant operating instructions required that an 'Operating Reactivity Margin' (ORM) of an equivalent of at least 30 rods should be maintained at all times. The operator had discretion between an ORM of 30 and 15 rods, but below 15 rods, in the words of Valery Legasov, 'no-one in the whole world, including the President of the country, can allow operation. The reactor must be stopped'. Nonetheless, at the actual start of the experiment the ORM was between six and eight rods. Under this regime it was impossible to shut down the reactor quickly.
- The operators overrode the many 'trip' commands coming from various parts of the plant. Indeed, the whole process of keeping the reactor at low power, in

order to repeat the experiment if necessary, was not part of the test programme; the test could in fact have been carried out soon after switching the reactor off, using the decay heat of the fuel.

Organisational factors

The actions of the operators cannot really be explained in terms of 'error' but must be seen against the institutional background of the plant's operation.

In many senses nuclear technology, along with the space programme, was the most prestigious of the industries of the USSR. Chernobyl had been described as the flagship of the Soviet nuclear fleet. Perhaps these factors contributed to complacency about safety that was endemic in the station's operations, though one can detect a similar hubris in the Japanese 'nuclear safety myth' before Fukushima. For example, from 17 January 1986 until the day of the accident the reactor protection system had been taken out of service on six occasions without any very good reason – and a reason must be very good indeed to justify such an action. Between 1980 and 1986, there were 27 reports of equipment failure that had not been investigated at all. The operators seem to have believed that no matter what they did with the reactor an explosion was impossible.[64] There were no RBMK control room simulators for operator training anywhere in the former Soviet Union.

The kudos attached to nuclear power also affected the attitude of operators in a more subtle way. Operators were selected not only for their technical ability (typically it took a nuclear engineer seven years to get their first degree and four or five more for a PhD, in a highly competitive system) but also for their loyalty to the Communist Party – it was not possible even to enter university without being an unswerving Party member. Officially, operating procedures, derived more from the plant design than from operating experience, were to be adhered to 'by the book'; overtly to do otherwise would be to invite instant dismissal and a return to a 25-year waiting list for an apartment. In reality, though, operators were constantly being put into situations which conflicted with this imperative, e.g. the local mayor, a high Party official, demanding extra power output during a cold spell, something which would have to be done whatever 'the book' said. Thus the highly talented workforce was officially discouraged on a daily basis from using personal initiative and taking responsibility for it, while they were quite used to bending the rules covertly.

The experiment had originally been proposed for the similar reactors at the Kursk power station in Russia. However, the plant manager there, an experienced nuclear engineer, apparently appreciated the dangers and refused permission. There were reports that a sequence of events similar to the early stages of the Chernobyl accident had occurred at the Leningrad (St Petersburg) RBMK plant in 1982. It is not clear whether this information ever reached the Chernobyl site management but clearly there were considerable communication problems within the Soviet Ministry of Power and Electrification.

Technological factors

As noted previously, the Chernobyl accident occurred without any of the plant's components malfunctioning. In other words, though questions have been raised over the standard of construction and maintenance at the station, the accident was purely the result of poor plant design and the aforementioned organisational factors. One can point to facts such as the operator's capacity to disable the ECCS, which served a number of safety purposes, one of which was to prevent operation at very low power, or override various trip signals. One can also note that the emergency control rods entered the core mechanically, taking up to around 18 seconds, rather than by gravity. These rods were unable to enter the reactor at all during the accident, presumably because the initial release of power caused the control rod channels to buckle. A design fault which involved the attachment of a graphite 'rider' on the bottom of the rods meant that insertion of the rods could actually lead to an increase in the level of nuclear fission as it introduced more moderation into the picture. But these can be regarded as human or organisational factors just as readily as technological ones – the plant behaved exactly as its design dictated when it was treated in the way it was.

Consequences

Around six tonnes of highly radioactive material was released from the core of the Chernobyl reactor. Thirty people died as an 'immediate' result of the accident at Chernobyl, two from falling masonry, burns etc. during the accident and a further 28 from acute radiation syndrome (ARS) during the following few weeks. All of these people were on site either at the time of the explosion or in the course of the firefighting operation which immediately followed. A further 106 cases of ARS were then confirmed. All of these people recovered though a further 14 had died by the end of 1995. The survivors would be expected to have a greater likelihood of developing cancer in the future.

There were no immediate health effects among people off site at the time of the accident. However, many people received measurable doses of radiation as a result of the event. As noted in chapter 11, it is known that substantial doses of radiation above 100 millisieverts (mSv) are associated with an increased risk of developing cancer after a 'latency period', which varies for different forms of cancer. Leukaemia has a latency period of two to ten years, thyroid cancer of five years and above, and most solid cancers of at least ten years.

The World Health Organisation (WHO) set up the International Project on the Health Effects of the Chernobyl Accident (IPHECA). The results of the IPHECA's pilot projects, published in 1995, were as follows.[65]

- There had been a very significant increase in thyroid cancer in the affected areas of Belarus, Ukraine and Russia. By the end of 1995 over 600 cases had been identified, three of whom had died. Further excesses were expected over the following few years – by 2005 the figures had grown to 6,000 cases with 15 deaths. The levels of thyroid cancers being observed were consistent with previous understanding of the dose-response relationship for this disease.

- There was no evidence of increased levels of leukaemia among the population in contaminated areas.

- There was an apparent increase in mental retardation and behavioural problems among children who were in the womb at the time of the accident. This was difficult to evaluate and attempts were being made to identify adequate cohorts of the relevant children for examination in the three Republics.

- There was no evidence of increases in solid cancers.

However, the effects of the evacuation itself were more severe. As the WHO concluded:

When people are evacuated from their homes, they often suffer considerable stress because they do not have full information about what is going on, they undergo disruption in community infrastructure and social interaction, and they face uncertainty about housing and employment. Many evacuees who move to new settlements after the Chernobyl accident were particularly depressed in their new homes because of financial difficulties, fear of isolation and concern for the health of their children. The tense situation caused considerable stress which, combined with the constant fear of health damage from the radioactive fallout, led to a rising number of health disorders being reported to local outpatient clinics. Although the countermeasures following the accident reduced radiation doses, they increased tension and the upheavals resulted in significant psychological stress in the affected population.

Sadly none of these lessons were to be applied in response to the next major nuclear accident 25 years later.

Fukushima

The accident at the Fukushima Daiichi F1 nuclear complex differs from those at Chernobyl and Three Mile Island in that it was caused at least in part by a severe external onslaught to the plant, in this case as a result of the Great East Japan (Tohoku) earthquake, of magnitude 9.0, at 1446 on Friday 11 March 2011 and, more so, the 15-metre high tsunami that was created as a result. The tsunami flooded nearly 600km^2 and caused over 19,000 deaths.

On the affected coastline 11 reactors were operating at four sites (F1, Fukushima Daini or F2, Onagawa and Tokaimura), with another three shut down for maintenance. Units 1 to 3 at F1 were BWRs designed in the 1960s which came into operation between 1971 and 1975. Power from the grid was lost in the earthquake but in all of the reactors the fission processes were shut off automatically when the earthquake hit. However, when the tsunami struck it overwhelmed the sea defences, flooding the emergency power generators which were powering the pumps circulating water in reactor units 1 to 3 at F1 and also the seawater pumps which allowed the waste heat to be dumped into the ocean. The fission products were still giving off a large amount of heat and the final line of defence, the back-up batteries, were flooded out at units 1 and 2 and only lasted for a few hours at unit 3. The control room suffered a blackout and effectively all the instrumentation stopped working. Site personnel were literally in the dark as to what was happening, though one thing was certain – in the absence of cooling water being passed through the cores of reactors 1 to 3 the fuel would ultimately melt, after which it could make its way through the steel pressure vessel into the 2.6 metre thick concrete 'containment' shield designed for such events and, in the worst case, melt through that too. (Unit 4 at F1 did not contain any fuel: units 5 and 6, though not generating, did contain spent fuel but one of the emergency generators capable of serving those two units remained operational.)

It was well known that Japan was subject to large earthquakes and the word 'tsunami' comes from the Japanese language. Nuclear plants were designed with both of these threats in mind. The Tohoku earthquake actually exceeded the level against which the plants were designed, but the plants were able to withstand the shock. However, the design basis for F1 was a 3.1 metre tsunami so the plant had been built around ten metres above sea level with the seawater pumps four metres above sea level. This was despite recognition that in the previous century there had been eight tsunamis in that region which were ten metres high or more at the point they arose, i.e. the location of the earthquake: an earthquake in June 1896 had produced a tsunami around 38 metres in height, killing over 27,000 people. By the time of the latest accident an improved understanding of tsunamis had led to a recognition that a tsunami 15 metres or more in height was a real possibility, but plant operator TEPCO had not taken any action and the regulator had not insisted on any.

A nuclear emergency was declared at 2103 hours and by the end of the next day an evacuation zone of 20 km round the plant had been established. Inside the reactors water was boiling, after which the steam started to react with the zirconium cans which contained the fuel, producing hydrogen. This caused a build-up of pressure which could potentially have breached the pressure vessel and containment. It became necessary to vent the pressure into the concrete outer buildings of the reactors. On the Saturday afternoon the build-up of hydrogen in the building around reactor 1 caused an explosion which blew that outer building apart. It was not a nuclear explosion but it came to signify the horror of the situation.

Over the following days a variety of ad hoc measures were taken to manage the situation. Water was introduced into the pressure vessels using fire pumps; from the Saturday evening seawater was used in place of fresh water, far from ideal – it irrevocably ended the operating lives of the power plants, a significant financial decision – but there was simply no fresh water available. The Japanese prime minister ordered TEPCO to stop this operation at unit 1 but fortunately the operators ignored him, feeling that of the many undoubtedly admirable qualities necessary to raise one to such exalted office a detailed knowledge of nuclear plant physics and dynamics may not have been one. As the water passed through the core it boiled coming out as steam or leaked out of the bottom of the core, but in either case carrying with it considerable amounts of volatile radioactive material which escaped into the environment. Power was restored to the reactors on 22 March but radiation levels remained too high for normal human access for another three months.

It is now thought that the fuel in reactor 1 quickly melted its way out of the pressure vessel and through around 65 cm of the concrete containment; melting also occurred, though to a lesser degree, in units 2 and 3. Although unit 4 was defuelled, hydrogen from the venting of unit 3 found a way into its outer building which therefore also suffered an explosion. It was to take several months to bring the units into 'cold shutdown', a condition in which the temperature in the core has fallen to below 100°C so water passes through as a liquid without boiling and is easier to collect.

As the fight to control the reactor cores developed a further problem emerged. The F1 design included cooling ponds for spent fuel which were located inside and near the top of the outer concrete shell of the reactor buildings. These ponds needed a constant flow of cool water to remove the heat being released by the spent fuel as it waited to be relocated to the site's external wet or dry stores. The pond associated with unit 4 was especially full as it contained the full charge of the reactor which had been defuelled for maintenance plus fresh fuel for the next run. Water in these ponds began to evaporate, raising the spectre of the fuel melting and releasing large amounts of radioactivity and maybe even the fission processes starting up again – as the remaining water could act as a moderator. Attempts were made to drop water into the ponds using airborne equipment designed for fighting forest fires and eventually water was introduced into the ponds using pumps designed for distributing concrete.

Even when the acute problems were brought under reasonable control the plant still faced the problem of dealing with the very large volumes of water which remained on site – some of it heavily contaminated. Water leaking from the reactors had become mixed with tsunami water and was making its way back to the sea. A large volume of storage capacity was installed (over 1,000 tanks each holding around 1,200 m³) and decontamination equipment was set up to scrub the radionuclides from the water. Unfortunately, it proved politically difficult to return the decontaminated water to the seas as planned, resulting in much of the storage capacity being blocked with relatively clean water.

Although they were undamaged in the accident, in December 2013 TEPCO announced it was to decommission units 5 and 6. It also seems politically impossible for the four reactors at F2 to resume operation, despite their dealing well with the challenges created by the earthquake and tsunami. Although the tsunami's height was 'only' nine metres by the time it reached the plant there was still considerable flooding, yet cold shutdown was achieved within five days.

Even given the external events which triggered the Fukushima accident, organisational factors were of great significance. In late 2011 the National Diet of Japan set up a Nuclear Accident Independent Investigation Commission (NAIIC) which harshly criticised the government, the plant operator and the country's national culture. The commission's report, published in July 2012, concluded that the accident was a 'man-made disaster' – the result of collusion between the government, the regulators and TEPCO. It said the root causes were the organisational and regulatory systems that supported faulty rationales for decisions and actions and attacked the regulator for insufficiently maintaining independence from the industry in developing and enforcing safety regulations; the government for inadequate emergency preparedness and management; and TEPCO for its poor governance and lack of safety culture. The chairman wrote: 'What must be admitted – very painfully – is that this was a disaster "Made in Japan". Its fundamental causes are to be found in the ingrained conventions of Japanese culture: our reflexive obedience; our reluctance to question authority; our devotion to "sticking with the programme"; our groupism; and our insularity.' The mindset of government and industry led the country to avoid learning the lessons of the previous major nuclear accidents at Three Mile Island and Chernobyl. 'The consequences of negligence at Fukushima stand out as catastrophic but the mindset that supported it can be found across Japan. In recognising that fact each of us should reflect on our responsibility as individuals in a democratic society.' One could read this as a succinct criticism of the weaknesses of a 'phase 1' relationship among the scientific/technical, political and public realms, where decisions can be taken quickly but where there is insufficient scrutiny of their quality.

Three Mile Island

The 959 MW reactor Three Mile Island 2 was destroyed on 28 March 1979. The accident, which started at 0400 hours, is often described in terms of 'operator error'. On the face of it, after the reactor had tripped for another reason, i.e. water entering a dry line through a check valve which had stuck open, the operators misread the information being presented to them, believing that a valve known as the Power Operated Relief Valve (PORV) had stuck open. In their attempts to understand and respond to the plant parameters they then made a number of errors, the most important of which was to throttle the high pressure injection flow machinery which was providing cooling water to replace that which was escaping through the stuck valve and then to turn off the main circulating pumps. The result was that the core boiled dry and about 40 per cent of the fuel melted before the situation was recognised and controlled.

However, further analysis once again indicates that these 'errors' should be seen against a less than perfect organisational and technological background, and indeed it seems harsh to blame the individual operators in any sense. For example, in 1977, at the Davis–Besse plant in Ohio, which like TMI had been built by Babcock and Wilcox, a PORV had stuck open, and operators again responded to rises in pressuriser water levels by throttling water injection. Because the plant was at a low power of 9 per cent and the valve closed after 20 minutes no damage was done, but an internal Babcock and Wilcox analysis concluded that if the plant had been operating at full power 'it is quite possible, perhaps probable, that core uncovery and possible fuel damage would have occurred'. In January 1978 a Nuclear Regulatory Commission report concluded that if such a circumstance had arisen it was unlikely that the operators would have been able to analyse its causes and respond appropriately[66] (Kemeny *et al.*, 1979). Yet none of this information was passed on to other plant operators either by Babcock and Wilcox or by the NRC. Indeed, the failure of a PORV valve was not one of the 'analysed incidents' with which operators were familiarised during training.

Lessons

Both Three Mile Island and Chernobyl, unlike Fukushima, occurred because of events that originated in the plants themselves. Even at Fukushima, where the main cause was the earthquake and more importantly the tsunami, institutional and design factors were important, improvements to sea defences having been delayed despite concerns being raised by geologists and others.

In this sense, then, none of these accidents 'should have' happened. Safety and operational regimes were fatally flawed. This leads to an observation that, particularly with nuclear technology, the main challenge is not so much 'can you deal with what you know might happen' but rather 'how can you be sure you have thought of everything'. To which the answer is of course, 'we can't'. The next major release of radioactivity from a nuclear plant will not arise because of a stuck PORV, a runaway nuclear reaction because of running a plant with a positive void coefficient at very low power or the effects of a tsunami. It will be because of something else which has not yet been fully appreciated. But it will surely happen.

This being said, the Three Mile Island accident involved human and organisational weaknesses as severe as those at Chernobyl and Fukushima, and also involved component failure which was absent at Chernobyl. Yet the release of activity was negligible. Key design features of the plant prevented the radioactive leaks as a result of the incident from being sufficient to cause significant damage to the environment or to human health. Among these key design features include the physics of the plant which would cause it to trip under any exceptional conditions and prevent a power surge under any circumstances. The containment of the plant was able to prevent releases of significant amounts of radioactivity into the environment. In other words, the design of the plant was 'forgiving', tolerating considerable violations of operating and safety codes by operators without causing major releases of hazardous materials.

But at the end of the day the question 'how do we prevent another nuclear disaster' may not mean 'how do we prevent another major release of radioactivity' but 'next time it happens, how do we prevent the relatively insignificant health consequences of radiation exposure being overtaken by the very considerable health damage caused by fear, evacuation and other responses'. As discussed in chapter 11, that is a question to which nuclear physics and engineering alone can offer very little constructive response.

3.10 A NEW GENERATION OF NUCLEAR GENERATION?

There are as many ways of making energy from uranium as there are from water, say. Just as 'water power' covers a wide range of approaches – run-of-river hydropower, hydropower using dams, wave power, tidal power and so on – so does 'nuclear power'. Most of the world's nuclear power is created in reactor designs which grew up in the command-and-control atmosphere of the electricity supply systems before liberalisation. Very large so as to benefit from economies of scale, guaranteed to be called first by the system operator so always enjoying a market for their output and able to pass their investment costs on to captive customers over years or even decades, the plants came to resemble the large herbivores of the dinosaur age, reliable and successful in a stable environment but vulnerable to opportunistic attack should more agile predators evolve to exploit a new environment. These reactors were not well suited to liberalised markets – why would they be?. However, other potential approaches might be.

The IAEA defines 'small' as under 300 MW while units in the range 300–700 MW are described as 'medium'.[67] Under the definitions the Calder Hall reactors, for example, would be described as 'small' while the AGR units would be 'medium'.

A very early application of small reactors was in the nuclear submarine, where a source of energy that did not consume large volumes of oxygen nor require the submarine to surface regularly for refuelling had obvious attractions. The US launched the USS *Nautilus*, the first nuclear submarine, in 1954. Nautilus could remain underwater for up to four months without resurfacing. The reactor, a PWR, was built by Westinghouse and became hugely influential in the development of nuclear power technology. As the Cold War deepened, the USSR started testing its first propulsion reactor in 1956. Following the 1958 US–UK Mutual Defence Agreement, HMS *Dreadnought*, the UK's first nuclear-powered submarine was commissioned by the Royal Navy in 1963 and powered by a Westinghouse S5W reactor – with subsequent submarines being propelled by a British-developed variant, the PWR1, which was superseded by the PWR2 in the late 1980s.[68]

The motivation for developing such reactors today includes the high capital cost of large power reactors and the need to serve small electricity grids under 4 GW. Small and medium-sized reactors (SMRs – confusingly the acronym can also mean 'Small Modular Reactors') may be built as one or two units or as modules in a bigger complex. In principle the simplification of design and use of passive safety systems which rely on natural forces rather than the active components of engineered safety systems should significantly reduce the cost of nuclear plants without any negative implications for safety or plant reliability. Passive designs, so the argument goes, will contain significantly fewer pumps, pipes, valves and cables so there would be fewer items to install, inspect and maintain than in a traditional plant. From a safety perspective such plants would rely on naturally occurring phenomena such as gravity, natural circulation and condensation, guaranteeing (it was claimed) a safe shutdown of the plant even in the highly unlikely event of an accident. Similarly, the economies of scale enjoyed by large units may be to an extent offset by series economies coming from building larger numbers of identical smaller units.[69] A bank of fifteen 100 MW units would be more flexible than a single 1,500 MW reactor, there being the option of switching some units off to follow load rather than trying to vary the output of the single larger unit. Among SMRs operating globally by the 2010s were the four units of the Bilibino cogeneration plant in Siberia, each unit providing 62 MWth for district heating and electricity production (11 MWe per unit); the Indian 220 MWe pressurised heavy water reactors (PHWRs); and the Chinese 300–325 MWe PWR such as those built at Qinshan phase I (China) and Chashma (Pakistan).

However, although there was a move towards smaller units of around 600–700 MW output in the early 2000s it proved to be something of a false start. By the end of the decade it was the large 1,000 MW-plus units that were being hawked around once more, mainly for reasons of economies of scale. The huge inflation in installed capacity costs which characterised the 2000s led to a re-emergence of interest in SMRs in the 2010s. Work on designs for SMRs, with generating capacities ranging from tens to a few hundred megawatts, was carried out in several countries, including Argentina, China, Japan, South Korea, Russia, South Africa and the US. SMR designs encompass a range of technologies, some being variants of the six Generation IV systems selected by GIF, others are based on established LWR technology. So in March 2012 the US Department of Energy (DOE) signed agreements with three companies interested in constructing demonstration SMRs at the Savannah River Site in South Carolina, the designs in question ranging from 25 MW to 140 MW.[70] The most advanced modular project was in China, where Chinergy was constructing a 210 MW High Temperature Gas Reactor (HTR), consisting of two 105 MW units.[71] In 2013 Toshiba-Westinghouse and China's State Nuclear Power Technology Company (SNPTC) signed a memorandum of understanding to work together developing a PWR-type SMR based on Westinghouse's 225 MW design.[72] Rosatom progressed plans to build a 70 MW floating nuclear power plant.[73] There was also interest in very small fast reactors of output below 50 MW. In the autumn statement in 2015 the UK Chancellor pledged £250 million to research SMRs.

In practice, then, all of these 'new' approaches share two features. They are not so new – they have been around some time, though nothing like as long as the 'new' renewables; and previous attempts have been made to overcome their shortcomings. And what might prove to be wildly optimistic assessments of how soon they may be available, if ever, risks diverting attention away from the urgent need for new low-carbon capacity now.

Thorium

Thorium is a potential rival to uranium as the basic fuel for a nuclear programme. The metal was discovered in 1828 by Swedish chemist Jöns Jacob Berzelius, one of the 'fathers of modern chemistry'. Although thorium atoms do not undergo fission directly, researchers in the 1940s found that they absorb neutrons when irradiated. A small fraction of the thorium transmutes into uranium-233, which once separated out from the mixture by reprocessing can undergo fission and therefore be used in a reactor or a rather unstable bomb.[74]

On the face of it, thorium looks an attractive proposition and indeed it has many advocates.

- It is a plentiful mineral – as a resource it dwarfs thermal uranium (uranium235). Since effectively all of the thorium is thorium232 and can therefore be used as a source of energy, one tonne of thorium can produce around the same amount of energy as 200 tonnes of uranium or 3.5 million tonnes of coal. Global reserves exceed six million tonnes, with countries such as India (850,000 tonnes), Brazil, Australia, US, Egypt, Turkey and Venezuela all having over 300,000 tonnes.
- Because it does not produce plutonium it has apparent non-proliferation attractions.
- It produces smaller volumes of waste.

In 2014 China announced plans to build the first fully-functioning thorium reactor within ten years instead of 25 years as originally planned. The *Telegraph* said they 'may do the world a big favour. They may even help to close the era of fossil fuel hegemony'.[75]

However, as noted below the first experiments with thorium as a fuel were carried out in the 1950s. One reason that these were not pursued, it seems, was the limited prospects for using thorium for nuclear weapons. Unlike uranium, the use of thorium requires reprocessing. Some of the fission products emit very penetrating gamma radiation which is difficult to manage. In 2009 the US DOE reviewed thorium reactors and concluded: 'the choice between uranium-based fuel and thorium-based fuel is seen basically as one of preference, with no fundamental difference in addressing the nuclear power issues [of waste management, proliferation risk, safety, security, economics and sustainability].'

The Generation IV nuclear energy initiative

Generation I nuclear reactors have been defined as the early programmes such as Magnox; Generation II as the PWRs, BWRs and AGRs of the 1970s and 1980s; Generation III as the advanced reactors available by the 2010s, including EPR, AP1000, ACR and ABWR.

Essentially, all of these reactors have pursued the broad technological route established by the nuclear submarines of around the same time. In principle there are many other ways of liberating energy from uranium or other nuclear fuels. Work has therefore been proceeding investigating systems that might be available from 2030 onwards, dubbed 'Generation IV'.

New technology rarely catches on overnight. Initially its basic physics has to be demonstrated – in principle could it be made to work? Then comes the engineering – can the process be harnessed as a practical source of energy? Finally comes the economics – how much will it cost? It is of course a relatively easy task to present estimates as to how long these stages will take – every scheme comes complete with such predictions. What is more challenging is that these estimates rarely prove to be confirmed by subsequent reality. So, in addition to basic research, new power concepts generally need a series of pilot and demonstration plants to allow in-depth study of such factors as the effects of different types of fuel, the nature of the resulting waste and its disposal, and the potential effect of the fuel cycle associated with the envisaged reactor or assembly on weapons proliferation.

To explore opportunities for such research, nine countries (Argentina, Brazil, Canada, France, Japan, South Korea, South Africa, UK and US) formed the 'Generation IV International Forum' (GIF)[76], subsequently joined by Switzerland, EURATOM (on behalf of the European Union), China and Russia. The technological goals for the Generation IV study are:

- to provide sustainable energy generation that meets clean air objectives and promotes long-term availability of systems as well as effective fuel utilisation for worldwide energy production;
- to minimise and manage nuclear waste and notably reduce the burden of the long term stewardship of waste;
- to increase the assurance that the waste streams are very unattractive for diversion into weapons;
- to excel in safety and reliability;
- to have a very low likelihood and degree of reactor core damage;
- to eliminate the need for off-site emergency response;
- to have a clear life cycle cost advantage over other energy sources;
- to have financial risks comparable to other energy projects.

In 2002 the US DOE, which initiated the programme, provided Congress with a 'roadmap' for evaluating potential nuclear energy concepts, selecting the most promising line for further development and defining the required research and development to bring the project to commercialisation within the proposed time. The roadmap addresses the full nuclear fuel cycle, not just reactors, and is updated annually.[77]

Among the concepts being considered were:

- a High Temperature Gas Reactor (HTR) under development by the General Atomic Corporation in conjunction with France, Japan and Russia;
- advanced light water reactor systems, such as IRIS, a modular design of 100–300 MW unit capacity, longer core reloading schedule (five to eight years) and enhanced safety features and proliferation resistance;
- renewed interest in fast reactors (FRs), though not necessarily as breeders;
- substantial advances in the design and use of powerful accelerators, making it possible to consider accelerator driven systems (ADS);
- a number of more specialised approaches including high temperature reactors providing high temperature heat for the production of hydrogen from hydrocarbons and for use in water desalination (an approach which had been demonstrated by the BN-350 fast reactor in Kazakhstan), both such uses being likely to become important during the twenty-first century;
- very small reactors, possibly down to 15 MW, with sealed fuel lasting the lifetime of the reactor for use as the energy source in isolated areas – the Russians have been at the forefront of the concept as a possible way of providing electricity for isolated regions such as Siberia's mining operations.

Many of these concepts are not entirely new in principle but do involve quite radically different approaches from the current commercial mainstream. It has been claimed that new designs of FRs are simpler and safer than thermal reactors and that they can destroy plutonium (from civil or military origins) and other actinides better and more effectively than via a mixed oxide fuel route in light water reactors. A particular example, the Power Reactor Innovative Small Module (PRISM), was designed by GE-Hitachi and based on the Integral Fast Reactor which had been developed at Argonne between 1984 and 1994. PRISM consists of modules of 300 MW capacity with passive cooling systems and was one of the options being considered for destroying the UK's separated plutonium stockpile from 2011 onwards. PRISM is sodium-cooled but attention was also being paid to FRs using a lead–bismuth eutectic, originally developed for Russian submarine reactors. This coolant is much denser than sodium but is also much less chemically reactive – one of the persistent problems in the unsuccessful attempts to commercialise FRs has been air getting into the coolant, reacting to form an ionic salt and thereby fatally increasing the viscosity of the fluid and preventing it from being pumped round the core and to the heat exchangers.

Molten Salt Reactors (MSRs) use a fluid fuel in the form of very hot fluoride or chloride salt. This allows the fuel to act as its own coolant. One variant which was attracting attention was the Liquid Fluoride Thorium Reactor (LFTR), using thorium and uranium dissolved in a fluoride salt. Fast breeder fluoride MSRs using plutonium were also on the drawing board. MSRs run at higher temperatures than water-cooled reactors, thereby delivering higher thermal efficiency, without any danger of the fuel vapourising. Again the idea is not new: a 1954 US project to create a nuclear-powered jet engine in 1954 looked at a molten sodium-zirconium-uranium fluoride mixture as fuel, while a Molten Salt Reactor experiment ran from 1965 to 1969.

The idea of creating nuclear power without a chain reaction, instead bombarding uranium or other fissile material with an externally created beam of neutrons, stretches back to the early days of nuclear science. In accelerator driven systems (ADS) the neutrons are generated in equipment known as an accelerator and are directed at fissile material – no chain reaction is involved, so interrupting the neutron beam stops the fission. It is possible that such subcritical reactors may have advantages over 'critical' fast reactors especially with regard to fuel composition, although the external source of neutrons may be very expensive. ADSs are also attracting interest as a possible way of 'transmuting' long-lived radioactive waste and thereby making it more manageable.

In 2004 the US DOE sought a partner to develop the Next Generation Nuclear Plant (NGNP), a Generation IV reactor of some kind, as its leading concept for developing advanced power systems both for electricity and hydrogen production on a very large scale. A pilot plant demonstrating technical feasibility was envisaged by 2020 at Idaho National Laboratory, with international collaboration. If successful, the NGNP would be: 'smaller, safer, more flexible and more cost-effective than any commercial nuclear plant in history. The NGNP was to secure a major role for nuclear energy for the long-term future and also provide the US with a practical path towards replacing imported oil with domestically produced, clean and economic hydrogen.'[78] The DOE goals for a commercial NGNP were electricity at less than 1.5¢ per kWh, hydrogen at less than 40¢ per litre petrol equivalent and an overnight capital cost below $1,000 per kW ($1,250 in 2015 prices), dropping to half of this in due course.

In 2010 the European Commission launched the European Sustainable Nuclear Industrial Initiative (ESNII) to support three Generation IV fast reactor projects as part of the EU's plan to promote low-carbon energy technologies. The three designs chosen were the Astrid sodium-cooled fast reactor (SFR) proposed by France, the Allegro gas-cooled fast reactor (GFR) supported by central and eastern Europe, and the Myrrha lead-cooled fast reactor (LFR) technology pilot proposed by Belgium. The timescale was typically ambitious, optimistic or unrealistic, depending on one's viewpoint. A 250–600 MWe Astrid prototype was to be built at Marcoule and connected to the grid in 2022; a 50–80 MWth Allegro plant would go to the Czech Republic, Slovakia and Hungary to operate from 2025; a 50–80 MWth Myrrha project in Belgium was planned to produce power in 2023.

The 2014 update of the GIF technology roadmap confirmed the choice for the six systems and focused on the most relevant of their developments so as to define the R & D goals for the next decade. It suggested that the Generation IV technologies that were most likely to be deployed first were the sodium-cooled fast reactor, the lead-cooled fast reactor and the very high temperature reactor technologies. The molten salt reactor and the gas-cooled fast reactor were shown as furthest from demonstration.

Fusion

Looking even further into the future, there may come a time when nuclear fusion becomes a serious contender.

To slip into the quasi-religious symbolism which often surrounds science, fusion is the holy grail of energy research. Fusion works on a different principle from fission; instead of breaking up relatively large atoms to make energy, in fusion the nuclei of the smallest atoms, generally various types of hydrogen, are built up into bigger ones, creating significantly more energy per reaction. The most practical reaction for terrestrial use is likely to be that between the second and third isotopes of hydrogen (deuterium, one proton and one neutron; and tritium, one proton and two neutrons), which react to form helium and a neutron which carries most of the energy. A single fusion reaction produces vastly more energy than a single fission reaction. It is the energy of the stars.

The mechanics of the fusion reaction began to be realised in the 1920s, with Hans Bethe confirming it experimentally in 1939. The first thermonuclear device was tested in 1951, and the first controlled fusion reaction achieved at Los Alamos in 1958.

Fusion has a number of attractive features. It would in effect run on water and lithium – a plentiful resource. There would be no fission products nor long-lived actinides such as plutonium so waste management would be very much easier. The power stations would become radioactive themselves over a period of time but it is likely that materials can be developed which could be reused in further fusion reactors, perhaps after being stored for a few decades. And because there would only be a few grams of radioactive material in the reactor at any one time the possibility of a big accident like Chernobyl could not arise.

However, there are considerable technical challenges still to be overcome. The 'plasma' in which fusion takes place must be heated to 100–200 million°C which is around ten times the temperature of the sun. No solid substance could be used to contain such temperatures so the plasma would have to be trapped in a magnetic field, or 'bottle', requiring massive electrical fields. Converting the energy released into electricity would be a difficult practical issue.

Fusion research has been carried out for some decades, notably in the UK, which has hosted the Joint European Torus, JET, since 1978, and in the US and Japan. This research has confirmed that fusion can be made to happen on earth – in other words, the physics works; although no machine has yet been built which can produce a self-sustaining fusion reaction. The next stage will be to determine if plant can be engineered which could harness the energy produced, coming in the form of neutrons travelling at a speed not far short of the speed of light, to make electricity. Then comes a commercial demonstration reactor to determine the economics.

Fusion is a further example of how difficult it is to take decisions within any sensible timescale. In 1985 the Soviet Union suggested building a next generation fusion reactor with Europe, Japan and the US. Between 1988 and 1990 initial designs were drawn up for an International Thermonuclear Experimental Reactor (ITER). In 1998 the ITER Council approved the first comprehensive design of a fusion reactor based on well-established physics and technology, at an expected cost at the time of $6 billion. The US then pulled out of the project but rejoined in 2003 when China and South Korea also joined, though Canada terminated its interest. After a further period of indecision the partners agreed in mid-2005 to site ITER at Cadarache in southern France but with major concessions to the Japanese who had proposed Rokkasho as an alternative. The EU and France were to contribute half of the cost, which by then had reached $10 billion, with the other partners – Japan, China, South Korea, US and Russia – putting in 10 per cent each. Japan was to provide high-tech components, host a €1 billion materials testing facility and enjoy the right to host a subsequent demonstration fusion reactor. India joined the ITER project in late 2005.[79] Construction of the tokamak, the main reactor component of the 500MWth project, began in 2013, with an expected date for full deuterium-tritium operation of 2027. The cost had now reached an estimated $16 billion.

The jokers have it that fusion is 30 years away (or 40 years, or 50 years) – always had been and always will be. In 1970, for example, a General Atomics pamphlet stated that 'by the year 2000, several commercial fusion reactors are expected to be online.'[80] Such predictions have inevitably given the technology something of a bad name and in turn that has put further pressure on budgets. Even if fusion is now 30 years away, having been 40 years away in the 1950s, the inherent uncertainties of such a large and complex programme mean that it will be some time before a realistic assessment of the practical role that fusion might play can be made. Even if an economic technology should emerge from ITER with the subsequent commercial prototype it will presumably take some decades before fusion is making a major contribution to global electricity production.

3.11 BELIEVING ONE'S OWN HYPE

As with all such things, perhaps one should hope for the best but prepare for the worst. As early as 1953 Admiral Hyman Rickover, 'the father of the nuclear navy', distinguished between 'academic' and 'practical' reactor designs, in terms that can be fruitfully applied more widely to energy sources as a whole.

'An academic reactor or reactor plant almost always has the following basic characteristics:

1. It is simple.
2. It is small.
3. It is cheap.
4. It is light.

5. It can be built very quickly.

6. It is very flexible in purpose.

7. Very little development will be required: it will use off-the-shelf components.

8. The reactor is in the study phase, it is not being built now.

On the other hand a practical reactor can be distinguished by the following characteristics:

1. It is being built now.

2. It is behind schedule.

3. It requires an immense amount of development on apparently trivial items.

4. It is very expensive.

5. It takes a long time to build because of its engineering development problems.

6. It is large.

7. It is heavy.

8. It is complicated.'[81]

Studies suggest that radically new nuclear technologies may take up to 50 years to become established due to factors like the need for safety licensing, building (successful) prototypes, scaling those prototypes up to full-scale, planning and siting approvals, slow construction times, historically rising costs and potential public opposition.[82] While ongoing research into new approaches to nuclear energy is highly attractive, the overall picture is reminiscent of the situation with wind power and solar. These technologies were available a long time ago – some decades in the nuclear example, rather longer for wind and photovoltaics. One wonders why, if these technologies were so easy to deploy, this had not been achieved previously. In the nuclear case the answer may lie in part within the structure of electricity supply systems which promoted very large 1,000 MW-plus reactors to the detriment of smaller designs. Be that as it may, it would be a serious mistake to rely on radically different technologies being available within the timescale of that required to replace the current global nuclear fleet, if that is a desired action.

ENDNOTES

1 Grimston M and Beck P (2002), *Double or quits – the global future of nuclear energy*, Earthscan.

2 http://www.worldnuclearreport.org/IMG/pdf/201408msc-worldnuclearreport2014-hr-v4.pdf, Schneider M and Froggatt A (2014), 'The World Nuclear Industry Status Report 2014', WISE.

3 http://www.neimagazine.com/opinion/opinionthe-world-nuclear-industry-is-it-in-terminal-decline-4394815/, Kidd S, 'The world nuclear industry – is it in terminal decline?', *Nuclear Engineering International*, 6 October 2014.

4 Bolter H. (1996), *Inside Sellafield*, Quartet Books.

5 http://filestore.nationalarchives.gov.uk/pdfs/small/cab-129-73-c-55-31-31.pdf, Ministry of Fuel and Power (1955), 'A programme of nuclear power', Cmnd 9389, HMSO.

6 http://www.nationalgrid.com/uk/Electricity/Data/Demand+Data/, National Grid (2013), 'Metered half-hourly electricity demand.'

7 https://www.gov.uk/government/statistical-data-sets/historical-electricity-data-1920-to-2011, DECC (2014), 'Historical electricity data: 1920–2014.'

8 Ministry of Fuel and Power (1957), 'Capital investment in the coal, gas and electricity industries', Cmnd 132, HMSO.

9 http://www.ncbi.nlm.nih.gov/pmc/articles/PMC1241789/, Bell M, Davis D and Fletcher A, (2004), 'A retrospective assessment of mortality from the London smog episode of 1952: the role of influenza and pollution', *Environmental Health Perspectives*, **112** (1) 6–8.

10 Official Report, 25 May 1965, **713**, 237–8.

11 Strauss L, 'Our children will enjoy in their homes electrical energy too cheap to meter – speech to the National Association of Science Writers, New York City', *New York Times*, 17 September 1954.

12 http://www.thisdayinquotes.com/2009/09/too-cheap-to-meter-nuclear-quote-debate.html, 'Too cheap to meter – the infamous nuclear power misquote', *This Day in Quotes*.

13 Dean G (1954), *Report on the atom: what you should know about atomic energy*, Eyre & Spottiswoode, London.

14 http://www.rug.nl/bibliotheek/support/law-library/document-instructions/traites_1957_euratom_1_en_0001. pdf, 'Treaty establishing the European Atomic Energy Community (EURATOM)', 1957.

15 http://books.google.com.au/books?id=4zigZvQ1wm8C&dq=bodega+bay+nuclear&printsec=frontcover&so urce=in&hl=en&ei=oRUqSu_BE9iOkAXoj_H4Cg&sa=X&oi=book_result&ct=result&resnum=12#v=onep age&q=bodega%20bay%20nuclear&f=false, Wellock T (1998), *Critical masses: opposition to nuclear power in California*, University of Wisconsin Press.

16 IAEA (1999), 'Nuclear research reactors in the world', reference data series 3.

17 http://www.iaea.org/About/Policy/GC/GC48/Documents/gc48inf-4_ftn3.pdf, IAEA (2002), 'Fifty years of nuclear energy'.

18 Abramsky K (2010), Sparking a worldwide energy revolution: *Social struggles in the transition to a post-petrol world*, AK Press, 346.

19 US Energy Information Administration (1986), 'An analysis of nuclear power plant construction costs'.

20 http://energyeconomyonline.com/uploads/Business_Risks_and_Costs_of_New_Nuclear_Power_Reprint_-_ Jan_2__2009_Craig_A._Severance.pdf, Severance C (2009), 'Business risks and costs of new nuclear power', Energy Economy Online.

21 http://www.phyast.pitt.edu/~blc/book/index.html, Cohen B (1990), *The nuclear energy option*, Plenum, New York.

22 Spiewak I and Cope D (1980), 'Overview paper on nuclear power', Oak Ridge National Laboratory Report ORNL/TM-7425.

23 http://www.phyast.pitt.edu/~blc/book/index.html, Cohen B (1990), *The nuclear energy option*, Plenum, New York.

24 http://seattletimes.com/html/opinion/2008082460_nukeop31.html, Pope D, 'A northwest distaste for nuclear power', *The Seattle Times*, 31 July 2008.

25 Blair J, 'N-plants provide power only for our greed', *Evansville Courier and Press*, 23 June 2004.

26 Pope D (2008), *Nuclear implosions: the rise and fall of the Washington Public Power Supply System*, Cambridge University Press.

27 Cook J, 'Nuclear follies', *Forbes*, 11 February 1985.

28 The Belarus Committee on the Problems of the Consequences of the Catastrophe at the Chernobyl NPP (2001), '15 years after Chernobyl disaster'.

29 http://www.theengineer.co.uk/news/edf-to-acquire-british-energy-group/308101.article, 'EDF to acquire British Energy Group', *The Engineer*, 24 September 2008.

30 Borenstein S (2003), 'Next generation nuclear plants appear too pricey despite Senate push', Knight Ridder/ Tribune News Service, 17 June 2003.

31 http://web.mit.edu/jparsons/www/publications/2009-004.pdf, Du Y and Parsons E (2009), 'Update on the cost of nuclear power', Centre for Energy and Environmental Policy Research, MIT.

32 http://bulk.resource.org/gpo.gov/hearings/108s/93750.pdf, Committee on Energy and Natural Resources, United States Senate (2004), 'Hearing before the Sub-Committee on Energy to receive testament regarding new nuclear power generation in the United States'.

33 http://seekerblog.com/archives/20080827/cera-construction-costs-for-new-nuclear-plants-up-over-230- since-2000/, Darden S (2008), 'CERA: construction costs for new nuclear plants up over 230% since 2000', *Seeker* blog.

34 https://www.ihs.com/info/cera/ihsindexes/index.html, IHS-CERA (2008), 'IHS CERA Power Capital Costs Index shows power plant construction costs decreasing slightly.'

35 Moody's Investor Services (2007), 'New nuclear generation in the United States: keeping options open vs addressing an inevitable necessity'.

36 http://www.synapse-energy.com/Downloads/SynapsePaper.2008-07.0.Nuclear-Plant-Construction-Costs. A0022.pdf, Synapse Energy Economics (2008), 'Nuclear power plant construction costs'.

37 https://www.keystone.org/wp-content/uploads/2015/08/06-2007-Nuclear-Power-Join-Fact-Funding.pdf, The Keystone Center (2007), 'Nuclear power joint fact-finding'.

38 Moody's Investor Services (2007), 'New nuclear generation in the United States: keeping options open vs addressing an inevitable necessity'.

39 Scroggs S (2007), 'Direct Testimony and Exhibits on behalf of Florida Power and Light in Docket No. 07-0650'.

40 Nuclear Engineering International (2008), 'Power market developments – the American way', 18 June 2008.

41 http://www.climateark.org/shared/reader/welcome.aspx?linkid=99329, Smith R (2008), 'New wave of nuclear plants faces high costs', *The Wall Street Journal*, 12 May 2008.

42 http://uk.reuters.com/article/2014/02/28/tvo-olkiluoto-idUKL6N0LX3XQ20140228, 'Finnish nuclear plant delayed again as Areva, TVO bicker', *Reuters*, 28 February 2014.

43 http://www.world-nuclear-news.org/RS-Lungmen-1-passes-pre-operational-tests-0108144.html, 'Lungmen 1 passes pre-operational tests', *World Nuclear News*, 1 August 2014.

44 http://flyvbjerg.plan.aau.dk/COSTCAUSESASPUBLISHED.pdf, Flyvbjerg B, Holm M and Buhl S (2004), 'What causes cost overrun in transport infrastructure projects?', *Transport Reviews*, **24** (1) 3–18.

45 Sovacool B, Nugent D and Gilbert A (2014), 'Construction cost overruns and electricity infrastructure: an unavoidable risk?', *The Electricity Journal*, **27** (4) 112–120.

46 http://www.4coffshore.com/windfarms/greater-gabbard-united-kingdom-uk05.html, 4C Offshore Wind (2014), 'Greater Gabbard'.

47 http://www.reuters.com/article/2015/01/21/eu-nuclear-austria-idUSL6N0V039W20150121, Nasralla S, 'Austria again pledges to sue Brussels over UK nuclear plant plan', *Reuters*, 19 January 2015.

48 http://www.cogentskills.com/media/1087/next-generation-skills-for-new-build-nuclear.pdf, Cogent (2010), 'Next generation – skills for new build nuclear'.

49 http://www.guardian.co.uk/environment/2013/apr/23/edf-energy-nuclear-power-station, Carrington D (2013), 'EDF Energy to cut jobs to control cost of building nuclear power station', *The Guardian*, 23 April 2013.

50 OECD/IAEA (2014), 'Uranium 2014: resources, production and demand', OECD: Paris.

51 http://www.world-nuclear.org/info/Country-Profiles/Countries-G-N/Kazakhstan/, World Nuclear Association (2015), 'Uranium and nuclear power in Kazakhstan'.

52 See for example: http://nuclearinfo.net/twiki/pub/Nuclearpower/ WebHomeAvailabilityOfUsableUranium/2009_Tamada.pdf, Tamada M (2009), 'Current status of technology for collection of uranium from seawater', Japan Atomic Energy Agency: Gunma.

53 http://www.neimagazine.com/features/featureuranium-market-doldrums-continue-4390747/, Steyn J and Meade T (2014), 'Uranium market doldrums continue', *Nuclear Engineering International*, 1 October 2014.

54 https://www.iaea.org/publications/documents/infcircs/report-radiological-accident-southern-urals-29-september-1957, Nikipelov B *et al.* (1989), 'Report on a radiological accident in the southern Urals on 29 September 1957', IAEA.

55 http://www.mdpi.com/1660-4601/6/1/174, Standring W, Dowdall M and Strand P (2009), 'Overview of dose assessment developments and the health of riverside residents close to the Mayak PA facilities, Russia', *International Journal of Environmental Research and Public Health*, **6** (1) 174–199.

56 http://www.foe.co.uk/news/nuclear_40884, Childs M (2013), 'A hard-headed look at nuclear power', Friends of the Earth.

57 http://www.no2nuclearpower.org.uk/nuclearnews/NuClearNewsNo54.pdf, No2NuclearPower, *NuClear News* **No. 54**, September 2013'.

58 https://www.gov.uk/government/uploads/system/uploads/attachment_data/file/68927/7386.pdf, Defra/BERR/ Devolved Administrations of Scotland and Wales (2008), White Paper, 'Managing radioactive waste safely: a framework for implementing geological disposal'.

59 http://www.westcumbriamrws.org.uk/documents/240-letter_from_DECC_regarding_DtP_and_RoW_7_ Nov_2011.pdf, Hendry C (2011), Letter, 'Managing Radioactive Waste Safely (MRWS) decision making process'.

60 http://www.cumbria.gov.uk/news/2013/January/30_01_2013-150007.asp, Cumbria County Council (2013), 'Cumbria says 'no' to underground radioactive waste repository'.

61 Irving A (2013), 'Copeland may go it alone in n-search', *Whitehaven News*, 31 January 2013.

62 http://articles.philly.com/1986-08-22/news/26063183_1_chernobyl-reactor-valery-legasov-soviet-officials, Goldstein S, 'Soviets say six human errors led to Chernobyl accident', *philly.com*, 2 August 1986.

63 Ignatenko E, Voznyak V, Kovalenko A and Troitskii S (1989), *Chernobyl – events and lessons (questions and answers)*, Moscow Political Literature Publishing House.

64 Kovalenko, A, 1989, *Soviet Weekly*, 19 August 1989.

65 World Health Organisation (1995), 'Health consequences of the Chernobyl accident – results of the IPHECA pilot projects and related national programmes (summary report)', Geneva.

66 http://www.threemileisland.org/downloads/188.pdf, Kemeny J *et al.* (1979), Report of the President's Committee on the accident at Three Mile Island', US Government Printing Office, Washington DC.

67 http://www.iaea.org/NuclearPower/SMR/index.html, IAEA (2013), 'Small and medium sized reactors: development, assessment and deployment'.

68 Ed. Chumbley S (1996), *Conway's all the world's fighting ships 1947–1995*, US Naval Institute Press, Annapolis.

69 http://www2.ans.org/pi/smr/ans-smr-report.pdf, American Nuclear Society (2010), 'Interim report of the American Nuclear Society President's Special Committee on small and medium sized reactors: general licensing issues'.

70 http://energy.gov/articles/energy-department-announces-small-modular-reactor-technology-partnerships-savannah-river, USDOE (2012), 'Energy department announces small modular reactor technology partnerships at Savannah River Site', energy.gov.

71 http://www.iaea.org/NuclearPower/Downloads/Technology/meetings/2011-March-TWG-GCR/Day1/HTR-PM-Status-SYL-20110328.pdf, Sun Y (2011), 'HTR-PM project status and test programme', IAEA.

72 http://www.powerengineeringint.com/articles/2013/05/westinghouse-and-chinas-snptc-sign-smr-nuclear-pact.html, Ross K (2013), 'Westinghouse and China's SNPTC sign SMR nuclear pact', *Power Engineering International*.

73 http://csis.org/blog/russias-floating-nuclear-reactors, Friedman J (2011), 'Russia's floating nuclear stations', Centre for Strategic and International Studies.

74 http://thebulletin.org/thorium-wonder-fuel-wasnt7156, 'Thorium: the wonder fuel that wasn't', *Bulletin of Atomic Scientists*, 11 May 2014.

75 http://blogs.telegraph.co.uk/finance/ambroseevans-pritchard/100026863/china-going-for-broke-on-thorium-nuclear-power-and-good-luck-to-them/, *Telegraph*, 19 March 2014; and
    http://www.theguardian.com/world/2014/mar/19/china-uranium-nuclear-plants-smog-thorium, Duggan J, 'China working on uranium-free nuclear plants', *The Guardian*, 19 March 2014.

76 https://www.gen-4.org/gif/jcms/c_9260/public, Generation IV International Forum.

77 https://www.gen-4.org/gif/upload/docs/application/pdf/2014-03/gif-tru2014.pdf, Generation IV International Forum, 'Technology roadmap update for Generation IV nuclear energy systems'.

78 http://www.power-eng.com/articles/2004/02/doe-releases-draft-requests-for-proposals-to-establish-a-world-class-nuclear-technology-lab-in-idaho-and-accelerate-site-cleanup.html, 'DOE releases draft request for proposals to establish world class nuclear technology lab in Idaho', *Power Engineering*, 19 February 2004.

79 http://www.iter.org/, ITER.

80 Rajput S (2009), *Nuclear energy*, Mahaveer and Sons.

81 http://ecolo.org/documents/documents_in_english/Rickover.pdf, Rickover H (1953) in 'AEC Authorizing Legislation: Hearings before the Joint Committee on Atomic Energy', 1970.

82 MIT (2009), 'Update of the 2003 future of nuclear power'.

4 | RENEWABLES

In 2001, on the day after the general election of that year, the newly re-elected Labour government launched a review of energy policy, to be overseen by the Performance and Innovation Unit which reported directly to No. 10. A prominent member of the team which produced the report, published in 2002[1], was economist Professor Gordon MacKerron, who was subsequently the first chairman of the Committee on Radioactive Waste Management (CoRWM). A generation of thinking (though only six actual years) later, in January 2008, the nuclear White Paper, *Meeting the energy challenge*, designed to send the signal for a new programme of nuclear reactors in the UK, was published. At that point MacKerron expressed a concern that nuclear investments would ultimately stall but that the expectation that new reactors would be built would hold back investment in the alternatives. So we could get to 2020 and find that neither nuclear nor other forms of carbon abatement technology had been built.[2]

History was not unkind to MacKerron's foresight. However, one could argue the case in reverse. Perhaps the damage to the ability to take necessary decisions has also resulted from a lingering and unrealistic fantasy as to the role that renewables can play.

4.1 AN ENERGETIC IDYLL?

In its widest sense, the term 'renewables' – sources of energy that do not rely on irreplaceable fuels such as crude oil, coal, natural gas or uranium – includes the power from animal muscles, including those of humans. In more recent times, however, it has come to refer to forces of nature such as wind, water and sun alongside more inert biological supplies – biofuels – such as plant and animal waste matter.

The emotional attractiveness of renewables is undeniable. Sources of electricity that rely on 'the power of nature', which don't belch smoke out of huge chimneys and which bring to mind pictures of Dutch windmills or waterwheels attached to ramshackle farmhouses and driven by babbling brooks, have an appeal that certainly stretches back to the Romantic movement.

> Water and windmills, greenness, Islets green;
> Willows whose Trunks beside the shadows stood
> Of their own higher half, and willowy swamp:

Farmhouses that at anchor seem'd – in the inland sky
The fog-transfixing Spires –
Water, wide water, greenness and green banks,
And water seen.[3]

When long-time Nazi Party member and supporter Martin Heidegger was allowed to resume teaching some years after the Second World War, he praised not only the Romanticism of renewables but also their intermittency as promoting a more authentic, hand-to-mouth way of interacting with the world than is demanded by modern industrial society.

> Modern technology is a [challenge], which puts to nature the unreasonable demand that it supply energy which can be extracted and stored as such. But does not this hold true for the old windmill as well? No. Its sails do indeed turn in the wind; they are left entirely to the wind's blowing. But the windmill does not unlock the energy from the air currents in order to store it.[4]

A typical recent paean, and one not without merit, goes:

> Fossil fuels are limited and therefore one day coal, oil, and natural gas will be depleted but the same scenario won't happen with renewable energy sources because the sun will continue to shine, the wind will continue to blow, etc. It is difficult to say how long fossil fuels will still be able to satisfy a large part of global energy demand – some energy experts believe this is likely to last till the end of this century – but in any case once fossil fuels become depleted the world will need to have already established an alternative in the form of renewable energy. From the ecological point of view renewable energy has an extreme advantage over fossil fuels: renewable energy sources are clean and unlike coal, oil, and natural gas release none or negligible direct carbon emissions. Fossil fuels on the other hand when burning release harmful carbon emissions that not only pollute our planet but have contributed to the severity of climate change impacts … Many countries in the world rely on foreign oil imports and by developing their own domestic renewable energy sectors they would help decrease the dependence on importing oil from other countries and would diversify their own energy portfolio in the process. The development of renewable energy sectors can also create many economic benefits for countries, mostly in form of many new green jobs.[5]

Another source summarises the benefits as follows.[6]

- Renewable energy promises greater security of energy supply by utilising abundant, diverse and domestic (non-imported) sources. This eliminates the resource exhaustion constraint.

- When substituting fossil fuels with renewable energy, global and local greenhouse gas emissions will be substantially reduced.
- Innovation and technology development in this field enable enhanced opportunities for implementation, particularly in rural areas and in emerging countries.
- The exploitation of renewable energy promotes local and regional employment in the energy manufacturing sector, installation and maintenance.

Opinion polling suggests that people support the concept of renewables, as shown in figure 4.1.

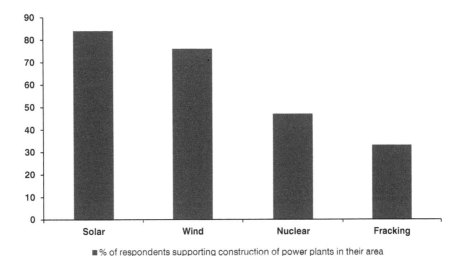

Figure 4.1 Percentage of respondents supporting construction of power plants in their area[7]

This is not a universal opinion, however. The twenty-first century in particular saw growing opposition to renewable generation, on both economic and environmental grounds. 'Energy insiders have long known that the notion of "renewable energy" is a romantic proposition – and an economic bust. But it is amazing what the lure of guaranteed "few strings attached" government subsidies can achieve. Even the Big Oil companies bought into the renewables revolution, albeit mostly for PR reasons. Like Shell, however, many quickly abandoned their fledgling renewable arms. Post-2008, they knew, the subsidy regimes could not last. Neither was the public buying into the new PR message.'[8]

The energy sources for renewables, as is often trumpeted, are free. But that is also true of oil, coal, gas and uranium – nobody had to invest a penny in putting these mineral resources into the ground. The challenges in all cases arise in trying to convert these free resources into electricity or other useful forms of practical energy and in this sense 'renewables' are highly resource intensive – perhaps 'fuel-free' would be a more accurate, if less PR-friendly, description. Alongside any benefits that renewables may bring in terms of low direct levels of emissions over their whole life cycle and the unlimited availability of fuel come two major disadvantages.

The power density of renewables is much less than that of a chemical fuel such as coal, or even more to a nuclear fuel such as uranium. Large amounts of machinery spread out over considerable areas of land or water, with concomitant environmental disruption, will therefore be necessary to collect significant amounts of power. DECC produced a graphic illustrating the point, contrasting the 430 acres needed for Hinkley Point C with 130,000 acres to produce a similar capacity of solar farms and 250,000 acres for onshore wind. (The graphic was removed from DECC's website very rapidly after the Renewable Energy Association claimed it was 'unhelpful'. The sway that the renewables lobby holds over some areas of government is perhaps illustrated by DECC's refusal to allow permission to use the simple figure for this book, despite it having been produced using taxpayers' money and being freely available online.)[9]

The extraction of very large volumes of aluminium for solar photovoltaics and nickel for the alloy used in wind turbines result in significant amounts of carbon emissions, resulting in an energy 'pay-back' period of 1 to 7 years for large solar installations and 1 to 12 years for wind farms.[10] The mining of rare metals such as neodymium for use in wind turbines may also not be sustainable.

Secondly, as renewable penetration grows the intermittency of the sources, especially wind and solar, may represent a considerable problem for managing a system where electricity supply and demand has to be matched precisely across fractions of seconds. This problem is easily manageable in technical terms when renewables play only a small part – though even then it reduces the benefits in terms of economics and environmental improvement compared to those predicted by a more simplistic approach – but can become highly challenging at high levels of penetration.

4.2 THE (NOT-SO-) 'NEW' RENEWABLES

The phrase 'new renewables', as these sources have (ironically, one assumes, or at least misleadingly) been dubbed by some of those who reap the benefits of the generous subsidies currently available, carries with it something of a hint that although humankind was dimly aware of water, wind and sunlight it had not realised that they could be used for energy until very recently. To quote singer Vanessa Williams, albeit in a different context, mankind tried out all of the nasty sources of energy first but decided to 'save the best till last'. In reality they have been the mainstay of human energy supplies for most of recorded history and well before. Use of wind to propel boats, sun to warm dwellings and of course to grow crops, water to drive equipment for grinding corn or aid irrigation have all been practiced for thousands of years.

Renewables were also among the earliest techniques for generating electricity. Faraday, 'the father of modern electricity', invented the first dynamo, or electricity generator, in 1831; the first dynamos capable of generating electricity for industry came in 1866, invented independently it seems by Charles Wheatstone (UK), Werner von Siemens (Germany) and Alfred Varley (UK) and mainly powered by steam engines fuelled by coal. Hydropower appeared in the 1870s. Electricity as a commercial venture made its first appearance in the remarkable decade of the 1880s which saw the first public power station, the Edison Electric Light Station, open at Holborn Viaduct in London, in January 1882, serving such auspicious customers as the City Temple, the Old Bailey law courts and the Telegraph Office of the General Post Office. A few months later the Wizard of Menlo Park, and 'father of' too many things to mention here, Thomas Edison, opened his celebrated Pearl Street Station generating electricity for lighting in lower Manhattan, New York. Within another five years power had been generated using both wind and solar power while wave and geothermal electricity emerged early in the twentieth century, though it was not until 1951 that they were joined by electricity from nuclear fission.

It is easy to forget just how rapidly electricity use grew in the years following its emergence, as can be seen in figure 4.2, and how urgent the need to find alternatives to coal appeared at the time.

Against such a background every potential alternative source of electricity – wind, solar, wave, geothermal, hydroelectric – received the same enthusiastic, inventive attention in the late nineteenth and early twentieth centuries as the other great achievements of the 'second industrial revolution' such as the internal combustion engine, air travel, electricity networks, refrigeration, the radio, recorded sound, the television, the plastics industry, materials such as alloy steels, sanitation, modern medicine and so on. Yet, despite early successes, only hydropower flourished to become a major power source by the middle of the twentieth century.

The numbers of patents granted to wave and tidal power, as sources of mechanical or electrical energy, in the UK serves to illustrate the rise and fall of many renewables over the last century and a half. After a steady increase from the first patent in 1850, interest evaporated almost overnight during the 1930s as the national supergrid emerged between 1927 and 1935, and the vast technical and economic superiority of a centralised power generation system made itself felt – see figure 4.3.

Before this point electricity had been dominated by small, 'embedded', localised generation systems, very similar in essence to those often advocated by Big Green today. By 1920 London alone had no fewer than 50 different systems of electricity supply with 24 different voltages and ten different frequencies. Moving home could mean having to change all the electrical appliances.

At Deptford operators were supplying London with 10,000 volt lighting supplies at 83.3 Hz (cycles per second); direct current at 460 V and 230 V for local industrial power and lighting; a three-phase 6,600 V 25 Hz system for large power consumers and traction; plus a single-phase 6,600 V 25 Hz system feeding the Brighton railway.

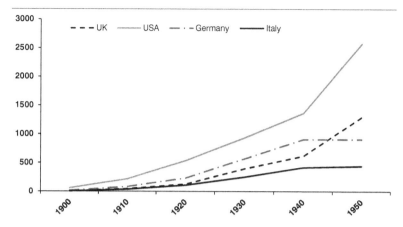

Figure 4.2 Electricity generated per head (kWh), 1900–50

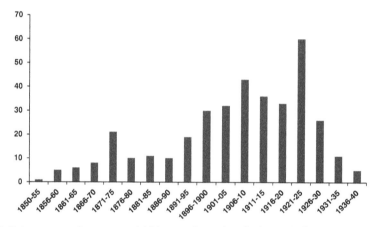

Figure 4.3 Patents granted to wave and tidal power inventions in the UK in five-year tranches, 1850–1940[11]

Deptford was one of the few 'large' stations of that time (some 160MW): most had a generating capacity of less than 5MW. Very few stations were interconnected, which meant that most needed to install their own reserve plant to cover breakdown and maintenance. Electricity was expensive for industry and a luxury that most people couldn't afford in their homes.

The rationalisation project was managed by the Central Electricity Board (CEB), established in December 1926. Its task was to build a 'gridiron' of high voltage transmission lines to link the most efficient stations, with the aim of providing abundant, reliable and cheaper supplies. At the time there was intense opposition both in Parliament and from electricity undertakings. As one power company chairman had it: 'The salvation of the industry is not to be brought about either by the waving of weird gridirons or the multiplication of authorities.' But the outcome was impressive. The 140 'selected stations' stayed in the ownership of the power companies and municipalities but CEB engineers controlled generation using the most efficient plant to achieve the lowest production costs. Crucially, because the interconnected stations could pool reserve plant there were even bigger savings. The average cost of electricity was halved. One company was offering domestic lighting supplies on a sliding scale from today's equivalent of 6.5p to 4.5p per unit and power for cooking or domestic chores at half that. Showrooms were hiring out cookers at six shillings a quarter. Industry was reaping equal benefits.[12]

The result was a new electricity revolution as improvements in energy efficiency resulted in enormous increases in demand (see figure 4.4). For the first time ordinary people could afford electricity.

The demise of the variable renewables in the middle of the twentieth century was not through a lack of interest or effort nor of demand for electricity. Contrary to the common impression, bolstered as it is by the propaganda value of the term 'new renewables', distributed energy powered by renewables did receive enormous attention and research effort around a century ago. Nor can its failure credibly be ascribed to a conspiracy by the fossil fuel industry, the nuclear industry or any other malevolent agent – hydropower thrived in the environment of the day while nuclear fission was not even discovered until 1938. A more likely explanation simply involves the disadvantages of low power density and variable output as outlined previously. When Nikola Tesla, variously called 'the father of electricity', the father of radio and even the father of modern technology', and George Westinghouse's AC distribution system won the 'war of the currents' over Edison's DC system in late nineteenth-century America, power stations could expand rapidly and export their power over much longer distances. The attractions of reliable,

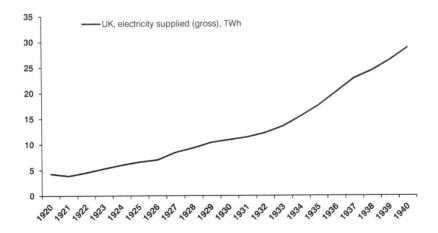

Figure 4.4 Annual gross electricity generation in the UK, 1920–40[13]

controllable electricity generated centrally, enjoying vast economies of scale and, crucially, not requiring huge battery capacity as back-up, were such that the variable renewables largely died out, despite the enthusiasm of advocates, until the post-Rio structure of widespread subsidies and preferential market treatment ushered in a new age. But this highly generous regime could not address the basic issues of unfavourable physics that had killed off the 'dash for renewables' in the 1920s and 1930s.

Let us start with an overview of the renewable technology that did succeed and then look at some of the others.

Hydropower

Evidence of the use of the potential energy stored in water and released as it falls, under gravity, from a higher level to a lower can be found in the artefacts of many ancient civilisations. The Roman Empire used water powered mills to grind corn for flour and to fashion wood and stone. The Han Dynasty in China (third century BC to third century AD) used watermills and also hydraulically powered pumps to raise water for irrigation; water power was also used in ancient India. It was perhaps the first driving force of the Industrial Revolution and though it was largely displaced by steam from burning coal and wood by the late eighteenth century in the UK, its use for small-scale applications survived through the nineteenth century.

British industrialist and armaments manufacturer William George Armstrong installed the world's first machinery to create electricity from hydropower at his house at Cragside, Northumberland, in 1878. It was used inter alia to power an arc lamp in his art gallery, to provide heating and hot water and to run a lift. In 1880 a brush arc light dynamo driven by a water turbine was used to provide lighting to a theatre and shops in Grand Rapids, Michigan; Schoellkopf power station near Niagara Falls opened in 1881; the following year the Fall River plant, later renamed the Appleton Edison Light Company, started operating in Appleton, Wisconsin with an output of 12 kW.[14] By 1886 there were 45 hydroelectric power stations in the US and Canada; by 1889 there were 200 in the US alone.[15]

The early development of hydropower and indeed of electricity itself illustrates the talent that the inventors, engineers, scientists and entrepreneurs of the late nineteenth- and early twentieth-century had for taking decisions and delivering on them, in sharp contrast to today's paralysis. By the beginning of the twentieth century small hydroelectric power stations were appearing in mountainous areas near many centres of population and in a range of countries.

- In Canada the first use of water to produce electricity was from a wheel built by the Ottawa Electric Light Company at Chaudieres Falls in 1881, used to power street lights and local mills. DeCew Falls 1, opened in Ontario in 1898 by the Cataract Power Company, was still in operation well over a century later. In 1945 hydropower provided 94 per cent of Canada's generating capacity.

- In 1882 a dynamo to provide electricity for lighting was installed at Ryds Bomullsspinneri, a spinning mill in Västergötland, Sweden, the power coming from the nearby Viskan River. The following year the Domnarvet ironworks installed an electric lighting system. The development of an AC system to replace the earlier inefficient DC approach allowed power to be transmitted from a power station near Lake Hällsjön in Dalarna to the mines in Grängesberg in 1893, following which three very large hydro stations were built.

- Hydropower came to Norway in 1885 when the outskirts of the town of Skien received a supply of hydroelectricity from a nearby wood-processing plant on the Skien River. Several hundred local or neighbourhood power systems were established over the period 1915 to 1930 and by 1929 a total of 1,452 hydropower plants were operating in the country.

- The first hydroelectric power station in New Zealand was established in 1885 to provide power for a stamping mill; by 1930, 92 per cent of the country's power was from hydroelectricity.

- At the beginning of the twentieth century Louis Tinchant built 14 hydropower plants in France, supplying electricity to 50,000 people. By 1960 more than half of France's power was from hydroelectric sources.

- Scotland's first hydropower scheme began generating at the Fort Augustus Abbey in the Highlands in 1891; rumour has it that when the monks played the electric organ the lights in the town went dim.

The year of 1922 saw the construction of the first hydropower plant specifically designed to provide peaking power. In 1925 an International Exhibition of Hydropower and Tourism held in Grenoble, France, attracted over one million visitors. A major milestone was the start of construction of the Hoover Dam in 1931 with capacity of 1,135 MW, the largest in the world at that time. Governmental funding was increasingly made available, notably the creation by the Roosevelt administration in 1933 of the Tennessee Valley Authority (TVA), which built dams and managed flood control and soil conservation programmes. In 1937 the Bonneville Power Administration (BPA) was established by an Act of Congress to market electricity from the Bonneville Dam on the Columbia River and create an associated transmission network. BPA is now based in Portland Oregon and is responsible for 31 federal hydropower dams. In 1942 the Grand Coulee Dam on the Columbia River became the world's largest with capacity 6,809 MW. The proportion of US electricity being generated using hydropower grew from 15 per cent in 1907 to 25 per cent in 1920 and 40 per cent in 1940. Hydropower became known as 'white coal': the appetite for practical renewable energy was apparently inexhaustible.

Although hydropower's share of global primary energy production dipped somewhat during the era of cheap oil, it edged up from 5.5 per cent in 1965 to 6.7 per cent in 2013. Absolute output from hydropower more than quadrupled during this period, as shown in figure 4.5.

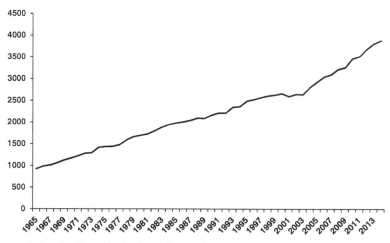

Figure 4.5 Global production of hydroelectricity (TWh), 1965–2014[16]

Hydropower succeeded because it does not suffer from the two main disadvantages that beset many of the other renewables to the same extent. Hydro dams, as opposed to 'run-of-the-river' hydro, in effect concentrate the energy source and therefore significantly increase the power density available. And since the dam can store water, which can be used for power production as the electricity is needed, variability of output is less of an issue – indeed, almost all of the world's power storage capacity is in the form of pumped storage schemes in which surplus electricity is used to pump water uphill and the resultant lake is used for a hydropower station when the electricity is needed.

This is not foolproof of course – some of the most serious acute power crises have been as a result of low rainfall in countries heavily dependent on hydropower, including Brazil (2000/1)[17], Chile (1998/9 and 2007/8)[18], China (2011)[19], New Zealand (2001, 2003 and 2008) and Norway (2003)[20]. It was also a factor in the California power crisis of 2000/1.

To take one example, hydropower has typically accounted for between 75 and 85 per cent of Brazil's total power generation, with fossil fuels producing about 10 per cent and the rest coming mainly from nuclear power and biomass. Among Brazil's hydropower capacity at the start of the twenty-first century was the world's biggest power station at the time, the 12,600 MW Itaipú Dam on the Parana River, co-owned with Paraguay and consisting of eighteen 700 MW turbines.[21] A combination of heavy reliance on hydropower and underinvestment in fossil fuel powered generating capacity left Brazil vulnerable to power outages during adverse weather conditions. In 2001 the country faced severe electricity shortages resulting from a prolonged period of lower than average rainfall and a particularly dry summer. Reservoir levels fell to 20 to 30 per cent of normal, representing just four months of reserve. It became clear that strict demand reduction programmes would be needed if widespread blackouts were to be avoided. These involved an overall target of a 20 per cent reduction in demand and a quota system within which each customer was required to reduce their consumption relative to a 'baseline', with financial penalties and disconnection for non-compliance. In early 2013 there was another dry period, during which short-term power prices were some 20 times the levels of the equivalent period in 2012, but there were no forced outages.

That being said, other dispatchable power sources – in effect those whose output does not in principle depend on random factors such as weather conditions – are also vulnerable to similar outages from time to time. These include nuclear power in France (2003 owing to a hot summer reducing availability of cooling water at inland reactors) and in Japan (2002/3 owing to a safety scandal and after Fukushima in 2011)[22]; coal in Australia (1949)[23] and the UK (1972 and 1974 and almost again in 1984/5)[24] owing to mining union strikes; gas owing to problems with pipelines, e.g. in Argentina in 2002.[25]

One of the major barriers to expansion of large scale hydropower, one of only two proven technologies that has made significant inroads into greenhouse gas emissions, has come from environmental protests. In 2014, for example, Greenpeace activists joined the Munduruku natives in the Amazon rainforest of Brazil to protest against the construction of an 8 GW power dam.[26] Similar protests have been provoked by the 2.75GW HidroAysén project in Chile, the 1.2GW Ilisu project in Turkey and the 700MW Karahnjukar plant in Iceland among many others. Further expansion of hydropower in Sweden is regarded as inconceivable: there are unused watercourses and Swedish rivers, including four large, entirely undeveloped Norrland rivers, which are protected by both Swedish law and the EU's Natura 2000. The major rivers have also been classified as national rivers with associated protection. Once again it is clear that Big Green has priorities higher than that of fighting climate change.

Wind power

There is evidence that wind was propelling boats on the Nile as early as 5,000 BC and was being used by the Babylonians and the Chinese to pump water for irrigating crops some 4,000 years ago. Babylonian emperor Hammurabi planned to use wind power for a large scale irrigation project in the seventeenth century BC.[27] The Persians used wind to pump water and grind grain between 900 and 500 BC, its use in food production spreading extensively to surrounding areas in the Middle East. Around AD 1,000 wind power technology spread north to European countries such as the Netherlands, which adapted windmills to help drain lakes and marches in the Rhine River delta; in the Middle Ages windmills became widespread across Europe for grinding corn for bread.[28]

The history of wind in the US and other industrialised and industrialising countries again gives lie to the idea that this is a new technology the potential of which has been criminally neglected by greedy oil producers or similar ne'er-do-wells. In the 1850s Daniel Halladay and John Burnham started the US Wind Engine Company and built the Halladay Windmill, designed for the landscape of the American West. The first windmill able to generate electricity was invented by Professor James Blyth of what is now the

University of Strathclyde in Scotland in 1887.[29] His ten-metre-high cloth-sailed device was used to power lighting at his holiday cottage in Marykirk – he offered the surplus output to the people of the village for street lighting but they rejected the offer in the belief that electricity was 'the work of the devil'. Blyth met less resistance from the altogether more sensible guests of the Montrose Lunatic Asylum, Infirmary and Dispensary where a later model windmill served as an emergency power source for the next 30 years.

However, Blyth's design was never economically viable, a major drawback being the inability to shut it down in high winds, risking severe damage. In Cleveland, Ohio, Charles F. Brush, 'the father of wind energy', constructed a more durable machine in the winter of 1887/8, which ran until 1900 and which could be shut off in high winds. It consisted of a 17-metre-diameter rotor mounted on an 18-metre tower and was rated at 12 kW. The dynamo was used to charge a bank of 12 batteries, thereby addressing the problem of intermittency, or to operate lighting and/or various motors in Brush's laboratory.

By the end of the nineteenth century American farmers and ranchers were employing wind power to pump water for irrigation and to generate electricity for homes and businesses. In 1893 the Chicago World's Fair showcased 15 windmill companies and their various designs. The invention of steel blades for the rotors in the late 1890s improved the efficiency and robustness of the process and by 1900 more than six million windmills had been erected through the US, mainly producing mechanical power but with 72 generating electricity by 1908.

In 1891 Danish physicist Poul la Cour constructed a wind turbine to generate electricity that was used to produce hydrogen by electrolysis, ushering in the 'hydrogen economy' which was still being talked about as an exciting new approach 125 years later. This could be stored for use in experiments and as a power source to light the local high school. La Cour made some headway in addressing the problem of intermittency in the very short term, i.e. over a period of seconds or less, by inventing a regulator which he called the Kratostate, and in 1895 he produced a prototype wind-powered electricity station which was used to light the village of Askov[30]. By 1900 there were around 2,500 windmills in Denmark producing an estimated combined output of some 30MW at peak.

Research into wind power allowed the technology's progress to continue through the first half of the twentieth century. In 1927 Joe and Marcellus Jacobs opened a factory in Minneapolis to manufacture wind turbine generators for lighting and battery charging on farms isolated from power grids. Over the next 30 years the firm produced around 30,000 small wind turbines, some of which ran for many years not only in the US but also, for example, in remote locations in Africa and even on Richard E. Byrd's expedition to Antarctica in 1934.[31]

Many other companies were also producing small-scale wind generators and by the 1930s wind turbines were being used across the US. These machines would typically have a capacity of a few hundred watts up to perhaps 25kW and would be used mainly to recharge batteries but also for more exotic purposes such as electrifying bridge structures to prevent corrosion. The most widely used small-scale wind generator during this period was a two-bladed horizontal-axis machine manufactured by the Wincharger Corporation which had a peak output of 200 W – these machines were still being manufactured in the US during the 1980s. In Australia the Dunlite Corporation built hundreds of small wind generators to provide power at isolated postal service stations and farms, such machines being manufactured from 1936 until 1970. The vertical axis wind turbine was invented in 1931. In the 1940s the then-largest wind turbine began operation in Vermont, rated at 1.25 MW and feeding electricity into the local transmission system for several months during the Second World War.

So a vast amount of commercial and research effort had been expended by the middle of the twentieth century on making electricity from the wind. Yet despite this long history wind never established a sustainable position in the electricity map of the US or indeed elsewhere. Electricity from wind was in effect a niche activity in rural America, in places where there was no access to what is now regarded as a conventional power supply. Windmills needed to be complemented by batteries to preserve some degree of security of supply, something which was just about feasible, though expensive, when farmers relied on electricity for a very small proportion of their energy. In 1935 Roosevelt succeeded in establishing the Rural Electrification Administration as part of the New Deal, as a result of which electrification of rural America became a public responsibility, very much against the American spirit of free enterprise but possible in the extraordinary circumstances of the Great Depression. As reliable and secure power supplies became available wind power died off, or at least slipped into suspended animation.

Solar energy

The sun, from the earth's point of view, is the most effective nuclear fusion reactor yet in existence, unlikely to be overtaken however successful earthbound experimentation into fusion might prove. It provides us with unimaginably large amounts of energy. Solar radiation reaching the earth could in theory power five 60 W light bulbs per square metre of surface. The total amount of power reaching the outer atmosphere from the sun is around 174 million GW, some 20,000 times as much as mankind uses. Further, the sun provides the energy which drives global weather systems and ocean currents on which wind, wave and tidal power depend. It evaporates water and causes the rainfall which creates the possibility of hydropower.

The indirect energy implications of solar power are hardly less important. Sunlight is vital for almost all plant life on which animals depend, the sources of the food which provides our personal energy as individual human beings, as well as biofuels such as wood and animal waste. The mortal remains of plants and animals over millions of years have mutated into the coal, oil and gas which still provide some 87 per cent of global traded primary energy. Indeed, one could come to the view that all energy is ultimately nuclear fusion energy, with the exception of fission and radioactive decay (which fuels geothermal energy), and even the uranium which powers nuclear energy was created as a result of fusion in ancient stars.

However, the development of solar engines and electricity generation, like other 'new' renewables, can be traced back to the second half of the nineteenth century.

Solar power looked especially attractive to those looking for new ways of making electricity. In a period spanning less than 50 years from the mid nineteenth century the early pioneers developed an array of innovative techniques for capturing solar radiation and using it to produce the steam that powered the machines of that era. All of the solar thermal conversion methods now being considered were well established by the outbreak of the First World War.[32]

Consideration of a few developments serves to give a flavour of the vigorous attention that solar power attracted in the late nineteenth and early twentieth centuries. The earliest example of the conversion of solar radiation into mechanical power – in effect a steam engine using sunlight rather than coal as the heat source – dates back to the 1860s, when French mathematician Auguste Mouchout was granted a patent for a motor running on solar power, a design he continued to improve until 1880. Sunlight was trapped by a glass-encased iron cauldron, resulting in the water within boiling. In 1865, by introducing reflectors to concentrate the solar radiation, he succeeded in using his apparatus to operate a small conventional steam engine. The following year he captured the enthusiasm of Emperor Napoleon III who provided financial assistance and by 1878 he had developed a solar-powered refrigerator. By 1881, however, the falling price of coal rendered Mouchout's device obsolete before it had really carved out a niche for itself.

In the mid-1880s Charles Tellier, 'the father of refrigeration', produced a solar collector which did not require the reflectors which had been severely limiting the potential scale of a solar engine. It was similar to today's roof panels for heating water and boiled liquid ammonia to power a water pump. He also further developed the concept of solar-powered refrigeration. However, there was a growing need for more compact refrigeration units which could be used on the ships transporting beef from South America to Europe. The area needed for solar refrigeration rendered this approach impractical – a manifestation of the problem of low power density.

During the same period work was being carried out into converting sunlight directly into electricity rather than using steam as an intermediary. The underlying science for photovoltaics had been developed by Antoine-César Becquerel (the actual grandfather of radiation pioneer Henri Becquerel) who in 1839 discovered the 'photovoltaic' effect in which a voltage develops when sunlight falls on certain materials. In 1883 American Charles Fritts constructed the first true solar cell using selenium and gold. These first devices were only capable of converting about 1 per cent of the solar radiation into electricity. Although another version of this device, using copper and copper oxide, was developed in the 1920s it was Russell Ohl's development of the silicon solar cell in

1941 which was to lead to a step change in performance. In 1954 three other American researchers, G.L. Pearson, Daryl Chapin and Calvin Fuller, demonstrated a silicon solar cell capable of a 6 per cent energy conversion efficiency when used in direct sunlight. (Silicon is the second most abundant element on earth, after oxygen. It is perhaps indicative of the inherent difficulties with solar photovoltaics as a useful source of electricity that it took so long to exploit it for power production.)

In 1904 Henry Willsie created the first solar device capable of operating at night using heat gathered during the day, in an attempt to address the problem of variability. Although he originally planned to market his device for desert irrigation he subsequently wrote that the invention was designed for furnishing power for electric light and power, for refrigeration and ice making, for milling and pumping at mines and for other purposes where large amounts of power are required. Unfortunately, no buyers emerged and Willsie's company disintegrated, like the earlier attempts to commercialise the technology.

In 1912, Frank Shuman, through the Sun Power Company, succeeded in installing a solar engine at Maadi outside Cairo, where the combination of intense solar irradiation and the costs of importing coal to the area made the economics look attractive. But two months after the final trials of the machine the First World War broke out, spreading rapidly to Europe's African colonies and resulting in the destruction of Shuman's device and the dissolution of his team. Shuman died in 1918 before the war had ended.

Between 1860 and 1914 some 50 patents were taken out for solar engines and scores of books and articles on the topic appeared. As was the case with wind power, the failure to make commercial progress was not through a lack of inventive effort or ingenuity. Developments in the use of oil and coal, sources with much higher power densities and which can be stored, outstripped those of solar power. By the early twentieth century oil and coal were beneficiaries of enormous infrastructures, stable markets and growing supplies as new resources were discovered and exploited.

It was only during the oil shocks of the 1970s that serious attention turned again to the inherently problematic solar technology. During the following two decades solar engineers tried myriad techniques to satisfy society's need for power, rediscovering many of the approaches of the early pioneers. Larger designs, in the 100 MW-plus range, were tested but once again the push collapsed under its own weaknesses when the oil price fell in the 1980s. In 1991 Luz, producer of more than 95 per cent of the world's solar electricity at the time, filed for bankruptcy. Chairman Becker claimed: 'The failure of the world's largest solar electric company was not due to technological or business judgment failures but rather to failures of government regulatory bodies to recognise the economic and environmental benefits of solar thermal generating plants.' Or to put it another way, it failed because it was unable to compete economically and unavoidably dependent on a degree of political support which by the 1990s was harder to come by. Other solar projects met a similar fate. Investors must fear that the latest round of institutional enthusiasm for solar power may ultimately prove just as ephemeral, especially if the fall in the oil price which began in mid-2014 were to be sustained. The withdrawal of subsidies for large-scale solar farms on agricultural land in the UK in 2015 will not have allayed such fears; neither will the comments by the environment secretary, Liz Truss, before the 2015 general election that they were a 'blight' on the countryside and that the sight of 'row upon row' of large-scale solar farms on once productive agricultural land 'makes the heart sink'.[33] Truss kept her job after that election.

As previously mentioned, hydropower faced a similar challenge from developments in fossil fuel technology but continued to flourish. The unavoidable conclusion is that far from being killed off until very recently by any lack of effort, solar power simply does not fit comfortably with the needs of a modern economy. Like wind power, the problem lies not in the effort or enthusiasm of the researchers but in the fundamental problems caused by the variability of the output and low power densities of these ancient 'new renewables'.

Wave power

The first known patent to use energy from ocean waves was filed in Paris in 1799 by a Monsieur Girard and his son. In their application 'to employ, by various means, the waves of the sea like motors', they rather poetically began:

> Waves, restless and unequal, rise up like mountains then collapse in a moment, resulting in the movement of all bodies which float, regardless of their weight or volume. The huge mass of a ship of the line, which no known power would be able to lift, obeys the slightest movement of the waves. Imagine, for a moment, such a ship suspended on the tip of a lever and one might get the idea of the most powerful machine that has ever existed.

Between 1855 and 1900 over 120 patents for energy-from-wave machines were issued in the UK alone.[34] Electricity was first generated from wave power in 1910, when Bochaux-Praceique generated around 1 kW to provide lighting and power to his house at Royan near Bordeaux. However, interest in wave (and tidal) power followed a similar pattern to that of other variable renewables, burgeoning until the 1930s and then fading away rapidly, despite the work of enthusiasts such as Yoshio Masuda in the 1940s.

Geothermal energy

Geothermal energy – energy from the earth's heat – is largely a result of radioactive decay deep underground. Its history perhaps goes back further than any source except for the sun, there being evidence that Native Americans were making settlements near hot springs some 10,000 years ago. The oldest known spa is a stone pool on China's Lisan Mountain built in the third century BC. The Romans established public baths at Bath (Aquae Sulis) in Somerset in the first century AD, while the world's oldest geothermal district heating scheme, at Chaudes-Aigues in southern France – the name derives from the Latin for 'hot waters' – has operated since the fourteenth century.

The first industrial use may have been in the extraction of boric acid from the Larderello Fields near Pisa, Italy, in the late eighteenth century. This location was also the site of the first generation of electricity from geothermal steam, in 1904. The first US geothermal generating plant, in 1922, produced some 250 kW for lighting street lamps and buildings in the local area. By the 1960s the US was operating an 11 MW geothermal facility at The Geysers in California. Nonetheless, by 1975 total global geothermal generating capacity was just 1,300 MW.[35]

4.3 THE RECENT SURGE

Most self-respecting environmental organisations – and all environmental organisations have enormous respect for themselves if not for people with different views – have produced reports claiming that the world could quite easily be powered entirely by renewables by 2050 (often dominated by hydropower). These reports generally assume significant reductions in energy use – the World Wide Fund for Nature (WWF), for example, predicts energy use in 2050 of around 280 EJ, against 340 EJ in 2010.[36] By contrast, the OECD International Energy Agency (IEA) expects energy use of 520 EJ in 2025, assuming that energy efficiency improves at more than double the historical (1980–2010) average of 0.9 per cent per year. WWF makes two conjectures, one of which is implausible – that energy efficiency will improve by 5 per cent per year throughout the period; and the other impossible – that there would be no rebound effect, i.e. a 50 per cent improvement in energy efficiency would lead to a 50 per cent reduction in energy use. The simplistic nature of such an approach is examined in chapter 5.

Be that as it may, renewables have certainly made a notable comeback in recent years. Before 1986, despite a century of research and attempted deployment and commercialisation, wind power globally generated less than 0.05 TWh of power per year – roughly equivalent to two days of output from a single decent-sized power station. By 1995 this had grown to 8.3 TWh, similar to the annual output of a 1.2 GW nuclear or gas power station such as say Sizewell B or Immingham. However, the growth in installed capacity among what might be termed the 'failed' renewables – in contrast to hydropower, the 'successful' renewable – over the two decades from the mid-1990s has been startling. Global cumulative wind capacity grew by a factor of 60 between 1996 and 2014, from 6 GW to 370 GW (see figure 4.6), while installed solar capacity grew by a factor of more than 100 between 2000 and 2014 (see figure 4.7).

For comparison, global installed nuclear capacity as of the end of 2014 was 375,500 MW, on the point of being overtaken by wind and little more than double that of solar. Electricity generated by wind and solar each increased 100-fold (see figures 4.8 and 4.9).

Technical and financial improvements were made during this period but the impressive growth did not occur primarily because of major new technical breakthroughs nor

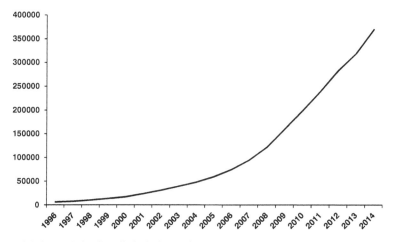

Figure 4.6 Global cumulative installed wind capacity (MW), 1996–2014[37]

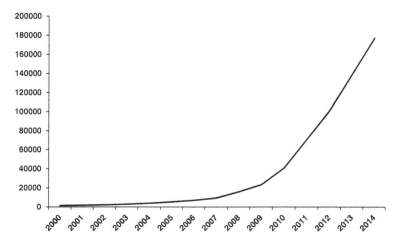

Figure 4.7 Global cumulative installed solar capacity (MW), 2000–14

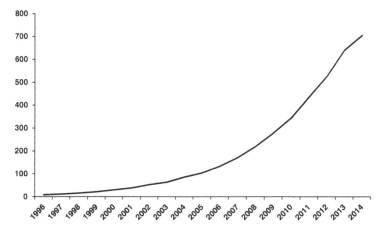

Figure 4.8 Global electricity generation from wind (TWh), 1996–2014[38]

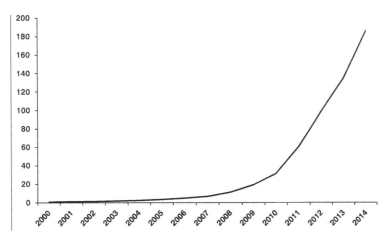

Figure 4.9 Global electricity generation from solar power (TWh), 2000–14

economies of scale, which are a predictable feature of any normal technical innovation and would have come as no surprise to those who devoted their lives to renewable research and development a century earlier. Unlike hydropower, the Orwellian 'new' renewables required growing levels of government subsidy to underpin their expansion. Estimates compiled by the Renewable Energy Foundation (highly sceptical of renewable sources) suggested that in the UK, for example, the direct subsidy to renewables, delivered through the Renewables Obligation, rose from below £500 million per year in 2002/3 to £2 billion per year in 2012/13, with an extra £500 million from 'feed-in tariffs'. Predictions based on government figures suggested this could reach £8 billion per year by 2020, though other estimates suggest it could be rather higher.[39] The Renewable Energy Foundation subsequently obtained DECC figures under a Freedom of Information request which suggested that household electricity prices could rise by 'as much as 40 per cent' by 2020.[40] As noted in chapter 8, renewables were still deemed to require guaranteed prices well above market rates when the 'strike prices' under the 2013 Electricity Act were announced.

4.4 A PRICE WORTH PAYING?

Nobody ever pretended that fighting climate change would be cheap. (Actually that is not true, many renewables advocates take time off from their arguments for high subsidies to foretell competitive costs just around the corner.) Investments of the size required, enormous though they may be, could surely be justified if they resulted in significant reductions in greenhouse gas emissions.

Yet despite the huge investment and the appearance of such quantities of renewable capacity, especially post-Kyoto, the world has not seen any noticeable reduction in the proportion of growing energy demand from fossil sources.

In 2013, after a review of nuclear power, Friends of the Earth UK (FoE) announced their primary objection to nuclear power was now because it was 'too slow to build and too costly' rather than on environmental grounds.[41] However, the only energy policy that has so far succeeded in rapidly displacing fossil fuels was in the 1970s and 1980s, the period when France, Sweden, Belgium, Canada, US, Germany, Japan, Switzerland, UK and others built or expanded their nuclear power fleets. The absence of further progress since 1995 shows clearly that very little overall decarbonisation had been achieved in the first 20 years of the post-Kyoto era of vast subsidies and other market support for renewables – see figure 4.10.

As noted in chapter 2, data show that for the last three decades the world has seen a halt in progress towards less carbon-intensive energy consumption, which has stalled at about 13 per cent of the total global supply. Ironically, but perhaps not surprisingly, the world was moving quite rapidly towards decarbonising its energy mix before climate policy became fashionable, only to stop when it became a policy priority. The main victim of the Big Green-inspired battle between renewables and nuclear power appears to have been the atmosphere.

The point is emphasised when one considers the speed with which various non-fossil fuel programmes have been brought into production. Renewable advocates prefer to cite capacity rather than per capita output, but a careful examination of this 'trick' reveals what a serious problem variable output is – see figure 4.11 (see colour section).

The growth in the rate of installing wind and solar looks slightly less impressive if one notes that the average 'load factor' of wind farms globally in 2013 was 23.9 per cent – in other words, if a wind farm had a rated capacity of 10 GW then on average some 7.6 GW of that output would have had to come from back-up capacity, almost certainly CCGT, with only 2.4 GW actually being generated by the wind. The situation with solar was twice as bad, with a global load factor of just 11.8 per cent. (The availability of

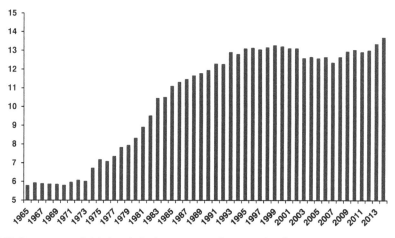

Figure 4.10 Percentage of global traded primary energy from non-fossil sources, 1965–2014[42]

nuclear power in 2013 was just over 75 per cent, a figure depressed by the continuing shutdown of Japan's reactors post-Fukushima.) As will be discussed later, this is just the start of the challenge, as the output of the variable renewables does not necessarily come at the time when the electricity is most needed, unlike the dispatchable power technologies, especially the more flexible ones.

To take one rather extreme example, at times of peak power demand (cold, cloudless, windless winter evenings) wind turbines sometimes become electricity consumers rather than generators. In these conditions offshore wind farms need to draw power from the National Grid to keep turning and prevent them from icing up in sub-zero temperatures and to power the hydraulic systems that turn the blades into the wind. While it is true that all power stations have power requirements when not generating, dispatchable capacity is generally shut down at times of modest demand (when its output is not required) rather than at times of peak demand.[43]

One might expect all this to be of at least interest if not concern to Big Green, but it seems not. The FoE review, carried out by the Tyndall Centre at the University of Manchester, chose not to examine these data.[44]

4.5 A GROWING BACKLASH

At the same time as Big Green began to lose its aura as the new religion of the age the concept of huge investment in renewables started to lose its shine. This shift was boosted by poor experience with large-scale renewable schemes, on both economic and environmental grounds, the growing failure of Energiewende in Germany, as discussed in chapter 6, and the prospect of lower fossil fuel prices.

One of the claims made for renewable technology is that it is 'simple' and 'straightforward'.[45] While this may be true in principle – as is nuclear fission for that matter – it is clear that practical deployment at large scale is a different matter. Time after time in all technical arenas one observes a simple failure to deliver technology up to the task, and renewables are no different.

In 2002 Denmark began installing generators at its 160 MW Horns Rev offshore wind farm 15 km west of Jutland in the North Sea. Project costs were of the order of €270 million, of which interconnection was to account for around €40 million. By February 2003 all 80 turbines were in place but over the next 18 months there was only one half-hour period in which all 80 were able to operate together. During this period there were some 75,000 maintenance trips. Failure of the transformers began in August 2003 owing to a variety of factors – manufacturing problems, weather conditions offshore, salt in the air (not really all that unexpected at sea one might have thought) and the aftereffects of having stored the transformers onshore. Eventually, in the spring of 2004 the responsible companies decided to take down the entire nacelles – the covers which house all of the generating compenents including the generator, gearbox, drive train and brake assembly – to be dismantled and rebuilt, while the blades were returned to the manufacturer to repair lightning damage. The facility, co-owned by Dong Energy and Vattenfall, continued to face problems for many years, for example being seriously affected by a generic problem identified in 2010 that resulted in the foundations of the turbines slipping, potentially limiting their operating lifetime and said to derive from a 'little-understood' physical property involving vibration of the towers. This technology had been adapted from the oil and gas industry, where it had worked effectively but where constant vibrations were less of an issue.[46]

Similarly, a dispute between Greater Gabbard Offshire Wind Ltd, a joint venture between SSE and RWE, and wind turbine manufacturer Fluor over the quality of turbine construction at the wind farm off the Suffolk coast took a year to resolve and led Fluor to take a pre-tax charge of $400 million to settle the matter. Germany's 400 MW wind energy project, BARD Offshore 1, met with a raft of technical, financial and legal problems resulting in a three-year schedule delay and major cost overruns. The farm was officially opened in August 2013 but had to shut down almost immediately because of technical faults. In March 2014 engineers tried once again to bring the facility online but failed as 'wild current' caused burnout within an offshore electrical converter station after just a few hours. Six months later engineers had still not solved the problem, the cost of lost output alone reaching €340 million. The inevitable legal wrangles ensued: 'Not only the engineers have been working feverishly on the repairs but also lawyers are now involved. In the meantime everything has turned to the question of who is responsible for the fiasco – and the costs.'[47] The problem was ascribed to the great distance the wind park is located from the coast which made it impossible to bring the power onshore with conventional technology. The power could not be transmitted through an AC underwater cable so had to be transmitted as DC current – back to Edison's approach which had lost out to Tesla's AC in the nineteenth century. Hans-Günter Eckel, Professor of Power Electronics at the University of Rostock, said: 'Most likely there isn't a single thing that is responsible but rather it's about a faulty total system. It's going to require patience. It's a completely new and complex technology.' For a considerable period Germany's wind turbines off the North Sea island of Borkum were rotating without being connected to the grid, leading to the extraordinary situation where the turbines had to be run as gigantic fans using diesel to prevent rusting when the wind dropped.[48]

It was ever thus, of course. New ventures rarely live up to the hype of their early enthusiasts and this in itself is no reason for not pursuing them. However, the economic failure did have inevitable repercussions for investment in new offshore wind capacity despite the very high levels of direct subsidy available. The European Wind Energy Association in 2014 revised down its 2009 forecast of an increase in offshore capacity by 2020 from 40 GW to 25 GW, though this still represented a significant increase from the 7 GW then installed.[49] In 2013 RWE abandoned its plans to invest £4 billion in its Atlantic Array project off the North Devon coast, blaming 'technical difficulties'. Some commentators felt it unlikely that any very major unexpected issues would have been encountered and ascribed the decision more to political uncertainties undermining investor confidence.[50] Centrica sold its interest in the proposed 580 MW Race Bank wind farm off the Wash in Norfolk for just £50 million[51] while SSE and RWE trimmed the size of their proposed Galloper offshore wind farm by a third. At the end of 2014 it was the turn of the First Flight Wind consortium to abandon plans for a 100-turbine, 600 MW, £1 billion wind farm off the County Down coast in Northern Ireland, blaming 'delays and red tape'.[52]

4.6 CHALLENGE TO THE LEFT OF THEM, CHALLENGE TO THE RIGHT

Criticisms of large-scale renewables have come from both the political Right and the political Left, albeit on different grounds. The Right, including financial institutions, tends to complain about the very high levels of taxpayer or billpayer subsidy when more economic options are available, a stance sometimes bolstered by varying degrees of scepticism about climate change and proposed responses. The Left objects to the transfer of large amounts of money from power consumers, including those who cannot afford to pay their bills as it is, to rich landowners and to the degree of local environmental degradation which inevitably arises from the low power density of the fuel sources themselves.

So the chief executive of insurance giant Legal & General, said that the green energy strategy would carpet the countryside with 'ugly modern windmills' and result in even higher bills for hard-pressed consumers: 'The Government is deluding itself that it is saving the world by building these ugly modern windmills. This is a very expensive and inefficient way of producing electricity. I fear we will end up having to import it from elsewhere.'[53] Right-leaning the *Daily Telegraph* published details of a Renewable Energy Foundation analysis suggesting that every job in the UK wind farm industry was subsidised to the tune of £100,000 per year.[54] Meanwhile liberal-minded *The Guardian* highlighted that the boom in onshore wind was being dominated by a small number of private landowners who were to share £1 billion in rental fees between 2012 and 2020, amounting to some 'risk free' £40,000 per year per turbine. Among those in line to benefit were senior members of the Royal Family and the Forestry Commission. These subsidies were on top of the farm subsidies the landowners automatically received from the EU. Individuals whose good fortune was singled out included the Earl of Moray (receiving some £2 million per year from 49 turbines on his Perthshire estate); the Duke of Roxburghe (£1.5 million); and the Liberal Democrat peer, the Earl of Glasgow (a mere £300,000). The Queen's cousin, the late Princess of Wales's brother and the prime minister's father-in-law were all recipients of electricity consumers' forced largesse.[55]

There was broad cross-party backing in Parliament for wind power and other renewables but this did not stifle dissent. Scottish Conservative MEP Struan Stevenson produced a booklet in which he argued that: 'Wind turbines violate the principle of fairness by transferring vast amounts of money from the poor to the rich. They despoil our unique landscape and environment; and through noise, the flicker effect and vibration, they abuse the health and welfare of people and animals which have to live near them. Far from being an eco-friendly answer to our energy needs, wind farms are an environmental disaster being inflicted on Britain that we will live to regret.'[56] Former Labour Party leader Neil Kinnock, as he went off to become European Transport Commissioner in 1995, said: 'My long-established view is that wind power can only provide a very small fraction of the output required to meet total energy needs and it unavoidably makes an unacceptable intrusion into the landscape.' Opposition to wind was a major plank of the emerging policy platform of the UK Independence Party (UKIP) in the 2010s.

The low power density of renewables inevitably brings with it significant local environmental degradation – visual intrusion, interruption in river courses, noise, effects on plant and animal life, perhaps especially birds, and so on. EDF Energy and Eneco, for example, scaled back their plans for 194 turbines in the English Channel known as Navitus Bay, moving it further offshore and reducing it to a maximum of 105 turbines. The scheme would have been visible from the Jurassic Coast, a UNESCO World Heritage site, and had met fierce local opposition.[57] (The government duly refused planning permission in September 2015 after five years of wrangling[58].) Many local groups sprang up to oppose wind generators – in 2012 national rural affairs campaigning group Country Guardian listed 285 such groups, covering small localities with such Betjeminesque names as Lochluichart, North Hambleton, Drumadarragh, Kelmarsh, Cefn Croes, Abbots Bromley, Criddling Stubbs and Hampole; as well as wider areas including Artists Against Wind Farms, and Scotland Against Wind and Salmon Farms; and boasting names such as STRAWS (Strathbogie Tourists and Residents Against Wind farm Stitch-up) and SHAFT (Stop the Higham and Fence Turbines).[59] A national group, National Opposition to Wind Farms, was launched by Liberal Democrat peer Lord Carlile in 2012. In 2011 Labour MP for Dumfries and Galloway, Russell Brown, attacked the then SNP leader saying: 'Alex Salmond once again brushed aside the concerns people have that we are sleepwalking into our region being carpeted in wind turbines.' In the same year Liberal Democrat Member of the Scottish Parliament, Jim Hume, celebrated the abandonment of plans for an offshore wind farm near Wigtown in south-west Scotland: 'I am pleased that this proposal will not go ahead. This is a victory for local campaigners who have fought hard to oppose the development. While these proposals were technically offshore, it was clear that any proposed wind farm would have been in extremely close proximity to the shore and would have therefore certainly dominated the horizon of Wigtown Bay.'

A further concern was prompted by regular blade shedding and other catastrophic plant failure, especially in high winds and occurring despite the turbines generally being shut down at wind speeds of around 55 mph. This led, for example, to the introduction of a '10H rule' in the Free State of Bavaria, whereby wind turbines cannot be constructed within a distance equivalent to ten times their height (ground to blade tip) from human dwellings. The calculation of the appropriate 'safe zone' was done on the basis of regarding projection from the tip of a blade as comparable to the ballistics of a stone or a projectile launched by a catapult, the centre being the axis of the rotor and the radius the length of the blade of the wind turbine. The Bavarian minister-president, Horst Seehofer, said: 'I won't let giant asparagus ruin Bavaria's nature and its wonderful landscapes.'[60]

The need to protect the interests of local residents against the renewable multinationals has brought an extra element of financial hardship to many communities through the costs of fighting planning applications. Powys Council estimated its costs in fighting a public inquiry into five wind farm applications in 2013 at £2.8 million.[61]

In 2014 Somerset cultural icons The Wurzels (whose classic 'Combine Harvester' had topped the UK music charts forty years earlier) released a song with these lyrics:

We started our own action group with Ernie, Jake and Sue
And sat around the pub all night deciding what to do.
We spoke to all the village folk who made it very clear,
They're noisy and they're ugly and nobody wants 'em here.
The planners are dictated to by him in Number 10,
Perhaps we'll get a subsidy and put one next to them.
I understand the argument for greener energy,
But save our countryside for you and me.

Calculations suggest that the 580 MW Altamont Pass wind farm kills some 2,700 birds each year, including 67 golden eagles, almost 200 red-tailed hawks and 350 American kestrels.[62] In Germany 32 protected white-tailed eagles were found killed by wind turbines, while in Australia 22 critically endangered Tasmanian eagles were killed by the Woolnorth wind farm between 2002 and 2010, an issue raised by the Green Party in the Tasmania State Parliament.[63]

James Lovelock, originator of the Gaia hypotheses and something of a maverick elder statesman to the genuine environmental movement, refused an invitation to open a wind farm at Delabole in Cornwall, saying: '[A while back] nobody was talking about a gigantic programme, getting 15 or 20 per cent of the country's energy from wind turbines. I think, now that I know as much as I do, I wouldn't have touched it with a bargepole. It has stolen up on us without any of us being aware of it.'[64]

Use of variable sources also has potential hidden effects on greenhouse gas emissions. The need in effect to double the amount of capacity – generating and transmission – to ensure secure supplies in a system with heavy use of variable renewables will have implications in terms of greenhouse gas emissions and other resource and environmental impacts simply from constructing the infrastructure itself, a further reason for questioning whether 'renewable' is really the right word for describing the practical effect of these technologies.

Even at high wind output, especially when electricity demand is also high, a substantial amount of spinning gas- or coal-fired capacity needs to be kept available to take over should the wind output drop rapidly. This reserve capacity produces carbon emissions as a direct consequence of the use of wind on the electricity supply system, although traditionally the emissions are not ascribed to the renewables in accounting or reporting terms but to the back-up capacity itself which is necessitated by the use of variable renewables.

Moreover, the thermal inefficiency introduced in fossil-fuelled generators by having to vary their output to compensate for changes in the output of variable renewables means more emissions of carbon dioxide per unit of electricity generated in these plants and potentially shortens their lifetime. In the same way a motor vehicle on the stop-start urban cycle uses more fuel and needs more maintenance per mile than one being used at a constant 56 mph on a motorway (if anyone ever drives at 56 mph on a motorway), so producing more emissions per mile travelled. Emissions from the entire electricity supply system will therefore be higher than one would believe by simply assuming that when renewables are generating they replace an equivalent amount of carbon-producing capacity with no further implications. In some cases, at least where coal is the main source of generation such as in Colorado, the thermal inefficiency losses may even outweigh the displacement advantages, resulting in higher emissions than if the wind farms were simply shut down.[65] In Colorado anyone who says wind is a very expensive way of reducing greenhouse gas emissions would seem to be mistaken – it actually appears to be a very expensive way of increasing greenhouse gas emissions. A 2011 MIT study put it as follows. 'While renewables can generate emissions-free electricity, the limited ability to store electricity, forecast renewable generation and control the availability of intermittent renewables forces the rest of the electric power system to adapt with less-efficient ramping and cycling operations. These operations potentially reduce the emissions benefits of renewables.'[66] To this may be added the shortening of the life of dispatchable plant which suffers more thermal stress owing to being forced to vary its output more frequently, reducing the period before a new plant has to be built and causing extra emissions during construction. Again these extra carbon emissions are not usually ascribed to the renewables, which continue to be described as 'low carbon'. Nonetheless, the complexities of the debate are gradually becoming recognised.

By the 2015 general election UKIP – the main party of protest in the UK in the mid-2010s, emerging as the largest party in the European Parliament elections in 2014, for example – and the right wing of the Conservative Party were both showing increasing scepticism towards renewables and indeed towards climate change in general.

Wind is not the only 'new' renewable to face opposition on environmental grounds. The 400 MW Ivanpah Solar Electric Generating System, covering five square miles of the Mojave Desert in California, consists of 350,000 large mirrors – to concentrate the sunlight – creating temperatures of over 500°C. State and federal regulators launched a two-year study of the effects on birds when the power plant was opened in 2014, while environmental groups questioned the value of cleaner power when native wildlife is being killed or injured. The Western Watersheds Project, a conservation group which works over six US states to protect wild environments, instituted a lawsuit against the federal agencies that reviewed the project, saying that alternatives to the site were not considered and serious environmental impacts, including fragmenting the tortoise population, were ignored. 'Do we really need to have these giant plants first or is it better to generate solar power on people's roofs, the place it's going to be used?'[67] During 2014 the plant's owners applied to increase the amount of gas it was allowed to burn to power boilers which are used to warm up the fluid in the turbines in the early morning to allow them to run for five hours a day, rather than the one hour which had been put forward during the licensing phase, and announced that output from January to August had been 0.25 TWh against the expected 1 TWh plus, owing to 'bad weather' and equipment failure. In the words of one solar advocate: 'The industry also has to come to terms with diminishing public support for solar and other renewable energies. To initiate public fervour in the first place and to encourage financial support, solar was inappropriately positioned as a saviour of all energy woes and represented as environmentally benign. This is not true … .'[68]

Studies also suggest that hydropower may have greater greenhouse gas implications than might at first seem to be the case. Carbon emissions vary from dam to dam but a study of the Curuá-Una Dam in Pará (Brazil), for example, suggested that total emissions were some three and a half times as great as would have been produced by generating the same amount of electricity from oil. This arose because large amounts of carbon tied up in trees and other plants would have been released when the reservoir was initially flooded and the plants rotted, followed by anaerobic decay of plant matter settling on the reservoir bottom creating methane which would dissolve in the reservoir and be released into the atmosphere when water passed through the dam's turbines.[69]

Of course, all sources of energy have their downsides. But the refusal of Big Green to engage with the wider concerns about how renewables, especially wind power, can work against social equity and justice is once again revealing and rather disappointing. One wonders how they might have responded to a logging firm or chemical company which received a critique like this: 'Landowners have woken up to the fact they can make a heck of a lot of money at the expense of those who have lived there for generations. There is a world of difference between a [wind farm] controlled by a local community and one imposed from outside by a landowner and a multinational company. The people benefiting are the ones who have always worked the subsidy system.'[70] Rather than uncritical support one might expect at least some questioning of this blatant transfer of money and social capital from poor to rich.

4.7 A RANGE OF OPINION

Big Green and its fellow travellers have not taken criticism of variable renewables lying down. The left-leaning Institute of Public Policy Research (IPPR) 'worked with' GL Garrad Hassan, a renewable energy consultancy which has carried out major pieces of work for Big Green (the funding for this project is not made clear), to produce a report on wind power which concluded that:

- it is inaccurate to describe the output from wind power as 'unpredictable';
- in the short term, wind power output is remarkably stable and increases and decreases only very slowly;
- the risks associated with 'long, cold, calm spells' have been overstated;
- in the UK, National Grid has reported that up to 30GW of wind power can be accommodated even if no changes are made to the way that the electricity system functions;
- in the longer term there are numerous technological options to facilitate much greater amounts of wind power, such as improved interconnection with other countries and intelligent management of supply and demand through a 'smart grid';
- for these reasons the authors conclude that wind power can play a major role in a secure and reliable future electricity system.

The slim (20-page) report can only address the issues in a relatively superficial way. Presumably for reasons of space it talks about the variability of wind speeds as though this were the only threat to secure supplies, without looking at the record of frequent equipment failure leading to unexpected outages. As noted previously, while multiple breakdowns of the generating units might be limited to extreme weather events, problems with the transmission of power from say an offshore wind farm to the national electricity grid have on many occasions knocked out a high proportion, if not all, of the output of a large array. The report does, however, note that even GL Garrad Hassan admits that output from wind farms can vary by as much as 20 per cent within half an hour; it also cites an example from Ireland in 2010 when wind production averaged less than 15 per cent of rated capacity for a period of a fortnight.

According to the IPPR this 'shows unequivocally that wind power can significantly reduce carbon emissions, is reliable, poses no threat to energy security and is technically capable of providing a significant proportion of the UK's electricity supply with minimal impact on the existing operation of the grid.'[71] The problem is that very many reports over the years have shown all sorts of things 'unequivocally' – one sees them in the early history of the nuclear industry and many other walks of life. This is no guarantee that those things actually come to pass.

In any case, IPPR's take is very different from that of E.ON Netz, the German network operating company responsible for integrating several GW of wind capacity into the power distribution system. Like IPPR and GL Garrad Hassan, E.ON Netz is 'actively committed to the use of renewable energy in the production of electricity' and 'supports the objective of making the production of electricity using renewable energy competitive as quickly as possible'.[72] Unlike IPPR and GL Garrad Hassan, however, E.ON Netz has both practical experience of and responsibility for making the power system actually work. E.ON Netz assumed that some 48 GW of wind capacity would be on the system in Germany by 2020. It concluded: 'In order to guarantee reliable electricity supplies when wind farms produce little or no power, traditional power station capacities must be available as a reserve. As wind power capacity rises, the lower availability of wind farms determines the reliability of the system as a whole to an ever increasing extent. As a result the relative contribution of wind power to the guaranteed capacity of our supply system up to the year 2020 will fall continuously to about 4 per cent. In concrete terms this means that in 2020, with a forecast wind power capacity of over 48 GW, 2 GW of traditional power production can be replaced by these wind farms.' In the UK context this implies that the dispatchable value of a wind park capable of fulfilling entire power demand when operating at full capacity would be little more than one decent-sized nuclear reactor.

4.8 MUDDLED THINKING

As discussed in chapter 11, arguments that simply do not stack up logically are not the sole preserve of the renewables industry. However, before considering the economics of renewable deployment in more detail, let us look in more depth at these apparent contradictions which are revealed in two rather strange arguments that underlie the Big Green position.

1. Renewables will be cheaper than the alternatives by 2020 so we should prolong and increase their subsidies beyond 2020 to make sure this happens.
2. Subsidies are only for mature sources; wind power (at least onshore) is a mature source; we should offer (onshore) wind more subsidy.

One can appreciate the attractiveness of each element of these arguments from the point of view of companies seeking to make money from wind power. If the population is convinced that wind power will be cheap very soon it may support it; companies want as much money for as little work as possible and so want higher subsidies – which depend on the technology being perceived as expensive. Danish firm Dong Energy claimed that offshore wind could cost £80 per MWh by 2020[73] – yet the statement was made in support of a UK government announcement (in December 2013) that the direct subsidies to offshore wind power for the fiscal year 2018/19 would fall from £155 per MWh to £140 per MWh, not £135 per MWh as had been previously implied. To the outsider these arguments do not make sense. Prima facie, either the wind industry is simply lying about how cheap it will become and/or how soon; or it is seeking to profiteer on the back of a naïve government that has not appreciated how far costs have fallen, at the expense of hard-pressed electricity consumers.

A similar tension underlies Big Green's commentary around 'maturity'. For example, in December 2013 Energy Fair, a hard line anti-nuclear group, published an open letter to European Competition Commissioner Joaquín Almunia before the Commission opened its investigation onto potential state aid for the Hinkley Point C nuclear project, saying: 'Nuclear power is a mature technology that should not require any subsidy. Subsidies are for newer technologies that are still finding their feet commercially.'[74] Again one can see the attractiveness of this message from the point of view of the renewables industry – nuclear should be excluded from a regime which offers guaranteed future prices at levels above those prevailing in today's market so more subsidy will flow into the pockets of the large multinational conglomerates and wealthy landowners profiting from renewables.

But the renewables industry also wants to give the impression that a century and a quarter of research into electricity from wind has not been entirely wasted and that it is a safe bet, a 'mature' technology – at least onshore. So the trade association for the industry, RenewableUK, says: 'The UK's first commercial wind farm was built in Delabole, Cornwall, in 1991. Since then, onshore wind energy has established itself as a mature, clean and productive technology.'[75] Other similar quotes:

- The EU (2009): 'Onshore wind energy is a near-mature technology.'[76]
- The US energy secretary (2009): 'Q: What can the government do to provide incentives for innovation in clean energy? A: In the case of a more mature technology like wind, what the government can best do is provide some stability so the companies can make long-term investments …'[77]
- UK Committee on Climate Change (2013): 'Pöyry's assessment suggests broadly comparable costs for mature technologies (i.e. onshore wind, nuclear, biomass conversion) …'[78]
- UK Green Party General Election manifesto (2015): 'We will concentrate on expanding mature renewable technologies such as wind energy and solar PV in the period until 2030.'[79]

The admittedly rather more modest Guernsey Renewable Energy Commission in 2012 even claimed that monopile foundations for offshore wind, drilled and driven into the rocky seabed, were a 'mature technology' and continued widespread deployment was developing their supply chain.'[80]

From a simple syllogistic point of view the premises lead to an inevitable conclusion.

- Premise: mature technologies should not receive subsidies.
- Premise: onshore wind is a mature technology.
- Conclusion: onshore wind should not receive subsidies.

The Conservative manifesto for the 2015 general election pledged to scrap subsidies for onshore wind and change the planning system to allow local councils in England and Wales to block any which did not already have planning consent, though subsidies already announced would continue and previously granted planning permission would be respected.[81] As the prime minister said, echoing RenewableUK and others: 'This is a mature industry which has had 20 years of subsidy.'[82]

One might have expected that such a policy, embracing the spirit of local control over energy and removing subsidies from a mature technology, would meet with a hearty cheer from Big Green. Oddly not. The author of the Energy Fair letter to Commissioner Almunia wailed: 'There seems to be little doubt this policy will eventually sink the onshore wind industry in this country. The potential for the technology to compete without subsidies is poor and the handing over of power to local authorities will likely be the final nail in the industry's coffin. [As an aside, the same author was passionately supportive of giving local authorities a veto over investigations into siting a radioactive waste repository.] Tories and many industry leaders are loath to admit this will put an end to onshore wind, saying that current capacity will be maintained and even grow for a time as projects granted permission before 2015 come online. But what industry survives without long-term growth? For onshore wind to continue to generate investment in research and development, and therefore to stay competitive, it will need to have the potential for new growth and projects.'[83] Jennifer Webber from RenewableUK said the policy would 'kill the industry dead'.[84]

The nuclear industry had promulgated similarly contradictory arguments a decade earlier when it had claimed that it would need no subsidy, only permission for firms to construct new reactors. When that permission was given a long debate followed about the market support, i.e. subsidies, that would be necessary. But while Big Green pounced upon the arguments as evidence that nuclear was not to be supported it was entirely unwilling to apply the same level of challenge and scrutiny to renewables. Whatever advantages this might appear to bring to renewable advocates in the short term, intellectual inconsistency bordering on dishonesty of this kind has a tendency of coming home to roost.

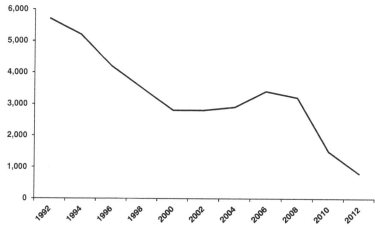

Figure 4.12 Cost of photovoltaic modules (US$ per kW)[87]

4.9 THE WIDER ECONOMIC PICTURE

The falling costs of renewables are widely feted by those who are making the running and the profits. 'Solar's' cost is falling so fast that the conventional energy sector does not realise how inexpensive it has become. Policy makers find it a challenge to keep pace with and find the comparison metrics challenging.'[85] 'The cost of wind power has decreased 58 per cent over the past five years.'[86]

As noted earlier, the costs of the generating units have indeed come down rapidly and at least in the case of solar continue to do so – see figure 4.12.

The rise in the mid2000s was owing to a silicon shortage; the rapid fall in the early 2010s because of overcapacity in China and the subsequent dumping of units into the European and US market, resulting in the imposition of anti-dumping tariffs.[88]

There were significant reductions in the levelised costs of onshore wind-generated electricity up to around 2005, when they began to rise again owing to increased commodity prices. In the US, for example, after hitting a low of roughly $1,200 per kW (in 2013 values) in the early 2000s, average installed turbine prices increased to more than $2,000 per kW by the end of 2008 before falling back to around $1,500 per kW in the mid-2010s – see figure 4.13.

In 2012 the Crown Estate, which as owner of the UK's seabeds and riverbeds is one of an array of very rich individuals and corporations which become considerably richer as wind farms are built, concluded: 'The costs of offshore wind in the UK have increased substantially since the first commercial scale wind farms were deployed in the early 2000s, driven both by underlying cost increases (commodity prices rises, currency fluctuations) and by more specific factors such as supply chain bottlenecks, sub-optimal reliability and the move to deeper water sites. Recent wind farm projects have indicated that costs have stabilised at around £140 per MWh for projects at Final Investment Decision in 2011.'[90]

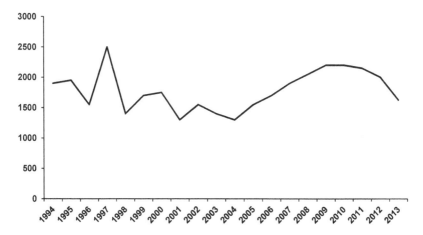

Figure 4.13 Wind capacity installation costs (US$ per kW), 1994–2013[89]

In fact the Crown Estate was using rather optimistic figures based on high availability factors of over 40 per cent; estimates carried out for DECC by consultants ARUP[91] and Mott MacDonald[92] put the levelised cost figure rather higher.[93]

However, the recent debate about costs of low-carbon electricity has tended to be based on a rather crude 'cost of a generating unit', which includes investment, operation, fuel and decommissioning costs, divided by the number of units of electricity produced approach. When discussing the costs of generation advocates of renewables in particular talk about 'grid parity', a term as powerfully misleading as 'new renewables' itself. The potentially vast costs associated with managing the inherent intermittency of some renewables, notably wind and solar, are not at present fully reflected either in the stated costs of these energy sources or in the quoted 'subsidies' they are receiving. These costs, unlike economies of scale in manufacturing, tend to rise as more variable renewable capacity is employed. Non-plant system-level costs are potentially very large and include:

- the need for grid strengthening, i.e. 'grid-level system costs' – in the UK the regulator Ofgem approved a £38.2 billion upgrade to the national electricity and gas grids (equivalent to around £12 per year on a domestic electricity bill) in 2012, saying: 'The energy mix has changed and the electricity transmission network needs to be extended to connect new sources of electricity: the investment will help to connect renewable sources of electricity to the main electricity transmission network' – the renewables industry had been arguing for a sum of £45.4 billion[94];
- the effect on the economics of other generators, leading to higher prices (or threats to security of supply, in itself a huge potential cost to consumers);

- the greenhouse gas emissions caused by the need to vary the output of dispatchable fossil-fuelled power plants, thereby reducing their thermal efficiency.

A rational approach to allocating limited resources to deliver maximum supply security and emission reduction would involve ascribing all costs and environmental effects, direct and indirect, to the source which has ultimately given rise to them. Instead of discussing 'grid parity', the relevant measure should be 'system-level parity', including all of these costs which, however they are allocated, will have to be met by consumers and/or taxpayers. This would lead to a very different discussion regarding the way of achieving economic, security of supply and environmental goals – even perhaps reappraising the extent to which, in practice, variable renewables can be described as 'low carbon'. In late 2015, UK Energy and Climate Change Secretary Amber Rudd, in a speech widely interpereted as 'resetting' the government's approach to energy said: "In the same way generators should pay the cost of pollution, we also want intermittent genrators to be responsible for the pressures they add to the system when the wind does not blow or the sun does not shine. Only when different technologies face their full costs can we achieve a more competitive market".[95] The subsequent squeals of protest from renewables firms showed they were all aware of the implications of being made accountable for at least some of their system-level costs.

4.10 THE CHALLENGE OF VARIABILITY

Despite over a century of research electricity can't be stored in significant amounts. In late 2014, for example, a 6 MW, 10 MWh battery storage scheme was installed near London at a cost of £18.7 million, with a hope that it might save some £6 million on traditional network reinforcement methods required to sustain a growing proportion of renewables. Given that the UK uses approximately 1 TWh of electricity per day this scheme could store power for just short of one second. On this basis, to store power for the whole day would cost some £1.8 trillion, rather more than the UK national debt, although admittedly the saved cost on network improvements would knock nearly £0.6 trillion[96] off this figure. But in the absence of such investment electricity supply and demand has to be balanced in real time, to prevent both shortages and surges in the amount of power available. In the absence of storage networks, the like of which are presently unfeasible, dispatchable plant has formed the basis of electricity supply systems globally ever since long-distance power transmission systems became available in the 1920s and 1930s.

The recent enormous growth in the output of variable renewables has complicated matters considerably. Leaving aside the possibility of equipment failure which renewables share with other sources of electricity, between 2010 and 2014 the 'load factor' for onshore wind in the UK varied between 21.8 and 28.8 per cent, while that of offshore wind varied between 30.6 and 39.1 per cent, the averages being 26.2% and 36.0% respectively.[97]

Though techniques for forecasting wind speeds have been improving, it remains difficult to predict with perfect accuracy, or very long in advance, when and how strongly the wind will blow – indeed, wind speeds can vary rapidly over a few seconds or minutes on a 'gusty' day – see figure 4.14 (see colour section).

Although the most rapid variations are to some extent compensated for by the inertia of the wind turbine rotor, the phenomenon still creates technical, financial and economic problems. Technically, intermittency necessitates the provision of a great deal more wire capacity in order to accommodate all power being produced when the wind is blowing but then very rapidly switch supply to another area of the country when the wind drops and other types of power plant are required. It also presents challenges to maintaining system frequency and imposes wear and tear on the wind turbines. Wind turbine towers are, however, usually tall enough to avoid the greater wind turbulence encountered close to ground or sea level.

Evidence suggests that low wind speed tends to be weakly correlated with high power demand (cold windless winter evenings and hot windless summer days), further exacerbating the challenges.[98] At other times wind or solar generators may be producing close to 100 per cent of their rated output, risking overloading the system and obliging other power sources to close down.

The rapidity of changes in wind output can be demonstrated by two 48-hour periods in 2013, at a time when some 8 GW of wind capacity was sold through the UK balancing market – see Figure 4.15. On 11–12 September wind output dropped practically to zero, while on 28–29 November output, having been very low during the period of peak demand on 28 November, increased by over 4 GW (i.e. 50 per cent of rated capacity) in a four-hour period overnight when system demand was low.

In China in 2010 about half of the energy produced in wind farms was wasted be-cause the grid connections were not yet in place.[100] By 2014 this had dropped to about one-fifth, representing almost 11 TWh or around 40 days'worth of electricity supply for Beijing. One issue was the rush by developers to lock projects in to generous subsidy rates ahead of an official cut in 2016 while grid operators were unwilling to come up with the relatively unprofitable investment compared to the more reliable coal-generated electricity.[101]

Managing a system which includes a lot of variable renewables therefore presents a number of challenges, depending on whether say the wind is blowing or is not.

In order to prevent excess supply causing damage to the transmission wires or un-acceptably high frequencies when the wind is blowing strongly or the sun is out or the tide is in, especially at times of low power demand, quite a lot of that energy has to be wasted. Alternatively, and/or in addition, some companies have had to cut their output to compensate. In the UK in recent years wind generators have often been paid quite large sums of money to stop generating electricity when the wind is good – see figure 4.16.

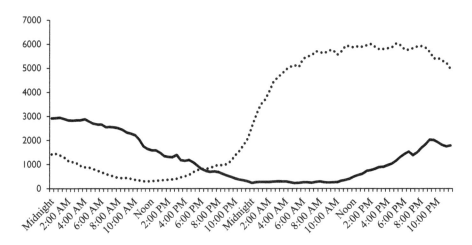

Figure 4.15 UK wind output (balancing market only), 11–12 September and 28–29 November 2013[99]

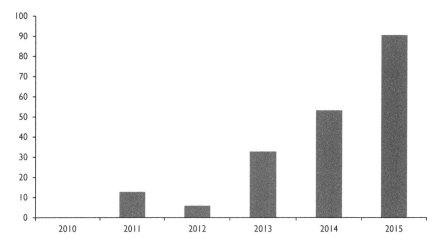

Figure 4.16 Constraint payments to UK wind generators (£ million)[102]

The amount of output being constrained was growing even more rapidly, as the average buyout cost has been falling, from 976MWh in 2010 to 658,611MWh in 2014; in the first six months of 2015 the figure was 582,352 – see figure 4.17.

This undermines the profitability of the gas-fired or coal-fired plants which are having their market share cut from under them. RWE, for example, closed about one-fifth of its gas-fired capacity in Germany in 2013 as it could not cover its costs.[103] The economics of nuclear power means that it would be the last to be withdrawn if there was an excess of variable output at any particular time, but even nuclear plants in Germany, prior to their early demise, were moving away from the traditional and efficient role of operating at stable levels close to full capacity, as the introduction of

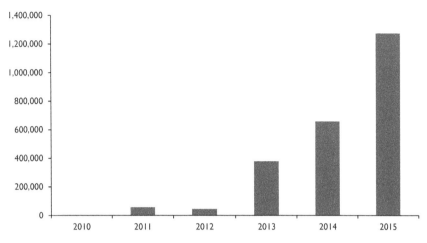

Figure 4.17 Output constraints to UK wind generation (MWh)

large amounts of variable renewables repeatedly led to prices below the marginal costs of nuclear, including several instances of negative prices. But when the wind is not blowing we still need almost as many gas- and coal-fired power stations as we would have needed if we had never built the wind farms in the first place. So some way has to be found of compensating the fossil fuel power stations for the market they lose when the wind is blowing. If this is not done they might be closed down or mothballed, and then not be available for when the wind drops. As noted in chapter 8, this problem has already been seen in countries such as Germany and Spain which have a lot of wind energy.[104] Pumped storage is also disadvantaged by the convergence of peak and off-peak wholesale power prices.

The situation is exacerbated in Britain and Ireland by the relative lack of interconnection to the continent, owing to its geographical position on the edge of Europe and the high costs of subsea interconnection, with just 3 GW in place – with France and with the Netherlands.[105] Germany, by contrast, has interconnection with nine neighbouring countries – counting the two Danish grids separately – amounting to some 17.3 GW of import capacity and 13.9 GW of export capacity.[106] This gives it more flexibility in managing its variable renewables. It can import power (if any is available) from a wide range of locations when the weather does not favour wind and solar plants. By the same token it can avoid potential damage caused by excess generation by in effect 'dumping' this excess electricity onto its neighbours' grids, thereby exporting the problem of potential power surges and reducing the need to shut off dispatchable capacity at home.[107] The UK and Ireland do not have these options to anything like the same extent and are therefore less able to manage highly variable renewable capacity. (A rather amusing comment in a report that GL Garrad Hassan produced for FoE Scotland on the implications of a high-renewables penetration policy for Scotland says: 'Security of supply will be satisfactory, provided the British electricity system itself is secure and provided there is sufficient interconnection capacity to Scotland.'[108])

As discussed in chapter 8, the Energy Act of 2013 included a number of measures designed to attract investment in new low-carbon capacity. However, if these measures were to work and more variable renewables were brought online then, as noted earlier, the economic case for investing in new CCGT or other dispatchable technology would be weakened, since these plants would be left without an income during those times when the wind was blowing at the right speed. This is even more the case with the more heavily capital-intensive sources such as coal with CCS, or nuclear: as their unavoidable costs are higher their economics are more seriously harmed by being taken offline, losing income but not significantly reducing costs.[109] But at some point new CCGTs or other flexible dispatchable capacity will be needed to provide power when the wind drops, otherwise 'the lights will go out'. Given this dilemma, 'energy-only' electricity markets need to be supplemented with 'capacity markets', whereby generators are paid to keep capacity available even if it is not being used, if sufficient dispatchable capacity is to remain in the market to provide back-up for when the wind is not blowing at the right speeds. The Energy Act of 2013 introduced capacity payments, in effect to persuade investors to build the CCGTs they would have built anyway had the measures to promote low-carbon capacity not been imposed. However, the first payments, amounting to almost £1 billion for the fiscal year 2018/19, went almost exclusively to existing plants in order to persuade them not to shut down: hardly any of it went to stimulate new build dispatchable capacity which was becoming increasingly urgent in the face of impending coal and nuclear plant closures.

The vast costs incurred by the need to manage the inherent intermittency of renewables are not at present taken into account when calculating the true costs of renewables and the full subsidies they receive. The German energy and environment minister, Peter Altmaier, has estimated the cost of Germany's transition to renewables at up to €1 trillion.[110]

The point was emphasised in a table from the UK government's National Policy Statement on Energy of 2011[111] – see table 4.1. When variable capacity was included the UK was nominally running a generous capacity margin of over 40 per cent. However, this figure was misleading because a growing proportion of the capacity could not be relied upon to help meet peak demand. The projected need for generating capacity in 2025 was no less than 113 GW, to cover peak demand unlikely to be more than around 60 GW in that year even with good economic recovery. That would seem to represent a capacity margin of over 70 per cent, vast by historical standards. However, from the point of view of secure supplies the 33 GW plus of new variable renewable capacity would have to be largely ignored. Whether the cost of building and maintaining such enormous redundancy should be allocated to the renewables or apportioned to other players (by distorting the market via 'must take' contracts for renewable output, for example, as discussed in the next section) it will ultimately land on the backs of consumers and/or the taxpayer.

Table 4.1 UK capacity data (GW)

Total current generating capacity	85
Peak electricity demand now and 2020	60
Average demand	43
Closure of coal plants by 2015 owing to the Large Combustion Plant Directive	12
Nuclear closures over next 20 years	10
Generating capacity required in 2025	113
Of which new build	59
Of which renewables	33
For industry to determine	26
Non-nuclear already under construction	8
Current proposals for new reactors	16

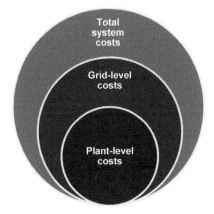

Figure 4.18 Relationships between different levels of cost in power generation

So when it comes to looking at the costs of various sources of electricity, it is highly misleading simply to consider the cost of installing the wind generator or solar panel divided by the number of units of electricity it produces. Some sources of electricity, notably the variable renewables, impose huge costs on the system as a whole.

4.11 SYSTEM-LEVEL COSTS

The issue is considered in depth in a major OECD/NEA study.[112] The report looks at the way in which various methods of generating electricity interact with each other. Such external costs of particular sources, or 'system costs', are defined as the total costs above plant-level costs to supply electricity at a given load and given level of security of supply. They can take the form of intermittency, network congestion, greater instability (i.e. higher risk of interruption to supply), etc. The focus of the report is on system costs associated with nuclear power and renewables such as wind and solar photovoltaics. In particular, the report considers 'grid-level system costs', a subset of overall system costs that consists of the costs of network connection, extension and reinforcement, short-term balancing and long-term adequacy in order to ensure continuous matching of supply and demand under all circumstances (see figure 4.18).

The report considers the complex issues arising from the integration of significant amounts of variable renewables, which profoundly affects the structure, financing and operation of electricity systems, thereby having economic and financial implications well above the 'plant-level' costs of these sources. It includes the first quantitative study of grid-level system costs in six countries (Finland, France, Germany, South Korea, UK and US).

The most important effects of a large variable renewable component in a particular electricity system include:

- lower and more volatile electricity prices in wholesale markets due to the influx of variable renewables with very low marginal costs (including zero fuel costs), leading to the closure or mothballing of existing plant needed to maintain secure supplies;
- the reduction of load factors of dispatchable power generators (the 'compression effect') as renewables with very low marginal cost are given priority over dispatchable supply;
- the introduction of inefficiencies in existing plants coupled with an influx of renewables, implying an increasing gap between the costs of producing electricity and prices on electricity wholesale markets;
- greater physical wear and tear to thermal power plants owing to the greater stress on components, especially as metal expands and contracts alongside unnecessary increases and decreases in output and hence core operating temperature, making it more likely to crack, thereby shortening the life and/or increasing the maintenance costs of these plants.

The financial implications of variable renewables are therefore potentially profound in both the short term and the long term.

In most jurisdictions where there is a political imperative to promote the growth of renewable generation, renewables enjoy a 'must take' arrangement, whereby if they have output available the system operator is obliged to take that output or pay the generating company compensation for not doing so.[113] Where markets are constructed in this way, dispatchable power technologies suffer owing to the compression effect as defined above.

The 'compression effect' affects different dispatchable technologies differently. For existing plant gas-fired capacity is most severely affected. A higher proportion of its costs are accounted for by the fuel, so greater savings are made by taking them offline when the wind is blowing or the sun is out. As noted earlier, gas-fired power plants are already experiencing substantial declines in profitability in many countries which have high shares of variable renewables. Coal does rather better, generally having lower operating costs. This leads to the undesirable outcome that the average carbon output from non-renewable generation per MWh generated increases, as discussed in chapter 6.

In the short term the effect on nuclear power may be less still as it generally has very low avoidable costs, including fuel. It is therefore likely to continue to run when large amounts of renewable electricity are available – unless, of course, there is sufficient renewable electricity available to fulfil total system demand – as there is little economic advantage in taking them offline.

It is of course possible to design an electricity market without including the must-take provisions for renewables. Since consumers place a high value on reliability of supply, such a market might leave generating companies whose output was dominated by variable renewables in the unenviable position of having to dip into the 'real-time' market to fulfil their contract obligations to customers during those times when the wind was still and night had fallen. When those times coincided with high system demand (as they often do in the UK, for example), the price of this replacement output might be very high indeed. This would clearly act as a severe disincentive to potential investors in renewables and therefore generally has not been pursued.

In the long term the position may be reversed. New build nuclear power may be more seriously affected than gas or coal by the advantages of access enjoyed by renewables, owing to the higher capital investment costs and hence higher risks in volatile low-wholesale price environments. In the UK's Electricity Market Reform this risk was addressed by introducing long-term Contracts for Difference – in effect guaranteed prices – for nuclear and renewable output, giving investors some confidence that encroachment of large volumes of renewable output when weather conditions are suitable would not reduce the price they could command for their output at those times.

MIT looked at the challenges presented by growing amounts of variable capacity to the concept of system reliability, analysed in terms of two attributes: security, which describes the ability of the system to withstand disturbances; and adequacy, which represents the ability of the system to meet the aggregate power and energy requirement of all consumers at all times.[114] The researchers noted that the debate on the implications of introducing renewables alarmingly lacked depth. Among their conclusions were the following points.

- Policies have been adopted around the world to promote deployment of renewable generation. These policies have been successful in increasing the volume of wind and solar generation capacity in various national systems but the cost and operating implications of these policies are not fully appreciated. The characteristics of intermittent sources require system operators to adopt different, and more costly, measures to balance load and generation and maintain system reliability. It is clear that policies that regulate investment, operations and rates will undergo significant change. It is becoming clear that the total costs and consequences of these policies were not fully understood. In order to ensure the goals of reliability and economic efficiency while simultaneously lowering carbon emissions, substantial regulatory changes are needed.

- Intermittent renewables present integration challenges at all timescales for the power system. As renewable penetration increases, system stability on the timescales of fractions of a second will increasingly matter as much as back-up capacity at the minutes to hours scales.

- Too much electricity generation from intermittent renewables is as much of a problem as too little generation. Frequently, wind integration problems involve having too much wind during low demand periods; many renewables mandates require the dispatch of wind energy, regardless of demand.

- Wind is highly variable and difficult to predict. Additionally, peak onshore wind does not usually coincide with peak electricity demand. As wind penetration increases in a power system, changes in the wind will have a larger effect on the net load. To account for this increasing uncertainty, the percentage of wind's capacity that a system operator considers 'firm' generally decreases as wind's share of the generation mix increases.

- The costs to thermal plant operators of dealing with increased ramping and cycling requirements at different timescales remain to be understood in detail.

Similar points have been made by the European Nuclear Energy Forum in its work on competitiveness.[115] It too notes that power development coupled with a large share of renewables implies structural changes to the electricity grid with large additional capital costs, including:

- transmission extension causing large additional capital costs for the power sector;

- implications for distribution networks which are not addressed in detail by most of the studies;

- a lack of detail surrounding the claims made in many studies for technologically advanced smart grids and super smart grids including smart metering;

- the absence of any clear vision on the possible total 'decentralisation' of power generation.

A more specific worked example was published in 2014 with regard to the effect of use of wind power in Ontario on system costs.[116] (The Fraser Institute is a free market think-tank whose vision is 'a free and prosperous world where individuals benefit from greater choice, competitive markets and personal responsibility', aims which in practice may prove to be incompatible with large-scale renewable investment.) The study took as its starting point the above-inflation growth in Ontario's electricity costs despite a decline in wholesale market prices for power. The authors concluded that solar and wind systems, which provided less than 4 per cent of Ontario's power, accounted for about 20 per cent of the average end-user price of electricity. They called for a moratorium on new renewable power facilities, pursuit of regulatory and legislative options to reduce the amount of installed renewable capacity, restarting four of twelve coal-burning units that, they argued, could operate as cleanly as natural gas plants, suspending conservation programmes when the province has surplus baseload and exploring the option of large-scale imports of power from Hydro Quebec until nuclear plants became available after refurbishment.

4.12 TENTATIVE NUMBERS

The OECD/NEA study attempts to quantify the grid-level system costs of various electricity sources in the six countries named earlier. There is quite a degree of variation among these countries, reflecting such factors as the siting of plants with respect to demand, the overall mix, the quality of wind cover and the levels of security of supply demanded. Taking the UK as an example, the calculated grid-level system costs of various sources of electricity are as shown in table 4.2. (The figures are cited for cases where the technology in question provides 10 and 30 per cent of total electricity and are in US$ per MWh.)

This implies that introducing variable renewables to provide up to 10 per cent of the total electricity supply will increase the 'per MWh' costs by between 5 and 50 per cent (depending on the country involved); whereas if the penetration level is 30 per cent this may increase 'per MWh' costs by between 16 and 180 per cent (the last figure referring to solar PV in Finland).

Table 4.2 Grid-level cost implications of renewable penetration

Technology	Nuclear		Coal		Gas		Offshore wind		Onshore wind		Solar	
Penetration level	10%	30%	10%	30%	10%	30%	10%	30%	10%	30%	10%	30%
Total grid-level system costs (US$ per MWh)	3.10	2.76	1.34	1.34	0.56	0.56	34.0	45.4	18.6	30.2	57.9	71.7

The study also looks at the effect on the profitability of dispatchable technology – and therefore the effect on the incentives for companies to invest in new plant – of having wind and solar at penetration levels of 10 or 30 per cent. Profitability of these plants is affected both by being taken offline, so losing direct income, and by the very low market price of electricity at times when significant amounts of renewables are available. The results are striking, as shown in table 4.3.

Table 4.3 Effects of renewable penetration on economics of dispatchable technologies

Penetration level		10%		30%	
Technology		Wind	Solar	Wind	Solar
Load losses	CCGT (%)	-34	-26	-71	-43
	Coal (%)	-27	-28	-62	-44
	Nuclear (%)	-4	-5	-20	-23
Profitability losses	CCGT (%)	-42	-31	-79	-46
	Coal (%)	-35	-30	-69	-46
	Nuclear (%)	-24	-23	-55	-39
Electricity price variation (%)		-14	-13	-33	-23

4.13 ALLOCATION OF COSTS

However the pricing system is structured, the full system costs of power production will need to be met. An economically rational system would place the system costs of renewables on the renewable technologies themselves, by mandating variable renewables to compete in the market on equal grounds to dispatchable technologies. In this way the link between the policies pursued and the costs to the consumer would be strong and transparent and economically rational decisions could be taken on the mix of generating plant required to deliver on environmental and other goals. As noted earlier, in reality renewables are generally shielded by transferring many of their cost implications on to the system as a whole. These market arrangements are just as important, if not more so, as the direct subsidies which have been offered to renewable investors through Renewable Obligation Certificates and subsequently by feed-in tariffs.

'Smart grids', which can, for example, switch fridges or heating systems off for one or two hours when supply drops, may help to mitigate some of these issues to some extent. However, it is more difficult to see how such an approach could revolutionise practice in industry, where shutting down the power supply would potentially leave a workforce idle.

There seem then to be three potential broad solutions to the challenge of moving towards a market structure which appropriately differentiates between a reliable unit of output and an unreliable one and apportions all of the associated costs – plant-level, grid-level and system-level.

- Capacity payments or markets with capacity obligations, in which variable producers need to buy 'adequacy services' from dispatchable providers, which would thus earn additional revenues.
- Long-term, fixed-price contracts subscribed by governments for guaranteed portions of the output of dispatchable plants whether in the form of Contracts for Difference or feed-in tariffs.

- The gradual phase-out of subsidies to variable renewables, the discontinuation of grid priority and a more direct allocation of additional grid costs to the sources which cause them – this would slow down deployment of renewables, hence reduce costs considerably, but would also force the internalisation of grid and balancing costs.

What is unlikely to lead to a return of effective decision-making in the field is prolonging the meaningless use of terms such as 'grid parity' and hiding the implications of variable output from sources with low power density in a raft of obscure market measures.

The NEA report makes four recommendations.

1. Increasing the transparency of power generation market costs at the system level, so it is clear to decision-makers the full economic costs of variable renewables.
2. Preparation of regulatory frameworks that minimise system costs and favour their internalisation – i.e. the costs fall where they are incurred – addressing four points:
   - the decrease in revenues for operators of dispatchable capacity owing to the compression effect;
   - the need to internalise the system costs for balancing and maintain supply adequacy effectively;
   - allocation of costs to the appropriate technology, insofar as it is possible;
   - the need for careful monitoring and internalisation of the carbon implications of the requirement for back-up, through a carbon tax again imposed on the causes of the emissions, not necessarily simply the plants that are producing them.
3. Better recognition of the value of dispatchable low-carbon technologies in complementing the introduction of variable renewables.
4. Development of flexibility resources for future low-carbon systems, e.g. working on increasing the extent to which nuclear plants can follow load, greater storage capacity and increasing international interconnection (a greater challenge for islands such as the UK or Japan than it is for the continental nations of Germany or Switzerland).

Whether or not these recommendations are followed, it is sensible for the full costs of use of various sources of electricity to be taken into account when planning a system that needs to balance cost alongside system security and environmental implications if the best policies are to be followed. It is of course still a perfectly defensible stance for government to take to argue that certain non-financial advantages of renewables, whatever such advantages might be, merit a very steep increase in power bills or a great increase in the national debt. But such statements must be made against a realistic assessment of what those costs are and what are the associated reductions, if any, in greenhouse gas emissions from the system as a whole, rather than the wind turbine or solar panel taken in isolation. Otherwise we may find consumers paying vastly inflated bills on the basis of promises which cannot be delivered on technically.

To fail to deal with these problems is likely further to delay necessary decisions. Unrealistic expectations concerning the potential of renewables, the maintenance of the fiction that variable renewables and delocalised power generation are new and exciting ideas rather than approaches which have been thoroughly tried and found wanting, and an unsustainable subsidy regime have all combined to divert attention away from practical ways of addressing the challenges of generating vast quantities of low-carbon electricity.

Endnotes

1   http://tna.europarchive.org/20080527124022/http://www.cabinetoffice.gov.uk/strategy/work_areas/~/media/assets/www.cabinetoffice.gov.uk/strategy/theenergyreview%20pdf.ashx, Performance and Innovation Unit (2002), *The energy review*, Cabinet Office.

2   http://www.independent.co.uk/voices/commentators/gordon-mackerron-this-way-is-more-likely-to-leave-us-in-the-dark-770005.html, MacKerron G, 'Gordon MacKerron: this way is more likely to leave us in the dark,' *The Independent*, 13 January 2008.

3   Coleridge S (1828), 'Netherlands fragment' in Holmes R (1996), *Samuel Taylor Coleridge: selected poetry*, Penguin Books.

4   Heidegger M (1954), 'The question concerning technology' in ed. Farrell Krell D, trans. Lovit W (1977), *Martin Heidegger: basic writings*, New York: Harper & Row, 1977.

5   http://www.renewables-info.com/drawbacks_and_benefits/renewable_energy_advantages_and_disadvantages.html, Haluzan N (2013), 'Renewable energy – advantages and disadvantages'.

6   http://www.th-nuernberg.de/fileadmin/Fachbereiche/bw/studienschwerpunkte/international_business/Master/CAIFD/SeminarPapers/HydropowerNorway_SeminarPaper.pdf, Gonzalez D, Kilinc A and Weidmann N (2011), 'Renewable energy development: hydropower in Norway', Georg Simon Ohm University of Applied Sciences, Nuremburg.

7   http://www.rtcc.org/2013/09/05/bbc-survey-highlights-uk-public-support-for-wind-and-solar/, 'BBC survey highlights UK public support for wind and solar', *Climate Home*, 5 September 2014.

8   http://www.thecommentator.com/article/3827/the_great_renewables_scam_unravels, Glover P, 'The great renewables scam unravels', *The Commentator*, 21 June 2013.

9   http://breakthrough.turing.com/images/elements/Footprint_17.jpg, Lovering J, Trembath A, Swain M and Lavin L (2012), 'Renewables and nuclear at a glance', Breakthrough Institute.

10 http://www.scientificamerican.com/article/renewable-energys-hidden-costs/, 'Matson J (2013), 'Renewable energy's hidden costs', *Scientific American*, **309** (4).

11 http://www.homepages.ed.ac.uk/v1ewaveg/0-Archive/EWPP%20archive/1976%20Leishman%20and%20Scobie%20NEL.pdf, Leishman J and Scobie G (1976), 'The development of wave power – a techno-economic study', National Engineering Laboratory, East Kilbride.

12 http://www.cegbmidreg.co.uk/cegbstory/frontouter.htm, Cochrane R and Schaefer M (1990), *The CEGB story*, CEGB, London.

13 https://www.gov.uk/government/statistical-data-sets/historical-electricity-data-1920-to-2011, DECC (2014), 'Historical electricity data: 1920 to 2014', The Stationery Office.

14 http://www.americaslibrary.gov/jb/gilded/jb_gilded_hydro_1.html, US Library of Congress, 'The world's first hydroelectric power plant began operation, 30 September 1882'.

15 http://energy.gov/eere/water/history-hydropower, US Department of Energy, 'History of hydropower'.

16 http://www.bp.com/en/global/corporate/about-bp/energy-economics/statistical-review-of-world-energy.html, BP (2015), 'BP statistical review of world energy 2015'.

17 Simoes S and Barros A (2007), 'Regional hydroclimatic variability and Brazil's 2001 energy crisis', *Management of Environmental Quality: An International Journal*, **18** (3) 263–73.

18 Bauer C (2009), 'Dams and markets: rivers and electric power in Chile', *Nat. Resources J.* **49** 583.

19 http://www.reuters.com/article/2011/05/25/us-china-drought-hydropower-idUSTRE74O1BK20110525, Stanway D, 'China power crunch to worsen as drought slashes hydro', *Reuters*, 25 May 2011.

20 http://docs.business.auckland.ac.nz/Doc/fficiency-and-fuel-switching-in-Norway-and-NZs-residential-electricity-sector-during-droughts.pdf, van Campen B (2010), 'Efficiency and fuel switching in Norway and New Zealand's residential electricity sector during droughts', IAEE: Cleveland, Ohio.

21 http://www.solar.coppe.ufrj.br/itaipu.html, Krauter S (1998), 'Itaipú dam'.

22 Salagnac J (2007), 'Lessons from the 2003 heat wave: a French perspective', *Building Research & Information*, **35** (4) 450–7

23 Deery P (ed.) (1978), *Labour in conflict: the 1949 coal strike*, Australian Society for the Study of Labour History: Sydney.

24 http://libcom.org/library/so-near-so-far-selective-history-british-miners, Dennis J et al. (2005), 'So near, so far: aspects of a history of the British miners', *libcom.org*.

25 http://www.bnamericas.com/news/oilandgas/Landslide_Damages_Norandino_Pipeline, Walter M, 'Landslide damages Norandino pipeline', *BNamericas*, 30 January 2002.

26 http://amazonwatch.org/news/2014/1127-activists-join-indigenous-people-to-protest-construction-of-amazon-mega-dam, *Amazon Watch*, 27 November 2014.

27 Sathyajith M (2006), *Wind energy: fundamentals, resource analysis and economics*, Springer Berlin Heidelberg.

28 http://www.energy.gov/eere/wind/history-wind-energy, US Department of Energy (2014), 'History of wind energy'.

29 Price T (2005), 'James Blyth – Britain's first modern wind power engineer', *Wind Engineering*, **29** (3) 191–200.

30 http://windowstoworldhistory.weebly.com/poul-la-cour-pioneered-wind-power-in-denmark.html, Warnes K (2005), 'Poul la Cour pioneered wind power in Denmark', *Windows to World History*.

31 'History of wind energy', *Energy Encyclopedia*, **6** 422.

32 http://www.theecoexperts.co.uk/the-history-of-solar-energy, Smith C (1995), 'The history of solar energy: revisiting solar power's past', *Technology Review*.

33 http://www.telegraph.co.uk/news/earth/environment/11312765/Liz-Truss-Solar-farms-in-rural-areas-make-the-heart-sink.html, Swinford S, 'Liz Truss: solar farms in rural areas make the heart sink', *Telegraph*, 27 December 2014.

34 http://www.homepages.ed.ac.uk/v1ewaveg/0-Archive/EWPP%20archive/1976%20Leishman%20and%20Scobie%20NEL.pdf, Leishman J and Scobie G (1976), 'The development of wave power – a techno-economic study', National Engineering Laboratory, East Kilbride.

35 https://pangea.stanford.edu/ERE/pdf/IGAstandard/EGC/2007/083.pdf, Bertani R (2007), 'World geothermal generation in 2007, *GHG Bulletin*.

36 http://assets.wwf.org.uk/downloads/2011_02_02_the_energy_report_full.pdf?_ga=1.48300078.944273677.141 8649672, WWF (2011), 'The energy report: 100% renewable energy by 2050'.

37 http://www.gwec.net/wp-content/uploads/2015/02/GWEC_GlobalWindStats2014_FINAL_10.2.2015.pdf, GWEC (2015), 'Global wind statistics 2014'.

38 http://www.bp.com/en/global/corporate/about-bp/energy-economics/statistical-review-of-world-energy.html, BP (2015), 'BP statistical review of world energy 2015'.

39 http://www.policyexchange.org.uk/media-centre/blogs/category/item/dear-energy-secretary-i-m-afraid-to-tell-you-there-is-no-money, Howard R (2015), 'Dear Energy Secretary, I'm afraid to tell you there is no money', *Policy Exchange*.

40 http://www.telegraph.co.uk/news/earth/energy/11292367/Green-policies-to-add-up-to-40pc-to-cost-of-household-electricity.html, Mendick R, 'Green policies to add up to 40% to the cost of household electricity', *Telegraph*, 14 December 2014.

41 http://www.theguardian.com/environment/2014/sep/10/friends-of-the-earth-nuclear-power-bbc-report, Vaughan A, 'Friends of the Earth denies dropping nuclear power opposition', *The Guardian*, 10 September 2014.

42 http://www.bp.com/en/global/corporate/about-bp/energy-economics/statistical-review-of-world-energy.html, BP (2015), 'BP statistical review of world energy 2015'.

43 http://www.telegraph.co.uk/news/earth/energy/windpower/11319763/Offshore-wind-farms-drawing-electricity-from-grid-to-keep-turning-in-icy-conditions.html, Rayner G, 'Offshore wind farms drawing electricity from grid to keep turning in icy conditions', *Telegraph*, 31 December 2014.

44 http://www.foe.co.uk/sites/default/files/downloads/tyndall_evidence.pdf, Tyndall Centre, 'A review of research relevant to new build nuclear power plants in the UK', January 2013.

45 E.g. http://www.sut.org/educational-support-fund/information-for-careers-in-underwater-technology-and-science/for-school-leavers-and-beyond/offshore-renewable-energy/, Society for Underwater Technology, 'Offshore renewable energy'.

46 http://www.windpowermonthly.com/article/1006708/slipping-turbines-trigger-investigation-design, deVries E, 'Slipping turbines trigger investigation into design', *Wind Power Monthly*, 1 June 2010.

47 http://www.abindieerde.de/2014-08-Doku/SPIEGEL_2014_35-Offshore.pdf, Brendel M and Traufetter G (2014), 'Knall auf hoher See', *Der Spiegel*.

48 http://www.spiegel.de/international/germany/high-costs-and-errors-of-german-transition-to-renewable-energy-a-920288.html, 'Germany's energy poverty: how electricity became a luxury good', *Der Spiegel*, 4 September 2013.

49 http://www.reuters.com/assets/print?aid=USL6NOT14F620141114, Steitz C and de Clercq G, 'Offshore wind industry races to cut costs as subsidies drop', *Reuters*, 14 November 2014.

50 http://www.bbc.co.uk/news/uk-england-devon-25095868, 'Atlantic Array wind farm dropped by developer', *BBC News website*, 26 November 2013.

51 http://www.4coffshore.com/windfarms/race-bank-united-kingdom-uk18.html, 4C Offshore (2013), 'Race Bank'.

52 http://www.belfasttelegraph.co.uk/news/local-national/northern-ireland/blown-away-1bn-wind-farm-project-for-northern-ireland-axed-30789867.html, Stewart L, 'Blown away: £1bn wind farm project for Northern Ireland axed', *Belfast Telegraph*, 2 December 2014.

53 http://www.thisismoney.co.uk/money/news/article-2385991/Plans-build-wind-farms-deluded.html, Salmon J, 'Plans to build more wind farms are deluded, L&G boss warns the government', *This is Money*, 9 August 2013.

54 http://www.telegraph.co.uk/news/earth/energy/windpower/10122850/True-cost-of-Britains-wind-farm-industry-revealed.html, Mendick R and Malnick E, 'True cost of Britain's wind farm industry revealed', *Daily Telegraph*, 15 June 2013.

55 http://www.theguardian.com/environment/2012/feb/28/windfarms-risk-free-millions-for-landowners, Vidal J, 'Wind turbines bring in 'risk-free' millions for rich landowners', *The Guardian*, 28 February 2012.

56 http://epaw.org/documents.php?lang=en&article=book5, Stevenson S (2011), 'The rape of Britain: Wind farms and the destruction of our environment', European Platform Against Windfarms.

57 http://www.telegraph.co.uk/finance/newsbysector/energy/11234261/Plans-for-3bn-Navitus-Bay-wind-farm-off-Jurassic-Coast-in-doubt.html, Gosden E, 'Plans for £3bn Navitus Bay wind farm off Jurassic Coast in doubt', *Daily Telegraph*, 17 November 2014.

58 http://www.telegraph.co.uk/news/earth/energy/windpower/11859736/Jurassic-Coast-saved-from-industrialisation-as-Navitus-Bay-wind-farm-rejected.html, 'Jurassic Coast saved from industrialisation as Navitus Bay wind farm rejected', *Daily Telegraph*, 11 September 2015.

59 http://www.countryguardian.net/campaign%20windfarm%20action%20groups.htm, Country Guardian (2012), 'Campaign against windfarms'.

60 http://en.friends-against-wind.org/realities/10h-regel, 'Setback distance between wind turbines and dwellings: the new 10H rule protects residents in Bavaria', Friends against wind, 28 December 2014.

61 http://www.bbc.co.uk/news/uk-wales-mid-wales-20727030, 'Powys windfarms: council faces £2.8m public inquiry bill', *BBC News website*, 14 December 2012.

62 http://www.biologicaldiversity.org/campaigns/protecting_birds_of_prey_at_altamont_pass/pdfs/Smallwood_2008-Altamont_mortality_estimates.pdf, Smallwood K and Thelander C (2007), 'Bird Mortality in the Altamont Pass wind resource area, California', *Biological Diversity*.

63 http://elementalpower.com.au/news/?p=1599, 'Renewable news 17th November', Elemental Power Industries, 17 November 2010.

64 http://ecolo.org/lovelock/lovelock-wind-power.html' Kuhn A, 'Wind power is just a gesture', ecolo.org, 3 February 2004.

65 http://docs.wind-watch.org/BENTEK-How-Less-Became-More.pdf, Bentek (2010), How less became more – wind, power and unintended consequences in the Colorado energy market.

66 http://mitei.mit.edu/system/files/intermittent-renewables-full.pdf, MIT (2011), 'Managing large-scale penetration of intermittent renewables'.

67 http://www.dailymail.co.uk/news/article-2560494/Worlds-largest-solar-farm-SCORCHING-BIRDS-fly-it.html, 'Horror at the world's largest solar farm days after it opens as it is revealed panels are scorching birds that fly over them', *Daily Mail*, 16 February 2014.

68 http://www.solar-international.net/pdf/magazines/2013_Issue_III.pdf, Ridsdale D (2013), 'An industry in denial', *International Solar*, issue III.

69 http://www.researchgate.net/publication/225823021_Do_Hydroelectric_Dams_Mitigate_Global_Warming_The_Case_of_Brazil's_Curu-una_Dam, Fearnside P (2005), 'Do hydroelectric dams mitigate global warming? The case of Brazil's Curuá-Una dam', *Mitigation and Adaptation Strategies for Global Change* (2005), 10 675–91.

70 http://www.theguardian.com/environment/2012/feb/28/windfarms-risk-free-millions-for-landowners, Vidal J, 'Windfarms bring in risk-free millions for landowners', *The Guardian*, 28 February 2012.

71 http://www.ippr.org/publications/beyond-the-bluster-why-wind-power-is-an-effective-technology, Platt R, Gardner P and Fitch-Roy O (2012), *Beyond the bluster – why wind is an effective technology*, IPPR.

72 http://docs.wind-watch.org/eonwindreport2005.pdf , E.ON Netz (2005), 'Wind report 2005'.

73 http://www.theguardian.com/environment/2013/dec/08/dong-energy-offshore-wind-power-subsidies-north-sea, Macalister T, 'Dong Energy upbeat about offshore wind power thanks to higher subsidy', *The Guardian*, 8 December 2013.

74 http://www.theecologist.org/blogs_and_comments/commentators/2160503/nuclear_subsidies_open_letter_to_commissioner_almunia.html, 'Nuclear subsidies – open letter to Commissioner Almunia', *Ecologist*, 15 November 2013.

75 http://www.renewableuk.com/en/renewable-energy/wind-energy/onshore-wind/index.cfm, 'Onshore wind', RenewableUK.

76 http://ec.europa.eu/research/energy/eu/index_en.cfm?pg=research-wind-background, European Union Research and Innovation (2009), 'Wind energy – technical background'.

77 http://www.nytimes.com/2009/04/19/magazine/19wwln-q4-t.html?_r=2&ref=todayspaper&, Solomon D, Interview with US Energy Secretary Steven Chu, *New York Times Magazine*, 9 April 2009.

78 http://www.theccc.org.uk/wp-content/uploads/2013/05/1720_EMR_report_web.pdf, Committee on Climate Change (2013), 'Next steps on Electricity Market Reform – securing the benefits of low-carbon investment'.

79 https://www.greenparty.org.uk/assets/files/manifesto/Green_Party_2015_General_Election_Manifesto_Searchable.pdf, UK Green Party, 'For the common good: General Election manifesto 2015'.

80 http://www.guernseyrenewableenergy.com/documents/managed/1.PDF, Lee O (2012), *Feasibility of offshore wind in Guernsey waters*, Guernsey Renewable Energy Commission.

81 http://www.telegraph.co.uk/earth/energy/windpower/10783823/No-more-onshore-wind-farms-if-Conservatives-win-2015-election.html, *Daily Telegraph*, 23 April 2014.

82 http://www.thetimes.co.uk/tto/news/politics/article4071244.ece, 'Cameron pledges to halt public aid for windfarms', *The Times*, 24 April 2014.

83 http://www.no2nuclearpower.org.uk/nuclearnews/NuClearNewsNo62.pdf, No2Nuclear Power, *NuClear News* **No. 62**, May 2014.

84 http://www.theguardian.com/environment/2014/apr/24/will-tory-plans-kill-onshore-wind-in-the-uk, Mathiesen K, 'Will Tory plans kill onshore wind in the UK?', *The Guardian*, 24 April 2014.

85 http://www.livosenergy.co.uk/solar-energys-falling-costs-can-cut-politics-renewables/, Hartnell G, 'Solar energy's falling costs can cut politics from renewables', Livos Energy, 14 April 2014.

86 http://w3.windfair.net/wind-energy/news/16557-awea-blog-falling-costs-for-wind-and-other-top-5-takeaways-from-new-wall-street-report, 'Falling costs for wind …', American Wind Energy Association blog, 23 September 2014.

87 http://solarcellcentral.com/cost_page.html, Solar Cell Central (2014), 'Solar electricity costs'.

88 http://www.reuters.com/article/2014/07/28/us-china-usa-solar-idUSKBN0FX09220140728, Martina M, 'China condemns US anti-dumping duties on solar imports', *Reuters*, 28 July 2014.

89 http://emp.lbl.gov/sites/all/files/2013_Wind_Technologies_Market_Report_Final3.pdf, US Department of Energy (2014), 'Wind technologies market report 2013'.

90 http://www.thecrownestate.co.uk/media/5493/ei-offshore-wind-cost-reduction-pathways-study.pdf, The Crown Estate (2012), 'Offshore wind cost reduction: Pathways study'.

91 https://www.gov.uk/government/uploads/system/uploads/attachment_data/file/42843/3237-cons-ro-banding-arup-report.pdf, ARUP (2011), 'Review of the generation costs and deployment potential of renewable electricity technologies in the UK', DECC.

92 https://www.gov.uk/government/uploads/system/uploads/attachment_data/file/65716/71-uk-electricity-generation-costs-update-.pdf, Mott MacDonald (2010), 'UK electricity generation costs update', DECC.

93 Element Energy (2013), 'The economic impact of the renewable path'.

94 https://www.ofgem.gov.uk/ofgem-publications/76242/20121217-press-release-riio.pdf, Ofgem, 'Ofgem announces major investment to upgrade Britain's gas and electricity network,' 17 December 2012.

95 https://www.gov.uk/government/speeches/amber-rudds-speech-on-a-new-direction-for-uk-energy-policy, 'Amber Rudd's speech on a new direction for UK energy policy', UK Government, November 18 2015.

96 http://www.pv-magazine.com/news/details/beitrag/uk-unveils-europes-largest-battery-storage-project_10 0017518/#axzz3NhL3Rw00, Clover I, 'UK unveils Europe's largest battery storage project', pv magazine, 15 December 2014.

97 https://www.gov.uk/government/uploads/system/uploads/attachment_data/file/337684/chapter_6.pdf, DECC (2014), 'Digest of UK Energy Statistics, Chapter 6: Renewable sources of energy'.

98 http://www.johnmuirtrust.org/about/resources/594-analysis-of-uk-wind-power-generation-november-2008-to-december-2010, Young S (2011), 'Analysis of UK wind power generation November 2008 to December 2010', John Muir Trust.

99 http://www.ref.org.uk/fuel/, Renewable Energy Foundation, 'British Electricity Generation (Balancing Mechanism) Fuel Mix'.

100 http://www.chinadaily.com.cn/bizchina/2011-02/15/content_12019641.htm, Song J (2011), '2.8 billion kWh of wind power wasted', *China Daily*, 15 February 2011.

101 http://www.reuters.com/article/2015/05/17/china-windpower-idUSL3N0Y24DM20150517, Wong S-L and Zhu C (2015), 'Chinese wind earnings under pressure with fifth of farms idle', *Reuters*.

102 http://www.ref.org.uk/constraints/index.php, Renewable Energy Foundation (2015), 'Balancing Mechanism windfarm constraint payments'.

103 http://www.bloomberg.com/news/2013-08-12/rwe-shuts-fifth-of-german-gas-power-capacity-as-profit-drops-1-.html, Mengewein J, 'RWE shuts fifth of German gas power capacity as profit drops', *Bloomberg Business News*, 12August 2013.

104 http://www.bloomberg.com/news/articles/2013-01-23/eon-rwe-may-have-to-close-down-unprofitable-gas-power-plants, Andresen T and Nicola S, 'EON, RWE may have to close down unprofitable gas power plants', *Bloomberg Business News*, 23 January 2013.

105 http://www.publications.parliament.uk/pa/cm201012/cmselect/cmenergy/1040/104007.htm, Energy and Climate Change Select Committee (2011), 'Seventh Report – a European Supergrid', parliament UK.

106 http://mainstream-downloads.opendebate.co.uk/downloads/11032013-Friends-of-the-Supergrid---Germany.pdf, PWC (2013), 'Supergrid in Germany – potential social, environmental and economical benefits'.

107 http://www.bloomberg.com/news/2012-10-25/windmills-overload-east-europe-s-grid-risking-blackout-energy.html, Bauerova L and Andreson T, 'Windmills overload East Europe's grid risking blackout', *Bloomberg Business News*, 26 October 2012.

108 GL Garrad Hassan (2010), 'Options for coping with high-renewables penetration in Scotland'.

109 http://www.oecd-nea.org/ndd/reports/2011/load-following-npp.pdf, OECD/NEA (2011), 'Technical and economic aspects of load following with nuclear power plants'.

110 http://www.faz.net/aktuell/wirtschaft/wirtschaftspolitik/energiepolitik/umweltminister-altmaier-energiewende-koennte-bis-zu-einer-billion-euro-kosten-12086525.html, 'Umweltminister Altmaier, Energiewende könnte bis zu einer Billion Euro kosten', *Frankfurter Allgemeine*, 19 February 2013.

111 https://www.gov.uk/government/uploads/system/uploads/attachment_data/file/47854/1938-overarching-nps-for-energy-en1.pdf, DECC (2011), 'Overarching National Policy Statement for Energy (EN-1)'.

112 http://www.oecd-nea.org/ndd/reports/2012/system-effects-exec-sum.pdf, OECD/NEA (2012), 'Nuclear energy and renewables: system effects in low-carbon electricity systems'.

113 http://www.dailymail.co.uk/news/article-2388417/Wind-farms-paid-30-million-year-stand-idle-grid-cope-energy-produce.html, McDermott N, 'Wind farms paid £30 million a year to stand idle because the grid can't cope with all the energy they produce', *Daily Mail*, 9 August 2013.

114 http://mitei.mit.edu/system/files/intermittent-renewables-full.pdf, MIT (2011), 'Managing large-scale penetration of intermittent renewables'.

115 http://www.snetp.eu/wp-content/uploads/2014/04/sessioni.speaker2.mr_.jppreplc-opportunitiesenefwg.pdf, European Nuclear Energy Forum (2010), 'SWOT analysis'.

116 https://www.fraserinstitute.org/research/what-goes-upontarios-soaring-electricity-prices-and-how-get-them-down, McKitrick R and Adams T (2014), 'What goes up … Ontario's soaring electricity prices and how to get them down', Fraser Institute.

5 | ENERGY EFFICIENCY

5.1 AN OLD, OLD SONG

Those who might want to prevent investment in reliable new sources of electricity generation have found a powerful ally in the alleged simplicity and cost-effectiveness of improving energy efficiency as a way of reducing energy use. Arguing that a particular sum of money spent on energy efficiency measures, rather than on nuclear or fossil fuel power plants (though interestingly not usually renewables), could produce several times as much reduction in carbon emissions and offer more robust security of supply is often used as a relatively straightforward way of arguing against investing in new generating capacity. For Greenpeace:

> The way we use energy is shockingly wasteful. Every year, we throw away more than eight times the amount of energy supplied by all of the UK's nuclear power stations combined. Yet implementing energy efficiency saves more money than it costs. The government's own research put the economic potential for saving energy across all sectors, including heat, transport and electricity, at 30 per cent. The same research found that introducing these measures would save consumers £12 billion a year in reduced bills. Meanwhile, researchers in the US have found that for each pound invested, efficiency measures are ten times more effective than nuclear power at reducing carbon dioxide emissions and closing the energy gap.[1]

To the Association for the Conservation of Energy (ACE), which in 1981 misleadingly coined the phrase 'the fifth fuel' for energy efficiency:

> Energy efficiency is now universally regarded as the first fuel. It is the key to reducing carbon dioxide emissions at least cost, it creates sustainable employment where and when we need it most, it saves consumers money and helps the 56 million Europeans who cannot afford to heat their homes adequately, it improves the state of public finances, enhances competitiveness and increases GDP. More pertinently than ever, it is the best way to enhance Europe's energy security, by reducing the need to import fossil fuels from dangerous places. It is the one sanction on [Russian President] Putin that has real bite and does not hurt Europe's own economic interests.'[2]

Quantitative arguments typically go something like this analysis from anti-nuclear website No2NuclearPower (N2NP). UK electricity demand in 2010 was around 370 TWh and an energy scenario by Friends of the Earth saw it increasing to 470 TWh by 2030.[3] However, a July 2012 draft (unrefereed) report for the Department for Energy and Climate Change by consultancy McKinsey & Co. argued that 155 TWh of electricity demand reduction was possible, 140 TWh of it at negative cost.[4] McKinsey said that, excluding the impact of any policies, underlying electricity demand was likely to grow to 411 TWh by 2030. Electrification of vehicles and heating could add an additional 5 to 15 per cent to electricity demand in 2030, i.e. increasing demand to between 432 TWh and 473 TWh. However, in the residential sector potential efficiency savings could reach 66 TWh. Current government policies expected to save 44 TWh of this, more than half coming from more efficient appliances, with institutional measures, such as the government's Green Deal and Energy Companies Obligation, being the second-largest contributor. Overall there were 100 TWh of potential efficiency savings which the UK government was failing to capture. This would be more than enough to replace the existing nuclear programme (70 TWh) or enough to replace four power stations the size of Hinkley Point C. (N2NP chooses to ignore that no fewer than four times in the report McKinsey says that its calculations 'take no account of rebound effects' as discussed later in this chapter.)

Common sense – not always a good guide one has to admit – would suggest that it is strange that a measure that would deliver cost savings to businesses and perhaps even to households, which tend to display more inertia in these matters, needs any encouragement from government or anyone else. Most future projections of demand start with an assumption that companies will behave rationally in their own financial interests. At the very least it seems unlikely that offering financial inducements to companies will make much of a difference – while such gifts might be happily accepted by companies or households that were going to carry out efficiency improvements anyway, financial incentives will not be effective in cases where the barriers to improving efficiency are not financial.

The idea that efficiency is not only a 'free lunch' but also one that 'you get paid to eat', in the words of US energy efficiency consultant Amory Lovins, one of Big Green's heroes in this field, has exercised an almost mesmeric influence over the environmental industry, politicians and business consultants alike. Indeed, any neutral observer would have to accept that this looks like a revelation. Even given Big Green's tendency only to quote reports which support its own political bias, one might ask in despair why, if energy efficiency is so simple that it can actually reduce emissions and deliver cost reductions at the same time, nobody has thought of it before.

In reality improving energy efficiency, like using the 'new renewables', is far from being a recent discovery. Consider the following: 'The objective [of government energy policy] is to change fundamentally the attitudes and behaviour of all energy users, on the assumption that energy for the foreseeable future is going to become increasingly scarce and expensive.' This was how journalist David Fishlock summarised the 'Save It' campaign introduced in October 1974 in the UK to respond to the twin challenges of the first OPEC oil price hike and the coal mining strike which had brought down the Conservative government of Edward Heath earlier that year. Within three years voices were being raised that the campaign was too restrictive and there was no need to be so prescriptive.[5]

Ten years later the UK was engaging in 'Energy Efficiency Year' following the setting up of the Energy Efficiency Office in 1986. In 1987 the Brundtland Report on sustainable development said: 'the Commission believes that energy efficiency should be the cutting edge of national energy policies for sustainable development' and praised progress on energy efficiency: 'Impressive gains in energy efficiency have been made since the first oil price shock in the 1970s. During the past 13 years, many industrial countries saw the energy content of growth fall significantly as a result of increases in energy efficiency averaging 1.7 per cent annually between 1973 and 1983.'

And so on. As one commentator put it:

Every ten years or so, Energy Secretaries feel the need to roll out an energy saving initiative of their own devising, backed by a high visibility media campaign. Sometimes they work, sometimes they don't. Posters, booklets and TV ads featuring [celebrity chef] Delia Smith urged the populace to Save It! in 1976, with a memorable rubber stamped logo … Ten years later they tried something similar but got a bit too clever by branding it 'Monergy'. 'Get more for your money! More out of energy! More for your Monergy!' was the clumsy explanation. This time the populace just shrugged a collective 'eh?' and went off to run a very deep bath for one.[6]

If one simply looks at UK primary energy use over the period from 1970 it is hard to guess the points at which the UK was engaged in an energy efficiency push and those when it was not. Indeed, the launches of Save It! (1976) and Monergy (1986) seemed if anything to be associated with reverses in falling energy use trends and a resumption of growth, as can be seen in figure 5.1.

As always, it is interesting to compare pronouncements from the leading lights of the energy efficiency cult with how reality actually panned out. In 1984 Amory Lovins told *Business Week*: 'We will never get, we suspect, to a high enough price to justify building centralised thermal power plants again. That era is over.' As often befalls Lovins's prognostications, between 1984 and 2000 US annual electricity usage duly grew from 2,400 TWh to 4,000 TWh, generated by a huge programme of new power stations fuelled largely by coal and gas.

Energy efficiency gains are an inherent feature of market economies, which have a financial vested interest in cutting waste where possible, certainly if to do so has a

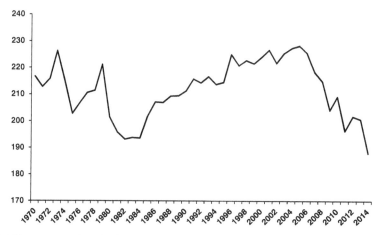

Figure 5.1 UK energy use (mtoe), 1970–2014

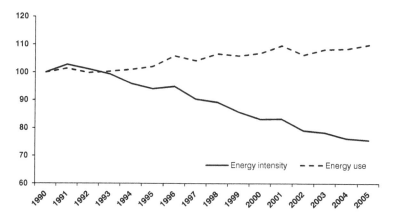

Figure 5.2 Changes in UK energy intensity and energy use, 1990–2005 (1990 = 100)[7]

quick payback period. Any company deploying cost-efficient technologies or techniques which are more energy efficient than those of its competitors will, all else being equal, enjoy a competitive advantage. The figures for the UK before the onset of the economic crisis in the mid-2000s shows that between 1990 and 2005 the energy intensity, i.e. the amount of energy consumed per unit of economic output and a crude proxy for energy efficiency, of the economy fell by nearly 25 per cent while energy use rose by 10 per cent – see figure 5.2.

Even given that recessions tend to deliver a windfall improvement in energy efficiency – older, less efficient plant tends to be rested or retired first when the economy is not operating at full capacity – the reduction in UK energy use of 18 per cent in the decade starting in 2005 is impressive.

However, two questions arise. Why have improvements in the efficiency of energy use in the past not resulted in the kinds of reductions in demand that current commentators predict? And what if anything is different this time round which might result in a different outcome in the future?

The apostles of energy efficiency claim that improvements in energy efficiency have at least allowed the economy to grow without expanding its use of energy as much as would otherwise have been the case. The reasoning behind such claims seems to be along the following lines. If we had not developed the light bulb – first incandescence and more recently LED – then if viewed from space the earth would still be producing just as much artificial light but it would be coming largely from candles. On this assumption the development of electric lighting has indeed cut energy use enormously. Similarly the development of the highly efficient LCD screen has reduced energy consumption enormously as long as we assume that several people would now have 50-inch cathode-ray tube TVs instead. The development of the jet engine has slashed energy use for long haul air travel because otherwise we would be undertaking some 3.5 billion passenger flights each year by turboprop or maybe even Gypsy Moth.

This argument is based on the notion that energy efficiency is a 'fuel' in competition say with using more gas – in other words that improving energy efficiency by 10 per cent results in a reduction in the demand for real fuels by 10 per cent. In fact, however, because energy efficiency affects the demand side it has entirely different macroeconomic consequences to increasing the availability of a fuel. If energy efficiency improves then the activity that is using the energy becomes more attractive because energy costs fall. So people will either carry out more of the activity or alternatively spend their savings on other goods or services, which in turn require energy in manufacture and use. So at least some of the benefit of improving energy efficiency is taken in increased economic output.

Once again this is not a recent discovery. W. Stanley Jevons, the great Liberal economist and 'father of neoclassical economics', first examined the apparent paradox in his 1865 book *The Coal Question.*[8] Jevons noted that England's consumption of coal had grown enormously since the 1780s when James Watt introduced a steam engine with much greater thermal efficiency than the design Thomas Newcomen had invented around 1710. (During the century from 1725 onwards alone, the efficiency of Newcomen engines improved from around 45 lb of coal per horsepower-hour to around 15 lb per horsepower-hour, while the introduction of the Watt engine had brought the figure down to below 5 lb per horsepower-hour, a combined improvement of a factor of around 10.[9] This represented an increase in thermal efficiency from 0.4 to 4 per cent – a modern CCGT such as those being forced off the system in Germany in favour of coal can convert chemical energy to electricity with about 60 per cent efficiency.)

At that time there were fears that Britain would run out of coal but some 'experts' (perhaps anticipating arguments similar to those currently presented by ACE, Greenpeace and others) argued that more energy efficient technology would reduce coal consumption, thereby extending the lifetime of the reserves. Jevons argued this was wrong – further increases in efficiency would actually result in an increase in the use of coal. Watt's innovations were making coal a more cost-effective power source, leading to the use of steam engines spreading to a wide variety of industries. So although the useful energy produced by a tonne of coal increased considerably, this had the effect of hugely expanding the activities for which coal power was used. The net effect was a growth in demand. 'It is a confusion of ideas to suppose that the economical use of fuel is equivalent to diminished consumption. The very contrary is the truth.'

To take an example, let us say that improvements in energy efficiency reduce the amount of energy needed to make a certain amount of steel by say 10 per cent. The apparent reduction in energy demand could be offset in a number of ways. For example, cheaper steel prices could result in more steel production as goods using steel become cheaper. If cars and lorries become cheaper more people might be able to afford them, resulting in more motor vehicle travel. In turn this might reduce transport costs for all sorts of goods, reducing their price and stimulating demand. Whether the overall effect, once such rebounds have been taken into account, is a reduction in energy use say of 7 per cent, or 2 per cent, no reduction at all, or even an increase of 2 per cent, will depend on a range of factors but the reduction will not be as great as 10 per cent.

History has certainly shown Jevons to be right. Bizarrely, however, there remain some who seem to continue to argue, or rather simply to state, that improvements in efficiency will deliver pro rata reductions in energy consumption. For example, the European Commission, in its '20-20-20' climate and energy policy set a target of 20 per cent improvement in the EU's energy efficiency by 2020 against the level that had been predicted on 'business as usual' scenarios.[10] However, in 2011 the European Parliament issued a report expressing 'deep concern that the EU is not on track to meet targets to reduce energy consumption by 20 per cent as compared with the projections for 2020', clearly conflating the concept of energy efficiency and reductions in energy use.[11] Similarly President Obama said: 'I think we have to have a strong push toward energy efficiency. We know that's the low-hanging fruit, we can save as much as 30 per cent of our current energy usage without changing our quality of life.'[12]

A considerable body of recent literature has analysed the 'Jevons paradox', also known as the rebound effect or the Khazzoom–Brookes postulate. The Breakthrough Institute is an environmental think tank which, while campaigning, for example, for measures to counter climate change, takes a rather different view from Big Green on issues where science and ideology clash. It states it is 'committed to modernising environmentalism for the twenty-first century', its mission being to 'accelerate the transition to a future where all the world's inhabitants can enjoy secure, free and prosperous lives on an ecologically vibrant planet'. In 2011 it produced a major report on the rebound effect[13] which referred to 'strong evidence' in academic literature that below-cost energy efficiency measures, i.e. measures that will pay for themselves, drive a rebound in energy consumption that erodes much and in some cases all of the expected energy savings and also to the widespread tendency to ignore this evidence in either politically- or commercially-driven statements on the subject.

One of the basic features of economic growth is that increasing the productivity of any single factor of production – labour, capital or energy – increases demand for all of those factors. The original generations of Luddites, for example, who feared there would be fewer jobs with the emergence of weaving looms, were proved wrong. Lower prices for woven clothing brought it within the price reach of many more people resulting in enormous increases in demand, while the associated profits stimulated demand for other goods and services elsewhere in the economy.

To offer an analogy, very few economists would argue that say a 10 per cent improvement in the productivity of labour, or the 'labour efficiency' of a particular activity, would result in a 10 per cent reduction in employment – or to put it another way, that the way to safeguard British jobs is to become less productive. More typical are statements such as: 'Labour productivity is an important factor in determining the productive potential of the economy. Countries with strong labour productivity growth tend to benefit from high rates of growth, strong export demand and low inflation. Increased labour productivity can enable a higher long run trend rate of growth.'[14] The stagnation in UK labour productivity after the 2008 recession was a topic of much concern. Evidence strongly implies that energy is no different, as per capita energy consumption is growing in most countries despite economies becoming more efficient year on year. This observation is weakest in the developed world, where the UK was not alone in seeing reductions in total energy use over the decade from 2005. Primary energy use in the EU fell by 11.6 per cent between 2005 and 2014 and in the OECD member countries by 3.3 per cent. This implies that rebound effects are weakest where economies are already well developed: there may be something of a ceiling to the amount of energy an individual desires or is even capable of consuming in such societies. However, even in these areas the fall in energy use was significantly less than the improvement in energy efficiency, while global energy use continued to grow, crossing the 1.3 billion tonnes of oil equivalent mark in 2014.

The idea that energy efficiency technologies could massively reduce macroeconomic energy demand re-emerged in the 1970s as part of the campaign against nuclear power, in the shape of the simplistic argument (still heard today, as noted before) that greater energy efficiency would remove the need for nuclear new build. In 1988 Lovins said: 'The concept of a nontrivial rebound effect is without basis in either theory or experience. It is, I believe, now widely accepted to be a fallacy whose tedious repetition ill serves rational discourse and sound public policy.'

Although the claim soon became a mantra among the denizens of Big Green pastures it met with concerted scepticism from the academic world. A 2007 study from UKERC (UK Energy Research Centre)[15] reviewed over 500 studies in the field and concluded:

'Rebound effects have been neglected by both experts and policymakers – for example, they do not feature in the recent Stern[16] and IPCC (Intergovernmental Panel on Climate Change) [AR4] reports or in the [UK] Government's 2007 Energy White Paper. This is a mistake. If we do not make sufficient allowance for rebound effects we will overestimate the contribution that energy efficiency can make to reducing carbon emissions.'

UKECR looked at both 'direct' and 'indirect' rebound effects. Direct effects relate to individual energy services such as heating, lighting and refrigeration and are confined to the energy required to provide that service. Improved energy efficiency will decrease the marginal cost of supplying that service and should therefore lead to an increase in consumption of the service, at least unless demand is already saturated. For example, consumers may choose to drive farther following the purchase of an energy efficient car because the cost per kilometre has fallen. The resulting increase in energy service consumption will tend somewhat to offset the expected reduction in energy use that might be expected by a crude 'fifth fuel' type of argument. They concluded that these direct effects were probably of the order of 30 per cent, i.e. an improvement in energy efficiency of 10 per cent would typically result in a 7 per cent reduction in energy use. 'Indirect' effects derive from two sources: the energy required to produce and install the measures that improve energy efficiency, such as thermal insulation; and the indirect energy consumption that results from such improvements. So the former relates to energy consumption that occurs prior to the energy efficiency improvement while the second relates to energy consumption that follows the improvement. Here UKERC looks at 'computable general equilibrium' models of the macroeconomy which simulate the full range of mechanisms responsible for rebound effects. These models were, and are, at a relatively early stage of development, being perhaps unlikely to be any more accurate than say models of the climate. Only a handful of studies had been published using these resources. However, the lowest value yielded by these models was a rebound effect of 37 per cent, with half of the models suggesting that when the widest effects of energy efficiency measures were taken into account the rebound effect could exceed 100 per cent – in other words, the overall consequence of the installation of energy efficiency measures was to boost energy use, not to reduce it. (Historically, this has patently been the case concerning, for example, lighting and air travel.) In a similar vein a 2009 Cambridge study suggested that if the International Energy Agency's recommendations for efficiency measures were followed in full in the following decades, the total rebound effect globally could be 31 per cent by 2020 and about 52 per cent by 2030.[17]

The Breakthrough Institute reviewers concluded:

For every two steps forward we take with below-cost efficiency, rebound effects mean we take one or more steps backwards, sometimes enough to completely erode the initial gains made. Rebound and backfire [the term they use for efficiency effects which result in an increase in energy use] could be mitigated through raising the price of energy. However, given the tight relationship between energy consumption and economic growth, climate change mitigation must focus on cutting the relationship between energy consumption and emissions, which means moving to low-cost, zero-carbon energy sources.

Several studies suggest that increasing the price of energy at the same rate at which energy efficiency improvements are achieved can deliver genuine reductions in energy use, although UK experience with the 'fuel price escalator' whereby petrol duty was increased by firstly 3 per cent, then 6 per cent, and then 1 per cent before being effectively abandoned, suggests that such measures can be difficult to sustain politically.

As previously mentioned, critics of the rebound theorists often argue that there is a ceiling to the demand for energy use – that 'nobody will vacuum more because their vacuum cleaner is more efficient', or 'people won't drive their Prius much more because petrol costs per mile come down'. Undoubtedly there is truth in such an argument but it may be simplistic in at least four ways.

First, even in developed countries demand in some areas of life may not be saturated. If improving energy efficiency is advocated as a way of tackling fuel poverty, for example, then it is surely the case that the main effect of improving insulation, say, at least in the first instance would involve people using the same amount of energy – as much as they could afford – but enjoying a warmer house in the winter (and perhaps a cooler house in the summer) as a result. This is clearly of great social benefit for the families involved but would significantly reduce the degree of energy saving compared to the improvement in energy efficiency. A related observation is that of the improvements in efficiency of car engines in the US between 1980 and 2006, estimated at some 60 per cent, three-quarters of the benefit went on improved performance, i.e. bigger and more powerful cars, with the result that the average amount of petrol used per mile driven fell by only 15 per cent.[18]

Secondly, and more importantly, although reducing the cost of running the car may not result in a great increase in miles driven it would liberate money which the car owners could spend on other goods or services – another overseas holiday, more electrical equipment, exotic imported foods or a new car sooner than they could otherwise have afforded – each of which would have energy implications in manufacture, delivery and/or use.

Thirdly, the highest rebound in energy use from efficiency may occur in the energy-intensive sectors of the economy (industry and commerce), which consume about two-thirds of the energy used by a typical developed economy. As already argued, improving the energy efficiency of a steel plant, for example, may result in lower prices of and therefore greater demand for steel, thereby creating higher economic growth and driving significant rebound in energy use.

Fourthly, in many developing countries energy demand is far from saturated. Worryingly, the effort to help countries such as China to adopt the most energy efficient technology may therefore backfire. Clearly such measures are good for the Chinese economy, but the scope for individuals using much more energy may well result in rebound effects considerably greater in these countries than in developed economies, as increases in the supply of energy and energy services are key drivers and enablers of economic growth. Efficiency efforts in rapidly developing nations will improve economic welfare but may ultimately increase rather than decrease energy usage over the levels that would have developed in the absence of aggressive energy efficiency measures. The history of the former Soviet bloc, where chronic energy inefficiency caused in part by heavily subsidised energy prices was a major barrier to economic development, is a case in point.

So efficiency measures reduce the price of the goods and services, leading to a number of potential behaviour changes among energy users.

- They may consume more of an energy service as its price falls.
- Producers may rearrange the services and goods they consume by substituting now cheaper energy services for other goods and services (i.e. materials, labour or capital).
- Any remaining savings in energy costs may be spent elsewhere, increasing demand for goods and services that in turn rely on energy to provide.
- Energy efficiency upgrades themselves require energy to produce and install.
- At a macroeconomic level, energy efficiency improves the productivity of the economy, promoting economic growth and in turn driving up energy demand.

Gradually the importance of rebound effects is becoming recognised despite the lack of interest shown by Big Green, which still of course tends to dismiss them. For example, in its Fifth Assessment Report of 2014 the IPCC, which previously had effectively ignored the phenomenon, said: 'Rebound effects cannot be ignored. They can erode roughly 20 to 60 per cent of the energy savings from cost-saving energy efficiency measures in rich, developed countries and rebounds are likely larger in emerging economies with a greater appetite for energy.'[19] The report accepts that the majority of studies show rebound effects for end-use energy services such as heating, cooling, and lighting in the region of 20 to 45 per cent, meaning that efficiency measures achieve only 55 to 80 per cent of their original desired purpose. For transportation the IPCC notes that 'there are some studies that support higher rebounds,' with one study finding rebounds in transportation eroded more than half of the original energy savings.

Similar acknowledgments have come from the Environment Directorate-General of the European Commission[20] which noted that the evidence and stakeholders showed clear recognition of the rebound effect's existence. (The Commission adds a rather strange statement that the fact that the rebound effect could be difficult to quantify might explain why it was simply ignored in a number of studies purporting to offer advice on policy responses to climate change, e.g. from the IPCC and the Stern Report. Similar difficulties in quantifying the expected temperature rise from various levels of atmospheric greenhouse gas concentrations would hardly be viewed as an acceptable reason for ignoring the possibility of climate change.) In the EC's view, claims by the likes of Lovins that rebound effects are generally small or non-existent were unjustifiable, as were suggestions that rebound effects were often above 100 per cent. It noted that direct rebound for household energy services in the developed world probably lay in the 10 to 30 per cent range, while the direct rebound effect of fuel efficiency in commercial transport was 30 to 80 per cent because, as discussed earlier, fuel efficiency lowers the cost for freight transport, making cost-efficient transportation possible for more goods over longer distances and more frequently.

But a search of the Greenpeace and Association for the Conservation of Energy websites reveals no reference at all to rebound effects. (Friends of the Earth makes a single reference to 'rebound effects where despite technological innovations in factors like efficiency the overall impacts of consumption actually increase' but offers no analysis or discussion of implications.) Once again a very significant but inconvenient truth, one which is of no use to Big Green's world view, is roundly ignored. Rather than seeking to engage with the questions here posed – what has driven the association between improved and increased energy use since the start of the industrial revolution and more importantly why should we believe that things will be different now rather than, for example, when Amory Lovins was making the same claims 30 years ago – Big Green seems to prefer simply to pretend that the issue does not exist in the hope that it will go away. In the early 2000s it looked like they might get away with that; this seemed much less credible in the early 2010s as higher power prices brought energy efficiency under more scrutiny.

Of course, for supporters of economic growth 'below-cost' energy efficiency remains a very attractive policy, creating jobs and improving material wellbeing and wealth, perhaps especially in the developing countries. Rebound rates below 100 per cent still deliver some reduction in energy use over the level which it would otherwise have reached. But in the absence of some very strong arguments to the opposite, including a reason for believing why such well-established laws of economics should suddenly stop applying, it is cloud cuckoo land to imagine that global energy use will fall significantly, if at all, over the next decades. The opposite is overwhelmingly more likely and any responsible plans to steer the world through the difficulties of the near and medium-term future must accept this simple fact. Such acceptance would remove one of the fictitious claims which has worked against investment in necessary power production facilities.

ENDNOTES

1 http://www.greenpeace.org.uk/climate/solutions/energy-efficiency, Greenpeace (2007), 'Energy efficiency'.
2 http://www.ukace.org/category/perspective/articles-and-blog/ , Warren A, 'Blog', Association for the Conservation of Energy

3 http://www.foe.co.uk/resource/briefings/plan_cbe_report.pdf, Friends of the Earth (2012), *A plan for clean British energy.*

4 http://www.decc.gov.uk/assets/decc/11/cutting-emissions/5776-capturing-the-full-electricity-efficiency-potentia.pdf, McKinsey & Co. (2012), *Capturing the full electricity efficiency potential of the UK* – draft report.

5 http://www.emeraldinsight.com/doi/abs/10.1108/eb056783?journalCode=im, Snobel A (1977), 'Energy – save it or create it?', *Industrial Management,* **77** (9) 21–23.

6 http://www.tvcream.co.uk/?p=14320, 'Energy savings campaigns', *TV cream.*

7 http://www.ons.gov.uk/ons/rel/environmental/environmental-accounts/2011/energy-consumption.html, Office for National Statistics (2010).

8 Jevons W S (1865), *The Coal Question* (2nd ed.), London: Macmillan.

9 http://www.lowcarbonpathways.org.uk/lowcarbon/publications/TP_Historical_Analysis_Workshop_Pearson. pdf, Pearson P (2010), 'Past and prospective UK energy transitions: insights from historical experience'.

10 http://ec.europa.eu/clima/policies/strategies/2020/index_en.htm, Europa, 2014, 'The 2020 climate and energy package'.

11 http://www.europarl.europa.eu/sides/getDoc.do?type=REPORT&reference=A7-2011-0219&language=EN, 'Report on the analysis of options to move beyond 20% greenhouse gas emission reductions and assessing the risk of carbon leakage', European Parliament, 1 June 2011.

12 http://www.nytimes.com/2009/06/29/us/politics/29climate-text.html, 'Interview with President Obama on climate bill', *New York Times,* 28 June 2009.

13 http://thebreakthrough.org/blog/Energy_Emergence.pdf, Jenkins J, Nordhaus T and Shellenberger M, 'Energy emergence – rebound and backfire as emergent phenomena', Breakthrough Institute.

14 http://www.economicshelp.org/blog/5887/economics/uk-labour-productivity/, Pettinger T (2014), 'UK labour productivity', *economicshelp.*

15 http://www.ukerc.ac.uk/programmes/technology-and-policy-assessment/the-rebound-effect-report.html, Sorrell S (2007), 'The rebound effect – an assessment of the evidence for economy-wide energy savings from improved energy efficiency', UKERC.

16 Stern N (2006), *Stern review – the economics of climate change,* HM Treasury.

17 Barker T, Dagoumas A and Rubin J (2009), 'The macroeconomic rebound effect and the world economy', *Energy Efficiency,* **2** (4) 411–427.

18 http://newsoffice.mit.edu/2011/cars-on-steroids-0104, Dizikes P (2012), 'The case of the missing gas mileage', MIT.

19 http://mitigation2014.org/, 'Climate change: mitigation of climate change', IPCC Working Group III (2014), Assessment Report 5, IPCC.

20 http://ec.europa.eu/environment/eussd/pdf/rebound_effect_report.pdf, Maxwell D et al. (2011), 'Addressing rebound effects', European Commission DG ENV.

6 | ENERGIEWENDE

The Energiewende, or clean energy transition, is the term used for the fundamental shift in Germany's energy supply.[1] The stated aim is for the current energy system, dominated by coal, oil, gas and nuclear power, to be replaced by a new energy supply based on renewables, mainly wind and solar but also hydropower, biomass and geothermal.

Energy systems, of course, are always in a state of transition, or at least flux. Germany's energy transition can be traced back at least as far as the 1991 'Electricity Feed-In Act', which mandated grid operators to pay a premium of 80 per cent of average historical electricity retail prices as feed-in tariffs for electricity generated by certain renewable sources. Furthermore, and crucially, it required electricity suppliers to accept any renewable electricity fed into the grid, often referred to as 'must-take' contracts. In other words if the wind happened to be blowing at a suitable speed suppliers would have to take that output, thereby forcing the shutdown of other more reliable power sources, or pay the wind generating company to stop generating. The very considerable financial consequences of this arrangement are not generally apportioned to the renewable sources themselves and the result is rarely described as what it is, i.e. a very large effective subsidy, albeit at the cost of the electricity consumer not the taxpayer. Nonetheless, as discussed in chapter 4, this unfair grid access, referred to by Big Green as 'fair grid access', remains probably the most important distortion which supports variable renewable output in most markets.

German public opinion on nuclear power has historically been among the least supportive in the nations which adopted the technology. Radioactive waste management in particular has proved an issue of enormous controversy, provoking large-scale and sometimes violent public demonstrations. Early in the chancellorship of Gerhard Schröder, who led an SPD (Social Democrats)/Green Party coalition federal government from 1998 to 2005, Germany had adopted a nuclear phase-out policy, to be completed by 2022. In 2000 the 1991 Act was replaced by the Renewable Energy Act (EEG).[2] In addition to supporting the nuclear phase-out it was also designed to replace coal- and gas-fired generation with renewables. Renewable plant operators were to receive a 20-year, technology-specific, guaranteed payment for their electricity generation, although over time the feed-in tariff would reduce to stimulate innovation.

Any excess costs would once again be picked up by the electricity consumer. However, it was made clear to the German people that this would not matter. In 2004 Jürgen Trittin, the Green Party's environment minister in the coalition government, promised: 'payments for renewable energy will cost each household on average around €1 per month – that's as much as one scoop of ice cream.'[3] Oddly, Big Green rarely gives this relatively recent comment as much publicity as it does Lewis Strauss's 'too cheap to meter' reference to nuclear fusion fifty years previously.

The term 'Energiewende' emerged as the German response to the Fukushima accident. While most European countries with nuclear plants had reversed their anti-nuclear stances by the end of the 2000s and had announced intentions to build new nuclear plants, Schröder's successor Angela Merkel announced a 12-year 'suspension' of the phase-out policy only in late 2010, less than six months before the Japanese accident. The Fukushima emergency began on 11 March 2011, just two weeks before important state elections in Germany in which the Greens were offering a serious challenge to the Christian Democrat Party (CDP)-led government of Mrs Merkel. A media frenzy over the accident, quite different, for example, from the more measured approach taken by the media in the UK, put heavy pressure on Merkel to step away from her supportive position towards nuclear power. On 27 March 2011 the CDP lost control of the industrial state of Baden-Württemberg, which it had ruled for 58 consecutive years, to the SPD/Greens. In May it also suffered its worst result for 52 years in the city state of Bremen, falling into third place behind the SPD and the Greens for the first time. At the end of May Merkel duly announced that Germany would readopt the 2022 deadline for phase-out of its nuclear plants, closing the oldest seven of the 17 operational plants immediately. However, Merkel said: 'not only do we want to renounce nuclear energy by 2022, we also want to reduce our CO_2 emissions by 40 per cent and double our share of renewable energies from about 17 per cent today to 35 per cent', asserting that Germany's energy policy would be safe, reliable and independent from imports, with affordable prices for both consumers and industry.

These were lofty goals – indeed, if they could be achieved simultaneously merely by chancellorial decree one might ask why the policy had not been pursued previously, Fukushima or no Fukushima. Targets were set for both renewable energy output and reductions in energy consumption (see table 6.1), although the latter was sometimes confused with 'improvements in energy efficiency' which is quite a different matter, as discussed in the previous chapter.

- Renewable energy: the electricity supply was to consist of at least an 80 per cent share of renewable energies by 2050, with intermediate targets of at least 35 per cent share by 2020, at least 50 per cent by 2030 and at least 65 per cent by 2040.
- Energy efficiency: energy consumption was to be reduced by 10 per cent by 2020 and 25 per cent by 2050 compared to 2008 levels.

Both the phase-out of nuclear energy by 2022 and the specific targets for renewable energy development in the electricity sector were fixed by law.

Table 6.1 Targets for renewable energy output and reduced energy consumption, 2015–50

	Reduction of greenhouse gases compared to 1990	Renewable energies		Energy efficiency			Nuclear	
		Minimum share of energy consu-mption	Minimum share of electricity consumption	Reduction in primary energy consump-tion compared to 2008	Reduction in electricity consump-tion compared to 2008	Reduction in energy consump-tion for transport compared to 2008	Reduction in consum-ption of electricity compared to 2008	Reduction in nuclear pro-duction compared to 2010
2015								-47%
2017								-54%
2019								-60%
2020	-40%	18%	35%	-20%	-20%	-10%	-10%	
2021								-80%
2022								-100%
2030	-55%	30%	50%					
2040	-70%	45%	65%					
2050	-80% to -95%	60%	80%	-50%	-80%	-40%	-25%	

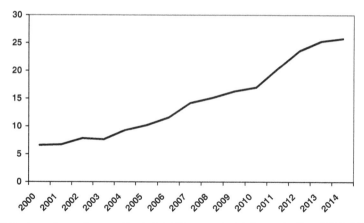

Figure 6.1 Percentage share of renewables in total German electricity production, 2000–14[4]

German experience concerning renewable energy on the one hand and energy consumption on the other had been very different. Generation of electricity from wind, solar, water and biomass had been increasing steadily since 1990, crossing the 25 per cent in 2013, as shown in figure 6.1.

However, Germany burned more lignite, possibly the dirtiest fuel of all, in 2013 than in any year since 1990 and consolidated its position as the world's largest producer, accounting for about one-sixth of global production. In 2012 two new lignite-fired power stations were connected to the grid, with a combined output of 2.7 GW. Hard coal's share also rose slightly between 2010 and 2014. In total over 10 GW of new coal-fired capacity began operating between 2012 and 2015. Coal imports in 2014 were the highest on record, at 67.2 million tonnes.[5] Figure 6.2 indicates the associated increase in the use of coal.

The main change was in the market share of nuclear power which halved over the period. The seven nuclear stations which were closed between 2011 and 2014 amounted to some 8.5 GW while the nine stations which remained in operation at that date had a combined capacity of 12 GW.

The importance of portraying the right message was illustrated by the way the figures were presented. By lumping all the renewables together while treating coal as two different fuels, Agora Energiewende – a consultancy set up to 'support' Energiewende – was able to claim that 'renewables win first place in German electricity generation', as shown in figure 6.3. Separating the renewables would give a rather different picture, as indicated in figure 6.4, as would combining the two types of coal.

German policy since Fukushima, then, has effectively been to preserve the market share of brown coal (lignite) and other fossil fuels rather than to retain nuclear plants. The period from 2011 onwards also saw a significant switch from gas to coal. As can be seen in figure 6.5, in 2010 gas was generating 14.1 per cent of Germany's electricity, falling to 10.5 per cent in 2013, while hard coal's share grew from 18.5 to 19.6 per cent and lignite's from 23.0 to 25.6 per cent. Some advocates of renewables argued that this was solely as a result of falling

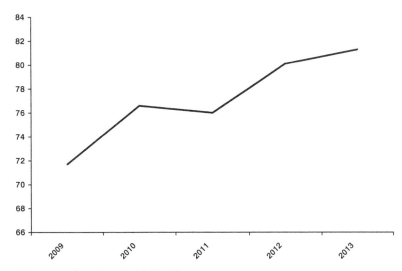

Figure 6.2 German coal use (mtoe), 2009–13[6]

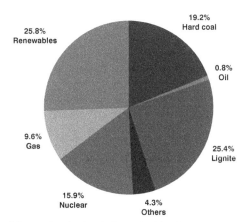

Figure 6.3 German electricity production by fuel type in 2014[7]

coal prices but others accepted that the greater flexibility of gas-fired plant and the higher proportion of their cost which is accounted for by the fuel means that variable renewable output differentially displaces gas rather than coal. This results in the average carbon emissions per unit of power generated by non-renewables or nuclear sources increasing.

This growing dependence on coal resulted in the peculiar spectacle of Germany, seeking to portray itself as a nation that took climate change seriously, begging Sweden not to pursue its national policy of divesting itself from coal. Sigmar Gabriel became leader of the SPD in 2009, and vice chancellor and economics minister in 2013, having previously held the environment portfolio in Merkel's first cabinet from 2005 to 2009. As an early supporter of the International Renewable Energy Agency his broad views on the matter are not a secret. However, in late 2014 he warned Sweden's new prime minister,

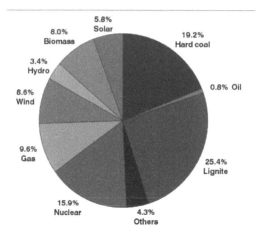

Figure 6.4 German electricity production by fuel in 2014 (renewables treated separately)

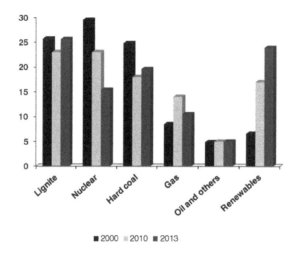

Figure 6.5 Percentage market share for different sources of electricity in Germany: 2000, 2010, 2013

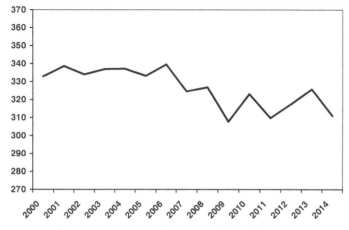

Figure 6.6 Germany primary energy consumption (mtoe), 2000–14[10]

Stefan Löfven, that there would be 'serious consequences' for electricity supplies and jobs if Sweden's state-owned utility Vattenfall abandoned its plans to expand two coal mines in the north-east of Germany.[8] Undoubtedly negotiations on the matter were complicated by the fact that Vattenfall had been forced to file a lawsuit against Germany seeking €4.7 billion in compensation over Berlin's decision forcibly to close the company's nuclear assets in Germany.

Somewhat hypocritically, in November 2014 Greenpeace activists dumped eight tonnes of coal in front of the German Ministry for Economic Affairs and Energy to protest at the growing use of coal, demanding Berlin shut down all German coal-fired power plants. This may appear akin to their forcing someone to throw half a dozen eggs off the top of the Fernsehturm, then dumping a big omelette outside the German Ministry of Food and Agriculture for allowing the eggs to break. Gabriel responded: 'When Greenpeace and others demand that all German coal power plants be shut down it is hardly more than mass disinformation of the people which will not lead to a single tonne of carbon dioxide being saved for the global climate.'[9]

In contrast to the growth of renewables, energy consumption, targets for which were not enshrined in law, made somewhat less headway, being just 6.5 per cent lower in 2014 than in 2000 and aided by a particularly warm winter – see figure 6.6.

6.1 ENERGIEWENDE – AN EVALUATION

It is difficult to exaggerate how important perceptions as to whether Energiewende is a success or not have been and will be in the energy field. Certainly there are powerful vested interests in the fossil fuel, nuclear and centralised electricity utility sectors which would not be heartbroken if it failed. Equally, Big Green and the renewables industry cannot contemplate failure or allow any actual failure to become recognised. If a renewables-first low-carbon economy cannot be created in Germany, a rich country surrounded by many other national grids on which it can dump excess output or suck in power at times of shortage, with a united political establishment prepared to put electricity consumers' money where its mouth is, then one strongly suspects that it cannot be created anywhere.

It is unchallengeable that Energiewende involves some very large companies making very large profits on the back of many poorly off electricity users. Further, even if the longer-term targets for renewables should eventually be met, the Germans will still be using very large amounts of coal-fired capacity well into the 2020s. Had they chosen to run their nuclear stations to the end of their economic and technical lives instead of phasing them out early, very many lives would have been saved both in coal mines and by avoiding the detrimental effects of particulates and other emissions on human health, while significant volumes of greenhouse gas emissions would have been avoided.

In normal circumstances such policies would be expected to bring Big Green to the barricades in defence of 'the people' against 'industry', especially if they were accompanied by a series of broken promises and targets. But Energiewende does not conform to the laws of politics any more than it is consistent with the laws of physics. Big Green does not seem even to have attempted to explain why it believes that increasing numbers of deaths and increasing levels of emissions would be a better outcome than allowing the nuclear plants to continue to operate until the end of their technical lives.

A typical Big Green take on the whole project comes from Amory Lovins, some of whose earlier pronouncements are examined in the chapter 5.

The March 2011 Fukushima accident destroyed four and closed all of Japan's 54 nuclear plants. Japan replaced nuclear energy with discomfort, sacrifice and costly fossil fuels because utility oligopolies suppressed renewable competitors and national energy efficiency languished. Two and a half years later, power reserves, though easing, remain tight, fuel bills exorbitant and carbon emissions elevated. In contrast, Germany, the world's fourth-largest economy, launched a purposeful, careful and successful energy transition in 1991 and agreed in 2000 to shut down nuclear power over a couple of decades. In 2011 all political parties agreed to shut down the oldest 41 per cent of Germany's nuclear capacity immediately, the rest by 2022. But unlike Japan, Germany offset its entire 2011 nuclear loss in the same year – 60 per cent by added renewable power – remaining a net exporter of competitive electricity. In 2011 Germany's economy grew by 3 per cent and stayed Europe's strongest, boosted by 382,000 renewable energy jobs, many for exporters. Power reliability remained the best in Europe, about ten times better than America's. Efficiency gains (plus a mild winter) cut Germany's 2011 energy use [by] 5.3 per cent, electricity consumption 1.4 per cent, carbon emissions 2.8 per cent and wholesale electricity prices 10 to 15 per cent.

Repeating 2011's renewable additions for a few more years would displace Germany's entire pre-Fukushima nuclear output. With one-ninth Japan's high-quality renewable resources per hectare, Germany achieved nine times Japan's renewable share of power generation. How? By giving renewables fair grid access, promoting competition, weakening monopolies and encouraging citizen and local ownership (now two-thirds of renewable capacity, which rivals peak demand). Germany's pump-priming investments triggered global scaling-up that Deutsche Bank predicts by mid-2015 will let solar power compete without subsidy in threequarters of global markets. With 30GW of solar rooftops German installed prices last year were half those of America.

In 2012 Germany's nuclear generation reached a 20-year low while net power exports hit a new high. Only Germany consistently exports net power to France. Renewables rose to 23 per cent of generation, passing every rival except lignite (soon to fade as its subsidies end). Wholesale prices plummeted 30 per cent in two years to near 8-year lows, attracting energy-intensive industries. Real GDP, damped by the Eurozone crisis, grew just 0.7 per cent but electricity use fell 1.3

per cent. Power plants and industries emitted no more carbon. Weather-adjusted total German carbon emissions fell, though a frigid winter raised absolute emissions 1.6 per cent.

Giant utilities, their profits squeezed and business models upended by the renewables they bet against, mounted a vigorous disinformation campaign. Critics claimed Germany had replaced lost nuclear power with coal. Actually, the brief and modest coal upturn was due to pricier gas and a flood of American coal displaced by cheaper gas in the United States (and, nearly twice as importantly in 2012, more efficient energy use). German coal-burning remained below its 1990–2007 high and no new coal plants were ordered.

Critics claimed German industry was fleeing (it wasn't) and renewables were destabilising the grid (they're not). Big German industries continued to enjoy highly reliable and ever cheaper electricity: their political patrons generously exempted them both from grid fees and from paying for the renewable expansion that slashed wholesale power prices. This favoritism modestly raised the renewable surcharge on households (whose bills are half taxes), inspiring the absurd election-season fiction that renewables have made electricity a luxury good and tipped Germany's poor into energy penury. All untrue – yet on 18 September 2013, even *The New York Times* echoed *Der Spiegel*'s breathless fabrications.[11]

Lovins goes on to speculate, or as he calls it 'prove', that a US economy 2.6 times its current size could eliminate coal, oil and nuclear energy and reduce natural gas use by one-third, treble energy efficiency, shift from one-tenth to three-quarters renewable supply, emit 82 to 86 per cent less carbon, and save $5 trillion. This could all be achieved without new inventions or Acts of Congress, the transition led by business for profit. That blueprint (Lovins cites one of his own books in support) and the European Climate Foundation's similar Roadmap 2050 show how climate change, energy insecurity, energy poverty and nuclear proliferation are artefacts of not choosing the best buys first.

In fact it is not clear, for example, that lignite is subsidised in Germany[12] – hard coal is and those subsidies are expected to end in 2018 – but in any case in 2013 Germany imported almost half of its coal requirements. But leaving such details aside, clearly if Lovins's description is basically correct – that this one policy is delivering on massive financial savings for all involved and improving security of electricity supplies, with a slashing of greenhouse gas emissions thrown in – then there really is little room for argument. As it will be able to compete without subsidies, governments could remove all support for solar power, for example, and sit back because the huge monetary bonanza from cheap renewables would be all the incentive that anyone needed. As another renewables advocate, Bob Johnstone, put it: 'The role of government is merely to ensure that producers of clean energy are appropriately compensated: not a penny of taxpayers' money is involved. This appeals not just to people who want to save the planet but also to people who want to make money.' The implication that those who pay the enormous subsidies through their power bills are not people to whom making, or saving, money is important is an interesting one for which Johnstone offers no supporting evidence.

How soundly based are such views? Sadly, a review of the first three years of Energiewende does not reveal such an emerging paradise. Let us look at the main claims separately.

- Costs – has Energiewende indeed cost no more than a scoop of ice cream or a little more?
- Greenhouse gas emissions – have these been minimised or have they increased more than would have been necessary had different policies been pursued?[13]
- Grid stability – has this proved a non-issue or is it a bit more complex?

6.2 COSTS

Power markets are complex beasts. As discussed in more detail in the section on the full costs of electricity generation in chapter 4, it is highly misleading, for example, to look at the wholesale price of power in a particular market and presume that it determines the price being paid by consumers.

Figure 6.7[14] confirms that wholesale power prices in Germany have been falling. Advocates of the Energiewende, and of renewables in general, often state something along the lines: 'The lower wholesale electricity price in Germany results from developing indigenous renewable power resources'[15]

They are right. However, such analyses are of little consolation to the electricity consumer in Germany's homes, to whom the size of the monthly power bill is the main consideration. In the decade and a half following the introduction of the EEG, household power bills in Germany doubled, German power prices becoming the second highest in Europe after Denmark – which also suffers from high levels of wind power.

Danish statistician and political scientist Professor Bjørn Lomborg is not regarded by Big Green as a reliable commentator on Green issues. His celebrated and notorious 2004 book *The Skeptical Environmentalist* was met with a failed attempt to have it declared flawed and misleading by the 'Danish Inquisition', in the form of the Danish Committees on Scientific Dishonesty. (Oddly the DCSD, in the words of the Danish government, attacked Lomborg's accuracy in general but 'did not provide specific statements on actual errors' and 'did not document where Lomborg has allegedly been biased in his choice of data and in his argumentation', adding that the negative ruling on Lomborg's work was 'completely void of argumentation for why the DCSD find that the complainants are right in their criticisms of [his] working methods. It is not sufficient that the criticisms of a researcher's working methods exist; the DCSD must consider the criticisms and take a position on whether or not the criticisms are justified, and why.')[16]

Such differences of opinion notwithstanding, Lomborg argues for an outcome which Big Green also sometimes claims to desire – access to the benefits of energy for the more deprived members of society.

The German government recently said that 6.9 million households live in energy poverty, defined as spending more than 10 per cent of their income on energy. This is partly a result of Energiewende. This year alone (2013) German consumers are expected to subsidise green energy to the tune of a whopping €23.6 billion on top of their normal electricity bills for the so-called 'renewable energies reallocation charge'. Since 2008, this charge has increasingly reallocated money from the poor to the rich, e.g. from poor tenants in the Ruhr area to wealthy homeowners in Bavaria who put solar panels on their roofs. The charge has skyrocketed from 1.15€¢ per kWh in 2008 to 6.24€¢ per kWh this year. Since then, another 1.4 million households slipped into energy poverty. German consumers have already paid €109 billion for renewable energies since 2000, with greater costs looming on the horizon. Between 2000 and 2013, real German electricity prices for households have increased 80 per cent. About one-quarter of household electricity costs now stem directly from renewable energy. [17]

The 'breathless' *Der Spiegel* article to which Lovins refers (as mentioned in the previous section) develops this point,[19] with reports of consumer advocates and aid organisations saying that more than 300,000 households a year were having their power disconnected because of unpaid bills. Once the power has been interrupted it is a lengthy and costly business to have supplies restored. Customers have to negotiate a payment plan and are also charged a reconnection fee of up to €100. By 2014 an average three-person household was spending around €90 a month for electricity, roughly double the figure in 2000 –

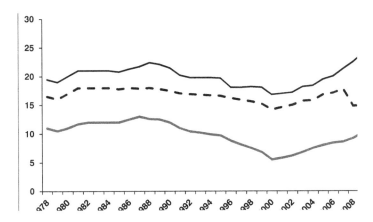

Figure 6.8 Electricity prices for households in Germany, 1978–2013 (2013 eurocents per kWh)[18]

gourmet ice cream indeed – with two-thirds of the price increase being accounted for by new government fees, surcharges and taxes. Government pensions and social welfare payments had not been adjusted to reflect these changes.

Der Spiegel, citing problems with the Borkum offshore wind farm, noted that by 2020 offshore wind turbines were required to be generating up to 10 GW of electricity if the whole Energiewende project were to remain on course. To attract investors the government was offering €190 per MWh, some 50 per cent more than onshore wind, with the government also assuming the liability risk. Furthermore, since there was little demand for electricity in thinly populated coastal regions new high voltage power lines would be needed to transport the energy to industrial centres in western and southern Germany. The German government estimated the costs of expanding the grid at €20 billion, excluding the additional ocean cables for offshore wind power. Big Green put an interesting spin on this.

> As long as nuclear remained an option for the Big Four power companies, a new grid was neither necessary nor desirable. Now the matter is urgent, particularly in light of the fact that in 2017 further nuclear reactors will go offline. Over the next decade, more than 2,300 miles of state-of-the-art, high-voltage cables will be laid at a cost of nearly $25 billion. Some of it will stretch over a hundred miles into the North Sea, a technological feat never attempted before on so large a scale.[20]

So the big power companies were in effect to be criticised for not making a vast investment which would have been unnecessary if the previous policy of large centralised power stations had been retained, thereby standing in the way of renewables. No quarter is given to the alternative view that these excess grid costs should be laid at the door of the renewables as they would not otherwise be necessary.

However it is dressed up, electricity rates for households rose from 13.94€¢ to 28.50€¢ per kWh between 2000 and 2012, a 104 per cent increase (see figure 6.9), with industrial rates increasing by 150 per cent from 6.05€¢ to 15.10€¢ per kWh. Taxes and charges rose by 171 per cent from 5.3€¢ per kWh in 2000 to what Big Green would undoubtedly in any other context call an 'eye-watering' 14.4€¢ per kWh in 2013.[21] One study for the federal government suggested that electricity would cost up to 40€¢ per kWh by 2020, some 40 per cent above 2012 prices.

The apparent paradox – that Energiewende is resulting in both much lower wholesale power prices and much higher electricity bills – is easily explained. While renewable output enjoys the privilege of a guaranteed market at guaranteed prices, periods of time during which solar and wind output is high result in lower demand for dispatchable electricity. Clearly whenever supply outstrips demand then prices fall. Generators are keen to cover their short-run marginal costs if possible, even if in total they are losing money, because to stop generating would increase losses even more, unless of course they liquidated the 'asset' entirely.

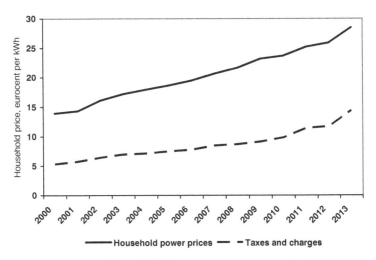

Figure 6.9 Electricity price for households in Germany (eurocents per kWh), 2000–13

What is less clear is why anyone should regard this as a good thing. Moving electricity supply out of the competitive part of the generation market to the subsidised part inevitably raises bills. When supply outstrips demand by a certain amount, electricity becomes a good of negative value owing to the damage the excess energy could cause to domestic or industrial appliances or to the transmission network. Prices on the intraday energy spot market in Germany regularly fall well below -€50 per MWh at times of high renewable output.[22] Even the most ardent advocate of renewables must suspect that there is something strange going on when people have to be paid to take more electricity than they need or renewable generators have to be paid to stop generating. In such circumstances it is more than a little misleading to refer to the dumping of electricity into neighbouring grids at negative prices as 'increased exports'.

Perhaps even more seriously, as discussed in chapter 4, this undermines the economics of dispatchable capacity, making it more likely that existing plants will close and much less likely that new plants will be built when needed. Yet when the wind and sun are not obligingly matching their output to demand, say on a still cold winter night, those dispatchable generators are still just as necessary as they would have been had no solar cells or wind generators ever been installed. The requirement arises, then, to compensate the dispatchable generators for their more or less random loss of market if they are to remain available when needed.

Furthermore, when there is a danger of electricity shortfall intensive users such as the ArcelorMittal steel mill in Hamburg might be asked to shut down production to protect the grid. Ordinary electricity customers (or taxpayers) are then expected to pay for the compensation these businesses are entitled to for lost profits.

Der Spiegel reported that the German government was predicting that the renewable energy surcharge added to consumers' electricity bills was to increase by 20 per cent from 5.3€¢ per kWh to between 6.2€¢ and 6.5€¢ per kWh owing to uncontrolled costs of the Energiewende. 'Electricity is becoming a luxury good in Germany and one of the country's most important future-oriented projects is acutely at risk.' Trittin's successor as environment and energy minister, Peter Altmaier, admitted that consumers were actually paying enough to 'eat everything on the ice cream menu'. In 2013 solar, wind and biogas plants were paid some €20 billion for electricity with a market value of just over €3 billion. In a separate interview, Altmaier said the total cost of Energiewende could reach €1 trillion by the end of the 2030s, four times more than the Greek euro bailout of the mid-2010s. Feed-in tariffs would account for more than half of the total, while improvements to the German grid would cost between an estimated €27.5 billion and €42.5 billion[23], with over 8,000km of new or upgraded transmission lines being required, and with an associated environmental impact. Grid operators said: 'The investments required for expanding the transmission network only represent a fraction of the energy switchover's cost but they are essential for its successful implementation.'[24]

Alongside the key issue of what the actual costs would prove to be – Trittin's scoop of ice cream or something rather less digestible – was the question of why the social campaigners of the Green movement seemed so indifferent to the scourge of fuel poverty while renewable generators creamed the huge profits off the backs off poorer billpayers, beyond occasionally criticising the German government for protecting the fuel bills of Big Business at the expense of domestic consumers.

That said, although German businesses had been wary of the growing costs that Energiewende was imposing on them, consumers had been largely stoic, protesting when the government seemed to waver in its commitment even as the cost of power climbed. However, this support may not be unconditional. Grid operators pointed out that if major new power lines were not built the stress on the existing power grid – already under heavy strain owing to the swings in power between sunny, windy days when renewable energy surged and dark, still winter days when renewables were all but absent – would be enormous, pushing the country back towards more coal or, less feasible politically, nuclear power.[25] SuedLink is an 800 km long and 1 km wide transmission project, one of four high voltage direct current lines needed to carry wind-generated power from the north to the south. Once its route was announced, however, there was a major outbreak of Nimbyism, with dozens of protest groups springing up along its path. Citizens living in the areas proposed for the lines said they were worried about the effect on the value of their property and health threats posed by high voltage power lines. A protest group called BI Fuldatal, dedicated to stopping SuedLink which became known locally as the Monster Line, saw its membership grow from five to over 400 within a week of its launch. Residents from different communities all argued that the link should blight somebody else and some called for the line to go through a nature reserve in what was once the 'death strip' along the Cold War boundary between former West and East Germany.

6.3 GREENHOUSE GAS EMISSIONS AND OTHER ENVIRONMENTAL ISSUES

Q Won't switching from nuclear to natural gas increase carbon emissions?

A Yes, but if you are concerned about climate change and support the Kyoto protocol you will have to admit that Germany actually has the *right* to increase its carbon emissions since it completely blew past its Kyoto target of a 21 per cent reduction. In August [2012] the German Environmental Ministry reported that the country had actually reduced its emissions by 28.7 per cent … If you are worried about carbon emissions, no industrial country had a more ambitious target than Germany, lots of countries (like the US) did not sign on to the Kyoto Protocol at all, and almost all of those who did missed their targets (like Canada).[26]

Bearing in mind that in the 2010s Germany's carbon emissions per head were some one-third higher than those of the UK or the EU as a whole and double the global average one might think the author of such a statement was on dangerous ground. However, a variety of players might have an interest in getting such an argument into play.

First suspects might be the companies using lignite. There were many examples illustrating how the continuing rise of brown coal in recent times had had environmental effects well beyond increased greenhouse gas emissions. Vattenfall planned to relocate the community of Atterwasch, a small village near the Polish border, in order to create a major new network of opencast lignite mines.[27] Similarly more than 800 residents were to be removed from Proschim in the former East Germany for another lignite complex:[28] Proschim was a particularly ironic example as about one-third of its houses had been fitted with solar roof panels, several fields boasted wind generators and one of the local farms had started converting agricultural waste into biogas. The Garzweiler lignite complex in North Rhine–Westphalia (see figure 6.10) was named after a town which was demolished in 2003; the first phase of mining ran from 1983 until it was exhausted in 2006, when work then started on phase II. At that time, the mining was expected to continue until 2045.

As one report put it: 'Germany, traditionally seen as one of the leaders in fighting carbon dioxide emissions as it tries to wean itself from fossil fuels and nuclear power, seems to have grown tired of leading by example and it is about to erase entire towns off its map because they are sitting on vast coal deposits.'

Another possibility could be a member of the Conservative-led German government, becoming increasingly embarrassed by the direction in which Energiewende was taking the country. Sigmar Gabriel said: 'We need strategic reserves of gas and coal power for the times when the wind doesn't blow and the sun doesn't shine.' In October 2014 a spokesperson for the Ministry of Economics said: 'For a country like Germany with a strong industrial base, exiting nuclear and coal-fired power generation at the same time would not be possible.'

Alternatively, it may have been a representative of some major German industrial company, horrified at escalating energy costs and seeking to damp down criticism of the practical outcome of the nuclear phase-out renewables in case the government imposed even higher taxes to support yet more renewables.

Whatever the source one would surely expect a howl of protest from German Big Green at such cavalier acceptance, almost advocacy, of increasing greenhouse gas emissions. However, the quote actually came from a publication from the Heinrich Böll Stiftung, described as 'The Green Political Foundation' and affiliated to the German Green Party.[29]

The German Green Party was not the only apologist for the growing greenhouse gas emissions from Germany. Other commentators within the renewable energy industry, who usually claim that greenhouse gas increases are a bad thing, seemed perfectly relaxed, indeed enthusiastic, about policies which increase such emissions as long as they supported their anti-nuclear ideologies (or commercial prospects).[30] Estimates suggested that by 2020 Germany would have produced an extra 300 million tonnes of carbon dioxide as a result of the nuclear closure, equivalent to almost all the savings that would be made in the 28 member states if the EU's energy efficiency directive were delivered.[31] Germany would still be using electricity generated with fossil fuels, whether from inside or outside its own borders, in 2030 which could be produced using its existing nuclear power stations.

The German data on greenhouse gas emissions were not encouraging. Germany's 1990 baseline emissions under the Kyoto protocol were inflated by the recent reunification of Germany, bringing with it large volumes of very inefficient and dirty communist-era East German industry. Closure of this industry and replacement of its production with cleaner West German technology gave a quick and easy reduction in greenhouse gas emissions, which fell from 1,248 million tonnes of carbon dioxide ($MtCO_2$) equivalent in 1990 to 1,041 $MtCO_2$ equivalent in 2000.[32] Indeed, more than half of the reduction in German greenhouse gas emissions between 1990 and 2013 happened between 1990 and 1994 purely because of this effect. Like the UK, which took credit for falling carbon emissions caused by a fortuitous side-effect of another policy, the 'dash for gas', Germany's claimed leadership in reducing emissions, as trumpeted by the Heinrich Böll Stiftung, actually had relatively little to do with direct policy initiatives and in the nuclear phase-out case happened despite those policies, not because of them.

Emissions continued to fall slowly until Energiewende was introduced in 2011 and then started to rise again. This was true even of the emissions from the 'basket' of greenhouse gases on which Lovins preferred to focus, although emissions of the other greenhouse gases proved much easier to reduce than did carbon dioxide emissions. For example, in the UK carbon dioxide emissions fell by 21.5 per cent between 1990 and

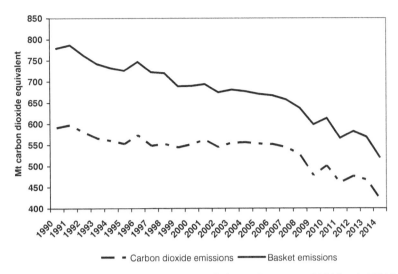

Figure 6.11 UK carbon dioxide and greenhouse gas emissions, 1990–2014 (1990 levels 591 MtCO$_2$ equivalent and 780 MtCO$_2$ equivalent respectively)

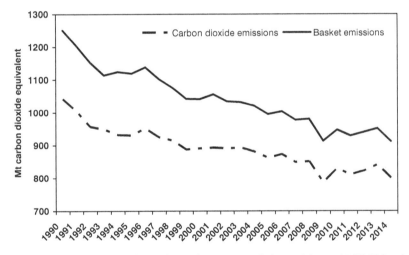

Figure 6.12 Germany carbon dioxide and greenhouse gas emissions, 1990–2014 (1990 levels 1,042 MtCO$_2$ equivalent and 1,251 MtCO$_2$ equivalent respectively)[33]

2013 but the carbon dioxide equivalent of the other five gases in the greenhouse gas basket fell by 43.8 per cent, as indicated in figure 6.11.

By contrast, the first two years of Energiewende saw German greenhouse gas emissions rise by 2.4 per cent with carbon dioxide emissions, mainly from energy production, up by 3.0 per cent, as shown in figure 6.12. Even the warm winter of 2014 was insufficient to return emissions to their 2009 level.

Agora Energiewende, despite its pro-Energiewende raison d'etre, offered a very different analysis from Lovins. It identified what it called an 'energy transition paradox', though others might call it an 'energy transition inevitability': greenhouse gas emissions rising despite an increasing share of power generation from renewables.[34] Given its commercial interests it can perhaps be forgiven for choosing to claim that the rising German emissions were not attributable to the country's decision to phase out nuclear power but were attributable to the increasing generation of electricity from lignite and hard coal instead of natural gas, as well as a lack of carbon dioxide emission reductions in the heating, transportation and industrial sectors. However, it is obvious that if the government had pursued the other measures within Energiewende but maintained its nuclear stations, instead closing lignite power plant, emissions would have continued to fall significantly. Just because the decrease in nuclear output and the increase in renewable output broadly cancelled each other out, this did not invalidate an alternative and indeed more plausible ascription that the problem had been an increase in the use of coal in place of nuclear. Coal and nuclear both occupy the baseload sector of the market to a greater or lesser extent, something which renewables cannot do, at least in the absence of large-scale storage.

Leaving aside such matters of how the issue is presented, Agora Energiewende ascribed the resumption in emissions increases from 2011 onwards to two factors. Although overall electricity consumption fell, at the same time the electricity market witnessed a shift towards lignite and hard coal power plants and away from cleaner natural gas power plants, while electricity 'exports' (or 'dumping') increased. As noted earlier, in 2012 Germany commissioned 2.7 GW of new lignite-fired capacity. Peter Altmaier proudly said: 'If one builds a new state-of-the-art lignite power plant to replace several older and much less efficient plants then I feel this should also be acknowledged as a contribution to our climate protection efforts.' A further 8 GW of hard coal capacity was to come online between 2013 and 2015. This was to be compared with the 8.3 GW of nuclear capacity permanently closed by the German government in 2011 in its reaction to Fukushima.

The second factor, the matter of reducing power consumption, as always proved easier to say than to do. Progress in reducing emissions from the building, transportation and industrial sectors largely stalled, in part owing to slow progress in the insulation of Germany's building stock.

6.4 GRID STABILITY

In 2012 *Der Spiegel* reported how, at 0300 hours on a typical Wednesday morning, machines at the Hydro Aluminium works in Hamburg suddenly shut down.[35] The rolling mill's monitoring system stopped production so abruptly that aluminium belts snagged, hitting machines, destroying part of the mill and causing damage amounting to some €10,000. The reported cause was a millisecond-long weakening of the electricity grid voltage. In the following three weeks the voltage weakened on two more occasions at the factory, each time for a fraction of second but fortunately at times when the machines were not operating. The company responded by investing some €150,000 to set up its own battery-based emergency power supply to protect itself from future damage. Joachim Pfeiffer, a Christian Democratic Union economic policy spokesman in the Bundestag, was reported as saying: 'You can hardly find a company that isn't worrying about its power supply.' A survey of members of the Association of German Industrial Energy Companies (VIK) revealed that the number of short interruptions to the German electricity grid had grown by 29 per cent over the previous three years, while the number of service failures had grown by 31 per cent percent and almost half of these failures had led to production stoppages. Company information suggested that damages caused by each instance could range between €10,000 and hundreds of thousands of euros. Pfeiffer said: 'In the long run, if we can't guarantee a stable grid companies will leave Germany. As a centre of industry we can't afford that.' Concerns also grew at the vulnerability of the grid to cyberhacking, it being relatively easy for a hacker to send messages containing incorrect information about the output from many small sources of generation.[36]

Challenges to grid security in Germany are to a certain extent mitigated by its location. Germany has interconnection with nine neighbouring countries – counting the two Danish grids separately – amounting to some 17.3 GW of import capacity and 13.9 GW of export capacity.[37] This gives it considerable flexibility in managing variable renewables by being able both to import power (if any is available) when the weather does not favour wind and solar plants and to avoid the potential damage caused by excess generation when wind and solar are generating strongly by transferring this excess electricity to its neighbours, thereby exporting to their grids the problem of potential power surges and minimising the need to shut down dispatchable capacity.[38] The UK and Ireland electricity market, by contrast, has limited options for increasing its degree of interconnection with continental Europe owing to its geographical position on the edge of Europe and the high costs of subsea interconnection, with just 3 GW in place (with France and with the Netherlands).[39] The UK and Ireland are therefore less able to manage highly variable renewable capacity.

In another rather effective piece of Big Green spin, the word used for this expulsion of unwanted excess electricity is 'exports'. So Energiewende is said to have been a boon for neighbouring countries as they benefit from the low wholesale prices that renewables bring while Germany has 'increased its exports' which of course must be a sign that things are going well.

This rosy perspective is not necessarily shared by the countries on the receiving end.

Utilities including CEZ (Czech Republic) and PGE (Poland) are on occasion forced to disconnect coal-fired plants in the western parts of these countries because of excess power flowing from Germany, causing strains both on the economic stability of their systems and on their wires. Both countries installed phase-shifter transformers in the transborder area with Germany to regulate power flows and protect their transmission networks.[40] CEPS, the Czech grid operator, took the decision to go ahead with constructing the equipment after failing to persuade Germany either to offer fair compensation for the damage it was causing or to allow its grid operators to switch off renewable sources at times of excess production.[41]

In reality the problem is that the detrimental effects of overgeneration of variable electricity on the economics of dispatchable plants, and hence the economics of the whole system, do not disappear simply because they are hidden in neighbouring grids. As with any 'exports' of goods below total cost, individuals in the receiving country may benefit in the short term but at the long-term expense of the integrity and ultimately the costs of their own industries. Here are two definitions:

- Export: to send a product to be sold in another country;
- Dumping: to sell a product in a foreign country for less than the cost of making the product.

In a functional cross-border electricity market with broadly predictable supply and demand the former definition is a perfectly good description of electricity transactions – for example, as the peak demand moves across Europe it makes sense for countries to import electricity at peak demand and to export power as its demand falls and that of its neighbours rises. Such arrangements tend to be symmetrical and mutually beneficial as the electricity has a positive value at all stages of the process. This is very different though from a country regularly having too much power for its own needs, facing problems such as surges on the grid or wires melting if it cannot remove that excess and paying an inadequate sum to land those problems on weaker neighbours, while still having the financial muscle to pay high prices for 'imports' when supply is weak. At least in December 2015 Germany had the good grace to pay Denmark to shut down some of its wind generators as prices slumped to -€117 per MWh.'[42]

Of course, a step change in the availability, cost and efficiency of electricity storage could in principle revolutionise the value of renewable output. Germany installed its first pumped storage scheme, i.e. using excess electricity to pump water uphill and then using the facility as a hydropower station when power is required, in the late 1920s at Niederwartha near Dresden. The scheme consisted of two lakes, one some 140 metres above the other. The nation now has around 7 GW of pumped storage capacity, representing up 40 GWh of storage, this being roughly equal to the energy stored in the high-level water when the upper lakes are full.

However, the German target of 60 to 80 per cent of energy from renewables by 2030 would require an estimated 8,400 GWh of storage capacity, over 200 times the 2013 capacity. Other estimates suggested that by 2050 the requirement could be 20,000 GWh to 30,000 GWh. Yet by the mid-2010s the growth of renewables – ironically given their long-term reliance on storage to manage their variability – was damaging the economics of pumped storage in the same way as it was undermining other forms of dispatchable capacity. In 2013 Vattenfall announced plans to close the two remaining operational units at Niederwartha while RWE and EnBW shelved plans to build a large pumped storage power station in the southern Black Forest and Trianel, an association of about 100 municipal utilities, withdrew from a similar project at Rursee Lake in the western Eifel Mountains. In 2014 E.ON announced that it was putting its plan to expand the company's Waldeck II pumped storage facility by 300 MW on hold: when the project had first been announced in 2010 the new capacity had been expected to be available in 2016.[43] Solar power in particular was eating into peak load hours, on which pumped storage schemes rely to cover their costs and deliver an acceptable rate of return, and narrowing the differential between peak and off-peak wholesale prices, thereby undermining the economics of the whole business.

While there is much bullish talk of a revolution in battery technology as a means of storing very large amounts of electricity it should be noted that some 99 per cent of the world's electricity storage capacity consists of pumped storage while batteries account for about 0.2 per cent of total storage capacity.[44] Interestingly even advocates of Energiewende are sceptical as to whether storage will be as cost-effective as varying the output of fossil fuel capacity or relying on neighbouring countries to accept excess output when renewables are overgenerating and provide input when the weather turns negative.[45]

6.5 REFLECTIONS

Even before 2011 German energy and environmental policy had had limited success in reducing greenhouse gas emissions compared to its neighbour France. Carbon dioxide releases from electricity and heating followed broadly the same path in the two countries until around 1980, after which the growing French nuclear fleet brought emissions down to effectively the same level or slightly below those of the early 1960s. German emissions have remained double their 1961 level, falling less than 1.5 per cent in the decade after the EEG Act in 2000 and, as noted previously, starting to rise again after the decision to close its nuclear plants – see figure 6.13.

In April 2014 Sigmar Gabriel told a solar industry meeting that the Energiewende was 'on the verge of failure'.[46] 'The truth is that in all fields we underestimated the complexity of the Energiewende. Those of you who are the engines of the transformation to renewable energies don't see how close we are to the failure of the energy transformation. The noble aspiration of a decentralised energy supply, of self-sufficiency! This is of course utter madness. Anyway, most other countries in Europe think we are crazy.'

This is not the only area in which other countries take a different view of Energiewende from the Germans themselves. A poll for *Time* magazine, which surveyed residents of Germany, US, Brazil, Turkey, India and South Korea, found that only in Germany itself did more respondents believe that Germany was part of the climate change solution rather than part of the problem and where a majority felt that Germany had a 'mostly' or 'somewhat positive' role in combating global warming.[47]

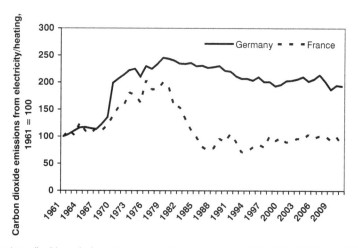

Figure 6.13 Carbon dioxide emissions from electricity and heating, 1961–2011 (1961 = 100)[48]

Given the self-righteous bluster of some of those in the Energiewende entourage it is interesting that Gabriel's comments were widely reported on anti-business websites including the 'SGT report: the corporate propaganda antidote'. In more usual circumstances Big Green might be expected to woo and support such attacks on Big Business but not on this occasion. Indeed, criticism of the Energiewende progressively spread from the political right, where it might have been expected, to the political left, for example with left-leaning Dissent Magazine publishing in 2013 a long expose of the failure of the Energiewende to reduce carbon emissions, concluding that Germany's enormous investments in renewables, together with plans to phase out its nuclear fleet, would cost the nation a generation in the fight against global warming.[49]

Big Green remains entirely unconcerned by these observations – indeed, the mantra remains that Energiewende is going well. The closest Big Green has come to criticism may have been from Greenpeace's Karsten Smid when he said: 'Germany is making itself a laughing stock because it hasn't set limits on brown coal.' But in the wider German society things may be changing – a hint perhaps of 'phase 3' thinking in the sense discussed in chapter 10. The way in which heretics are treated is often illustrative of changing societal values. In March 2014 ZDF German television broadcast a sketch parodying Energiewende and its acolytes. The piece started with the Merkel government's refusal to accept the findings of an expert commission which gives the Energiewende project a grade of F minus. At one point, one of the protagonists confirmed that the Energiewende makes no sense but claimed that this is all beside the point. It is something 'you just have to believe in! The Energiewende works only if we all really believe in it!' Another exhorted non-believers to stop questioning the policies. 'Thinking about them just gives you a headache! Just pay and be happy!' In a country which is sometimes (unfairly) derided for its sense of humour this demonstration of the decline in deference towards the Green priesthood may be a significant moment.

ENDNOTES

1 http://www.agora-energiewende.org/topics/the-energiewende/germanys-clean-energy-transition-what-is-the-energiewende/, Agora Energiewende (2014), 'Germany's clean energy transition: what is the Energiewende?', Mercator Foundation/European Climate Foundation.

2 http://www.worldfuturecouncil.org/fileadmin/user_upload/Miguel/feed-in_systems_spain_germany_long_en.pdf, Ragwitz M and Huber C (2003), 'Feed-in systems in Germany and Spain and a comparison', Fraunhofer Institute.

3 http://www.bmub.bund.de/presse/pressemitteilungen/pm/artikel/erneuerbare-energien-gesetz-tritt-in-kraft/, Trittin J (2004), 'Erneuerbare-Energien-Gesetz tritt in Kraft', Government of Germany, 30 July 2004.

4 http://www.agora-energiewende.de/fileadmin/downloads/publikationen/Analysen/Trends_im_deutschen_Stromsektor/Analysis_Energiewende_Paradox_web_EN.pdf, Graichen P and Redl C (2014), 'The German Energiewende and its climate paradox: an analysis of power sector trends for renewables, coal, gas, nuclear power and CO_2 emissions 2010– 2030', Agora Energiewende.

5 http://www.platts.com/latest-news/coal/london/germanys-2014-coal-imports-hit-all-time-high-26044121, Platts, 'Germany's 2014 coal imports hit all-time high of 56.2 mil mt: VDKi', 20 March 2015.

6 http://www.bp.com/en/global/corporate/about-bp/energy-economics/statistical-review-of-world-energy.html, BP (2014), 'BP statistical review of world energy'.

7 http://www.agora-energiewende.de/fileadmin/downloads/publikationen/Analysen/Jahresauswertung_2014/Agora_Energiewende_Review_2014_EN.pdf, Agora Energiewende (2015), 'The Energiewende in the power sector: state of affairs 2014: a review of the significant developments and an outlook for 2015'.

8 http://www.mining.com/german-begs-sweden-for-more-coal-90155/, Els F, 'Germany begs Sweden for more coal', *mining.com*, 24 November 2014.

9 http://notrickszone.com/2014/11/09/federal-minister-of-economics-sigmar-gabriel-ending-coal-is-mass-disinformation-of-the-people-an-illusion/#sthash.AL6CizK4.dpbs, Gosselin P, 'Federal Minister Of Economics Sigmar Gabriel: ending coal is mass disinformation of the people', *NoTricksZone*, 9 November 2014.

10 http://www.bp.com/en/global/corporate/about-bp/energy-economics/statistical-review-of-world-energy.html, BP (2014), 'BP statistical review of world energy 2014'.

11 http://breakingenergy.com/2013/12/06/germanys-revolution-in-efficiency-and-renewable-energy/, Lovins A (2013), 'Germany's revolution in efficiency and renewable energy', *Breaking Energy*, 6 December 2013.

12 http://www.rwe.com/web/cms/en/1754836/rwe-generation-se/about-us/energy-mix/lignite/, RWE (2015), 'Lignite-fired power plants'.

13 http://www.agora-energiewende.de/fileadmin/downloads/publikationen/Analysen/Trends_im_deutschen_Stromsektor/Analysis_Energiewende_Paradox_web_EN.pdf, Graichen P and Redl C (2014), 'The German Energiewende and its climate paradox: an analysis of power sector trends for renewables, coal, gas, nuclear power and CO_2 emissions 2010- 2030', Agora Energiewende.

14 http://reneweconomy.com.au/2013/energiewende-part-3-the cost-of-germanys-energy-vision-70701, Hope M (2013), 'Energiewende – Part 3: the costs of Germany's energy vision', *Renew Economy*, 2 August 2013.

15 http://www.theecologist.org/blogs_and_comments/commentators/2489204/germanys_renewable_revolution_shows_the_way.html, Barnham K (2014), 'Germany's renewable revolution shows the way', Ecologist, 26 July 2014.

16 http://www.economist.com/node/2299989, 'A reprieve for free speech', The Economist, 18 December 2003.

17 https://www.linkedin.com/pulse/20140321133218-322580126-german-energy-policy-is-failing-the-poor-while-being-a-poor-way-to-help-the-climate, Lomborg B (2014), 'Germany's energy policy is failing the poor, while being a poor way to help the climate'.

18 https://www.linkedin.com/pulse/20140321133218-322580126-german-energy-policy-is-failing-the-poor-while-being-a-poor-way-to-help-the-climate, Lomborg B (2014), 'Germany's energy policy is failing the poor, while being a poor way to help the climate'.

19 http://www.spiegel.de/international/germany/high-costs-and-errors-of-german-transition-to-renewable-energy-a-920288.html, 'Germany's energy poverty: How electricity became a luxury good', *Spiegel online*, 4 September 2013.

20 http://www.emagazine.com/author/guest/paul-hockenos/, Hockenos P, 'Germany's renewable energy gamble', *The Environmental Magazine*, 1 September 2012.

21 http://bdew.de/internet.nsf/id/17DF3FA36BF264EBC1257B0A003EE8B8/$file/Energieinfo_EE-und-das-EEG-Januar-2013.pdf, BDEW (2013) 'Erneuerbare Energien und das EEG: Zahlen, Fakten, Grafiken'.

22 http://www.pv-magazine.com/news/details/beitrag/german-renewables-pushing-wholesale-electricity-further-into-negative-territory_100019141/, Hall M (2015), 'German renewables pushing wholesale electricity further into negative territory', *pv magazine*.

23 http://www.faz.net/aktuell/wirtschaft/wirtschaftspolitik/energiepolitik/umweltminister-altmaier-energiewende-koennte-bis-zu-einer-billion-euro-kosten-12086525.html, 'Umweltminister Altmaier, Energiewende könnte bis zu einer Billion Euro kosten', *Frankfurter Allgemeine*, 19 February 2013.

24 http://bigstory.ap.org/content/grid-operators-say-germany-must-invest-25-billion, Baetz J,'Grid operators say Germany must invest $25 billion', *The Big Story*, 30 May 2012.

25 http://www.nytimes.com/2014/12/25/world/europe/germans-balk-at-plan-for-wind-power-lines.html?_r=0, Eddy M, 'Germans balk at plan for wind power lines', *New York Times*, 24 December 2014.

26 http://www.renewablesinternational.net/the-german-switch-from-nuclear-to-renewables-myths-and-facts/150/537/33308/, ,The German switch from nuclear to renewable – myths and facts', *Renewables International*, 9 March 2012.

27 http://www.mining.com/green-germany-returning-to-brown-coal-villages-in-limbo/, Jamasmie C, 'Green Germany returning to brown coal, villages in limbo', *mining.com*, 11 April 2014.

28 http://www.independent.co.uk/environment/green-living/green-village-to-be-bulldozed-and-mined-for-lignite-in-germanys-quest-for-nonnuclear-fuel-9760091.html, Paterson T, 'Green village to be bulldozed and mined for lignite in Germany's quest for non-nuclear fuel', *The Independent*, 28 September 2014.

29 http://www.go100percent.org/cms/index.php?id=45&tx_ttnews%5Btt_news%5D=152&cHash=762928366b28de1ca4e33086fb81f7a5, Morris C, 'Myths and facts: the German switch from nuclear to renewables', for the Heinrich Böll Stiftung Foundation, 9 March 2012.

30 http://www.jeremyleggett.net/2011/05/germanys-nuclear-phase-out-will-prove-easier-than-many-think/, Leggett J (2011), 'German nuclear phase-out easier than many think'.

31 http://www.newscientist.com/article/mg21128236.300-the-carbon-cost-of-germanys-nuclear-nein-danke.html?full=true, Strahan D, 'The carbon cost of Germany's nuclear Nein danke!', *New Scientist*, 30 July 2011.

32 http://www.umweltbundesamt.de/sites/default/files/medien/376/bilder/dateien/greenhouse_gas_emissions_in_germany_1990_-_forecast_2013_0.pdf, Unweltbundesamt (2014), 'Greenhouse gas emissions in Germany 1990-2013 forecast'.

33 http://edgar.jrc.ec.europa.eu/news_docs/jrc-2014-trends-in-global-co2-emissions-2014-report-93171.pdf, Olivier J et al., *Trends in global CO$_2$ emissions*: 2014 report, European Commission.

34 http://www.agora-energiewende.de/fileadmin/downloads/publikationen/Analysen/Trends_im_deutschen_Stromsektor/Analysis_Energiewende_Paradox_web_EN.pdf, Graichen P and Redl C (2014), 'The German Energiewende and its climate paradox: an analysis of power sector trends for renewables, coal, gas, nuclear power and CO$_2$ emissions 2010- 2030', Agora Energiewende.

35 http://www.spiegel.de/international/germany/instability-in-power-grid-comes-at-high-cost-for-german-industry-a-850419.html, Schröder C, 'Energy revolution hiccups:Grid instability has industry scrambling for solutions', *Der Spiegel*, 16 August 2012.

36 http://www.smartgridtoday.com/public/Grid-security-experts-warn-of-renewables-risks-2.cfm, 'Grid security experts warn of renewables risk', *Smart Grid Today*, 6 May 2015.

37 http://mainstream-downloads.opendebate.co.uk/downloads/11032013-Friends-of-the-Supergrid---Germany.pdf, PWC (2013), 'Supergrid in Germany – potential social, environmental and economical benefits.'

38 http://www.bloomberg.com/news/2012-10-25/windmills-overload-east-europe-s-grid-risking-blackout-energy.html, Bauerova L and Andreson T, 'Windmills overload East Europe's grid risking blackout', *Bloomberg Business News*, 26 October 2012.

39 http://www.publications.parliament.uk/pa/cm201012/cmselect/cmenergy/1040/104007.htm, Energy and Climate Change Select Committee (2011), 'Seventh Report – a European Supergrid', parliament.uk.

40 Toomey J (2014), *An unworthy future*, Archway Publishing.

41 http://uk.reuters.com/article/2013/04/17/czech-germany-grid-idUKL5N0D43LA20130417, Lopatka J, 'Czech grid acts to guard against German wind power surges', *Reuters*, April 17 2013.

42 http://www.bloomberg.com/news/articles/2015-12-01/german-wind-power-surplus-spurs-cash-for-neighbour-to-switch-off, Starn J and Zha W (2015), 'Germany pays to hault Danish wind power to protect its own input', *Bloomberg Business News*, 1 December 2015.

43 http://www.icis.com/resources/news/2014/09/03/9817356/german-pumped-storage-in-crisis-as-solar-crushes-economics/, ICIS, 'German pumped-storage in crisis as solar crushes economics', 3 September 2014.

44 http://www.leonardo-energy.org/all-hope-relies-electrical-energy-storage, de Wachter B (2013), 'All hope relies on electrical energy storage', *Leonardo Energy*.

45 http://www.agora-energiewende.de/fileadmin/downloads/publikationen/Studien/Speicher_in_der_Energiewende/Speicherstudie_Preface_and_summary_english_preliminary_english_version.pdf, Agora Energiewende, 'Electricity storage in the German energy transition'.

46 http://srsroccoreport.com/germany-death-of-renewable-energy-bring-on-the-dirty-coal-monsters/germany-death-of-renewable-energy-bring-on-the-dirty-coal-monsters/, St Angelo S (2014), 'Germany: renewable energy policy "complete failure", bring on the coal monsters', SRSrocco Report.

47 http://time.com/3028723/germany-climate-change-coal-poll/, Nicks D, 'Poll: only Germans think they are helping to fix global warming', *Time*, 24 June 2014.

48 http://api.worldbank.org/v2/en/topic/19?downloadformat=excel, World Bank (2014), 'Climate change knowledge portal'

49 http://www.dissentmagazine.org/article/green-energy-bust-in-germany, Boisvert W, 'Green energy bust in Germany', Dissent, Summer 2013.

7 | Japan – a worked example of medium-term power capacity shortage

The most feared outcome of a failure to take appropriate decisions in the energy field is probably 'the lights going out'. As discussed in chapter 8, this rather gentle-sounding phrase does not really capture the huge disruption, perhaps even devastation, that a prolonged power outage could wreak.

But how realistic is this eventuality, even in a country such as the UK which seems determined to delay necessary infrastructure decisions for as long as possible? Some pointers may be drawn from the experience of Japan, which found itself facing a sudden and substantial reduction in available power capacity after the Great East Japan earthquake and tsunami in March 2011 and the consequent accident at the Fukushima Daiichi nuclear station.

Although there have been many examples of countries running short of power for relatively short periods and for a variety of reasons, Japan's situation is unusual in its combination of two factors which combine to make it perhaps a realistic surrogate for a country which has not invested in enough capacity. First, a considerable contribution to Japan's electricity generation became effectively unavailable quite rapidly and remained so for over four years – the UK is facing a considerable amount of capacity coming offline not overnight but in a relatively short period of time as existing plant ages. In the year before the tsunami Japan had the world's third-largest nuclear fleet: 54 reactors generating 30 per cent of its power needs. Secondly, Japan was (and is) the most heavily dependent of the world's economies on energy imports, a situation exacerbated by its island status which severely limits opportunities for electricity interconnection to neighbouring countries. Again there are parallels with the UK as North Sea fossil fuel reserves run short and interconnection opportunities remain inevitably limited owing to its geographical location.

The Japanese situation created both a power challenge and an energy challenge – i.e. risks that peak demand could not be fulfilled, requiring efforts to shift peak demand to other times of day, and risks that continuing to fulfil total electricity and energy demand over the day and year might prove unsustainable for reasons perhaps of plant breakdown, high costs or environmental emissions.

7.1 Before the earthquake

Japan, with a population of around 126 million people, is the most heavily industrialised country of the Asia–Pacific region. In the mid-2010s it was the world's third biggest economy after the US and China, and fifth biggest energy user after China, the US, the Russian Federation and India. However, it had practically no indigenous energy resources. Japan was therefore the most heavily dependent of the G7/G8 countries on energy imports, which in January 2011 were accounting for some 88.8 per cent of primary energy consumption (nuclear being counted as indigenous owing to its small fuel requirements).[1] Before the 2011 earthquake Japan was already the world's largest liquefied natural gas (LNG) importer, some 85 mtoe in 2010; the second largest coal importer, 123 mtoe in 2010; and the third largest oil importer, 204 mtoe in 2010. High energy prices resulted in Japan becoming one of the most energy-efficient countries in the world (see figure 7.1) – energy intensity improving by some 37 per cent between 1973 and 2006[2] and standing at 92 kgoe per thousand euros gross domestic product (GDP) in 2009.[3]

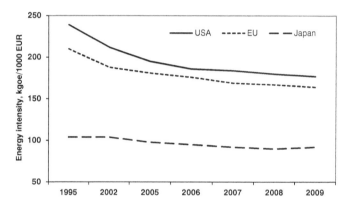

Figure 7.1 Energy intensity in the EU, US and Japan (kgoe/€1000), 1995–2009[4]

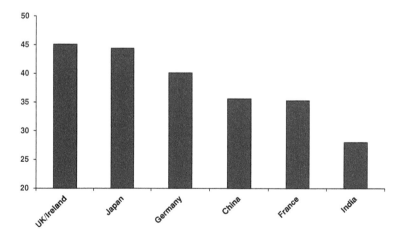

Figure 7.2 Percentage average thermal efficiency of fossil-fired power production – selected countries, 2010[5]

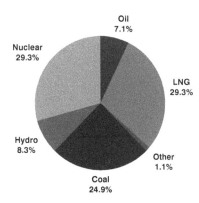

Figure 7.3 Japanese electricity production, FY2009[7]

The average thermal efficiency of its fossil-fired power stations in 2010, at 44.4 per cent, was among the highest in the word (see figure 7.2) – its coal-fired and oil-fired stations were the world's most efficient (with average thermal efficiency of 41.3 and 42.0 per cent respectively), with its gas-fired stations among the top five (at 48.3 per cent).

Electricity accounts for more than 40 per cent of Japan's primary energy delivery, one of the highest proportions in the world.[6] In 2009, the last full fiscal year (FY) before the 2011 earthquake and tsunami, its electricity was largely generated by LNG (281 TWh, 29.3 per cent), nuclear power (280 TWh, 29.3 per cent) and coal (238 TWh, 24.9 per cent) in roughly equal proportions, as shown in figure 7.3.

Energy growth stagnated after the mid-1990s owing to the country's economic recession and the wider Asian economic crisis of 2009. Nonetheless, in 2011 electricity consumption per capita stood at 7,800 kWh per year, the same as the OECD average and one-third higher than the EU average.[8] Electricity Power Companies (EPCOs), the ten vertically integrated private utilities which had monopoly powers over electricity supply to small power users, aimed to maintain a capacity margin of between 8 and 10 per cent.[9] The market was divided between larger 'deregulated' customers using more than 50 kW, representing some 62 per cent of the market by volume, who could shop around among the EPCOs for their supplies, and small 'regulated' customers over whom the local EPCO had monopoly powers. Transmission and distribution grids are owned by the EPCOs, affording them considerable opportunity to exclude competitors.

Under the Kyoto Protocol Japan had a target of a 6 per cent reduction in 1990 greenhouse gas emission levels by the 2008–12 commitment period. In 2009 the prime minister, Yukio Hatoyama, committed the country to a 25 per cent reduction in carbon dioxide emissions by 2020.[10] According to EIA figures the use of nuclear power in Japan pre-March 2011 had been reducing Japan's carbon dioxide emissions by some 14 per cent per year, representing over 200 million tonnes and in theory saving some

$3.4 billion to buy international carbon emissions permits.[11] The government expressed its intention to increase the nuclear share of total electricity generation from 24 per cent in 2008 to 40 per cent by 2017 and 50 per cent by 2030, requiring 14 more plants, according to the Ministry of Economy, Trade and Industry (METI).[12] Among Japan's pre-March 2011 climate change goals was a 20 per cent carbon intensity reduction from 1990 levels to an average of 334 g carbon dioxide (CO_2) per kWh during the 2008–12 Kyoto commitment period. From 2008 to 2011 Japan had been 7 per cent above this target at 358 g CO_2 per kWh.[13]

7.2 THE EARTHQUAKE AND TSUNAMI

Straight after the tsunami the main problem for power production was a severe loss of available capacity in the regions served by TEPCO (Tokyo Electric Power Company) and the Tohoku EPCO – not only nuclear output but also a considerable amount of fossil-fuelled generation capacity damaged in the earthquake – with the rest of the country largely unaffected. Over time, however, capacity in the earthquake region recovered somewhat. By the summer TEPCO had raised generating capacity to 45 GW from a low of 31 GW immediately after the earthquake, still well below the 65 GW it had available before, by restoring damaged thermal plants and bringing those mothballed back into service.[14]

However, the problem slowly spread across the whole nation. Ten of the 37 Japanese nuclear plants which had been online in March 2011 were shut down as a direct result of the earthquake, tsunami and subsequent crisis at the Fukushima Daiichi complex. Over the following months reactors across Japan were refused permission to resume operations as they progressively closed for their routine 13-month maintenance and inspection outages. (Permission to restart reactors after these routine outages lay with the local prefecture as well as with the regulator.) As a result, in May 2012 Japan was without nuclear generation for the first time in over 40 years. Over the following year only Kansai EPCO (KEPCO)'s Ohi Units 3 and 4 were given permission to restart, in June 2012.[15] When they reached their next scheduled maintenance shutdown in September 2013 Japan was again without operating nuclear reactors, a situation that persisted for two further years until the first plant received permission to restart in September 2015.

The focus in Japan, then, has largely turned to preserving power supplies in the medium to long term with only a proportion of the former nuclear fleet returning to service, rather than getting through an acute crisis until 'business as usual' can be resumed.

7.3 REDUCTIONS IN DEMAND

The first thing to note is that in these stressed conditions Japanese industry and the wider public responded by reducing their demand for electricity, in a way that seemed that it

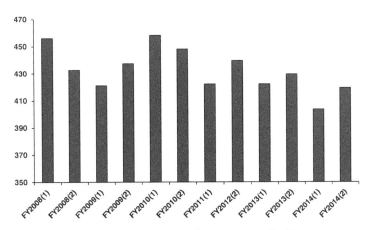

Figure 7.4 Japan electricity sales (TWh), fiscal years (first half/second half) 2008–14[16]

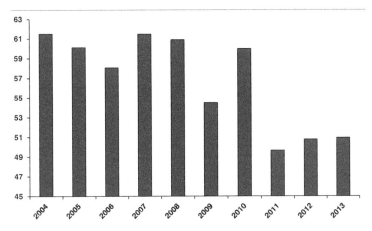

Figure 7.5 Peak demand in TEPCO region (GW), 2004–13[18]

might be sustainable, at least over the four years following the emergency. Electricity sales in the first half of fiscal year 2010 (before the earthquake) stood at 458 TWh; in the corresponding period in 2014 the figure was 404 TWh, a reduction of some 11.9 per cent, as shown in figure 7.4.

Peak demand also fell from 178 GW in 2010 to 155 GW in 2011.[17] In the TEPCO region the fall was even more precipitate, peak demand falling by 17.2 per cent between FY2010 and FY2011 – see figure 7.5.

In FY2011 the national government sought a number of electricity reduction outcomes:[19]

- a uniform 15 per cent cut in peak power consumption in the TEPCO and Tohoku EPCO service areas;
- a 10 per cent cut in the Kansai service area, owing to KEPCO's high dependence on nuclear power (later extended to Kyushu EPCO for similar reasons);

- a legally binding National Power-Saving Edict requiring a 15 per cent reduction in power use for all customers contracted for 500kW or more by 1 July.

METI played a central role in providing public information, providing ¥8.7 billion ($93 million) for an electricity saving campaign targeting consumers and industries and for projects to support electricity savings by targeting small businesses.[20] In June 2011 the Ministry of Environment launched the 'Super Cool Biz' campaign, with full-page newspaper advertisements and photographs of ministry workers at their desks wearing polo shirts and colourful short-sleeved 'kariyushi' shirts. The campaign also encouraged workplaces to set air conditioning at 28°C, switch off computers when not in use, switch working hours to the morning and take more summer holidays.[21]

The Electricity Conservation Ratio (ECR) – in effect the real reduction in power demand correcting for temperature and other factors – fell in the directly affected area but rose elsewhere in Japan as the year proceeded,[22] as shown in table 7.1.

Table 7.1 Electricity Conservation Ratios, 2011

	Japan	TEPCO	Tohoku EPCO	KEPCO	Others
April 2011	7.4%	11.7%	8.9%	3.6%	4.5%
Summer 2011	6.7%	8.1%	5.4%	7.6%	5.0%

Nine months after the earthquake the average 2011 ECR for Japan was 6.2 per cent, i.e. demand was 6.2 per cent lower than it would have been in a similar year before the earthquake. Research suggested that the main actions taken by households to reduce their electricity use were with regard to air conditioning – almost everyone reported shortening the hours of cooling at night and setting the temperature higher, with a majority also refraining from cooling during the day – and lighting, followed by installing a sunshade, reducing hot water use and raising the refrigerator temperature setting. Other actions included reducing electric heating (in the winter) and television use, and decreasing water heater temperature settings. The government target of a 15 per cent cut in electricity consumption was achieved by 30 per cent of households in the TEPCO area, with 17 per cent of households exceeding 25 per cent savings. A small but growing number of consumers adopted more severe measures, such as lowering the capacity of their circuit breakers, thereby reducing the amount of electricity they could use before a fuse blows.[23]

Lessons learned from this period include:[24]

- when measures are mandatory, more electricity may be saved than the set targets;

- even when only a voluntary target is set some electricity saving can be expected simply due to having the target;
- to minimise negative impacts on economic activity, it is necessary to request businesses, particularly in the service sector, to undertake detailed efforts to conserve electricity, taking into account the differing circumstances of each company.

By the end of FY2011 the overall situation appeared to have eased and in January 2012 the trade minister, Yukio Edano, predicted, correctly, that there was a good chance of Japan coping without mandatory cuts on electricity usage during the following summer even if all the nuclear reactors were shut down.[25] The government had considered reintroducing formal rationing and a programme of planned outages, with railways, hospitals and fire departments to be exempt, but eventually they did not feel there was sufficient need to do so.[26] No power disruptions were faced during FY2012. Overall power use continued to fall marginally, sales by the ten regional utilities dropping by 1 per cent from the previous year (to 852 TWh). The Federation of Electric Power Companies of Japan (FEPC) ascribed the reduction to a decline in manufacturing activity and greater energy conservation efforts.[27] In surveys some 47 per cent of people had said they intended to continue with actions they had started in 2011 into 2012 and beyond, 18 per cent said they would not continue in 2012 and 33 per cent did not know or were not taking measures.[28]

Power prices, especially to larger users, increased considerably as a result of higher generating costs. For example, on 26 November 2012 Kansai EPCO raised rates by 11.9 per cent in the regulated (household) sector and 19.2 per cent in the deregulated sector.[29] In the TEPCO and KEPCO service areas 'demand-controlled power supply contracts' were introduced in December 2012, designed to encourage industrial consumers to reduce electricity usage during hours of high demand.[30]

By 2013 fears were being voiced about the operational state of some of the (non-nuclear) generating capacity. Since March 2011 the power companies had been operating under emergency plans which involved putting off periodic inspections of their thermal power plants and restarting those that had been shut down, whether or not maintenance had been completed.[31] The breakdown rate of plant over 40 years old increased significantly after the earthquake, although that of plant less than 40 years old did not change significantly – see figure 7.6.

New fossil fuel capacity entered into commercial operation: a 600 MW CCGT in Chubu and two new coal-fired power stations, with a combined capacity of 1,600 MW, by TEPCO.[32]

Unlike in 2012 the government did not prescribe any specific targets for energy to be saved during the summer. METI described its strategy for the summer as follows. 'Power conservation will be requested in a manner ensuring that expected power-saving efforts are made without undue stress and with minimum effects on people's lives and economic activities. Without setting numerical targets, the government will present the expected

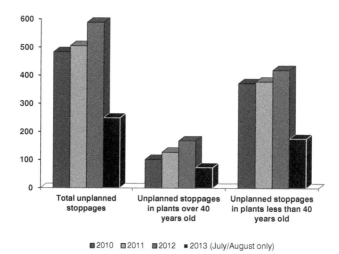

Figure 7.6 Summer and winter unplanned stoppage rates (July/August and December/February)

reduction amount in power consumption in each service area as guidelines for power conservation and promote power conservation.'[33]

On the face of it the overall situation was quite extraordinary. Despite losing almost 30 per cent of its power production, Japan was able to maintain secure power supplies for a full four years, with little reason to believe that this fundamental achievement would not last despite the problems already discussed. However, this should not be taken as a further excuse for the UK ducking or needlessly delaying necessary investment decisions. The costs to Japan have been extremely high.

7.4 THE FINANCIAL COSTS OF JAPAN'S POWER SUPPLIES

The financial consequences of Japan's failure to resume operations in its undamaged nuclear plants have been significant. Japan became even more dependent on imported fossil fuels for its electricity generation.[34] Imports of LNG by the ten regional utilities were some 39 per cent higher in January 2012 than in January 2011 (5.2 million tonnes against 3.7 million tonnes), while imports of fuel oil and crude oil both nearly tripled – as much of the available underused generating capacity was oil-fired, dating back to before the 1973 oil price shock.[35] The increased demand for LNG coincided with a weakening of the yen and an increase in LNG unit costs, the latter in part caused by this greater demand. The result was a 77.5 per cent increase in overall LNG importation costs. By the end of 2013 increased imports of oil, gas and coal had cost Japan an extra ¥9.2 trillion (around $93 billion).[36] The nine biggest EPCOs cited fuel costs as the main reason for their combined losses of ¥1.59 trillion in FY2012.[37]

In response, the Japanese government stepped up imports of coal. The ten EPCOs, which accounted for half of Japan's coal use, consumed nearly 16 per cent more coal in the first ten months of 2013 compared with the same period a year previously and imported nearly 11 per cent more.[38] The overall costs of power generation, based on the June 2013 financial report of the electricity utilities, rose from ¥7.5 trillion in 2010 to ¥9.5 trillion in 2011 and ¥10.6 trillion in 2012 (unit costs following a similar trend as shown in figure 7.7). Fuel costs for thermal power generation rose from ¥3.7 trillion in 2010 to ¥6.1 trillion in 2011 and ¥7.3 trillion in 2012. Some ¥1.2 trillion could be ascribed to rising energy prices, the rest to increased volumes of fossil fuel imports.

The burden of increased costs associated with the need to import more fossil fuel for electricity production was not to be spread evenly. For example, in late 2012 TEPCO increased its prices to its larger 'deregulated' customers by 14 per cent. However, its request for a similar increase for its 'regulated' customers was turned down by the regulator, who capped rises in this segment of the market to 8 per cent. As a result household and small business consumers were to an extent cushioned from the financial effects of the increased fuel costs and could largely offset the tariff increases by reducing their demand. So while large industry lobbied heavily for the nuclear plants to be returned to service, small consumers, perhaps especially those who did not work outside the home (and so to an extent did not perceive the problems that high prices were causing to larger businesses), more often took the view that the modest increase in unit prices was worth paying to keep nuclear plants offline. The third group to suffer from higher fuel costs were the EPCOs themselves. Even ignoring the special situation in which TEPCO found itself – having to deal not only with increased fuel costs but also the direct costs of the rescue operation at Fukushima Daiichi – which had led the Japanese government to buy more than half of the company's shares, other EPCOs, notably those most reliant on nuclear power, also faced severe financial challenges. In mid-2014 Hokkaido Electric Power announced that it was seeking a ¥50 billion ($484 million) rescue package from the Development Bank of Japan (DBJ), having suffered a net loss of some ¥60 billion in FY2013 – its third consecutive year in the red. Kyushu EPCO said it was also in discussions over a possible capital infusion – its net debt of ¥2.54 trillion in FY2013 was equal to 28 times its estimated earnings before interest, tax, depreciation and amortisation. It had not been in profit since April 2011 and recorded a net loss of ¥59 billion in the nine months to 31 December 2013. DBJ also issued a loan of ¥45 billion to Kansai EPCO.[40] Clearly continued losses at EPCOs are not sustainable – unless they return to profitability they may be unable to continue to operate even their existing plants, let alone make the heavy investment in plant maintenance and construction required to bring Japan's energy situation onto a more stable footing.

Projected long-term effects on Japan's GDP and balance of payments were severe. The failure to return any nuclear plants into operation before 2020 could reduce GDP in that year by an estimated 7.6 per cent against the 'reference scenario' whereby nuclear had continued to be used as per the pre-2011 plans – see figure 7.8.

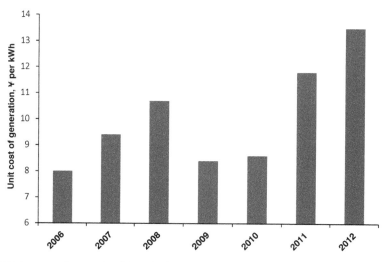

Figure 7.7 Unit costs of generation (¥ per kWh), 2006–12[39]

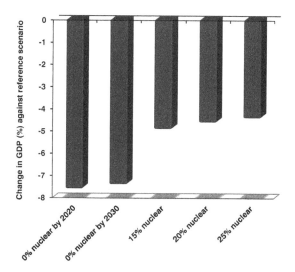

Figure 7.8 Impact on Japan's GDP in 2030 of lower nuclear output[41]

The increased imports had a significant effect on Japan's balance of payments, which went into deficit in FY2011 for the first time in 31 years, the deficit growing year-on-year to reach over $138 billion in FY2014 – see figure 7.9.

7.5 SACRIFICING THE ENVIRONMENT

Before the earthquake Japan was well on course to meet its Kyoto commitments, which it duly did. However, the closure of nuclear plants resulted in an increase in the carbon intensity of Japanese energy supplies, i.e. the quantity of greenhouse gas emitted per unit

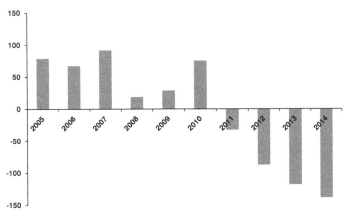

Figure 7.9 Japan trade balance (US$ billion), 2005–14[42]

of energy supplied, expressed as tonnes of carbon dioxide equivalent emitted per tonne of oil equivalent supplied (tCO_2 per toe).

Two trends were identifiable a year after the disaster: a fall in energy use and an increase in the carbon intensity of the economy. Carbon intensity, which had fallen by over 3 per cent between January 2008 and March 2011, rose by almost 5 per cent in the eight months after the earthquake – the increase between November 2010 and November 2011 was some 15 per cent, from 2.01 tCO_2 per toe in November 2010 to 2.31 tCO_2 per toe in November 2011. A comparison between the eight month period after the earthquake (April to November 2011) with the same period one year earlier reveals an increase in average carbon intensity of 7.6 per cent, from 2.07 tCO_2 per toe to 2.22 tCO_2 per toe – see figure 7.10 (see colour section).

Despite the steep drop in energy consumption, Japan's heightened dependence on fossil fuels and the consequent sharp increase in carbon intensity drove Japan's carbon dioxide emissions up by 3.8 per cent in November 2011 relative to the same month in the previous year. The combined effect of the two trends, after the period pre-2011 in which the carbon intensity of Japanese energy supply had been falling, is striking – see figure 7.11 (see colour section).

With regard to electricity, in FY2010 (which ended three weeks after the earthquake) Japan's power stations produced an average of 350 g of CO_2 per kWh of electricity generated, against its 20 per cent carbon intensity reduction target of 334gCO_2 per kWh during the 2008–12 Kyoto commitment period. In FY2011 the figure was no less than 36 per cent higher, at 476 g CO_2 per kWh.[43] The carbon intensity of power production rose further in FY2012 to 487gCO_2 per kWh, 39 per cent higher than 2010.[44] The final 2008–12 average was therefore 406 g CO_2 per kWh, some 22 per cent above target. In FY2012 the carbon intensity of electricity production was 17 per cent above 1990 levels, representing in absolute terms an increase of some 100 million tonnes of carbon dioxide above typical annual emission levels of 1,250 million tonnes.[45]

As in the German Energiewende, the environment was therefore the first casualty of the political decision to withdraw from nuclear power. In April 2013 METI relaxed the environmental assessment requirements for new coal-fired generating plant, limiting carbon dioxide restrictions only to the extent that requesting businesses use the 'best available technology' for currently operating power plants and to observe national objectives and policies.[46] Japan's national leaders accepted that there was no hope of meeting the nation's near-term targets and refused to sign any new binding emissions reduction commitments, notably stage 2 of Kyoto. At the November 2013 United Nations Climate Change Conference in Warsaw, Japan formally abandoned its plans to cut greenhouse gas emissions by 25 per cent by 2020, instead saying they would in effect allow an increase of 3.1 per cent.[47] The new Basic Energy Plan published in April 2014 did refer to an expansion of renewable energy (to 20 per cent of total electricity generation by 2030, this figure to include hydropower), but designated coal as an important long-term electricity source and did not set specific targets for wind, solar or geothermal. Far from expressing disapproval, Big Green has consistently opposed the restarting of Japanese nuclear plants and has said very little about Japan's burgeoning greenhouse gas emissions.

Japan's renewable energy sector was very small, with renewable sources such as wind, solar, and geothermal collectively providing less than 3 per cent of the nation's electricity supply in FY2012. The government did state an intention to increase the contribution of renewable energy: a feed-in tariff (FiT) regime for independently generated renewable electricity was introduced in 2012.[48] As of January 2014 a total of 33.2 GW of potential installations had received recognition: almost all (31 GW) were solar.[49] However, replacing the zero-carbon electricity Japan had previously planned to derive from nuclear with electricity from renewable energy sources would require a near 50-fold increase in the electricity provided by wind, solar and geothermal to the national energy system, from just over 8 TWh in 2009 to 399 TWh in 2030. Managing variable renewables is a particular challenge given the fragmentary nature of Japan's grid system. The east of the country operates a separate grid with a different frequency from that in the centre and the west, with very limited frequency conversion capacity between the two, and, unlike Germany, Japan has no interconnection capacity with neighbouring countries, leaving it without the option of using its economic muscle to dump excess renewable energy on its neighbours or outbid them when power in the region is short.[50]

7.6 A SUSTAINABLE CHANGE?

Three years after the emergency there was some mild evidence that a sense of complacency might be creeping in. Forecasts, or perhaps even assumptions, of doom for the country's future without nuclear plants had been expressed both before the accident (not least by the nuclear industry) and afterwards. Soon after the earthquake, for example, the Japan Centre for Economic Research published a report stating that the country would face a serious electricity shortage if all its nuclear reactors stopped operating because the country would be unable to bridge the gap with only fossil fuels[51] – but, as mentioned previously, no power outages had ensued. When asked to consider a range of motives and actions associated with saving energy – 'good for society', awareness of government publicity, cost, to avoid personal hardship, trying new ways to keep up the effort, becoming more proactive – consumers reported that only cost was more important to them in FY2013 than it had been in FY2012. Awareness of the energy implications of using various appliances and of actions actually taken to reduce demand showed year-on-year decline between 2011 and 2013 – see figures 7.12 and 7.13.

Some of the reductions in energy demand resulted in genuine inconvenience, e.g. the higher temperatures endured in shops and small business places and perhaps even the minor inconvenience of unplugging electronic equipment rather than leaving them in 'standby' mode. It is at least possible that a return to 'business as usual' and a perception that the capacity crisis was over, say, because of a new fleet of coal-fired or gas-fired power stations being built or a return to operation of some of the nuclear plants, would be accompanied by an increase in demand for these creature comforts.

Such a perception was aided in 2014 by three factors. The rapid and significant fall in the oil price in the second half of the year benefited Japan more than any other economy, by both lowering its costs for industry and reducing its balance of payments deficit. (There was some countervailing concern that the potential slowdown of the economies of oil-producing nations might represent a threat to exports, especially from the Japanese motor industry.) In 2013 Japan had imported some ¥15 trillion worth of oil; halving this figure would clearly be a major economic boon and would be likely to delay the point at which the financial burden of the Japanese situation became unbearable.

Secondly, there was increasing talk of some of the nuclear plants being brought back into service as they began to meet the new regulatory requirements.

Thirdly, and again as was being seen in the German Energiewende, the main winner of the situation was the coal industry. In July 2014 the Electric Power Development Company (J-Power) announced it was to seek permission to expand its Takasago coal-fired power plant from 500 MW to 1,200 MW, having already published plans for a new 640 MW coal power plant northeast of Tokyo. At the same time TEPCO said it wished to build a 1,000 MW coal power plant at Yokosuka. In the final quarter of FY2014 alone seven large new coal-fired power stations were ordered, with a combined capacity of 7.26GW.[54] Having in effect decided that reducing greenhouse emissions was a luxury Japan could not afford if it wished to keep most of its nuclear plants offline, this response was perhaps inevitable.

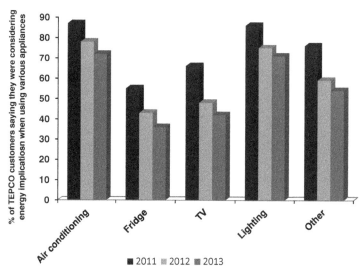

Figure 7.12 Percentage of TEPCO customers reporting that they were reflecting on energy usage when using various appliances, fiscal years 2011–13[52]

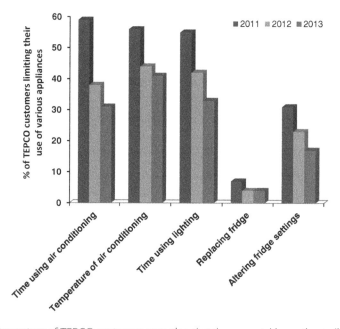

Figure 7.13 Percentage of TEPCO customers reporting that they were taking action to limit their use of various appliances, fiscal years 2011–13[53]

7.7 REFLECTIONS

The Japanese situation arose as a result of an unexpected external emergency rather than through a chronic failure to take decisions – the partial liberalisation of the Japanese power system did leave enough authority with the EPCOs to allow them to consider suitable investments in a reasonable time. Ironically, one of the consequences of the Fukushima accident has been the further pressure to break up their vertical integration and especially to divest them of their grid capacity by 2020, despite the dubious record of timely investment in new capacity in liberalised power markets. Nonetheless, the outcome may have been similar – an apparent shortage of capacity with a severe threat of ongoing power outages.

A number of actions are available to countries to respond in these circumstances. They include:

- use of price signals (to encourage both reductions in demand in the short term and investment in new capacity in the medium to long term);
- information campaigns;
- technology replacement;
- rationing;
- market mechanisms.[55]

Japan used all of these.

- TEPCO introduced higher tariffs during peak demand times to encourage people to switch the timings of their electricity use where possible, thereby reducing peak demand.
- A vigorous campaign of public information was launched in 2011 and continued for the two following summers. It included providing energy savings advice to small and medium enterprises, a website for households to check their energy use, and the promotion of cooler business wear.
- Residents took a number of 'mechanical' measures to reduce their energy consumption, for example by buying LED light bulbs.
- The government introduced quotas for various sections of the economy, e.g. mandatory rationing of 15 per cent for industry.
- Programmes such as TEPCO's 'Demand Diet Plan' and KEPCO's 'Demand Cut Plan' offered discounts to industrial users for reducing consumption during times of peak demand.

The situation four years on from the accident was not necessarily a stable one. Old fossil-fuelled generation which had ambled into semi-retirement as peaking plant was brought back into use for baseload production, without a major programme of refurbishment and maintenance, bring a growing threat of major breakdown. People were perhaps becoming slightly complacent and lowering the priority they placed on reducing energy use by behavioural change.

Nonetheless, the lights and to a lesser extent the air conditioning had stayed on despite the loss of a source of electricity which had been generating almost 30 per cent of Japan's total power before the earthquake. Part of the solution has lain in lower power demand, in part caused by and in turn a cause of the economic slowdown caused by the earthquake and its effect on many industries and businesses; part because of the success of energy conservation programmes of various kinds, including behaviour change. The country seemed to be quietly preparing for the possibility of a very low or non-nuclear future. Heavy government subsidies were being offered to small-scale generators, notably solar, and autogeneration, while plans were being prepared for a major build programme of gas-fired and coal-fired capacity.

So a tentative conclusion would seem to be that power generation systems may be considerably more robust to major shocks than they might appear on casual observation. Governments and individuals, at least in the Japanese case, can work together to mitigate both the likelihood of major power outages and the effects of power shortages. However, in order to preserve these supplies the other legs of the traditional energy 'trilemma' are clearly prone to be sacrificed.

Greenhouse gas emissions in FY2013 stood at near-record levels, 8.5 per cent higher than in 2010 and 10.6 per cent above the Kyoto base year of 1990 – see figure 7.14.

In the absence of a robust global greenhouse regime in the years after the first Kyoto compliance period this has proved little more than a minor embarrassment. Big Green has certainly not been critical of Japan's position, falling back on the fiction that Japan could more or less overnight cut its demand by a further 20 per cent to compensate for the loss of nuclear output. This may be more difficult to maintain after the Paris agreement of 2015.

If environmental protection is so easily dispensable as a political goal, higher prices are more of a challenge. Japan's response was to pile the cost increases disproportionally onto industry rather than household consumers and to allow power generators to drift towards bankruptcy while providing large loans to slow the process. This approach may not be sustainable in the long term but undoubtedly delivered more immediate political benefits.

This gives rise to an interesting policy perspective. If a country faces loss of a significant proportion, say 25 per cent, of its power generation capacity over a relatively short period, with little prospect of its return in the medium to long term, its system operator, alongside regulators and legislators, may be faced with two broad choices. The first would be to plan for a programme of rolling blackouts, with the associated danger that the system operator itself, and by proxy the government of the day, becomes the target for public anger and protest. The second would be to make appeals to the public to reduce its energy and power demands by say 15 per cent, trust that this would happen and therefore institute no public plans for outages. In the latter scenario one would expect that the system operator and the government would be seen more to be in collaboration with consumers in overcoming the crisis and thereby less likely to become the focus for discontent – in

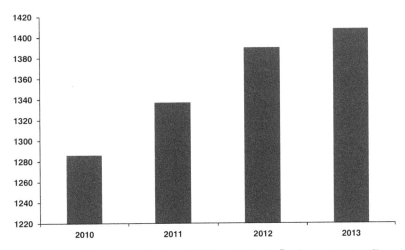

Figure 7.14 Japan's greenhouse gas emissions (MtCO$_2$ equivalent), fiscal years 2010–13[56]

effect the policy interest and the political interest coincide. Japan's experience post-March 2011 suggests that at least for a reasonable period of time policymakers can afford to be bold in this respect. Japan's consumers have indeed reduced their demand sufficiently to allow other emergency measures to keep supplies flowing.

Perhaps the key message though is that the energy tetralemma is not a brotherhood of equals. Ultimately, in Japan at least, political factors seem to trump all else. The political decision, based on very heavy social pressure and expressed indirectly in terms of an overriding 'safety' imperative, not to operate the available nuclear plants has placed considerable pressures on the goals of security of supply, economics and reducing emissions of greenhouse gases, all of which would significantly benefit from the restart of as many nuclear plants as feasible. Security of supply, the factor most closely allied to the political decision, comes either second or equal first depending on one's point of view. Next comes economics, with attempts to reduce costs, for example by encouraging the construction of a new generation of gas-fired and coal-fired capacity, which is cheaper than running, maintaining and refurbishing old oil-fired plants. Last in line comes climate change commitments, abandoned with barely more than a passing comment even from non-governmental organisations.

This would seem to imply that any attempt to put the environment at the sole heart of policymaking, as for example the European Union arguably did in the early 2000s, may ultimately be self-defeating. As one analyst has put it:

Modern governments have three drivers shaping their energy policy: tackling climate change, affordability, and security of supply. Most governments try to find a balance but the EU has given overwhelming weight to tackling climate change. But the implementation of this policy stance is proving hugely expensive and politicians are finding that in the real world affordability matters, a lot ... All of a sudden the affordability of power for both residential and industrial consumers has become a major political issue in many countries. EU politicians are finding out that in the real world, the issue of affordability is not after all the minor factor they assumed it to be back in the mid-2000s.[57]

To maintain social and, therefore, political commitments to reducing greenhouse gas emissions may require ensuring that the other elements of the traditional trilemma – security of supply and affordable economics – are satisfied first.

Any underlying assumption that the lights will stay on almost whatever, as an excuse for not taking timely decisions on new infrastructure, may be seductive and more justified than one might initially expect. But the consequences for the environment and for the economy could be dire.

Some of the material in this chapter derives from unpublished work by Malcolm Grimston with Steve Ashley and Bill Nuttall of the Open University, supported by a grant from the Engineering and Physical Sciences Research Council, UK via the 'NREFS' project led by Philip Thomas, City University, London, UK.

ENDNOTES

1 http://data.worldbank.org/indicator/EG.IMP.CONS.ZS, The World Bank, 'Energy imports, net (% of energy use)'.

2 http://www.carnegiecouncil.org/publications/articles_papers_reports/0036.html/_res/id=sa_File1/Japanchapter. pdf, Stewart D, 'Japan –the power of efficiency', in ed. Luft G and Korin A, *Energy security challenges for the 21st century*, Praeger: Santa Barbara CA, 2009.

3 http://www.eia.gov/cfapps/ipdbproject/IEDIndex3.cfm?tid=92&pid=46&aid=2, US EIA (2013), 'Energy intensity using market exchange rates'.

4 http://vmisenergy.com/2011/10/27/energy-intensity-in-europe-the-us-and-japan/, VMIS Energy (2011), 'Energy Intensity in Europe, the US and Japan'.

5 http://www.ecofys.com/files/files/ecofys-2014-international-comparison-fossil-power-efficiency.pdf, Ecofys (2014), 'International comparison of fossil power efficiency and CO_2 intensity – update 2014'.

6 http://www.fepc.or.jp/english/environment/asia-pacific/green_handbook_peer/__icsFiles/afieldfile/2008/10/20/ chapter1_1.pdf, FEPC (2007), 'Green Handbook Peer Review'.

7 http://www.fepc.or.jp/english/library/electricity_eview_japan/__icsFiles/afieldfile/2011/01/28/ERJ2011_04.pdf, FEPC (2011), 'Electricity review Japan 2011', Tokyo.

8 http://data.worldbank.org/indicator/EG.USE.ELEC.KH.PC, World Bank, 'Electric power consumption (kWh per capita)'.

9 http://www.chuden.co.jp/english/resource/corporate/ecsr_report_2013_2.pdf, Chubu Electric Power Company Group, 'Annual Report 2013 –measures to ensure supply and demand outlook for summer 2013'.

10 http://www.theecologist.org/News/news_round_up/315886/japan_commits_to_25_per_cent_cut_in_co2_ emissions.html, 'Japan commits to 25% cut in CO_2 emissions', *Ecologist*, 7 September 2009.

11 http://www.ensec.org/index.php?option=com_content&id=335:japans-energy-security-predicament-in-the-aftermath-of-the-fukushima-disaster&catid=121:contentenergysecurity1111&Itemid=386, Vivoda V, 'Japan's energy security predicament in the aftermath of the Fukushima disaster', *Journal of Energy Security*, 14 December 2011.

12 http://climateobserver.org/wp-content/uploads/2015/03/strategic-energy-plan.pdf, METI (June 2010), 'The Strategic Energy Plan of Japan –meeting global challenges and securing energy futures'.

13 http://www.fepc.or.jp/english/news/message/__icsFiles/afieldfile/2013/08/02/20130731_e_CO2.pdf, FEPC (2013), 'Environmental Action Plan by the Japanese electric utility industry: CO_2 emissions 2008–2012'.

14 https://www.semiconportal.com/en/archive/news/main-news/110414-electricity-saving-25percent-summer. html, 'This summer's big challenge: cut peak electricity demand by 25%', *Semiconportal*, 15 April 2011.

15 http://www.kepco.co.jp/english/corporate/pr/2012/__icsFiles/afieldfile/2012/06/20/2012_jun_1.pdf, Kansai EPCO, 'Future processes involved in restarting Ohi Power Station Units 3 and 4', 16 June 2012.

16 http://www.fepc.or.jp/english/news/demand/__icsFiles/afieldfile/2015/05/01/juyou_k_e_fy2014.pdf, FEPC (2015), 'Electricity demand in fiscal 2014'.

17 http://fepc-dp.jp/pdf/06_infobase.pdf, FEPC (2013), FEPC Infobase.

18 http://www.tepco.co.jp/en/corpinfo/illustrated/power-demand/peak-demand-recent-e.html#, TEPCO Illustrated (2014), 'Peak demand for recent years'.

19 http://www.aceee.org/files/proceedings/2012/data/papers/0193-000219.pdf, Murakoshi C, Nakagami H and Hirayama S (2012), 'Electricity crisis and behaviour change in the residential sector: Tokyo before and after the Great East Japan earthquake', American Council for an Energy-Efficient Economy.

20 http://www.japanfs.org/en/news/archives/news_id031101.html, 'METI addresses summertime power problem with supplementary budget', *Japan for Sustainability*, 24 July 2011.

21 http://www.japantimes.co.jp/opinion/2011/06/12/editorials/super-cool-biz/, 'Super cool biz', *The Japan Times*, 12 June 2011.

22 http://www.aceee.org/files/proceedings/2012/data/papers/0193-000219.pdf, Murakoshi C, Nakagami H and Hirayama S (2012), 'Electricity crisis and behaviour change in the residential sector: Tokyo before and after the Great East Japan earthquake', American Council for an Energy-Efficient Economy.

23 http://online.wsj.com/news/articles/SB10000872396390443720204578003524193492696, Dvorak P, 'In Japan, people getting charged up about amping down', *The Wall Street Journal*, 3 October 2012.

24 http://www.japanfs.org/en/news/archives/news_id031565.html, 'Outcomes and lessons learned from measures in Japan to save electricity in summer 2011', JFS Newsletter No. 112, *Japan for Sustainability*, 24 January 2012.

25 http://eandt.theiet.org/news/2012/jan/japan-power.cfm, Fergusson R, 'Japan will avoid power cuts this summer', *Engineering & Technology Magazine*, 27 January 2012.

26 http://asian-power.com/environment/news/japan-might-ration-electricity-summer, 'Japan might ration electricity this summer', *Asian Power*, 31 May 2012.

27 http://www.thefreelibrary.com/FY+2012+electricity+sales+in+Japan+slip+for+2nd+straight+year.-a0327207064, FEPCO, 'FY electricity sales in Japan slip for second straight year', 24 April 2013.

28 http://www.stanford.edu/group/peec/cgi-bin/docs/events/2011/becc/presentations/1%20-%204E%20%20The%20Worst%20Eelectricity%20Crisis%20Ever%20-Sho%20Hirayama.pdf, Hirayama S (2011), 'The worst electricity crisis ever: how Tokyo cut its electricity use by 15%', Jyukankyo Research Institute Inc.

29 http://www.kepco.co.jp/english/corporate/list/report/pdf/e2013.pdf, Kansai EPCO, 'Kansai Electric Power Group Report 2013'.

30 http://www.rieti.go.jp/en/columns/a01_0368.html, Ohashi H, 'How to control peak electricity demand; policy evaluation', RIETI, 23 April 2013.

31 http://www.fepc.or.jp/english/news/conference/__icsFiles/afieldfile/2013/04/25/kaiken_e_20130419.pdf, FEPC, 'Summary of press conference comments made by Makoto Yagi, FEPC Chairman, on 19 April 2013'.

32 http://www.tepco.co.jp/en/press/corp-com/release/2013/1232692_5130.html, TEPCO, 'Commercial operation commencement of Unit 6 of Hirono Thermal; Power Station', 3 December 2013; and http://www.tepco.co.jp/en/press/corp-com/release/2013/1233043_5130.html, TEPCO, 'Commercial operation commencement of Unit 2 of Hitachinaka Thermal Power Station', 18 December 2013.

33 http://www.meti.go.jp/english/earthquake/electricity/pdf/20130610_01.pdf, METI (April 2013), 'Electricity supply-demand outlook and measures for the summer of FY2013'.

34 http://www.eia.gov/todayinenergy/detail.cfm?id=10391, EIA, 'Japan's fossil-fuelled generation remains high because of continued nuclear plant outages', 15 March 2013.

35 http://www.bloomberg.com/news/2012-02-13/japanese-power-utilities-lng-imports-rise-to-5-2-million-tons-in-january.html, Adelman J, 'Japanese power utilities import 39% more LNG in January', *Bloomberg Business News*, 13 February 2012.

36 http://www.economist.com/node/21586570/print, 'Electricity in Japan – power struggle', *The Economist*, 21 September 2013.

37 http://www.bloomberg.com/news/2013-05-31/japan-pays-record-price-for-lng-imports-in-april-as-yen-weakens.html, 'Japan pays record price for LNG imports in April as yen weakens', *Bloomberg Business News*, 31 May 2013.

38 http://www.brecorder.com/markets/commodities/asia/146463-japans-soaring-coal-use-may-push-down-lng-imports-this-year.html?tmpl=component&page, 'Japan's soaring coal use may push down LNG imports this year', *Reuters*, 29 November 2013.

39 http://eneken.ieej.or.jp/data/5252.pdf, Matsuo Y and Yamaguchi Y (2013), 'The rise in cost of power generation in Japan after the Fukushima Daiichi accident and its impact on the finances of the electric power utilities', *IEEJ*, November 2013.

40 http://www.ft.com/cms/s/0/0241b30a-ba41-11e3-aeb0-00144feabdc0.html#axzz3mUQAGl1c, Thompson J and Soble J, 'Bailout requests colour Japan's nuclear debate', *Financial Times*, 2 April 2014.

41 Akimoto K (2014), 'Energy strategy perspectives in Japan after the Fukushima–Daiichi nuclear power accident', Research Institute of Innovative Technology for the Earth (RITE).

42 http://www.tradingeconomics.com/japan/balance-of-trade, Trading Economics (2015), 'Japan balance of trade 1963-2015'.

43 http://www.world-nuclear-news.org/EE_Climate_targets_missed_in_Japan_0108131.html, 'Climate targets blown in Japan', *World Nuclear News*, 1 August 2013.

44 http://www.fepc.or.jp/english/news/message/__icsFiles/afieldfile/2013/08/02/20130731_e_CO2.pdf, FEPC (2013), 'Environmental Action Plan by the Japanese electric utility industry: CO_2 emissions 2008–2012'.

45 http://www.world-nuclear-news.org/EE_Climate_targets_missed_in_Japan_0108131.html, 'Climate targets blown in Japan', *World Nuclear News*, 1 August 2013.

46 http://www.meti.go.jp/press/2013/04/20130426003/20130426003-3.pdf, METI, 'Summary of a meeting by relevant directors on TEPCO's bid for thermal power', 25 April 2013.

47 http://wwwp.dailyclimate.org/tdc-newsroom/2013/11/japan-emissions-walkback, Brown P, 'Dismay in Warsaw as Japan abandons greenhouse targets', *The Daily Climate*, 16 November 2013.

48 http://www.meti.go.jp/english/policy/energy_environment/renewable/pdf/summary201209.pdf, METI, Agency for Natural Resources and Energy (2011), 'Feed-in tariff scheme for renewables energy', Tokyo.

49 http://www.meti.go.jp/english/press/2014/0418_01.html, ANRE (2014), 'Announcement regarding the present status of introduction of facilities generating renewable energy as of January 31', METI, Tokyo.

50 http://mainstream-downloads.opendebate.co.uk/downloads/11032013-Friends-of-the-Supergrid---Germany.pdf, PWC (2013), 'Supergrid in Germany – potential social, environmental and economical benefits'.

51 http://www.reuters.com/article/2011/07/15/us-japan-nuclear-capacity-idUSTRE76E1XQ20110715, Mogi C, 'Fossil fuels can't save Japan from power shortages – report', *Reuters*, 15 July 2011.

52 Nishio K and Ofuji K (2014), 'Ex-post analysis of electricity saving measures in the residential sector in the summer of 2013', CRIEPI.

53 Nishio K and Ofuji K (2014), 'Ex-post analysis of electricity saving measures in the residential sector in the summer of 2013', CRIEPI.

54 http://www.wsj.com/articles/japan-continues-to-re-embrace-coal-1426162227, Iwati M, 'Japan continues to embrace coal', *The Wall Street Journal*, 12 March 2015.

55 http://www.iea.org/media/training/presentations/Day_4_Session_2_Saving_Electricity_in_a_Hurry.pdf, Pasquier S and Tromop R, 'Saving electricity in a hurry', IEA, 11 April 2013.

56 http://www.env.go.jp/press/files/en/600.pdf, Japan Ministry of the Environment (2015), 'Japan's national greenhouse gas emissions in FY2013 (final figures)'.

57 www.eurelectric.org/Download/Download.aspx?DocumentFileID=72013, Atherton P et al., 'A very hostile political environment', Citi Group, 13 September 2011.

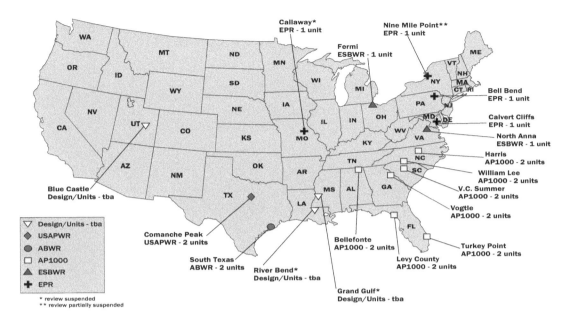

Figure 3.13 Location of projected new nuclear reactors in the US, 2009

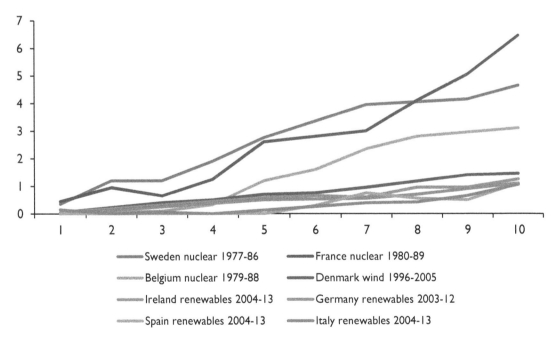

Figure 4.11 Added electrical energy per capita (MWh/yr), 10-year periods (http://thebreakthrough.org/index.php/programs/energy-and-climate/nuclear-has-scaled-far-more-rapidly-than-renewables, Russell G, 'Nuclear has scaled far more rapidly than renewables', Breakthrough Institute, 20 June 2013)

Figure 3.2 UK nuclear sites

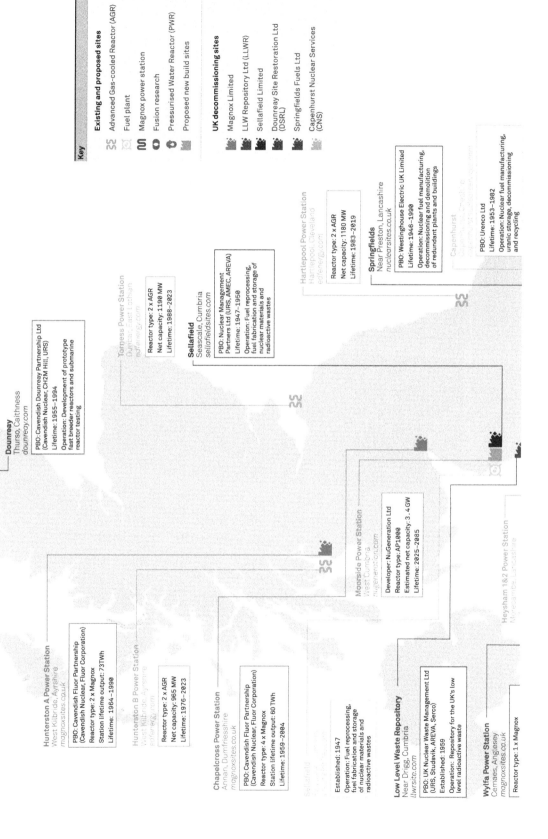

Dounreay
Thurso, Caithness
dounreay.com

PBO: Cavendish Dounreay Partnership Ltd (Cavendish Nuclear, CH2M Hill, URS)

Lifetime: 1955–1994

Operation: Development of prototype fast breeder reactors and submarine reactor testing

Torness Power Station
Dunbar, East Lothian
edfenergy.com

Reactor type: 2 x AGR
Net capacity: 1190 MW
Lifetime: 1988–2023

Sellafield
Seascale, Cumbria
sellafieldsites.com

PBO: Nuclear Management Partners Ltd (URS, AMEC, AREVA)

Lifetime: 1947–1950

Operation: Fuel reprocessing, fuel fabrication and storage of nuclear materials and radioactive wastes

Hartlepool Power Station
Hartlepool, Cleveland
edfenergy.com

Reactor type: 2 x AGR
Net capacity: 1180 MW
Lifetime: 1983–2019

Springfields
Near Preston, Lancashire
nuclearsites.co.uk

PBO: Westinghouse Electric UK Limited

Lifetime: 1946–1990

Operation: Nuclear fuel manufacturing, decommissioning and demolition of redundant plants and buildings

Capenhurst

PBO: Urenco Ltd

Lifetime: 1953–1982

Operation: Nuclear fuel manufacturing, uranic storage, decommissioning and recycling

Hunterston A Power Station
West Kilbride, Ayrshire
magnoxsites.co.uk

PBO: Cavendish Fluor Partnership (Cavendish Nuclear, Fluor Corporation)

Reactor type: 2 x Magnox
Station lifetime output: 73TWh
Lifetime: 1964–1990

Hunterston B Power Station
West Kilbride, Ayrshire
edfenergy.com

Reactor type: 2 x AGR
Net capacity: 965 MW
Lifetime: 1976–2023

Chapelcross Power Station
Annan, Dumfriesshire
magnoxsites.co.uk

PBO: Cavendish Fluor Partnership (Cavendish Nuclear, Fluor Corporation)

Reactor type: 4 x Magnox
Station lifetime output: 60 TWh
Lifetime: 1959–2004

Moorside Power Station
West Cumbria
nugeneration.com

Developer: NuGeneration Ltd
Reactor type: AP1000
Estimated net capacity: 3.4 GW
Lifetime: 2025–2085

Sellafield

Established: 1947

Operation: Fuel reprocessing, fuel fabrication and storage of nuclear materials and radioactive wastes

Low Level Waste Repository
Near Drigg, Cumbria
llwrsite.com

PBO: UK Nuclear Waste Management Ltd (URS, Studsvik, AREVA, Serco)

Established: 1959

Operation: Repository for the UK's low level radioactive waste

Heysham 1&2 Power Station
Morecambe, Lancashire

Wylfa Power Station
Cemaes, Anglesey
magnoxsites.co.uk

Reactor type: 1 x Magnox

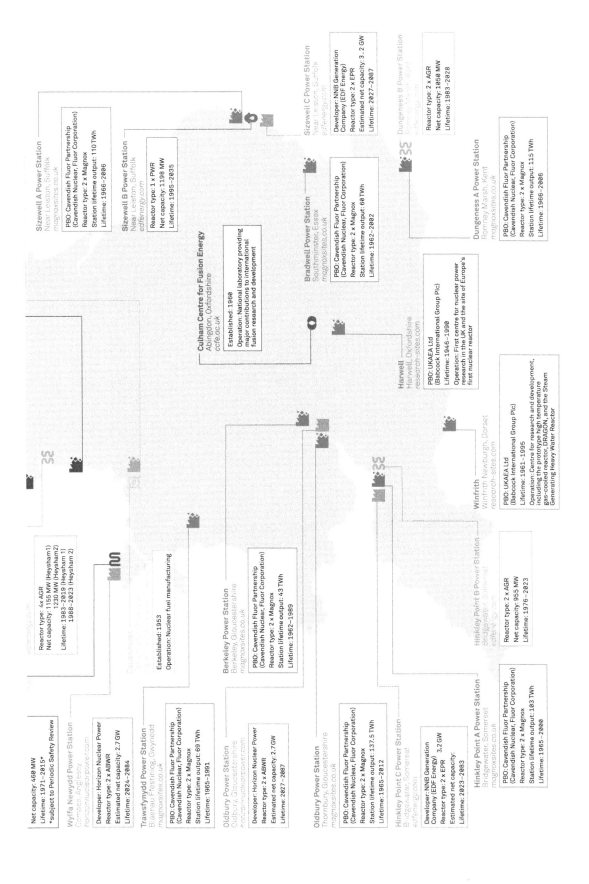

Net capacity: 460 MW
Lifetime: 1971–2015*
*subject to Periodic Safety Review

Reactor type: 4x AGR
Net capacity: 1155 MW (Heysham1)
1230 MW (Heysham2)
Lifetime: 1983–2019 (Heysham 1)
1988–2023 (Heysham 2)

Wylfa Newydd Power Station
Cemaes, Anglesey
horizonnuclear.com

Developer: Horizon Nuclear Power
Reactor type: 2 x ABWR
Estimated net capacity: 2.7 GW
Lifetime: 2024–2084

Trawsfynydd Power Station
Blaenau Ffestiniog, Gwynedd
magnoxsites.co.uk

PBO: Cavendish Fluor Partnership
(Cavendish Nuclear, Fluor Corporation)
Reactor type: 2 x Magnox
Station lifetime output: 69 TWh
Lifetime: 1965–1991

Established: 1953
Operation: Nuclear fuel manufacturing

Berkeley Power Station
Berkeley, Gloucestershire
magnoxsites.co.uk

PBO: Cavendish Fluor Partnership
(Cavendish Nuclear, Fluor Corporation)
Reactor type: 2 x Magnox
Station lifetime output: 43 TWh
Lifetime: 1962–1989

Oldbury Power Station
Oldbury, Gloucestershire
horizonnuclearpower.com

Developer: Horizon Nuclear Power
Reactor type: 2 x ABWR
Estimated net capacity: 2.7 GW
Lifetime: 2027–2087

Oldbury Power Station
Thornbury, Gloucestershire
magnoxsites.co.uk

PBO: Cavendish Fluor Partnership
(Cavendish Nuclear, Fluor Corporation)
Reactor type: 2 x Magnox
Station lifetime output: 137.5 TWh
Lifetime: 1965–2012

Hinkley Point C Power Station
Bridgewater, Somerset
edfenergy.com

Developer: NNB Generation
Company (EDF Energy)   3.2 GW
Reactor type: 2 x EPR
Estimated net capacity:
Lifetime: 2023–2083

Hinkley Point B Power Station
Bridgwater, Somerset
edfenergy.com

Reactor type: 2 x AGR
Net capacity: 955 MW
Lifetime: 1976–2023

PBO: Cavendish Fluor Partnership
(Cavendish Nuclear, Fluor Corporation)
Reactor type: 2 x Magnox
Station lifetime output: 103 TWh
Lifetime: 1965–2000

Hinkley Point A Power Station
Bridgewater, Somerset
magnoxsites.co.uk

Sizewell A Power Station
Near Leiston, Suffolk
magnoxsites.co.uk

PBO: Cavendish Fluor Partnership
(Cavendish Nuclear, Fluor Corporation)
Reactor type: 2 x Magnox
Station lifetime output: 110 TWh
Lifetime: 1966–2006

Sizewell B Power Station
Near Leiston, Suffolk
edfenergy.com

Reactor type: 1 x PWR
Net capacity: 1198 MW
Lifetime: 1995–2035

Sizewell C Power Station
Near Leiston, Suffolk
edfenergy.com

Developer: NNB Generation
Company (EDF Energy)
Reactor type: 2 x EPR
Estimated net capacity: 3.2 GW
Lifetime: 2027–2087

Dungeness B Power Station
Romney Marsh, Kent
edfenergy.com

Reactor type: 2 x AGR
Net capacity: 1050 MW
Lifetime: 1983–2028

Dungeness A Power Station
Romney Marsh, Kent
magnoxsites.co.uk

PBO: Cavendish Fluor Partnership
(Cavendish Nuclear, Fluor Corporation)
Reactor type: 2 x Magnox
Station lifetime output: 115 TWh
Lifetime: 1966–2006

Bradwell Power Station
Southminster, Essex
magnoxsites.co.uk

PBO: Cavendish Fluor Partnership
(Cavendish Nuclear, Fluor Corporation)
Reactor type: 2 x Magnox
Station lifetime output: 60 TWh
Lifetime: 1962–2002

Culham Centre for Fusion Energy
Abingdon, Oxfordshire
ccfe.ac.uk

Established: 1960
Operation: National laboratory providing
major contributions to international
fusion research and development

Harwell
Harwell, Oxfordshire
research-sites.com

PBO: UKAEA Ltd
(Babcock International Group Plc)
Lifetime: 1946–1990
Operation: First centre for nuclear power
research in the UK and the site of Europe's
first nuclear reactor

Winfrith
Winfrith Newburgh, Dorset
research-sites.com

PBO: UKAEA Ltd
(Babcock International Group Plc)
Lifetime: 1961–1995
Operation: Centre for research and development,
including the prototype high temperature
gas-cooled reactor, DRAGON, and the Steam
Generating Heavy Water Reactor

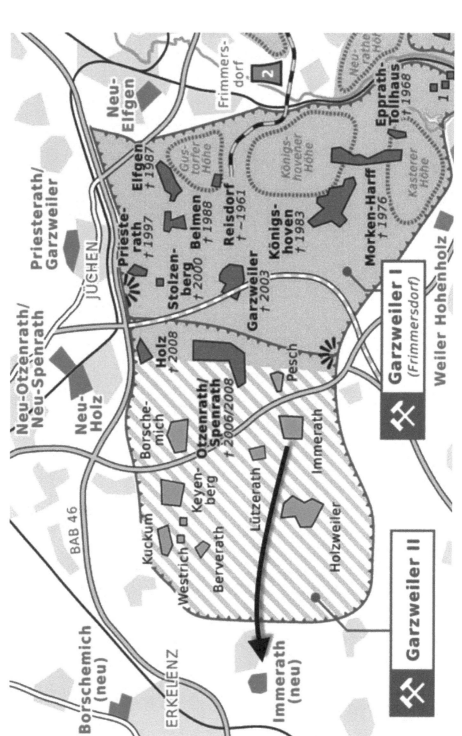

Figure 6.10 The Garzweiler lignite mining complex in western Germany

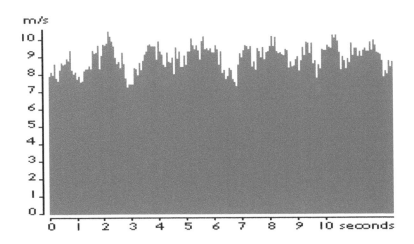

Figure 4.14 Short-term variability in wind speed (http://www.windpowerwiki.dk/index.php?title=Turbulence, Danish Wind Industry Association (1998), 'Turbulence'.)

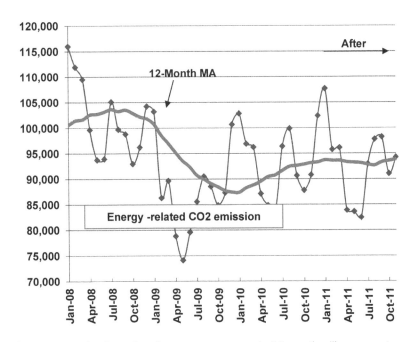

Figure 7.10 Carbon intensity of Japanese energy supply (12-month rolling average), November 2009 to November 2011. Redrawn from The Breakthrough Institute from an analysis by Mark Cain and Jesse Jenkins.

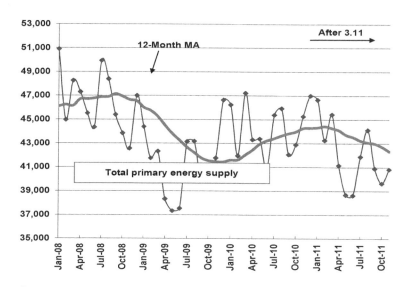

Figure 7.11 Japanese economy-wide energy use and carbon dioxide emissions
(12-month rolling average values), January 2008 to November 2011

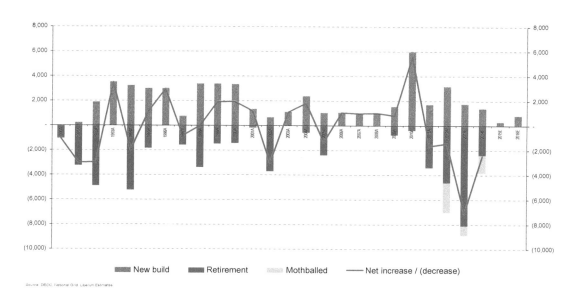

Figure 8.12 Short-term capacity challenge in UK

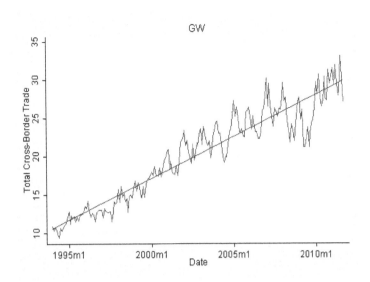

Figure 8.19 Cross-border trade in electricity in Europe. (http://dx.doi.
org/10.1787/5k4869cdwnzr-en, Bahar H and Sauvage J (2013) 'Cross-border
trade in electricity and the development of renewables-based electric power:
lessons from Europe', OECD Trade and Environment Working Papers.)

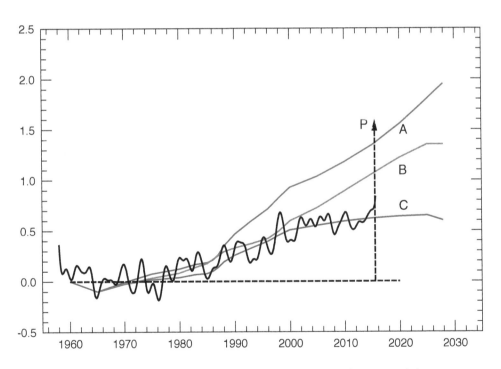

Figure 9.5 Comparison of Hansen's projections and observed outcomes (http://www.kaltesonne.
de/was-ist-eigentlich-aus-james-hansens-temperaturvorhersage-von-1988-geworden-zeit-fur-
eine-uberprufung/, Solheim J (2012), 'James Hansen's predictions versus observations'.)

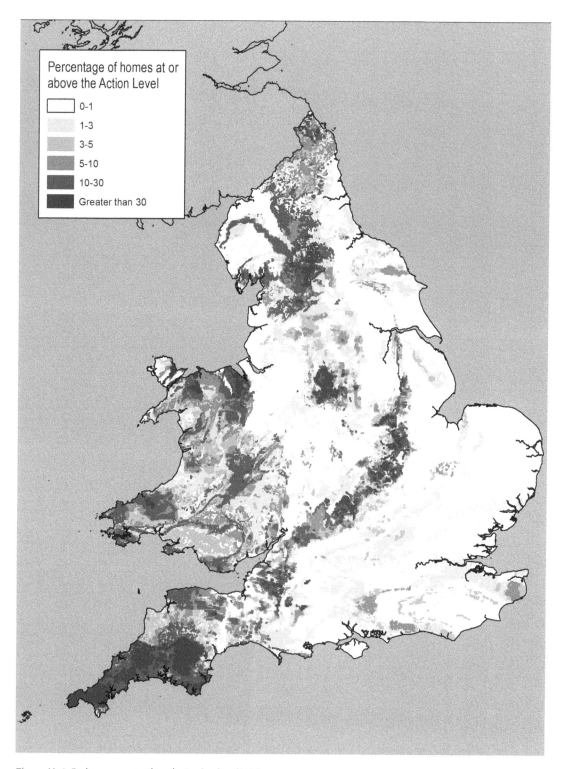

Figure 11.4 Radon concentrations in England and Wales

8 | THE LIBERALISATION OF POWER MARKETS AND ITS EFFECT ON INVESTMENT

When we are thinking about why it seems so difficult to take necessary decisions in the field of energy now compared say to the 1950s or 1960s, it is helpful to consider what is meant by a 'decision' and who might be responsible for taking it.

From the great nationalisation programme of the late 1940s to the great privatisation programme some 40 years later, decisions in many fields which were of considerable importance to the economy and which required large capital programmes were taken directly or indirectly by the government of the day. UK government and its agencies were responsible for steel, for telecommunications, for gas, electricity and large parts of the oil industry, for a proportion of motor car manufacture, for the railways, airports and the national airline.

In such an environment, taking a decision was in principle a relatively straightforward affair. The relevant government department would take a view as to the national interest in the field involved and weigh up the various projects that might deliver it, against the background of the Treasury acting as an arbiter of the dreams and desires of the different departments within the context of what the nation could afford. Although the agencies of government – British Telecommunications, the British Steel Corporation, the British Gas Corporation, British Leyland, the Central Electricity Generating Board, British Aerospace, British Sugar Corporation, Cable & Wireless, Britoil, BP and so on[1] – would have a certain amount of responsibility and therefore autonomy over how to deliver the national policy (and would take much of the flak if things did not work out), the will of the government would usually prevail.

Of course reality was not always quite so simple. Other agencies in society, for example powerful trade unions, may have different interests coupled with the political or industrial power to make things difficult. Many of the industries were led by individuals with a long history in the business, leading perhaps to something of a technocratic mindset and a sense of being product-oriented rather than consumer-oriented, the nuclear industry arguably being a particular example. In practice these nationalised industries were often subjected to heavy political interference depending on the policies or whims of the government of the day. Furthermore, when investment was necessary the agency in question would find itself in a queue competing with other government projects for funds, some of which may have a more rapid or greater political payback in terms of popularity, or avoided unpopularity, and hence of votes. Nonetheless, the lines of strategic and operational responsibility and accountability were reasonably clear and the government of the day could be judged on the success or otherwise of the projects involved.

For a variety of reasons, this centralised approach to decision-making was largely supplanted by a market-based approach as large nationalised industries were broken up and sold off, especially in the 1980s. State-owned monopolies in particular came to be regarded as inherently inefficient. Being shielded from competition they tended to become complacent and lack innovation. Being backed by government money the agencies in question were in effect prevented from failing – customers were trapped and government would bail out any financial losses. In extreme cases it could even be in the interests of the companies involved to be inefficient – a major project running over schedule and budget meant more work and wages for the state-employed workforce and their managers. BNFL, the monopolistic state-owned company responsible for the nuclear fuel cycle, contracted its services to the monopolistic CEGB on a 'cost-plus' basis, albeit diluted with incentive fee arrangements – in other words it would be paid a fixed rate of return on top of its costs. The higher the costs, the higher the returns. The government was shareholder of both sides of these contracts and took an interest in the relationship, but nonetheless a regime which rewards poor performance with higher profits is clearly flawed. It was only in 1992, more than 20 years after BNFL's creation, that arrangements shifted to fixed-price contracts with the nuclear generating companies, which though still state-owned were by then competing in a liberalised marketplace and looking forward to privatisation.[2]

The word 'privatisation' was actually used in two senses. The 'British Telecom-style' (BT) privatisation involved selling off not only the phones and wires but also the responsibility for providing telecommunications services. If the mobile phone network goes down nobody now blames the government – it is clearly the responsibility of the phone companies. The privatisation of British Steel and the car manufacturer British Leyland were other examples. However, there was also the 'British Rail-type' (BR) privatisation, where although the trains and tracks were sold off, the responsibility for the strategic rail network remained with the government, as was demonstrated when network operator Railtrack collapsed in 2002. The government was forced to transfer its assets into a state-owned successor, Network Rail, rather than simply allow it to go broke. Other examples might include the privatisation of the water industry and that of the electricity supply – if the taps run dry or the lights go out the government will clearly be held to account.

The situation is complicated where there is an element of 'natural monopoly' about the activity in question, or part of it. It makes no sense to have competing networks of water pipes, rails or electricity transmission wires (Finland once did have two national electricity grids but they were not directly in competition). The level of capital investment would be prohibitive and outweigh any operational benefits of competition. So these activities tend to be treated as regulated monopolies, allowed to make a modest rate of return on investment and to charge more or less fixed prices – often one or two percentage points below inflation to act as a surrogate for the benefits of competition – unless they can persuade their regulator of the need for investment and that it can be done efficiently.

Whether the sale of and introduction of competition into the great state-owned monopolies was a success or not largely depends on one's political tastes. In economic terms it may be argued that, at least in the early years following privatisation, those areas in which competition was feasible and government did not remain the 'guarantor of last resort' were the most successful. This is a fairly typical appraisal a decade or so after the big privatisations:

> BT was once memorably described, probably accurately, as the most hated institution in the land … Today we have better quality of service, more choice and lower prices in Britain than almost anywhere else in the world apart from the US. BT is also at the forefront of international developments in telecoms. Liberalisation and privatisation are the two key causes of this extraordinary turnaround. The lesson seems to be, therefore, that provided privatisation is also accompanied by liberalisation, initial teething difficulties are eventually overcome to produce publicly recognised advances. With water, electricity and rail, it is proving much more difficult to introduce competition into the market, virtually impossible in the case of water. As a result we have had to rely solely on regulation to protect the public interest. This regulation has frequently been seen as wanting. Perhaps more seriously, these industries have embraced wholeheartedly all the worst manifestations of corporate excess.[3]

Among the first of the major privatisations were British Telecom and British Gas. Each was privatised in effect as a monopoly, which came to be regarded as a mistake, or at least a wasted opportunity. By the time that electricity was privatised at the end of the 1980s the government had come to believe that introducing competition wherever possible was just as important as transferring state-owned assets into the private sector. The National Grid and the local distribution networks were treated as natural monopolies. However, competition could be introduced at the level both of generation and of supply.

Decisions about investment in generating capacity, then, became matters for private sector companies, or at least for companies not owned by the UK government, who were responsible to shareholders. However, the outcome of those decisions were of enormous interest to British society at large.

Normally, as has been understood since the time of Adam Smith, the eighteenth-century philosopher and 'father of economics', markets work well because the interests of producer and consumer coincide to a high degree. As Smith famously said in his seminal work *The Wealth of Nations*: 'It is not from the benevolence of the butcher, the brewer, or the baker that we expect our dinner, but from their regard to their own interest.' In principle, and largely in practice, the producer sells at a price where the money they take is of more value to them than the goods or services being sold, while the consumer buys at a price where the goods or services are worth more to them than the money. The actual price, at least in theory, is the one that maximises the benefit to both parties. This is not to say that governments have no part to play: government has a vital role in 'internalising externalities', most notably perhaps in making sure that firms cannot use the environment as a free dumping ground and have to cover the costs of any damage they may do, and be prevented from releasing emissions beyond a certain statutory limit. Regulation is also necessary to make sure that companies do not indulge in unfair competition, say by predatory pricing, by buying up all their competitors or by colluding with other firms to keep prices high. But notwithstanding these interventions, in general government can, indeed must, stand back and allow the market to follow its own logic. Effective and innovative companies thrive, inefficient and expensive ones do not and the customer reaps at least some of the benefit of this competition.

However, electricity is different from any other commodity in key senses, which contribute to the difficulty which has beset decision-making in competitive power markets.

8.1 THE UNIQUE CHALLENGES OF ELECTRICITY

Despite over two centuries of research – the first real battery was invented in 1800 by Alessandro Volta, 'the father of electrochemistry', while the production of hydrogen from surplus electricity for subsequent power production can be traced back to the late nineteenth century – electricity can't be stored in significant amounts. Pumped storage is feasible in some circumstances – indeed it dominates storage in the world today – but requires appropriate geology and major capital investment and suffers from losses of around 20 per cent of the power input.[4] In any case, pumped storage is generally regarded as an 'ancillary service' to manage grid stability in real time rather than as an indirect method of storing significant amounts of electricity per se. For some decades there has been considerable effort devoted to producing more effective batteries but it is unlikely that the ability to store trillions of kWh of electricity economically will appear soon, if at all. Indeed, countless breakthroughs have been announced over time but 'time and time again these advances have failed to translate into commercial batteries with anything like the promised improvements in cost and energy storage.'[5]

In the case of most commodities it is not crucial when the product is actually produced. Shoe manufacturers do not have to produce all their shoes during shopping hours nor do they have to throw away any shoes that are not sold at the moment they are cobbled together. In the energy field oil can be stored in barrels, gas in tanks, coal in yards and uranium in warehouses. But electricity is a process, not a material. Furthermore, there are few commodities where an interruption in supply for a short period, say a fraction of a second to a day or so, or indeed a significant surplus of supply over demand would be quite such a serious matter as is the case with electricity. So with electricity, unlike other commodities, we have to be able to make exactly the right amount to fulfil demand on a moment-by-moment basis.

A lot of discussion goes into the challenge of producing enough electricity to fulfil demand and prevent 'the lights going out' – a particularly insipid phrase to describe what would be a very serious problem for health, social order and the economy. Actually the challenge is more complex. A modest discrepancy between supply and demand can result in an increase or decrease in the 'frequency' and/or voltage of the power supply. In the UK the supergrid operator, National Grid, has a statutory duty to maintain the frequency in the range of 49.5 Hz to 50.5 Hz (cycles per second) and usually upholds it at between 49.8 Hz and 50.2 Hz.[6] More severe divergences can result in serious consequences.

If the frequency is too low then equipment will tend to underperform. In turn this may lead operators to increase their demand where they can, e.g. the driver of an electric train opening the throttle to compensate for the lower speed, while voltage stabilisers, which automatically increase current in equipment in response to drops in voltage or frequency, are readily available commercially and indeed are inbuilt in many appliances. Such measures may exacerbate the supply/demand discrepancy. Should supplies fall very short of demand then eventually power cuts may follow. Among the potential effects of prolonged power outages, for example, are failures in the transport systems, loss of a day's production in the workplace, loss of a freezer full of food, the inability to pump water into our homes or power our hospitals, and severe social disruption, e.g. looting and vandalism if a prolonged power cut occurs overnight, especially in an urban location – though the claim of a rise in the birth rate nine months after a major outage appears to be an urban myth. Even very short outages, down to fractions of seconds, can have a severe effect on electronic systems, for example requiring lengthy and expensive rebooting of computer networks or causing the loss of unsaved electronic data. In 2014 a UK government exercise examining what might happen should there be a major power outage that lasted a fortnight or so concluded that there would probably be a very rapid descent into public disorder unless the government could maintain the perception of security and that populations were far less resilient than they once were. The UK had not faced prolonged power outages for over 40 years. In the interim, the nature of the economy and especially the role that electricity plays within it had changed almost beyond recognition. One of the major problems identified by the exercise was that crucial fuel supplies, vital in the absence of power to run generators and emergency response vehicles, might not be accessible because petrol stations and even fuel bunkers now relied on electric pumps. There would be 'increased mortality rates which would put pressure on the practicalities of movement, storage and disposal of the deceased. Aside from the environmental health problems there is the cultural and social issue of ensuring dignity in death.' Transport systems would be paralysed as even diesel trains and road vehicles would be immobilised by fuel shortages and signal failures. Mobile phone coverage would start to drop out after two hours. Electronic tagging systems would fail, allowing high risk offenders in the community to 'disappear'. Fire and rescue services may struggle to cope after being inundated with automated alarm systems. Only one-third of staff might be able or willing to go in to work, further delaying attempts to recover the situation. Some types of sewage treatment works could cease to work after six hours and sewage would have to be discharged into water courses. Milk collection from dairy farms would fail, triggering an environmental emergency as it would have to be disposed of by spreading it over farms. Panic buying and hoarding would be triggered, casting doubt on assumptions over food supplies. Efforts to restore power would be hampered by significant metal theft from 'dead' circuits unless the military or emergency services patrolled power lines, diluting their abilities to maintain order in population centres.[7]

Rather less attention tends to be paid to the difficulties caused by there being too much power production for the demand at the time, initially pushing up the system frequency. If the frequency should become too high, direct damage to a variety of appliances and/or increased wear and tear can result. More significant, overgeneration can ultimately put strain on the wires which carry electricity from the power station to the customer and in extreme cases can lead to their deforming and finally melting, with serious consequences such as 'electrical arcing' or severe damage to transmission capacity and eventually to system collapse. The New York blackout of August 2003 was initiated by a 3,500 MW surge of power which caused wires to soften and sag, hitting trees and shorting out.[8] At that point the power tried to force its way down nearby wires, causing them to malfunction and forcing power stations to trip to avoid even more damage. The outcome was power cuts affecting 55 million people in eight US states plus Ontario in Canada.

Power demand varies considerably, both during the day and during the year. The difference between peak demand and lowest demand in the UK is roughly a factor of three (from 60 GW, or 60,000 MW, down to 20 GW) – see figure 8.1. Usually demand fluctuation is fairly predictable but variations can be very rapid. The biggest demand surge to date in the UK was some 2,800 MW after the penalty shoot-out in the 1990 football World Cup semi-final between England and West Germany, when supporters put the kettle on for a consolatory cup of tea. (Subsequent England managers seem to have been following policies to ensure that this would not be repeated.) Electricity planners needed to be ready for the moment when coverage of the Royal Wedding of Prince William to the Duchess of Cambridge in 2011 reverted to the TV studio, when demand surged by 2,400 MW.[9]

This balancing act creates considerable challenges, both technical and economic. In a functioning system the grid operator, who has operational (if not legal) responsibility for

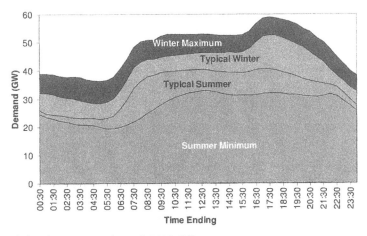

Figure 8.1 Variations in UK power demand, 2010–11[10]

keeping supply and demand in balance, has to be able to order up (or 'dispatch') enough reliable electricity to fulfil demand, but no more. In turn, to maintain secure supplies all year round some entity has to keep power stations available to fulfil any very high levels of demand, say on a cold, still, cloudless winter evening, knowing that these plants will not be called upon for most of the rest of the year when demand is lower. The costs to the consumer of failing to fulfil demand are very high – typically the 'Value of Lost Load' (VOLL) is estimated at between 50 and 350 times the price of a delivered unit of electricity.[11]

As a result of the interplay between these two unusual factors – the inability to store electricity and the very high costs to the consumer of supply interruptions – the interests of producer and consumer diverge in the case of electricity to a much greater degree than is the case with other commodities.

From the point of view of a producer, maintaining secure supplies may not always be the top priority. The costs of doing so would include the need to carry spare capacity that was available to generate whenever demand was at its highest. Such capacity would need to be maintained and staffed but may not get a market at all if peak demand should be lower than expected in any particular year. Not only is it a disadvantage to any company to keep such plant available, it is also to the disadvantage of every company for its rivals to do the same if it would result in a greater oversupply of capacity throughout the year, inevitably bringing prices down as companies compete simply to cover their marginal or 'avoidable' costs. However, if there are power cuts occasionally then while generating companies miss out on a few sales at the time this may well not outweigh the financial costs of maintaining the spare capacity. Indeed, at times of tight supply prices tend to rise strongly as consumers compete vigorously for what power there is, offering greater rewards for generators who are selling through the real-time market rather than on long-term fixed-price contracts, while long-term contracts become more valuable if future shortages are expected. To this one might add the highly capital-intensive nature of much electricity-related investment which means that investors need confidence that the rules will not be changed radically before they have got at least a good proportion of their money back with a decent return. These challenges can to an extent be addressed by 'capacity payments' – in effect paying producers to have idle plant available in case it is needed, rather than just paying for power actually generated and consumed, but such payments would have to be considerable to stimulate new build.

From the point of view of the consumer though things are simpler – power supply interruptions, especially unplanned ones, are extremely expensive. Some consumers may be prepared to accept occasional planned interruptions in supply but they would need to be heavily compensated to do so.

8.2 THE POWER OF CENTRAL GENERATION

The possibility of transmitting large amounts of electricity over long distances with relatively low losses effectively emerged in the 1920s and 1930s, the AC technology on which this depends having been invented in the late nineteenth century. As soon as national grids became available, the case for small-scale localised electricity systems embedded within the demand network, which was powered by variable renewables and the electricity usually transmitted by DC, effectively evaporated. Rather than each local consumption network requiring its own back-up capacity or having to cope with prolonged outages, a national system of large power stations which could be turned on and off or whose output could be varied in response to changes of demand without threatening security of supply was a great attraction.

As already mentioned, the 'baseload' – the demand which never goes away day or night, summer or winter – is roughly 18-20 GW in the UK. This is considerably more than the installed nuclear capacity of around 10 GW in the mid-2010s, meaning that when the system is being run rationally the nuclear stations, with their low fuel costs, tend to operate continuously whenever they are available. Fossil-fuelled power stations – particularly gas but also coal – are used to 'load-follow'. Switching such plant on and off, or varying its output, makes rather more sense because the cost of fuel saved is higher. The most flexible plant, in terms of load-following, tends to be the Open Cycle Gas Turbine (OCGT) and relatively small diesel generators, the latter in particular enjoying a considerable boost in the mid-2010s in the UK despite their greenhouse gas implications.[12] The important thing is that all of this plant is dispatchable – available for operation unless, of course, it is closed down for maintenance or has broken down unexpectedly. A national grid does not entirely solve the problem of managing peak demand – at times of very high demand the real-time power price increases enormously to compensate those companies which keep power stations ticking over to fulfil that very high demand – but overall the system is vastly more efficient than the decentralised constellation which it replaced in the 1920s and 1930s.

In order to maintain the required quality of supply and ultimately to keep the lights on in unforeseen circumstances – say a particularly cold snap or a time when two or three large power stations break down unexpectedly – grid operators seek to maintain what is called a 'capacity margin' over the system as a whole. So if the highest demand expected in a particular year is around 60,000 MW, it was thought the ideal would be to keep around 72,000 MW of power capacity ready for use, in case the demand should happen to be 65,000 MW and up to 7,000 MW of power plant was unavailable because of undergoing maintenance or breakdowns, a highly unlikely but not impossible scenario. A capacity margin of about 20 per cent was generally regarded as sufficient if not generous, though from time to time it exceeded and indeed fell below this value. When a large power station does break down or has to be taken offline for maintenance or refuelling it is in principle a relatively straightforward task to bring another large station online. There is an inevitable tension, however, between security of supply and cost – in principle, a big capacity margin can deliver a very robust system which may be resistant to power cuts in extreme circumstances, but is expensive to maintain.

8.3 THE COMMAND-AND-CONTROL MODEL

One possible way of approaching this challenge is to give some entity monopoly powers over electricity generation and allow it to spread the costs of maintaining and operating peaking capacity more or less evenly over all units of electricity sold. Electricity in the postwar period, following the 1947 Electricity Act, was a state-owned affair organised in effect as a national monopoly. In England and Wales, generation (with a few minor exceptions, e.g. the Calder Hall military/power nuclear reactors) and transmission was the responsibility of the Central Electricity Generating Board (CEGB), a state-owned monopoly created in 1957 which operated coal, oil, nuclear and hydropower stations and owned the supergrid. The CEGB had a 'duty to supply' electricity and could invest in a range of sources, some of which would not be economic at least in the short term. Distribution (taking electricity from the national grid and transporting it to the end user) and supply (the relationship with the customer – metering, billing and the like) were the task of twelve area boards which had a local monopoly over the relationship with the end-use customer.

In general it was accepted that the CEGB was good at maintaining secure supplies, especially keeping the grid in good condition – its performance in restoring supplies after the devastating storms of October 1987 was impressive – and in addressing environmental issues such as acid rain. In principle it could invest in whatever mix of generating plant it felt might be needed to ensure secure and environmentally acceptable supplies and pass on the cost of that investment to its customers, the area boards, which could in turn use their monopoly powers to pass costs on to the end user. In the case of the Sizewell B nuclear power project, as an example, the associated public inquiry did have to consider whether the project would offer economic benefits sufficient to outweigh any detriments including potential health and safety risks and damage to the local environment. The inquiry inspector, Sir Frank Layfield, concluded in his report that in his judgment, 'the expected national economic benefits are sufficient to justify the risks that would be incurred'.[13] However, the risk of failing to deliver on this aspiration largely lay with the electricity consumer, not the investor. The CEGB would in effect 'build and operate' the required plant – the Sizewell B project was managed by the CEGB Sizewell B project management team. There was no need to set long-term power prices because if the original economic appraisal proved to be wrong then power prices would go up to compensate and captive electricity users would not be allowed to switch to suppliers to avoid this.

The regulatory regime was perhaps captured by the technocrats running the CEGB and it was difficult for lay people or even politicians to offer effective challenge. This was perhaps especially the case with regard to the nuclear programme.

The result was that, in theory at least, the CEGB could take and deliver upon decisions that it perceived necessary, decisions that private companies would not take owing to the economic risk involved, but these decisions were not always well scrutinised so economic efficiency suffered. Further, there was a suspicion that in discharging its legal duty to supply the CEGB tended to 'gold-plate' security of supply, again resulting in higher costs than were entirely necessary.

In practice the CEGB was prone to political interference that could further undermine effective decision-making, most notably in being forced to source a large amount of its fuel from the UK coal industry. At the time this was dominated by a militant trade union, the National Union of Mineworkers (NUM), which proved very willing to take on the government (notably in 1971, 1973 and 1981), generally with devastating success. The situation became effectively self-sustaining – the NUM could use its ability to disrupt electricity supplies to prevent governments from diversifying the fuel mix, thereby perpetuating its position.

8.4 THE ADVENT OF COMPETITION

By the mid-1980s political fashion was changing as many in the UK came to believe that the post-war consensus had been associated with severe social problems and economic decline. A significant gulf opened up between the two major political parties as to how the problem should be addressed, driven by a shared recognition that 'something had to be done'. The large-scale privatisation programme of the Thatcher government, elected in 1979, had begun – by 1983 concerns such as British Aerospace, Cable & Wireless, the National Freight Corporation, Britoil and Associated British Ports had been sold, with British Telecom to follow the next year. In the years up to 1984 Thatcher, determined that her government should not be destroyed in the way that its predecessor had been a decade earlier, outmanoeuvred the NUM by building up coal stocks at the power stations and reviving oil-fuelled generating capacity. The government duly survived a year-long mining strike in 1984/5. The more moderate Union of Democratic Mineworkers split off from the NUM. Further undermining of the coal industry became a central, if unspoken, aim of the privatisation of the power industry.

As negotiations over privatisation progressed in the late 1980s it became clear that it would not be possible to impose a 'duty to supply', i.e. a legal responsibility to ensure sufficient supply was available to fulfil demand such as had been exercised by the CEGB, anywhere in the new system. Clearly, to give responsibility for 'keeping the lights on' to any one generating company would be to place it at a fatal competitive disadvantage. It would have to carry the costs of maintaining plant which might only be used for a few hours per year (if at all), while its competitors could focus on providing a smaller amount of capacity to serve baseload requirements and at times of relatively low demand, and could do this at a much lower average cost per MWh, thereby gaining a large price advantage. Nor could a way be found to give 'collective responsibility' to the generators as a whole. Any individual generator which chose to 'freeload' on the others would always be at a competitive advantage. In any case the government was hoping to promote a competitive market where no player was big enough to wield significant market power. Imposing a general duty to supply on the market as a whole appeared incoherent: who would be punished for a failure to fulfil this requirement?

The challenges of the financial risks associated with accommodating low-carbon electricity in a market context were apparent right from the start. To manage the size and implications of nuclear investment and decommissioning costs the generating side of the CEGB was initially to be broken into just two competing companies. 'Big G' would hold about 70 per cent of the fossil fuel generation plus the nuclear plants, while 'Little G' would hold the remaining 30 per cent. Yet even this glaring imbalance was not enough and it proved impossible to privatise the nuclear power stations, which were generating around 20 per cent of the UK's electricity in the late 1980s. The older stations (Magnox) were too close to the end of their lives (one, at Berkeley, had already closed) and private companies buying the Magnoxes would have demanded a very high guaranteed power price, or other financial inducement, in order to be sure they would not be left with a net liability at the time when costs of waste management and decommissioning would have to be met. The second programme of nuclear power in the UK, the Advanced Gas-cooled Reactors (AGRs), had a very poor operational record (with average load factors around 50 per cent at the time) and once again potential buyers wanted unacceptably high guaranteed prices in case their output record remained disappointing. The projected third programme of nuclear power, to be based around a Pressurised Water Reactor (PWR) design imported from the US, proved undeliverable because private companies were not prepared to put up the large capital sums required when they could build much cheaper Combined Cycle Gas Turbines (CCGT) and get their money back much more quickly. The Magnox programme was withdrawn from the privatisation process in July 1989,[14] to be joined by the AGRs and the construction project at Sizewell B that November (on the same day as the Berlin Wall fell). Although it was agreed that Sizewell B should be completed, coming online in 1995, a five-year moratorium was placed on any further nuclear new build.

Privatisation proceeded with the CEGB being broken into four parts. 'Big G' was christened National Power and still held about 70 per cent of the fossil fuel capacity of the CEGB, there being insufficient time to tear up the privatisation plans and start again. 'Little G', named PowerGen, still held the other 30 per cent. The nuclear plants were placed in a new public company, Nuclear Electric plc – a sister organisation, Scottish Nuclear Ltd, was created north of the Border – while ownership of National Grid was divided among the twelve area boards, who were renamed 'Regional Electricity Companies' (RECs) and privatised separately.

8.5 LIBERALISATION AND THE DASH FOR GAS

Dominating the story of UK electricity in the 1990s is the 'dash for gas' as liberalisation was rolled out in stages. Gas rose from a zero market share in 1990 to account for about 40 per cent of electricity generated in 2000 – see figures 8.2 and 8.3.

For a brief period lasting a little over a decade a single policy, the switch from coal to gas, was serving all the requirements of electricity policy – cheap (the global gas price had fallen dramatically, on the back of falling oil prices, from the highs of the 1970s), secure (North Sea gas was plentiful – the UK became a gas exporter – and a lot of new CCGT power capacity had been constructed), relatively clean (generating half as much carbon dioxide per unit of electricity as is generated using coal) and in general not causing public unrest or political jitters – indeed, it was extremely popular to supporters of the government as it significantly reduced the power of the NUM.

A White Paper in 1994 in effect confirmed that there was no case for nuclear new build, though it did set the stage for the privatisation in 1995 of the AGRs (including

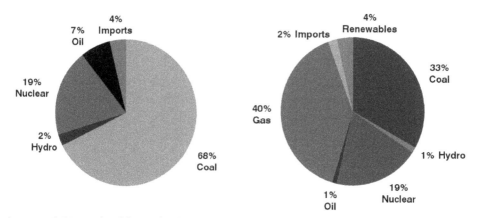

Figure 8.2 [left] UK electricity production by fuel, 1990
Figure 8.3 [right] UK electricity production by fuel, 2000[15]

Hunterston B and Torness in Scotland) and Sizewell B as British Energy, their technical performance having improved substantially.[16] Ownership of the Magnox stations, including Hunterston A in Scotland, were transferred to a new state-owned company, Magnox plc, later incorporated into BNFL.

Over the course of the 1990s the size, and consequent market power, of National Power and PowerGen was reduced through forced divestment of generating assets, while these companies were not allowed to buy the supply businesses of the RECs until well into the decade. By 2000 there were several dozen companies generating electricity for sale through the UK grid and there was a very healthy capacity margin (from the point of view of the consumer) – most of the generating capacity that had existed in 1990 was still available and it had been joined by a new tranche of CCGT.

8.6 THE INVESTMENT PROBLEM

As previously suggested, companies in a competitive marketplace have less of an incentive to invest in new power capacity ahead of supply shortages than monopolistic entities with a duty to supply.

Liberalisation also affects the relative attractiveness of different sources of power. Even where levelised costs – defined as the present value of the total cost of building and operating a generating plant over its economic life, converted to equal annual payments, adjusted to remove the impact of inflation and expressed in terms of cost per unit generated – are expected to be similar, in a competitive market CCGT has a very heavy advantage over competing technologies when it comes to new build. Its construction costs are lower, and it is quicker to build, than is the case with coal or nuclear power plants so the initial investment is recouped relatively rapidly. In 2012 DECC's calculations of the cost profiles of dispatchable capacity were as shown in figure 8.4.

It will be noted how influential assumptions about the value of carbon emissions are on levelised power costs. Ignoring carbon costs, the differences are even clearer, as shown in figure 8.5.

The disparity is a little less if carbon capture and storage (CCS) is required for the gas and coal plants, as this increases the capital costs of fossil-fired plants significantly, though still not to nuclear levels, as seen in figure 8.6.

For CCGT the key economic sensitivity is the gas price; but if gas prices do rise significantly so do power prices, especially for household consumers who have relatively little market power, so the risk is largely borne by the consumer rather than the investor. Further, CCGT output is reasonably flexible, unlike nuclear power, at least in the absence of the ability to store large quantities of electricity. At present nuclear is more or less stuck in the baseload – it makes little sense to load-follow as the total costs of a nuclear plant are all but independent of how much power it is generating, the fuel cost being a very small proportion of that total. So it effectively has to take what it can get rather than playing a role in setting the price like the more flexible generators can.

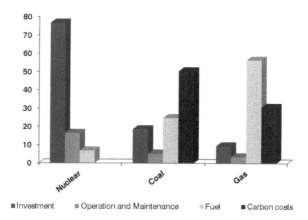

Figure 8.4 Percentage UK levelised costs of unabated methods of generating electricity at 10% rate of return for projects starting in 2018[17]

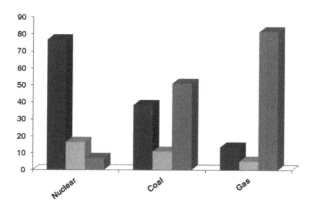

Figure 8.5 Percentage UK levelised costs of unabated methods of generating electricity at 10% rate of return for projects starting in 2018 – excluding carbon costs

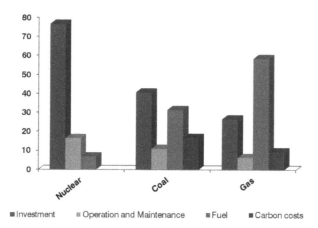

Figure 8.6 Percentage UK levelised costs of methods of generating electricity at 10% rate of return for projects starting in 2018 – including CCS and carbon costs

In principle, if nuclear electricity could be stored it could become almost as flexible as a gas turbine or coal-fired plant depending on how efficient the storage technology is. This would put nuclear in a much stronger market position with respect to the flexible technologies, while it remained reliably available when called upon. As noted earlier, large-scale storage has been the holy grail of the power industry since its invention and remains elusive.

Whether intentional or not from the government's point of view, the rise of CCGT further weakened the strength of the NUM by reducing reliance on domestically mined coal. CCGT new build was incentivised in the following fashion. The large new generating companies – National Power and PowerGen – were prevented from owning supply companies. However, contestability in supply was only fully in place by the 1998/9 financial year – before that date more than half of the market by volume was still franchised to the local REC, i.e. most electricity was still sold to the end user under the old monopoly powers. The RECs, responsible for distribution and supply, were allowed in effect to build their own CCGT generation capacity and to pass the construction costs on to their captive customers, i.e. those using less than 1 MW from 1990 to 1994 and those using less than 100 kW from 1994 to 1999. The result, by 2000, was an electricity market well stocked with capacity (older coal and nuclear power stations and new CCGTs), producing extremely cheap electricity with lower carbon emissions, thereby easily fulfilling the UK's commitments under the Rio Convention and subsequent Kyoto Protocol.

8.7 CHOPPING THE LEGS FROM UNDER THE GENERATING COMPANIES

In 2001 the government introduced the first major rewriting of the rules of the power market, having already tinkered with them on a number of occasions; for example, a ban on connecting new gas-fired stations to the grid in 1998 in order to protect the coal industry, a measure which was lifted after 18 months. The New Electricity Trading Arrangements (NETA) abolished the power 'pool' through which all sales were being made. Previously, the system operator, the independent National Grid Company, would take bids for output from any available capacity. National Grid would estimate demand in half-hour periods and call up enough capacity to fulfil that demand, starting with the cheapest bids. The pool price for a particular half-hour would be the bid price offered by the last (i.e. most expensive) generator to be called, plus an amount to cover transmission, distribution and other costs. The pool price would obviously be higher at times of high demand than at times of low demand. This meant that baseload power producers, notably nuclear plants, could bid at a very low price (often zero) and be guaranteed of receiving the market price pertaining, so guaranteeing themselves a market for their entire output at a reasonable price. In practice a significant proportion of the electricity delivered had been bought and sold on longer-term contracts, e.g. week-ahead or year-ahead, with 'Contracts for Difference' (CfD) working alongside the pool mechanism to give stable prices. However, unlike many European power markets the contract length in the UK remained relatively short, typically one year or less.[18]

NETA allowed for bilateral 'over-the-counter' trading, pushing power prices down as baseload providers would only be paid what they offered. The capacity payments which had been a feature of the pool market, to reward generators for keeping plant available as a backup against unexpected increases in demand or breakdowns, were also abolished – there was so much capacity scrambling for a market that they were unnecessary. In effect, NETA made it impossible for 'pure' power generating companies to make any return on the investment they had made in the 1990s or before. Many of the seeds of the subsequent impasse on investment decisions were sown in this period: if governments could change the rules so drastically to prevent companies making a fair rate of return once then maybe they would do so again.

In the early years of the new century the wholesale price of power fell to unsustainably low levels. By now the barriers to vertical integration – companies owning both generation and supply businesses – had been relaxed to allow National Power and PowerGen (and British Energy, though it did not really take up the offer) to enter the supply market. By now, however, their share of the generation market was very modest and their market power even less. The roots of the 'Big Six', which came to dominate the market by the 2010s, were set down as smaller companies could not sustain a presence in such a depressed market and the Americans, who had bought considerable amounts of generating capacity in the 1990s, began to quit the market altogether.

Those generators with domestic supply businesses were to a considerable extent insulated from the effects of the wholesale price collapse as domestic customers proved relatively unwilling to shop around among potential suppliers. As a result household power prices remained quite high – small consumers were behaving as though they were still captives of their local power supplier even when they were given the right to switch. Highly complex tariff structures and a series of selling scandals, whereby customers' accounts were fraudulently switched from one company to another, further undermined domestic customers' confidence in changing supplier.[19] Indeed, it can be argued that the case for retail competition was always fatally flawed. If the wholesale power market was working then everyone would be paying almost the same for their electricity. Distribution and transmission charges are standard and would be the same for all supply companies. A retailer could only gain advantages in the area of their own costs, which account for only about 10 per cent of the total cost being charged to consumers. The differentiation available would be unlikely to compensate consumers for the inconvenience and uncertainty involved in switching suppliers.[20] And competition itself brings an extra set of costs in administering whatever switching is occurring, marketing the service and so on, costs that will ultimately have to be paid through higher power bills. Of course, if the wholesale market is failing – if, for example, vertically integrated companies can charge their own supply businesses a lower wholesale price than they would charge their competitors –

then there could be a value in consumers switching, but any market in which one part can only succeed if another fails is clearly unlikely to be either efficient or sustainable.

By contrast, companies with big power bills shopped around vigorously and took full advantage of low wholesale prices. The outcome was that nobody made much money out of electricity, certainly not enough to persuade them to invest in new capacity – not that any new capacity was needed – but the generating companies which did not own supply companies were especially hard hit. In 2002/3 some 40 per cent of the generating capacity, including all independent power producers who did not have a retail supply business, was effectively bankrupted, including AES Drax, owner of the UK's largest coal-fired power station, and British Energy.

Unsurprisingly, a White Paper on energy, *Our energy future – creating a low carbon economy*, published in 2003, in effect dismissed nuclear power as a serious option, saying:

> Nuclear power is currently an important source of carbon-free electricity. However, its current economics make it an unattractive option for new, carbon-free generating capacity and there are also important issues of nuclear waste to be resolved ... This White Paper does not contain specific proposals for building new nuclear power stations. We do not rule out the possibility that at some point in the future new nuclear build might be necessary if we are to meet our carbon targets. Before any decision to proceed with the building of new nuclear power stations, there will need to be the fullest public consultation and the publication of a further White Paper setting out our proposals.[21]

Indeed, the whole tone of the White Paper was quite relaxed, if not complacent. The market was delivering secure and cheap electricity to customers while greenhouse gas emissions were falling – what need was there for radical decisions? What need was there for decisions of any description?

8.8 CHANGING PERCEPTIONS 2003–08

By the middle years of the decade things no longer looked quite so comfortable. North Sea gas was becoming depleted and concerns about the implications were exacerbated, for example, when Russia cut gas supplies to Ukraine on New Year's Day in 2005 and again in 2009,[22] a situation that became even more worrying when conflict between Russia and Ukraine erupted in 2014.

Awareness was also growing that the UK was not investing in new electricity generating capacity of any description, beyond a few heavily subsidised renewable schemes, while a lot of existing capacity would be soon coming offline. Many large coal-fired stations would have to be retired from 2015 as a result of the European Large Combustion Plant Directive aimed at reducing sulphur emissions, and nuclear power stations were also nearing the end of their lives. All of the Magnox nuclear reactors were closed by the end of 2015 and the AGR programme was expected to be largely over by the late 2020s even with ongoing lifetime extensions.

With regard to the economics of electricity production, coal and particularly gas prices increased enormously through the decade, as shown in figures 8.7 and 8.8 respectively.

Power prices increased in tandem (figure 8.9), causing considerable political and social disquiet.

Environmentally, the reduction in carbon dioxide emissions seen during the 1990s slowed as the dash for gas peaked – high gas prices saw increasing use of coal for electricity generation, though coal prices also rose during the period – and demand for electricity edged upwards.

In 2007 the government responded by publishing an energy White Paper, *Meeting the Energy Challenge*, with a noticeably more urgent tone than its predecessor just four years earlier.

> The UK will need around 30–35 GW of new electricity generation capacity over the next two decades and around two-thirds of this capacity by 2020. This is because many of our coal and most of our existing nuclear power stations are set to close. And energy demand will grow over time, despite increased energy efficiency, as the economy expands.[23]

The following year a nuclear White Paper was published. The contrast to 2003 was remarkable. In the foreword the prime minister, Gordon Brown, said:

> The government has today concluded that nuclear should have a role to play in the generation of electricity, alongside other low-carbon technologies. We have therefore decided that the electricity industry should, from now on, be allowed to build and operate new nuclear power stations, subject to meeting the normal planning and regulatory requirements. Nuclear power is a tried and tested technology. It has provided the UK with secure supplies of safe, low-carbon electricity for half a century. New nuclear power stations will be better designed and more efficient than those they will replace. More than ever before, nuclear power has a key role to play as part of the UK's energy mix. I am confident that nuclear power can and will make a real contribution to meeting our commitments to limit damaging climate change.[24]

Despite these clearly very positive comments, there remained a lack of clarity about what the government was actually saying. In stating that 'more than ever before' nuclear power should have a 'key role' to play in the generation of electricity, the prime minister was clearly saying that nuclear new build would be in the national interest, which would seem to imply that government's role was to ensure that it should happen. But in stating

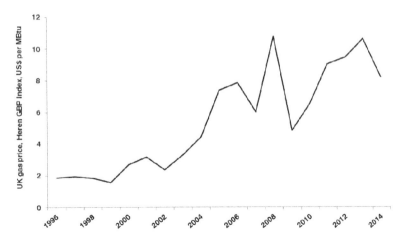

Figure 8.7 North-west Europe coal price (US$ per tonne), 1996–2014

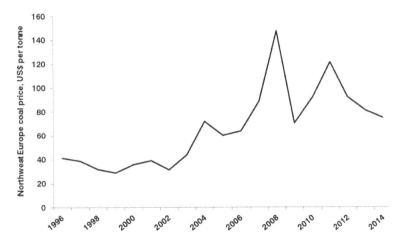

Figure 8.8 UK gas price, Heren NBP index (US$ per MMBtu), 1996–2014

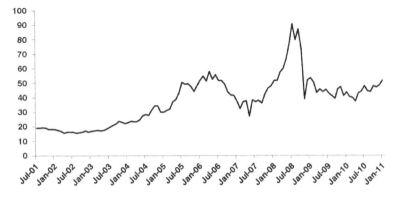

Figure 8.9 UK wholesale electricity price – October year ahead (£ per MWh)

that companies 'should be allowed' to build new nuclear reactors he seemed to be saying the opposite – that it was not the government's job to ensure the long-term robustness of UK electricity policy and generation.

Inherent in the concept of the command-and-control approach to electricity provision is the idea that electricity is basically a social and industrial service, of such importance to the economy, the environment and society at large that government has to play the role of guarantor of last resort. This is not of course to say that market forces may not be harnessed to improve value for money, for example by encouraging competitive tenders to build new coal-fuelled or nuclear power stations, but the outcome is determined ultimately by government decree not market logic – i.e. the ultimate responsibility for how much coal, nuclear, renewable or other plant should be built lies with the national government, not the individual decisions of 'private' investors or foreign governments.

Inherent in the concept of a more or less competitive market in which private companies take decisions on whether to invest in generating capacity or not is the idea that electricity is basically a commodity like any other. In the words of Nigel Lawson, the energy secretary in 1982:

> Energy is a traded good. I do not see the government's task as being to try to plan the future shape of energy production and consumption. It is not even primarily to try to balance UK demand and supply for energy. Our task is rather to set a framework which will ensure that the market operates in the energy sector with a minimum of distortion and energy is produced and consumed efficiently.[25]

The inescapable corollary is that it is up to investors to decide how and indeed whether to value security of supply and environmental emissions. Government can nudge those companies in the 'right' direction in the way it sets up the market. It can put a negative value on environmental emissions, for example, through carbon taxes or a regime of tradable emission permits. It can also decide whether to value security of supply – treat it as an externality – by introducing capacity payments to run alongside payments for energy output. But then it must essentially leave that market alone for very long periods of time – decades – if firms are to feel confident that they understand the environment in which their long-term investments will be operating.

Brown's foreword seemed to fall between two stools. In effect the government was simultaneously speaking as though electricity was a social and industrial service and that it was to be regarded as a commodity, with government's role being to set up a stable market which put a value on economic externalities but then leave it to the players in the marketplace to determine the final outcome.

One might think that is fair enough – after all electricity has features of both commodity and social service so why not treat it as both? The problem is that trying to pursue both of these contradictory philosophies at the same time can lead to a situation less satisfactory than either of the extremes. A command-and-control approach, not necessarily with public ownership but with central planning of the electricity supply system, can in principle deliver secure and environmentally sensitive supplies, though it may well fail to do so with economic perfection – it will doubtless pick many 'losers' alongside the technological 'winners' – and could be vulnerable to political interference. A stable market, where investors can have confidence that the rules will not frequently change in pursuit of short-term social goals or political fashions, e.g. caps on power prices at times of high demand, might well invest in a timely programme of new build, mainly CCGT unless carbon penalties were very high. But the UK government had had to divest itself of many of the managerial levers of intervention when setting up the market; it then regularly tried to intervene at the system design level, thereby undermining confidence in investing. Experience of the first quarter century of the liberalisation age suggests a major rewriting of the whole framework every 12 years (1989, 2001 and 2013), with tinkering of this subsidy or that market rule on an annual basis. The average tenure of office of an energy minister was one year or less – there were five incumbents between June 2012 and June 2015 though all belonged to the Conservative Party. This is not an environment to attract investment in technologies such as nuclear power or large-scale renewables with a financial horizon of several decades.

Nor was this a unique UK experience. One analyst listed 27 political interventions between January 2010 and September 2011 in Europe alone which had resulted in €200 billion in lost shareholder value in France, Germany, Spain, Portugal, Finland, Italy, the Czech Republic and the UK.[26]

In essence the tension arises over how risk is to be managed and apportioned. For reasons already discussed, investment in low-carbon energy is a much riskier business than is the case with CCGT. Private companies operating in a competitive marketplace, as opposed to regulated monopolies, demand considerably higher rates of return on capital invested to compensate for the possibility that they may fail to compete successfully. This acts to the disadvantage of sources of power which have high initial costs needing to be funded either through borrowing or through offering attractive dividends to shareholders. As if this were not sufficiently challenging, these high-capital low-operational cost sources, being more expensive and taking longer to build, require reliable income and a stable operating environment for much longer periods into the future than sources which can repay their initial investment more quickly. This means that investors require an even higher rate of return still for them to choose to invest in nuclear or renewables rather than CCGT or indeed relatively safe non-energy investments such as supermarkets or shoe manufacturers. As the costs of producing low-carbon electricity are not far short of being proportional to the rate of return on the initial investment, while as noted the sensitivity of the cost of CCGT-generated power is much less dependent on rates of return and much more on the cost of the fuel, this has an enormous effect on the relative economics of various ways of making electricity.

As a result neither nuclear power nor renewables have been able to stand on their own feet in largely liberalised markets. If governments wish to persuade investors to find such projects, then ways need to be found of transferring a considerable proportion of the economic risk to either the taxpayer or the electricity consumer. Of course in the command-and-control approach most of that risk lies with the government in the first instance and ultimately with the power consumer who must pay the rates set by the monopoly which serves them. The quid (or several billion quid) pro quo is that the nominal rate of return offered to investors, where there is private ownership, is much lower as the likelihood of failing to achieve that return is less.

The tension between viewing electricity as commodity or social service, and the effects of trying to mix the two, became very clear as the UK government tried to find ways of persuading 'private' investors – including agencies largely or totally owned by foreign governments – to provide funding for the renewables and nuclear new build programme that it deemed desirable.

8.9 A LIBERALISED SYSTEM IN CRISIS – CALIFORNIA 2000/1

15 years or more on from its power crisis California still stands as a beacon illuminating the risks of trying to mix and match markets and non-market interventionism by setting up a power market riven with regulation, perhaps most damagingly in the form of price caps, designed to drive the market where policymakers want it to go.

California is the most populous state in the US, with around 39 million people. Its economy – the eighth largest in the world – is dominated by technology industries, for many of which uninterrupted electricity supplies are of great importance. Installed electricity capacity is of the order of 80 GW and in 1998, the year in which liberalisation of power supplies was introduced, the state used 276 TWh of electricity, some 17 per cent of which was imported.[27] In 2002, just after the crisis, the mix of power sources, including imports (which accounted for 23 per cent of the total) was as shown in figure 8.10.

Power use had been increasing at a steady rate for some decades, rising by a total of over 40 per cent between 1982 and 2000. However, in common with the rest of the western region with which it is and was interconnected, commissioning of new capacity in California was slow in the 1990s. There were a number of contributory factors, many of them typical of a 'phase 2' society in the sense that is outlined in chapter 10 – a regulatory regime which made it hard to make reasonable profits from generating plant; difficulties in obtaining licences for environmental reasons; uncertainty caused by the announcement of impending market rule changes with the introduction of liberalisation. No applications for new plants of over 80 MW were filed between 1994 and 1998, a situation which was to find its parallel in the UK after 2000.

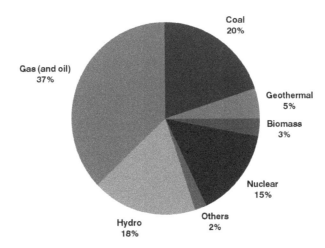

Figure 8.10 California electricity supply, 2002[28]

Before liberalisation, the Californian electricity system, like those in many US states, consisted mainly of vertically integrated investor-owned utilities. Approximately 80 per cent of the state's power was delivered by the three biggest utilities (referred to as simply the 'utilities' in the rest of this discussion):

- Pacific Gas and Electric (PG&E) in the northern half of the state, including San Francisco;
- Southern California Edison (SCE) in the southern half of the state;
- San Diego Gas and Electric (SDG&E) serving the city of San Diego at the southern tip of the state.

The utilities owned power plants plus transmission and distribution facilities and sold power to individual retail customers within their franchise areas at prices set by the California Public Utilities Commission (CPUC). The rest of the state's power was provided by several municipally owned utilities, the largest of which, indeed the largest in the US, was the Los Angeles Department of Water and Power (DWP). Although these were interconnected with the wider California grid systems they were under local democratic control and not subject to price regulation by the CPUC.

The main exception to the vertically integrated structure involved the independent power producers known as Qualifying Facilities (QFs) under the Public Utility Regulatory Policies Act (PURPA), introduced after the oil shocks of the 1970s to encourage investment in renewables. They were heavily subsidised in the form of long-term guaranteed prices at levels well above market rates, selling all of their output to the local utility. By 1992, 9.5 GW of QF capacity was generating over 25 per cent of the electricity requirements of the three utilities.[29]

As California entered the 1990s it had an overgenerous capacity margin of 40 per cent and relatively high power prices. Household power prices charged by the utilities were in the order of 9¢ to 10.5¢ per kWh, some 30 to 50 per cent above both the national average and the competitive rate for new supplies.[30] Deregulation in other industries such as trucking and telecommunications seemed to be having positive consequences.

In September 1992 the CPUC launched a review of the Californian power market.[31] As a result the utilities were strongly encouraged to divest themselves of their fossil fuel generating plant in return for a deal which would allow them to recoup their 'stranded costs', the debts which they had run up on the assumption that they would be able to recoup the investment from their captive customers in due course. These costs were estimated at between $21 billion and $25 billion at the time the new market was established.[32] This deal consisted of their being allowed to charge their customers rates well above the market price until 2002 or until the stranded costs were paid off, whichever came first. Customers could not escape the stranded cost repayments by switching supplier. In return an apparently generous retail price cap of $70 per MWh for industrial customers was imposed on the utilities.

Two new not-for-profit bodies were established:

- the Power Exchange (PX), which operated a day-ahead hourly spot market for auctions between generators and purchasers;
- the California Independent System Operator (CAISO) to manage system aspects.

The utilities were mandated to buy their electricity only from the PX – they were not allowed to sign bilateral 'vesting contracts' with their own generating output or independent producers – and had to sell their own real and virtual (QF) output through the PX. CAISO's responsibilities included making sure their transmission wires did not become overloaded, e.g. by making some north California plants shut down and commissioning power from more expensive southern ones on days where there was excess demand in the south of the state.

Consumers were eligible to switch their demand to alternative retail providers but for a variety of reasons, including the requirement that they should not be able to avoid making their contribution to paying off the utilities' stranded costs, very few did so. By early 2000 only 1 per cent of residential consumers and 20 per cent of industrial ones had switched supplier. Enron abandoned the household supply market within just three weeks.[33]

In order to gain public support for the whole scheme, the retail price charged by the utilities was to reduce by 10 per cent in the first four years of the liberalised system, which after some delay started in April 1998. It was assumed that the gap between the (capped) price charged to customers and the wholesale price the utilities would be paying through the PX (plus distribution charges) would be sufficient to allow for the stranded costs to be paid off within that four-year period.

The utilities duly divested themselves of their fossil fuel generating capacity, in fact even faster than they were mandated to do. SCE sold most of its thermal units in the two months following the opening of the market, PG&E sold most of its units in the first year and SDG&E divested its thermal units in mid-1999. Five companies – AES, Duke, Dynergy, Reliant and Southern/Mirant – each bought about one fifth of the divested capacity. The fact that these plants were sold for high prices – a total of some $3.1 billion against a book value of $1.8 billion[34] – suggests that the purchasers expected power prices to rise significantly in due course. The utilities retained responsibility for the long-term contracts with QFs, the nuclear plants running at baseload and the hydropower stations.

In its initial months the PX worked smoothly, with an average wholesale price of $26.2 per MWh in 1998/9 and $31.2 per MWh in 1999/2000. At these wholesale prices, well below the retail price cap of $70 per MWh, the utilities could make a considerable contribution to paying off their stranded costs, some $17 billion of debt being retired in those two years,[35] with SDG&E paying off all of these costs by the end of 1999. There were occasional shortages in CAISO's reserve market, in which power was bought and sold in real time rather than on fixed price contracts in order to balance supply and demand, and prices there reached $9,999 per MWh in July 1998. The institutional response was to impose a cap of $500 per MWh, raised to $750 per MWh in October 1999, which may have appeared a clever way of preventing 'profiteering' but in effect put a cap on the power exchange prices too. If utilities could buy as much power as they wanted in the real-time market they would have no incentive to pay higher prices in the day-ahead market – which one would only do precisely to avoid the danger of being caught with a price of several thousand dollars per MWh – introducing a serious level of uncertainty into system planning.[36]

Late in the spring of 2000 the market began to fail. In June wholesale prices averaged $132 per MWh. CAISO reduced the price cap in the day-ahead market to $500 per MWh on 1 July 2000 and to $250 per MWh in August.[37] Prices stabilised a little, only to spike dramatically in December (see figure 8.11), when prices averaged $386 per MWh as significant amounts of capacity were moved into CAISO's reserve market with its more generous $750 per MWh price cap.

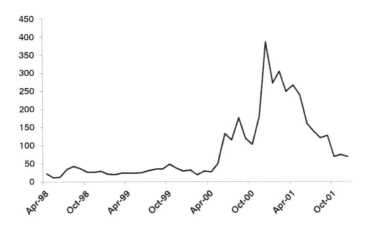

Figure 8.11 Average monthly wholesale power price in California ($ per MWh)[38]

By the end of January 2001 the collapse was complete. Blackouts occurred on eight days between June 2000 and May 2001, the PX suspended operations and the CAISO, SCE and PG&E were by now all insolvent.

The causes seem to be a mixture of rising costs to the utilities, naïve price caps (there may not be any other kind) and misbehaviour by many key market participants, itself a function of a complex regulatory structure constructed to deliver on market goals. As with any other welfare system (as this effectively was) this was full of potential for abuse, notably manipulation of the market price. Many billions of dollars in fines and indeed several prison sentences were subsequently to be imposed.

The position of the utilities was relatively straightforward. Being almost completely exposed to the spot market, they were facing the impossible situation of paying very high wholesale prices for power from the PX. A precipitate rise in the gas price, on which the marginal price for the whole system depended, caused in part by the breakdown of a major gas pipeline, and the shortage of hydropower pushed the average wholesale price above \$100 per MWh from June 2000 to September 2001. However, the price the utilities could charge their industrial customers was capped at \$70 per MWh (7¢ per kWh). Inevitably they became insolvent, with huge debts. With the benefit of hindsight, the utilities would have benefited considerably from being less exposed to the spot market by taking more forward positions but even this would simply have pushed the problem further upstream and threatened the solvency of gas producers. PG&E filed for bankruptcy in April 2001 and it took over two years for a compromise between CPUC and PG&E which allowed the utility to keep trading under state direction.

The situation with respect to generation was more complex. The first point to note is that only one of the eight blackouts occurred when demand exceeded 35 GW – one was even on a Sunday. The one in which demand did exceed this occurred because of transmission restraints rather than a shortage of available power capacity. Peak demand in 2000 (43,500 MW) was actually lower than in 1999 (45,600 MW).[39]

That said, low rainfall in the winters of 1999/2000 and 2000/1 significantly curtailed the availability of hydropower in the region, affecting both California's own capacity (the hourly average hydropower on the system fell from 4.4 GW in 1999 to 2.6 GW in 2000) and the ability of other states in the western region to export.[40] The shortage of investment in new capacity in the western region generally in the 1990s certainly did nothing to ease this. However, it is difficult to ascribe the problems of 2000/1 to a genuine shortage of electricity capacity – as noted earlier capacity margins were on the face of it more than adequate. The problem was that existing capacity was not available. It was more profitable for the generating companies to create power cuts than to fulfil demand.

A number of other factors were involved. There was evidence of abuse of market power and the generating companies seem to have deliberately precipitated the crisis in order to inflate their prices. Some generating companies, notably those trading through the PX which did not rely on gas or hydropower, made very high profits, as consumers were so dependent on electricity that they would pay extremely high prices to maintain supplies at times of shortage.

However, the whole fiasco cannot simply be ascribed to a few greedy power company executives. The price caps which pervaded the system had counterproductive outcomes in terms of removing price signals to consumers to reduce demand while forcing some players into insolvency when their input prices rose uncontrollably. Power companies had the option of exporting their output to neighbouring states which had not imposed price caps, either for use there or, even more bizarrely, to reimport into California as the CPUC had no powers to impose price caps on companies in other jurisdictions. (This became known by the delightful phrase 'megawatt laundering'.) The cost of using the wires for such an exercise, though substantial, was dwarfed by the profits available for selling the power at prices well above the market cap of $250 per MWh.

One further element in the withdrawal of plant involved the QFs. As soon as it was obvious that the utilities to which they sold their output were going bust and would be unable to pay their bills, many of the QFs simply stopped generating. During the March 2001 blackout, for example, about half of the 3,100 MW of capacity contracted to utilities by QFs was withdrawn as they had not been paid for three to four months.[41]

The final nail in the market's coffin came when the Washington-based Federal Energy Regulatory Commission (FERC) intervened in the wholesale market in late 2000.[42] Its badly designed interference reduced the volume of power going through the PX to practically zero, as most of the non-utility generators shifted their sales into the real-time market to avoid price capping. CAISO was taking a 'keep the lights on at all costs' approach which was leading it to pay extremely high prices in the reserve market. Of course anyone bidding into this market faced the possibility that they might not be called but if they were they could earn several times more than the $250 per MWh cap in the PX, making the gamble an attractive one. And of course the more plant that 'broke down' or was otherwise unavailable, the more the likelihood that what plant was available would be called and that prices would be extremely high. CAISO was therefore operating in violation of the new FERC 'soft cap'. PX suspended operations in January 2001 and declared itself bankrupt in March 2001. Meanwhile, PG&E and SCE defaulted on payments owed to the PX, causing the PX in turn to default on payments to CAISO for balancing generation. CAISO became insolvent as well: the whole market had collapsed. The only entity in a fit state to take over was the California Department of Water Resources (CDWR) which, via emergency legislation, was allowed to procure electricity under secret but no doubt very worthwhile terms, at least from the point of view of power generators. Even so power cuts persisted until May 2001.

While all this was going on, the municipally owned DWP in Los Angeles, with 3.8 million customers and which remained vertically integrated while SCE, PG&E and

SDG&E were divesting generating assets, benefited significantly by retaining an orderly supply for its own customers while profiting from sales to the rest of the state from capacity that it could bring out of mothballs for the purpose. Prior to 2000 the prices in Los Angeles were rather higher than those which were being paid through the PX, in order to compensate for unused reserve capacity[43], but in the particular circumstances that faced California in the five years following the introduction of the market reforms in 1998 it was clear that Los Angeles had done well to keep out of it.

In June 2001 the crisis dissolved almost as rapidly as it had emerged. The gas price, which had gone from around $2 to $3 per MMBtu before the trouble to $12 per MMBtu, fell back to $5 per MMBtu in June and normal levels by September, bringing down power generation costs for gas-fired generation and prices for the whole system because in general marginal gas plants set the price in the PX.[44] The FERC capped prices in the whole western region, thereby removing the incentive for megawatt laundering. Generators were mandated to offer all available capacity and the market price was once more set at the highest accepted bid.

Many of the tenets of the liberalisation of California's power markets were abandoned. The state tied itself into long-term power contracts at high prices by historical standards and prevented direct-access contracts between consumers and generators at lower prices. Price caps were relaxed and the utilities moved back into plant ownership.[45] Things settled down, Governor Gray Davis was 'recalled' and sacked by the electorate in 2003 and new Governor Arnold Schwarzenegger promised to reintroduce competition but 'make it work'.

Predictably, the crisis was blamed by those on the Left on too much liberalisation.

> We must face reality. California's deregulation scheme is a colossal and dangerous failure. It has not lowered consumer prices. And it has not increased supply. In fact, it has resulted in skyrocketing prices, price gouging and an unreliable supply of electricity. In short, an energy nightmare. We have lost control over our own power. We have surrendered the decisions about where electricity is sold, and for how much, to private companies with only one objective – maximising unheard-of profits.[46]

Equally predictably, those on the Right blamed too little liberalisation.

> Deregulation never really took place in California. Instead, political forces imposed a contrived market structure that made failure almost predictable. California's disaster was of its own making and largely avoidable.[47]

> The California electricity crisis is not really a story about environmentalists gone bad, deregulatory details ignored or capitalists running amok. It's a story about what happens when price controls are imposed on scarce goods.[48]

The last point is a serious one. Imposing price caps, or threatening to do so, is a tried and tested way of increasing the price of something, especially something rare enough for the price cap to matter. As discussed later, the Labour Party's announcement that if elected it would do the same, at an even lower level, appeared to deter generating companies from reducing their prices in the run-up to the 2015 election.

However, in a sense both sides of the argument are right. The experience of the unliberalised Los Angeles DWP showed that although power prices were higher than in the liberalised market areas for a couple of years after deregulation, the command-and-control system was in a much better position when it came to dealing with severe challenges when they arose. On the other hand, had the utilities been allowed to raise their power prices to sustainable levels when the wholesale power they were buying became vastly more expensive, rather than being forced into bankruptcy by the price caps under which they laboured, supplies would presumably have been maintained, there being no capacity-based reasons why not, and prices would therefore not have peaked as high nor remained there so long as they did. As a worked example of how toxic the mixture of a competitive marketplace and a regulatory system forcing the market to go where it doesn't want to go, California should not be forgotten.

8.10 THEORY INTO PRACTICE, UK 2008–13?

By the end of the first decade of the new century the size of the challenge was enormous. In 2011, for example, analysts were calculating that the UK would have to spend €320 billion over the following decade (the figure for all Europe being €938 billion) to replace its asset base in electricity and water and meet its environmental commitments, but that the maximum amount the investors would provide would be €181 billion. One view was expressed as a serious of 'even ifs':

> **even if** – the utility companies across Europe had the appetite to spend €938 billion they didn't have the organisational capacity to do so;
>
> **even if** – they had the organisational capacity to spend the money the supply chain couldn't provide the equipment;
>
> **even if** – the supply chain could provide the equipment the utility companies didn't have the balance sheet to finance the investment;
>
> **even if** – the utility companies could raise the equity they wouldn't be able to afford the cost; and
>
> **even if** – the utilities could finance the investment, the consumer wouldn't be able to afford their bills.[49]

As far as UK electricity capacity went, plant closures were predicted significantly to outstrip new build as coal and nuclear stations reached the end of their lives – see figure 8.12.

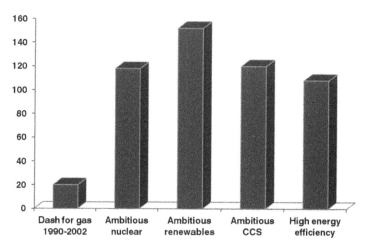

Figure 8.13 New capacity requirements in UK compared to the 'dash for gas' (£ billion), 2014–30[50]

As a result the gross new build requirements over the period up to 2030 dwarfed the investment which had been associated with dash for gas in the 1990s, as shown in figure 8.13.

The incoming Conservative–Liberal Democrat coalition government elected in 2010 adopted the policy stance of its predecessor with all of its lack of clarity. The coalition agreement talked of 'reforming electricity markets' but also lists a range of specific outcomes that it will achieve in terms of secure supplies, low prices and reductions in emissions. In a single interview in June 2012, Ed Davey, Secretary of State for Energy and Climate Change, made the following, apparently contradictory, statements.

- Britain could survive without nuclear power.
- Can nuclear consortia build power stations with no public subsidy? I don't know, we'll see.
- Nuclear power is an essential part of the country's energy mix.[51]

Like its predecessor, the new government sometimes spoke as though it regarded itself as the guarantor of last resort with respect to secure low-carbon power supplies, sometimes as though ultimately it was up to the market. What would a potential investor make of such apparently mixed messages? The issue was complicated yet further by a fudge necessitated by the need to create a workable policy out of the most instinctively pro-nuclear and anti-nuclear of the main parties (Conservative and Liberal Democrat respectively). It is often claimed that the wording alighted upon was that the new government would 'allow the replacement of existing nuclear power stations provided that … they receive no public subsidy.'[52] In fact this is wrong – the coalition agreement stated the difference between the Liberal Democrat and Conservative election manifestoes, acknowledging that the Liberal Democrats opposed building new nuclear reactors while Conservatives supported it provided that they received no public subsidy. However, there was no reference to either of these positions in the statement of the agreement, which said simply:

> We will implement a process allowing the Liberal Democrats to maintain their opposition to nuclear power while permitting the Government to bring forward the National Planning Statement for ratification by Parliament so that new nuclear construction becomes possible. This process will involve: the Government completing the drafting of a national planning statement and putting it before Parliament; specific agreement that a Liberal Democrat spokesperson will speak against the Planning Statement but that Liberal Democrat MPs will abstain; and clarity that this will not be regarded as an issue of confidence.

In any case opinion polling suggested that 43 per cent of the UK public was comfortable with government subsidies to support nuclear new build with only 28 per cent against.[53] The form of words was inevitably a recipe for further policy paralysis as the political debate got somewhat bogged down around defining the word 'subsidy'. As discussed earlier, the cost profile of nuclear would make it very difficult for funding to be found even if the projected levelised costs at constant rates of return over a 70-year project appeared likely to be much lower than alternatives. Even if investors were prepared to provide a healthy capacity margin, the level of risk associated with nuclear (and renewable) new build would inevitably be higher than would be the case for say CCGT. Whether this were viewed as an absolute barrier to investing or as requiring higher rates of return is effectively irrelevant, the outcome would be the same.

The government had to find a definition of the word 'subsidy' which would not include market measures such as guaranteed power prices and a floor price for carbon emissions designed to encourage the new electricity market to invest in low-carbon energy sources. Unsurprisingly, opponents of nuclear power sought to define the word in such a way as to include any and all measures that involved government.

8.11 ELECTRICITY MARKET REFORM (EMR)

Broadly speaking, experience suggests that competitive markets are good at 'sweating assets', i.e. getting best value for money out of existing power stations and networks, but are less good at sending signals for new investment to protect secure supplies of power in the medium to long term, and (arguably) at protecting environmental goals. Central planning perhaps tends to be the mirror image. But is there any way of creating a structure which can deliver on both of these goals? In pursuit of this end the main development in the UK electricity supply industry during the first half of the 2010s was the introduction of the Electricity Market Reform (EMR), published for consultation in 2010[54] and embodied in the Energy Act of December 2013.[55]

As the above analysis suggests, a key challenge for the power market was that of encouraging private investors (or foreign governments) to fund sufficient low-carbon technology to maintain secure supplies and move the UK towards its long-term carbon commitment, expressed through a series of four-year carbon 'budgets' culminating in a 60 per cent cut in emissions by 2050. It was calculated that in order to meet its goals the UK would need to invest some £100 billion in low-carbon technology by 2020 – about half in wind and half in new nuclear and other renewables – with another £90 billion through the 2020s.[56]

On the one hand, the Energy Act was responding to the likelihood that, left to a free market, investors would only fund new build CCGT – quick and cheap to build and maintain and quite appropriate for load-following, it being economically worthwhile to vary its output depending on the system demand. As noted previously the OCGT is even more flexible, though its overall thermal efficiency is lower. Although it is relatively expensive to run CCGT and OCGT unless gas is cheap, they represent relatively low economic risk investments because increases in the gas price can largely be passed on in higher electricity prices to customers who have nowhere else to go, at least in the short term. Growing dependence on imported gas in an increasingly competitive world would bring with it some concerns about both climate change goals and the reliability of supplies as China and India, in particular, increased their use of imported gas enormously. Indeed, the government's Gas Generation Strategy of 2012 foresaw this as a real possibility, stating there could be a need for at least 26 GW of new gas-fuelled power capacity by 2030.[57] Some commentators expressed great hope in shale gas but the evidence that this could be a game changer in the UK in the sense it appeared to be in the US was unconvincing, particularly given considerable public and Big Green objections to developing fracking wells. Nuclear and renewables, being much more capital-intensive, are inherently much riskier.

EMR sought to respond to this challenge by introducing a number of measures to dissuade the market from solely building CCGT. An escalating 'carbon floor price' was to be introduced to give a degree of surety to low-carbon generators that they would gain a predictable competitive benefit over higher-carbon rivals. The carbon price through the European Union Emissions Trading Scheme (ETS) had collapsed as the recession delivered sufficient carbon emission reductions from 'business as usual'. In the summer of 2008 the price peaked at a price of nearly €30 per tonne of carbon emitted but had fallen to €3 per tonne of carbon emitted in early 2013, recovering to around €8 per tonne at the end of 2015 (see figure 8.14[58]).

A 'carbon standard' for emissions was introduced which would in effect rule out new coal build without carbon capture and storage. Some commentators argued that the carbon floor price should be reinforced by a legally binding target for carbon intensity of 50g CO_2 per kWh.[59]

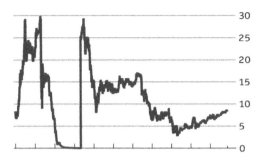

Figure 8.14 ETS carbon price, 2008–12

More importantly, low-carbon sources of electricity were to be offered CfDs which involved 'feed-in tariffs' at a negotiated 'strike price' – effectively a long-term guaranteed price for their output. DECC described the measure as:

A new mechanism to support investment in low-carbon electricity generation. The CfD works by stabilising revenues for generators at a fixed price level known as the 'strike price'. Generators will receive revenue from selling their electricity into the market as usual. However, when the 'market reference price' – in effect the wholesale power price at any particular time – is below the strike price they will also receive a top-up payment from suppliers for the additional amount. Conversely if the reference price is above the strike price the generator must pay back the difference.

The challenge was to find a figure – if such a figure existed – that would be high enough to be sufficiently attractive to potential investors in low-carbon new build but low enough to be acceptable to government and consumers, politically credible against future changes of government over subsequent decades and, in the case of nuclear, that would not fall foul of European Union competition directives. (As an hangover of the monolithically Big Green bias of EU policy in the 2000s, renewables could in practice receive as much help as a national government wished to give them without triggering unfair competition enquiries.)

In October 2013 figures for the strike prices eventually emerged. The strike prices were to be index-linked to inflation and to last for about two-thirds of the expected lifetime of the project in question, i.e. around 15 years for wind and 35 years for nuclear:[60]

nuclear   £89.50 per MWh (if Sizewell B were built) or £92.50 per MWh;

onshore wind   £95 per MWh, falling to £90 per MWh in 2018;

offshore wind   £155 per MWh, falling to £140 per MWh;

large solar   £120 per MWh, falling to £110 per MWh;

hydropower   £95 per MWh;

biomass	£105 per MWh;
tide/wave	£305 per MWh.

These figures are to be viewed against a wholesale price of power at the time which had stood, as shown in figure 8.15, at a quarterly average of between £40 and £60 per MWh throughout the period 2010 to early 2013.[61] There was an assumption, based on projections of fossil fuel prices, that this price, though considerably higher than it had been a decade earlier, was artificially low both because of the effects of the economic recession and because very little new capacity was being constructed. These conditions were thought unlikely to persist for long; for example, the 'peak oil' doomsayers were arguing that fossil fuel prices would rise enormously as supplies dried up.

In addition the Hinkley Point C nuclear deal included loan guarantees for 70 per cent of the construction cost, another measure to transfer risk from the investor to (in this case) the taxpayer.

Yet however generous these effective subsidies may appear, they could not guarantee that companies within a marketplace would conform with the country's requirements for investment in the relevant sources of energy. As the parliamentary Energy and Climate Change Select Committee put it in a report in March 2013, there did not seem to be a 'Plan B' in case the market did not come up with investment in nuclear power despite the inducements to do so in the Energy Act.[62]

8.12 THE PROBLEMS OF SUCCESS

The tortuous passage of Electricity Market Reform – over three years from first consultation to Act of Parliament, a longer period than it took to develop the atomic bomb after the first demonstration of nuclear fission in the 1940s – and the absence of any real reason for thinking they would succeed in stimulating the construction of sufficient new low-carbon capacity, was just one of the difficulties. Another would emerge if the measures were 'successful'.

As discussed in chapter 4 the presence of variable renewable output, which even when it is largely predictable is not generally well correlated with variations in demand, seriously damages the profitability of dispatchable generation. But the variability means that to maintain secure supplies the UK would still need almost as many gas, coal and nuclear power stations as would have been needed if the variable renewables had never been developed. Some 4.5 GW of coal-fired capacity and over 2 GW of oil-fired capacity had closed earlier than expected in 2012/13.[63] Germany and Spain, with high quantities of wind energy, were already facing such problems[64], resulting, for example, in the long-term mothballing of the world's most efficient CCGT plants, at Ingolstadt in southern Germany, in favour of older and more polluting coal-fired capacity with lower marginal costs.[65] At the same time UK capacity margins were projected to fall dramatically,

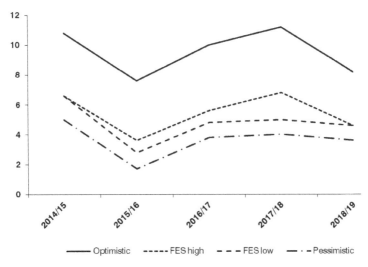

Figure 8.16 Projected UK capacity margins, 2014/15–2018/19

perhaps below 4 per cent by 2015/16, within the Future Energy Scenarios (FES) used by National Grid[66] – see figure 8.16.

There was a brief concern over the robustness of supply in the autumn of 2014. Five power stations with a combined capacity of 4.4 GW suffered unplanned outages. There were fires at the coal-fired plants at Ironbridge (370 MW was taken offline permanently as a result) and Ferrybridge (1,000 MW temporarily closed) and the CCGT at Didcot (700 MW temporarily). In addition the discovery of cracks in one of 16 boiler units at Heysham nuclear power station led to the temporary closure of this plant and its counterpart in Hartlepool while checks could be done on the other 31 units, losing some 2,330 MW of output. Fortunately the interruptions occurred at a time of relatively low demand during a remarkably warm October and November but National Grid responded by calling up mothballed capacity and opening discussions with major power users to accept voluntary disconnection should supplies get tight.[67] Similar concerns arose in November 2015.

So Electricity Market Reform had at its heart a rather intriguing paradox. If the measures designed to dissuade investors from building only CCGT were to prove successful and more variable renewables were brought online then the economic case for investing in, or even continuing to operate, CCGT or other dispatchable technologies would be weakened, as these plants would be left without an income during those times when the wind was blowing at the right speed. This is an even more acute problem with new build of more heavily capital-intensive sources such as coal with CCS or nuclear, as their economics are more seriously harmed by being taken offline, most of the cost being fixed and independent on the amount of power produced.[68] But eventually new CCGTs or other flexible dispatchable capacity would be needed to provide power when the wind dropped, or 'the lights would go out'. So the government had to (re)introduce other measures, notably capacity payments – paying companies to keep power capacity available even if it is not being used – to attract the investment in CCGT that would have happened but for the effects of the measures it introduced to deter CCGT in the first place.

The first round of capacity payments, announced in early 2015, resulted in the procurement of almost 50 GW of capacity at a clearing price of £19.40 per kW for delivery in 2018/19, a total of just under £1 billion.[69] Critics argued that the scheme would actively subsidise coal power stations to stay open that would otherwise have shut, claiming this would crowd out opportunities for new gas power stations to be built and discourage new demand-side response. They also argued that the payments would represent a huge 'windfall' for the nuclear plants. However, these were not unfortunate side-effects of the policy but the desired outcome: the capacity market was required precisely to compensate for the loss of market share of dispatchable capacity and as a hedge against insufficient new dispatchable capacity being committed in time. Later that year one of the few new build projects to receive support through the capacity auction, a 1.9 GW CCGT planned for Trafford in Greater Manchester, was declared to be 'in doubt' owing to 'continued direct subsidies for low-carbon technologies' forcing down the wholesale price. Researchers suggested that in order to stimulate investment in the new gas-fired power plants essential to back up variable renewables, capacity payments may need to double.[70]

8.13 PRESSURE UPON PRESSURE

The scale of the financial challenges facing the power industry, despite the levels of subsidy available, can be inferred from the underperformance of the sector in recent years. The German Energiewende has perhaps had the most damaging effect on the large utilities which in the mid-2010s were being called upon to make the many hundreds of billions of euros investment over the next 15 years. Between 2008 and 2014 the 'Big Four' German power utility companies lost an average of 50 per cent of their market capitalisation despite 2013 having been the best year for general equities since 1995 – for example as shown in figure 8.17.[71]

Utility companies in other EU28 countries typically lost 33 to 50 per cent of their stock value in the same period.[72]

The situation was exacerbated by a deterioration in the support the large utilities enjoyed from the political establishment. In Spain, Italy, France and Germany as well as the UK governments had generally supported, or at least did not obstruct, the creation of very large local energy utilities which had considerable market power and which in return were big enough to make the necessary investments to 'keep the lights on'.[73] The 1990s and, especially, 2000s were a time of consolidation and the creation of cross-European giants operating in many countries – see figure 8.18.

At the same time cross-border trading in electricity grew enormously (see figure 8.19), taking advantage of the longitudinal geography of Europe which means that peak

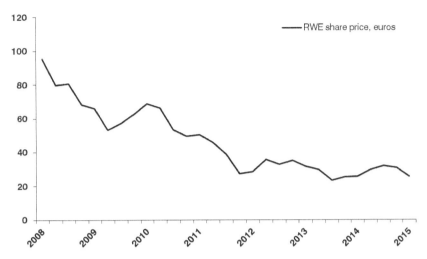

Figure 8.17 RWE share price, 2007–15

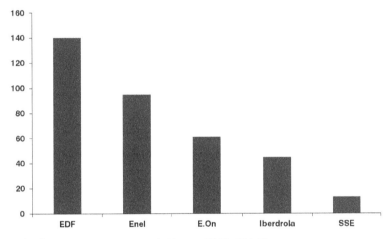

Figure 8.18 Five largest power generators in Europe (GW), 2014[74]

demand moves across the continent, reducing the need for each country to maintain its own capacity margin and also making it easier to dump excess production onto (sometimes unwilling) neighbours when necessary. In early 2014 the day-ahead power markets in 15 north-western European countries, from the UK to Finland, were linked with the aim of reducing the differences in price among the countries and thereby improve the workings of the cross-border market. Prior to this move traders selling power into another country had first to buy wire capacity and then make a separate trade for the power itself, exposing them to two sets of price risk. The coupling of the markets allowed traders to bid for energy on local exchanges, which would then automatically allocate cross-border transmission capacity.

The logic of moving to larger generating systems, in stark contrast to the Big Green dream of a return to pre-supergrid localised generation with all of its costs and inefficiencies, was irresistible.

However, as the economic crisis gathered force in the late 2000s so the effective compact between utilities and governments came under strain. Governments in general found themselves facing heavy criticism of the profits being made by power companies, focusing on the headline figures rather than the return on investment which those figures represented. This was especially the case in the UK. One of the consequences of the unsustainably low prices pertaining during the moderate market collapse of the first half of the 2000s was that, in the perception of the public, these levels became a benchmark of what power 'should' cost. Through the decade from 2005 onwards UK power prices increased by 120 per cent (see figure 8.20), far faster than the European Union average of just over 50 per cent.

In reality the UK price remained below the EU average and well below most of those in the western countries – see figure 8.21.

With the economy and a growing anti-business narrative being the dominant political themes throughout the 2010–15 parliament, it was inevitable that political capital would be sought by attacking the levels of profit being made by the privatised power industry. This is a case which is much more difficult to make against nationalised industries, however inefficient and therefore expensive they may be. The Left had made such attacks on many occasions – for example, during a power crisis in Norway in 2002/3 which was caused by low rainfall (hardly the fault of the ownership structure of the power industry one might think), Socialist Left Party MP Hallgeir Langeland said: 'What we've seen this winter is that the market doesn't function. The market works for companies but not for consumers who have to pay for electricity themselves.'[75] However, in previous crises there has usually been a countervailing voice within government and the Right supporting the players in the liberalised markets. Following the California crisis, for example, Bill Eastlake of the Idaho Public Utilities Commission said: 'If you believe in markets you can't blanch at the sight of victims.'[76] The UK regulator's version was: 'Failure can be a sign that competition is working effectively, because in many cases it is the degree of rivalry between companies and the extent to which customers exercise choice that inevitably leads to success for some and failure for others.'[77] By the mid-2010s the power companies had few such vocal friends in the political establishment. There was a great deal of talk about 'making the market work', as though this would bring prices down. In reality the returns which were being earned even from existing plant were insufficient to encourage investors to consider the levels of investment required to stave off a long-term crisis.

The most egregious version of this political game came from the opposition Labour Party in 2013, leader Ed Miliband telling his Party Conference that he would freeze domestic gas and electricity prices for 20 months if Labour were elected in 2015. The freeze was to be accompanied by measures to break up the vertical integration of the Big

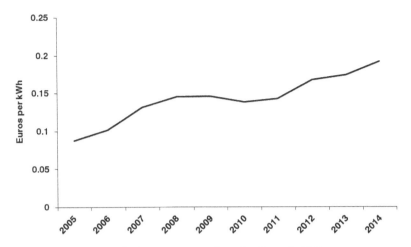

Figure 8.20 Electricity prices to final users – UK medium-sized households[78]

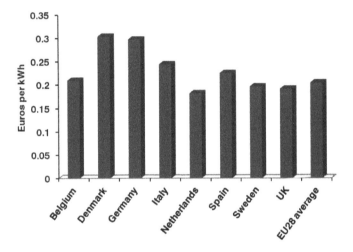

Figure 8.21 Average power prices for medium-sized households in selected European countries, 2014

Six, thereby removing their ability to hedge risk across the generation/supply interface and in effect introducing a very large new risk element which would need to be financed. Analysts attacked the proposals, describing them as a 'life threatening problem' because market participants would not be able to react to commodity prices (the California problem). The dangers in trying to make a market deliver on non-market goals were once again emphasised. 'There are reasons we have independent economic regulators and there are reasons why we have the competition and markets authority – it's to stop politically attractive but economically insane policies like this.'[79] After the policy was modified to refer to a 'cap' rather than a 'freeze', another researcher said: 'The price cap will either be so high it will be meaningless or it will bind and it will lead to shortages –

not necessarily in electricity but certainly in investment. It's not an economically sensible policy to help hard pressed consumers.'

The policy threatened at least four seriously negative outcomes. The first would be a rerun of California if gas prices in particular were to enjoy a significant upturn, leading inevitably to very large losses among generating companies and consequent interruptions in supply. The level of the cap would be far lower than had been the case in California, leaving companies exposed to even modest rises in gas or coal prices. In the event, the fall in fossil fuel prices from the middle of 2014 reduced this possibility, while of course also removing the relevance of freezing prices at a time when costs were falling. However, the second danger was the mirror image – fears among power companies of a potential upturn in fuel prices from mid-2015 to mid-2017 prevented companies cutting their prices as far as they would otherwise have done on the back of falling gas prices. RWE npower's UK chief executive explained: 'The political and media pressures at the moment make it more difficult to reduce prices and then increase them again next spring. Then we are acutely aware that if the Labour party were to implement their proposed price freeze, we will be living with the consequences of our standard rate tariff price for a very long time and beyond the level of risk that we could manage in the wholesale market.'[80] One does not have to take a view as to whether this is a real argument or a convenient one to aid higher profits – in either case the consumer suffers. Thirdly, inevitably if power companies are to reduce the risk that they may be unable to absorb increases in gas prices they have to try to hedge those risks by transferring some of them onto gas suppliers. As risk is expensive this pushes power prices up – power generators have to increase the proportion of their fuel supplies that are bought on long-term (say three-year) contracts which on average are more expensive than a mixture with a higher proportion of month-ahead and day-ahead purchases. Finally, and perhaps most damagingly of all, the perception that future governments will once more simply change the market rules almost on a whim to prevent companies from making a fair return on the huge investment they are being asked to make inevitably drives investment out, creates delays and/or increases the demanded returns to cope with this additional risk.

For such reasons the idea of price freezes, especially if they are made over a year in advance, have found very little favour among market practitioners or commentators. Yet despite these obvious and unavoidable problems caused simply by making the announcement itself, it proved to be a pledge that played well given the atmosphere of the time which tended to blame all economic problems on large companies, notably the banks, and which had once more started to view profit as 'a filthy word' in the phrase of one self-appointed apostle of the new social mores.[81]

8.14 IN THE EVENT – THE 2015 GENERAL ELECTION

Sadly from the point of view of the student of power markets, though probably in the best interests of everyone else, the price cap was not effected as the Conservative Party won an overall majority in the 2015 general election. However, it was widely felt that the failure of opinion pollsters to foresee this outcome had led the Conservatives to produce a manifesto which included a number of somewhat extreme statements which it would then put up for negotiation if say a second coalition with the Liberal Democrats were to be sought. Be that as it may, the unambiguous commitment to low prices in the short term risked severely limiting the new government's ability to drive the market into valuing the other goals of security of supply and falling environmental emissions – or increasing the cost of such intervention.

The 2015 Conservative manifesto, in its section on energy entitled 'Guaranteeing you clean, affordable and secure energy supplies', took the following line.

> Our commitment to you: affordable, reliable energy is critical to our economy, to our national security and to family budgets. We will:
> - keep your bills as low as possible and promote competition in the energy market;
> - ensure your homes and businesses have energy supplies they can rely on;
> - help you insulate your home;
> - halt the spread of subsidised onshore wind farms;
> - meet our climate change commitments, cutting carbon emissions as cheaply as possible, to save you money.

We need to secure clean but affordable energy supplies for generations to come. This means a significant expansion in new nuclear and gas; backing good-value green energy; and pushing for more new investment in UK energy sources. Healthy competition is the best way to secure a good deal for consumers. So we will keep on relentlessly pushing for more competition to keep bills low.

> - We will promote competition to keep your bills as low as possible.
> - We will secure your energy supplies.
> - We will halt the spread of onshore wind farms.
> - We will protect our planet for our children.
> - We want a better deal – and low bills – for hardworking families.

('Hardworking families' was the buzzword of the 2015 election generally, though little detail was given as to how the new government would identify these and make sure that the lazy families were excluded from such benefits.)

The need for investment was not entirely ignored.

Our long-term plan has unlocked £59 billion of investment in electricity. All parts of the UK will soon be helping to deliver secure, affordable and low-carbon energy, from the Hinkley Point nuclear power station, to offshore wind turbine manufacturing at the new Green Port in Hull, the next generation of pipelines West of Shetland and the Swansea tidal lagoon. Our tax cuts have encouraged record levels of investment in existing North Sea gas and the birth of a new industry, shale gas, which could create many thousands of jobs.

But even here the focus was unmistakeable.

We have delivered a better deal for consumers too. We have demanded that energy companies simplify their tariffs; encouraged more independent suppliers – which now account for 10 per cent of the household market; and made it much easier for people to switch energy providers.

The themes – notably the commitment to reduce prices at a time when there was a chronic shortage of long-term investment – were largely reflected in the manifestos of the other parties, albeit with different emphases:

- Labour – price freeze/cap; create an Energy Security Board to plan and deliver the energy mix we need, including renewables, nuclear, green gas, carbon capture and storage, and clean coal;
- Liberal Democrats – reduce energy demand by 50 per cent by 2030; expand the Green Investment Bank to provide more low-carbon energy; promote competition; accept that new nuclear power stations can play a role provided concerns about safety, disposal of waste, and cost are adequately addressed and without public subsidy for new build;
- SNP – lower energy bills; more renewables; no new nuclear; support for Longannet coal-fired power station; continue to argue for changes to ensure that Scottish renewables are not penalised because of their distance from markets in the south of England [i.e. penalise someone else for the extra costs of transporting their electricity to where it might be useful];
- UKIP – abolish DECC, Green subsidies and the carbon floor tax; withdraw from the EU Emissions Trading Scheme; repeal the Climate Change Act 2008; support a diverse energy market based on coal, nuclear, shale gas, conventional gas, oil, solar and hydro;
- Green Party – phase-out nuclear power by 2025; invest up to £35 billion of public money in renewable generation and in the National Grid *if necessary* [my emphasis – what a wonderful concept that it might not be necessary to wire

up these new wind farms and solar arrays so people shouldn't worry too much about the costs]; concentrate on expanding mature renewable technologies such as wind energy and solar PV. [The Greens had argued to the European Commission State Aid investigation in Hinkley Point C that subsidies should never be offered to 'mature' technologies, making this particular comment a real hostage to fortune.]

The 2015 election provided the relative stability of a majority government. However, the main party of opposition, Labour, responded by electing a leader who expressed a firm desire to renationalise the energy industries. Though as will be argued later there may be considerable merit in considering returning certain duties for electricity to the government, such an announcement may not have been taken as encouragement among non-UK government entities thinking of investing in the industry.

8.15 SUMMARY – HOW HAS MARKET STRUCTURE ADDED TO THE DECISION PARALYSIS?

As previously noted, in the early 1950s it took less than five years to move from a few flickering light bulbs powered by the world's first nuclear-generated electricity to the first output from a more or less commercial-scale nuclear power station. Sixty years later, eight years came and went without a decision over whether to act on a firm exhortation by government to build new nuclear capacity.

There are several reasons for this change but the transfer of decision-making from a centralised structure to a more devolved market-based one, while it has delivered considerable advantages in terms of efficient operation of existing assets, has been a particularly important factor in delaying investment in dispatchable generating capacity and other infrastructure assets. Underlying the phenomenon are two things that investors do not like: complexity and unpredictability.

Two questions emerge. To take nuclear power as an example – in order to secure the claimed social benefits of a new build programme would government have to intervene in the market so heavily that, in a way reminiscent of the banking crisis, the risks of investment were nationalised but the potential benefits privatised? And even if it did, how credible would potential investors regard a regime which in principle could be undermined by future changes in the market rules or by changing government attitudes, such as was seen in Germany after the Fukushima accident in 2011? Experience of the first 25 years after privatisation suggests market rules might be expected to undergo major changes roughly every 12 years while a nuclear plant ordered in 2015 might well still be generating electricity in 2085.

It may be that the challenges of providing secure, environmentally sensitive long-term electricity supplies are simply too complex to be delivered within a market context. Circumstances in the UK seem to be highly favourable towards new nuclear development – cross-party support in Parliament; successive governments who wished to launch a major infrastructure programme as a response to the recession; considerable public support; an improved planning and licensing regime and companies interested in making the investment (though it is going too far to describe it as 'streamlined', certainly compared to things in a 'phase 1' society). If despite such an environment a sufficiently large programme of low-carbon generation new build is not initiated then it may be time to reappraise the fundamental role of markets in the area.

For markets to work there must be both the opportunity to profit heavily if one makes successful investments and the risk of going broke if one does not. Neither of these would seem to be the case in power markets. Political pressure will always prevent companies making the kind of rate of return on capital which would send out the signals for new investment through the operation of the normal business cycle.

However, the size of the investments involved in electricity, and the difference in the capital/operational cost profile of the major options, seems to require that government (or the power consumer) has to take on a very large proportion of the risk of investment in low-carbon power or it simply will not happen. Electricity Market Reform appears to move quite close to a situation in which public sector policy was being pursued at private sector rates of return.

Perhaps, then, the challenge of finding a single set of policies that both delivers new build when required and promotes competitive operation of existing assets is an impossible one. Until liberalisation has completed a full cycle, returning to a plant mix which is similar in both quantity and age to that which pertained at the start of full liberalisation in 1998/9, it will be impossible to judge this issue or to decide if liberalisation was 'successful' but the signs in the mid-2010s were not good.

There may, however, be a way of sidestepping the matter. If governments are relatively good at safeguarding a healthy capacity margin by ensuring that new plant is built in time and that a suitable mix of different power capacity is maintained, then perhaps that role should be returned to them. After all, as already discussed, the level of political 'engagement' (or interference) in the marketplace in the first quarter century of the liberalisation experiment to promote non-market outcomes was profound and constant. If competitive markets are better at 'sweating assets', operating existing capacity efficiently and effectively, then maybe that role should remain with them. A system might then evolve whereby the government unequivocally retakes formal responsibility for the capacity mix and quantum – something which de facto it never really relinquished despite liberalisation – but once the plants are built they can be franchised out or sold on to private companies to operate within a competitive generating market.[82] (Whether it would then make any sense to persist with the faltering supply market, at least for household consumers, is another matter – EMR already moved some way towards a 'single buyer' model to support the CfD regime.) The most obvious political stumbling block to the command-and-control model working in the post-war period was perhaps the malign power of the mining unions, something which has long been consigned to the dustbin of history.

There are of course flaws in such an approach, as there are with any. The European Competition Directorate wouldn't like it, for example. But it would at least return us to the days when it was clear what a 'decision' would look like and who was ultimately responsible for taking it, something which at least in principle could lower one of the barriers to timely decision-making which has blighted the market for most of the twenty-first century.

Endnotes

1   http://hansard.millbanksystems.com/commons/1982/mar/16/amersham-international, 'Amersham International', *Hansard*, 16 March 1982.

2   http://www.independent.co.uk/news/business/bnfl-to-slash-3000-contractors-jobs-at-sellafield-1542733.html, Fagan M, 'BNFL to slash 3,000 contractors' jobs at Sellafield', *The Independent*, 27 August 1992.

3   http://www.independent.co.uk/news/business/why-privatisation-has-been-a-success-story-1281602.html, Warner J, 'Why privatisation has been a success story', *The Independent*, 4 January 1997.

4   http://energystorage.org/energy-storage/technologies/pumped-hydroelectric-storage, Energy Storage Association (2015), 'Pumped Hydroelectric Storage'.

5   http://www.technologyreview.com/review/534866/why-we-dont-have-battery-breakthroughs/, Bullis K, 'Why we don't have battery breakthroughs', *MIT Technology Review*, 10 February 2015.

6   http://www.nationalgrid.com/uk/Electricity/Data/Realtime/, National Grid (2013), 'Electricity – real time operational data'.

7   http://www.telegraph.co.uk/news/earth/energy/11314480/Blackouts-report-death-disorder-and-other-key-consequences.html, Gosden E, 'Blackouts report: death, disorder and other key consequences', *Telegraph*, 28 December 2014.

8   http://www.hks.harvard.edu/hepg/Papers/NYISO.blackout.report.8.Jan.04.pdf, New York Independent System Operator (2004), 'Interim report on the August 14, 2003 blackout'.

9   http://www.clickgreen.org.uk/analysis/general-analysis/122208-royal-wedding-triggered-record-energy-demand-on-uks-national-grid.html, 'Royal Wedding triggered record energy demand on National Grid', ClickGreen, 30 April 2011.

10   http://www.nationalgrid.com/NR/rdonlyres/D4D6B84C-7A9D-4E05-ACF6-D25BC8961915/47015/NETSSYS2011Chapter2.pdf, National Grid (2011), '2011 National Electricity Transmission System (NETS) seven year statement'.

11   Cramton P. and Lien J. (2000), 'Value of lost load', University of Maryland.

12   http://www.theguardian.com/environment/2015/may/06/uk-energy-bill-subsidies-driving-boom-in-polluting-diesel-farms, Macalister T, 'UK energy bill subsidies driving boom in polluting diesel farms', The Guardian, 6 May 2015.

13   Layfield F (1987), *Sizewell B Public Inquiry: Report*, HMSO.

14   http://hansard.millbanksystems.com/commons/1989/jul/24/electricity-privatisation-nuclear-power, Parkinson C, 'Electricity privatisation (nuclear power)', *Hansard*, 24 July 1989.

15   https://www.gov.uk/government/publications/digest-of-united-kingdom-energy-statistics-2012-internet-content-only, DECC (2013), 'Digest of UK Energy Statistics 2012.'

16   DTI/Scottish Office (1995), 'The prospects for nuclear power in the UK: conclusions of the government's nuclear review', Cm2860, HMSO.

17   https://www.gov.uk/government/uploads/system/uploads/attachment_data/file/65713/6883-electricity-generation-costs.pdf, DECC (2012), 'Electricity generating costs'.

18   http://www.publications.parliament.uk/pa/cm200506/cmselect/cmenvaud/584/5111612.htm, Select Committee on Environmental Audit (2005), 'Supplementary memorandum submitted by Ofgem', parliament.uk.

19   http://www.ofgem.gov.uk/Media/PressRel/Documents1/SSE%20Press%20Release.pdf, Ofgem (2013), 'Ofgem fines SSE £10.5 million for misselling'.

20   Thomas S (2014), 'Has the need for low-carbon generation killed the electricity market or was it dead already?', University of Greenwich.

21   http://webarchive.nationalarchives.gov.uk/+/http:/www.berr.gov.uk/files/file10719.pdf, DTI (2003), 'Our energy future: creating a low carbon economy'.

22   http://news.bbc.co.uk/1/hi/world/europe/7806870.stm, 'Russia shuts off gas to Ukraine', *BBC News website*, 1 January 2009.

23 http://webarchive.nationalarchives.gov.uk/20121205174605/http:/www.decc.gov.uk/assets/decc/publications/white_paper_07/file39387.pdf, DTI (2007), 'Meeting the energy challenge, a White Paper on energy', TSO.

24 http://webarchive.nationalarchives.gov.uk/+/http:/www.berr.gov.uk/files/file43006.pdf, BERR (2008), 'Meeting the energy challenge: a White Paper on nuclear power'.

25 Lawson N (1982), reproduced in Helm D, Kay J and Thompson D (1989), *The market for energy*, OUP.

26 Atherton P (2011), 'A very hostile political environment', Citigroup.

27 Weinstein S and Hall D (2001), 'The California electricity crisis – overview and international lessons', PSIRU: London.

28 http://energyalmanac.ca.gov/electricity/electricity_generation.html, California Energy Commission (2015), 'California electrical energy generation'.

29 http://www.ucei.berkeley.edu/PDF/csemwp103.pdf, Blumstein C, Friedman L and Green R (2002), 'The history of electricity restructuring in California', University of California Energy Institute.

30 http://www.eia.gov/totalenergy/data/annual/archive/038496.pdf, EIA (1997), 'Annual energy review 1996'.

31 http://www.hks.harvard.edu/hepg/Papers/Old_Papers/CAElecSvcIndus_1993.pdf, Dasovich J, Meyer W and Coe V (1993), 'California's electric services industry: perspectives on the past, strategies for the future', California Public Utilities Commission.

32 http://www.cato.org/sites/cato.org/files/serials/files/regulation/1997/4/reg20n2i.html, Michaels R (1997), 'Stranded in Sacramento: California tries legislating electricity competition', Cato Institute.

33 Flaim T, 'The big retail bust: what will it take to get to true competition?', *Electricity Journal*, March 2000.

34 http://www.salon.com/2001/01/30/deregulation_mess/, York A, 'The deregulation debacle', *Salon*, 30 January 2001.

35 Smith R and Emshwiller J, 'California's PG&E gropes for a way out of electricity squeeze,' *The Wall Street Journal*, 4 January 2001.

36 http://www.ucei.berkeley.edu/PDF/isomsc.pdf, Wolak F, Nordhaus R and Shapiro C (1998), 'Preliminary report on the operation of the ancillary services market of the California ISO'.

37 Joskow P and Kahn E (2002), 'A quantitative analysis of pricing behaviour in California's wholesale market during summer 2000'.

38 http://www.cpuc.ca.gov/PUC/energy/Electric+Rates/ENGRD/ratesNCharts_elect.htm, CPUC (2012), 'Average energy costs 2000–2011'.

39 http://enduse.lbl.gov/Info/LBNL-47992.pdf, Brown R and Koomey J (2002), 'Electricity usage in California: part trends and present usage patterns', *Energy Policy*.

40 Krapels E, 'Was gas to blame?', *Public Utilities Fortnightly*, 15 February 2001.

41 Purdum, T, 'Rolling blackout affects a million across California', *New York Times*, 20 March 2001.

42 http://www.ucei.berkeley.edu/PDF/csemwp103.pdf, Blumstein C, Friedman L and Green R (2002), 'The history of electricity restructuring in California', CSEM.

43 http://www.cato.org/pubs/pas/pa406.pdf, Taylor J and VanDoren P (2001), 'California's electricity crisis – what's going on, who's to blame and what to do', *Policy Analysis*, **406**.

44 Wilson J (2002), 'High natural gas prices in California, 2000-2001: causes and lessons', *Journal of Industry, Competition and Trade*.

45 http://pqasb.pqarchiver.com/latimes/370911591.html?did=370911591&FMT=ABS&FMTS=FT&date=Jul+18,+2003&author=Nancy+Rivera+Brooks&desc=New+Power+Plant+for+Edison, Brooks N, 'New power plant for Edison', *Los Angeles Times*, 18 July 2003.

46 Davis G (2001), 'State of the State Address'.

47 http://www.ncpa.org/pub/ba348/, Michaels R (2001), 'California's electrical mess: the deregulation that wasn't', National Center for Policy Analysis: Dallas.

48 http://www.cato.org/pubs/pas/pa406.pdf, Taylor J and VanDoren P (2001), 'California's electricity crisis – what's going on, who's to blame and what to do', *Policy Analysis*, **406**.

49 Atherton P, 'The €1 trillion decade revisited', Citigroup, 29 September 2010.

50 Atherton P (2014), 'So, will the light go out', Liberum Capital.

51 http://www.ft.com/cms/s/0/70f2a90e-b89e-11e1-a2d6-00144feabdc0.html#axzz2QMjZsYn7, 'We have to be green – and straight', *Financial Times*, 17 June 2012.

52 https://www.gov.uk/government/uploads/system/uploads/attachment_data/file/78977/coalition_programme_for_government.pdf, HM Government (2010), 'The Coalition: our programme for government'.

53 http://www.imeche.org/news/archives/2013/05/28/Public_backs_Government_subsidy_for_new_nuclear, Institution of Mechanical Engineers (2013), 'Public backs government subsidy for new nuclear'.

54 https://www.gov.uk/government/uploads/system/uploads/attachment_data/file/42636/1041-electricity-market-reform-condoc.pdf, DECC (2010), 'Electricity Market Reform consultation document', TSO.

55 http://www.legislation.gov.uk/ukpga/2013/32/contents/enacted, legislation.gov.uk (2013), 'Energy Act 2013', TSO.

56 http://www.theccc.org.uk/wp-content/uploads/2013/05/1720_EMR_report_web.pdf, Committee on Climate Change (2013), 'Next steps on Electricity Market Reform – securing the benefits of low-carbon investment'.

57 https://www.gov.uk/government/uploads/system/uploads/attachment_data/file/65654/7165-gas-generation-strategy.pdf, DECC (2012), 'Gas Generation Strategy'.

58 http://cen.acs.org/articles/91/i7/EU-Carbon-Emissions-Trading-Scheme.html, Scott A, 'EU Carbon Emissions Trading Scheme in freefall', *Chemical and Engineering News*, 18 February 2013.

59 http://www.economist.com/blogs/freeexchange/2015/schr-dinger-s-emissions-trading-system, van Benthem A and Martin R (2015), 'Europe's carbon-trading system is better than thought, and could be better still', Economist, 11 December 2015.

60 http://www.greenwisebusiness.co.uk/news/government-unveils-new-strike-prices-for-renewable-energy-4187. aspx, Bateman L, 'Government unveils new strike prices for renewable energy', *Greenwisebusiness*, 4 December 2013.

61 http://www.businesselectricityprices.org.uk/wholesale-business-electricity-prices/, Smith J. (2013), 'A guide to wholesale electricity pricing,' Business Electricity Prices.

62 http://www.publications.parliament.uk/pa/cm201213/cmselect/cmenergy/117/117.pdf, House of Commons Energy and Climate Change Committee (2013), 'Building new nuclear: the challenges ahead'.

63 http://www.argusmedia.com/pages/NewsBody.aspx?id=841211&menu=yes, 'Coal-fired plant closures to increase UK gas burn', *Argus Media*, 3 April 2013.

64 http://www.bloomberg.com/news/articles/2013-01-23/eon-rwe-may-have-to-close-down-unprofitable-gas-power-plants, Andresen T and Nicola S (2013), 'EON, RWE may have to close down unprofitable gas power plants', *Bloomberg Business News*, 23 January 2013.

65 http://notrickszone.com/2015/03/18/green-progress-worlds-most-efficient-gas-fired-turbines-to-get-shut-down-due-to-energiewende/#sthash.1DZgZRHt.dpbs, Gosselin P (2015), 'Green progress: world's most efficient gas-fired turbines to get shut down due to Energiewende', *NoTricksZone*.

66 https://www.ofgem.gov.uk/ofgem-publications/84728/electricitycapacityassessment2014-consultation.pdf, Ofgem (2014), 'National capacity assessment report 2014'.

67 http://www.thisismoney.co.uk/money/bills/article-2810732/Factories-asked-shut-help-lights-Britain-faces-tightest-energy-crunch-eight-years.html, Straus R, 'Factories could be asked to shut down to help keep the lights on as Britain faces tightest energy crunch in eight years', *This is money*, 28 October 2014.

68 http://www.oecd-nea.org/ndd/reports/2011/load-following-npp.pdf, OECD/NEA (2011), 'Technical and economic aspects of load following with nuclear power plants'.

69 https://www.gov.uk/government/news/first-capacity-market-auction-guarantees-security-of-supply-at-low-cost, DECC (2015), 'The first ever Capacity Market auction official results have been released today', gov.uk.

70 http://www.telegraph.co.uk/news/earth/energy/11925444/UK-energy-crisis-Trafford-gas-plant-in-doubt.html, Gosden E, 'Blow to UK energy plans as new gas plant in doubt', *Daily Telegraph*, 11 October 2015.

71 http://www.stockopedia.com/share-prices/rwe-ag-ETR:RWE/, 'RWE share price', *Stockopedia.com*.

72 http://www.rense.com/general96/euopres.html, McKillop A (2014), 'Europe's energy transition paradox', rense.com.

73 Atherton P, 'A very hostile political environment', Citigroup, 13 September 2011.

74 http://www.power-technology.com/features/featurethe-top-10-biggest-power-companies-of-2014-4385942/, The top 10 biggest power companies of 2014, *Power Technology*.

75 http://content.time.com/time/quotes/0,26174,425836,00.html, Wallace C, 'Consumers in the Nordic region are furious over soaring electricity prices – and they blame deregulation', *Time*, 23 February 2003.

76 'How State's consumers lost with electricity deregulation', *Los Angeles Times*, 9 December 2000.

77 Ofgem (2003), 'Consideration of responses to the consultation document: Supplier of Last Resort - security cover and levies'.

78 http://ec.europa.eu/eurostat/tgm/table.do?tab=table&init=1&language=en&pcode=ten00117&plugin=1, Eurostat (2015), 'Electricity price by type of user'.

79 http://www.utilityweek.co.uk/news/labours-price-freeze-slammed-as-%E2%80%98economically-insane%E2%80%99-at-party-conference/1054082, 'Labour's price freeze slammed as economically insane at party conference', *Utility Week*, 23 September 2014.

80 http://www.theguardian.com/money/2014/aug/22/npower-blames-labour-energy-price-freeze Hickey S, 'Npower chief blames high prices on Labour's threatened energy price freeze', *The Guardian*, 22 August 2014.

81 http://www.dailymail.co.uk/tvshowbiz/article-2476191/Russell-Brand-lays-political-revolution-Jeremy-Paxman-Newsnight.html, Flint H, 'Profit is a filthy word!: Russell Brand lays out his political revolution to Jeremy Paxman on Newsnight', *Daily Mail*, 25 October 2013.

82 http://www.ippr.org/assets/media/publications/pdf/when-the-levy-breaks_Jun2015.pdf?noredirect=1, Garman J and Aldridge J (2015), 'When the levy breaks: energy bills, green levies and a fairer low-carbon transition', IPPR: London.

9 | CLIMATE CHANGE

Just as the civilisation of ancient Rome was built on slaves, ours is built on fossil fuels. What happened in the beginning of the nineteenth century was not an 'industrial revolution' but an 'energy revolution'. Putting carbon into the atmosphere is what we do. What used to be the energy-intensive lifestyle of today's high-income countries has gone global. Economic convergence between emerging and high-income countries is increasing demand for energy faster than improved energy efficiency is reducing it. Not only aggregate carbon dioxide emissions but even emissions per head are rising.[1]

If the views of what appears to be the majority of scientists working in the field are to be believed, there is no area of policy where effective decision-making is more urgent than in reducing global emissions of greenhouse gases. One estimate, albeit from a source heavily associated with the solar industry, suggests that in order to limit the rise in global temperature as a result of human activity to 2°C a total of no more than 886 billion tonnes (Gt) of carbon dioxide equivalent can be released into the atmosphere. Between 2000 and 2010, 321 Gt of carbon dioxide were released and in the mid-2010s 38 Gt were being emitted each year. This would all seem to imply that should annual greenhouse gas emissions stabilise at their 2015 levels – and emissions of carbon dioxide from energy production did appear to stall in 2014 for the first time in the absence of an international recession[2] – the crisis point would be reached sometime between 2025 and 2030. The total known reserve of fossil fuels as of the mid-2010s, if burned for energy, would cause the emission of 2,800 Gt of carbon dioxide. Sixty-five per cent of this carbon is trapped in the world's coal reserves, with 22 per cent in oil and 13 per cent in gas.[3]

Yet very little progress is being made. Although many players would like to see reductions in greenhouse gas emissions, almost everyone seems to have a higher priority – a price they would not be prepared to pay to deliver on those reductions.

9.1 THE END OF SCIENCE?

By the mid-2010s the social and political environment in which climate science found itself was a complex one. Echoing the once popular view that the collapse of Communism and the victory of Western (neo) liberal democracy represents 'the end of history', supporters of the notion that climate change is real, serious and caused by human activity often sought to portray the scientific debate as concluded – there was no space for any views that might challenge the orthodox position. This was often propounded even by individuals who had no compunction about gainsaying the scientific consensus in other fields such as genetically modified foods or the health effects of low-level radiation.[4]

For example:

> We have good reason to consider the funding of climate denial to be criminally and morally negligent. The charge of criminal and moral negligence ought to extend to all activities of the climate deniers who receive funding as part of a sustained campaign to undermine the public's understanding of scientific consensus. Criminal negligence is normally understood to result from failures to avoid reasonably foreseeable harms or the threat of harms to public safety, consequent of certain activities. Those funding climate denial campaigns can reasonably predict the public's diminished ability to respond to climate change as a result of their behaviour. Indeed, public uncertainty regarding climate science and the resulting failure to respond to climate change is the intentional aim of politically and financially motivated denialists.[5]

Lawrence Torcello, the author of this proposal, does not seem even to consider the parallel criticism that those who receive research grants, media appearances or lucrative consultancies with environmental organisations or multinational renewable manufacturers have a direct vested interest in increasing public fear of climate change and thence political action, thereby feeding their research careers, egos and pockets. He does not try to explain why one group of scientists – defined in terms of their agreeing with his views – should be immune from such base motivations while others – those with whom he personally disagrees – are so wicked as to merit being silenced, by way of incarceration if necessary. Less judicially, the *Los Angeles Times* letters editor, Paul Thornton, stated flatly that he would not publish some letters from those sceptical of anthropogenic climate change as such views were 'factually inaccurate'.[6] In the UK the Green Party leader, Natalie Bennett, an agriculture and communications graduate, argued that any senior government advisor refusing to accept 'the scientific consensus on climate change' should be sacked – not just those with responsibility for environmental issues – although she seemed to value the freedom of speech of those who rejected the equally well-established understanding say of the health effects of radiation rather more generously.[7] She subsequently told her Party Conference that the 'the scientific debate [on climate change] is over'.

Yet despite such blandishments, opponents of large-scale and extremely expensive responses to the climate threat persisted. Their views ranged from broad acceptance of the science but concerns that the cost of some of the proposed mitigation measures may outweigh their potential benefit, to serious questioning of the science itself and in particular the probity of the institutions which promote it, not least the Intergovernmental Panel on Climate Change (IPCC). As with Energiewende and indeed every other field of science and policy it is effectively impossible to find neutral, disinterested participants – almost everyone has a vested interest of one sort or another in their point of view prevailing.

For the scientific generalist this kind of debate can be stimulating, intriguing and disconcerting. Stimulating in that the science itself, and particularly the observations and reasoning on which it is based, is complex and engaging, requiring an understanding of a number of different disciplines. Intriguing in that individuals and organisations who act as self-appointed scourges of the 'establishment' view in so many fields seem extraordinarily defensive about climate change and unwilling even to question how a conclusion is reached as long as it is the 'right' – i.e. useful – one. Disconcerting both because the stakes are so high – either in terms of damage done by climate change if the phenomenon is real or the potentially vast waste of resources if it is not – and because alarm bells always ring when anyone claims that in any complex scientific field we should stop looking for alternative explanations and counterexamples in the evidence.

This chapter is written from the standpoint of someone who is not a climate science specialist, who is generally sceptical of vested interests be they industrial or environmental but who is generally convinced by the anthropogenic climate change theory.

9.2 A BRIEF HISTORY OF CLIMATE CHANGE SCIENCE

Climate change as a major public and political issue may seem a rather recent phenomenon. The first United Nations Conference on Environment and Development (UNCED) was held in Rio de Janeiro in 1992, leading to the 'Rio Convention', or Framework Convention in Climate Change (UNFCCC) to give it its full title. The Kyoto Conference of Parties (CoP) in 1997 was a milestone in international cooperation (or at least discussions about international cooperation) in the field, spawning the eponymous Protocol which formed the basis of efforts to counter climate change, at least until its targets expired in 2012.

Yet just as 'new renewables' is a comically misleading way to describe sources of electricity generation such as wind and solar which were invented in the 1880s, so theorising about possible human influence on the climate has a long history. As early as 1824 – well before the possibility emerged of making direct measurement of the earth's temperature – Joseph Fourier, 'the father of modern engineering', performed calculations that implied that the earth seemed to be rather warmer than would be expected based on its distance from the sun. (Though the information would not be available until over 60 years after Fourier's death in 1830, the moon, which is on average the same distance away, seems to have an average temperature of around -77°C – the enormous variations make this a rather meaningless figure – while that of the earth is +14°C.[8]) Fourier considered various possible explanations; although eventually he settled on interstellar radiation he did also suggest that perhaps the earth's atmosphere could be acting as an insulator in some way. He postulated, correctly as it turns out, that the atmosphere was in effect transparent to visible light but when that light struck the earth its energy was reradiated as infrared radiation, which had been discovered in 1800: the atmosphere was to an extent opaque to this radiation so would not allow it to escape. The trapped heat resulted in the earth enjoying a higher temperature than would otherwise be the case.

It was already known that glass could behave in an analogous way, allowing visible light to pass through but then preventing heat from escaping. This observation would eventually lead to the phenomenon being called the 'greenhouse effect', although like in a greenhouse much of the heating mechanism is actually accounted for by convection rather than by the insulation. Life on earth owes its existence to this natural phenomenon.

Fourier's work was contemporary with a growing realisation that the climate of the earth was not stable but could, over long periods of time, alter quite radically. The term 'ice age' seems to have been coined in 1837 after decades of speculation as to whether glaciers could have been responsible for depositing huge rocks in valleys at various locations. Fourier speculated that human activity could influence climate, though his focus was on land use. In an 1827 paper he said:

> The establishment and progress of human societies, the action of natural forces, can notably change, and in vast regions, the state of the surface, the distribution of water and the great movements of the air. Such effects are able to make to vary, in the course of many centuries, the average degree of heat because the analytic expressions contain coefficients relating to the state of the surface and which greatly influence the temperature.[9]

The rest of the nineteenth century saw Fourier's speculation receive experimental support. In the 1860s physicist John Tyndall, 'the father of environmental monitoring', found that water vapour, carbon dioxide and hydrocarbons such as methane strongly trap infrared radiation. By now a wide range of theories was emerging to explain the widely accepted ice age idea. Changes to the atmosphere represented but one possibility – others included changes in levels of solar radiation, shifts in ocean current or migration of the axis of rotation of the earth.

A major step forward in the development of the atmospheric theory of climate change came with the great Swedish physicist/chemist Svante Arrhenius, referred to as both 'the father of physical chemistry' and 'the father of climate change'. In 1896 he calculated that the effect of doubling the carbon dioxide concentration of the atmosphere would be a mean temperature increase of between 5°C and 6°C, drawing on recent work by the American Samuel Pierpoint Langley who estimated the temperature of the moon and the brightness of infrared radiation coming from it. However, using calculations from another Swede,

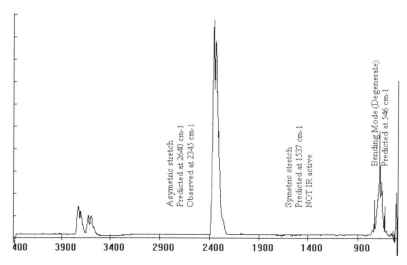

Figure 9.1 Infrared absorption spectrum for carbon dioxide (inverse wavelength, cm⁻¹)[10]

Arvid Högbom, on the level of human releases of carbon dioxide into the atmosphere at the time, he suggested that it would take some thousands of years for the atmospheric carbon dioxide levels to double and even then the overall effect might be beneficial.

Arrhenius's speculations did not quickly command widespread acceptance, opponents arguing, for example, that the oceans would absorb carbon dioxide and that in any case the gas was too weak an absorber of infrared radiation to account for major climatic changes. The 'solar cycle' explanation was popular in the 1920s and 1930s and, through a variant involving 'sunspots', still emerges from time to time today.

It was not until the mid-to-late 1950s that a number of observations resulted in Arrhenius's hypothesis being revived. It became clear that the ocean could not absorb carbon dioxide as rapidly as had previously been assumed. Better instrumentation revealed that carbon dioxide was a better absorber of infrared than had been thought. Carbon dioxide absorbs infrared in two particular wavelength bands, preventing heat at those frequencies from passing through, although not materially affecting the passage of heat of other wavelengths – see figure 9.1.

Doubling of carbon dioxide concentrations in the atmosphere would not therefore double the degree of global greenhouse forcing, i.e. the upward pressure on global temperatures. A huge increase in carbon dioxide levels would have relatively little effect on infrared radiation around the 2200 to 800 cm⁻¹ level for example, allowing this radiation to continue to pass through practically unimpeded. When it comes to radiation around the 2350 cm⁻¹ mark, by contrast, a significant amount of this is already absorbed by natural levels of carbon dioxide in the atmosphere so increasing those levels severalfold will have a proportionately lower effect. This was first noted by another Swedish scientist and contemporary of Arrhenius, Knut Ångström, 'the father of spectroscopy'. If, roughly speaking, 'natural' (i.e. pre-industrial revolution) levels of carbon dioxide were responsible for raising the global temperature by 10°C above what might otherwise be expected, doubling those levels would be expected to have a smaller effect, i.e. to raise temperatures by perhaps 2°C to 3°C. However, there are other gases, many of which are not found in significant quantities in the natural lower atmosphere, which absorb infrared radiation at frequencies which are practically unaffected by carbon dioxide. These include methane; ozone; chlorofluorocarbons (CFCs) and hydrofluorocarbons (HFCs) which are both gases formerly and currently used in activities such as refrigeration – CFCs also damage the ozone layer so have largely been replaced by HFCs; nitrous oxide; and sulphur hexafluoride. The warming potential of these gases, molecule for molecule, is therefore much greater than that of carbon dioxide (table 8.1), although calculations are complicated by the fact that they tend to stay in the atmosphere for relatively short periods of time, typically a few decades. That said, carbon dioxide remains responsible for some 75 per cent of global greenhouse forcing (excluding water vapour), with methane accounting for about 14 per cent and nitrous oxide for 8 per cent.

The other major greenhouse gas is water vapour. However, its concentration in the atmosphere is so great that it is unaffected by releases of water vapour as such, being almost entirely determined by the temperature and other weather or geographical conditions and therefore fluctuating widely.

Table 8.1 Global Warming Potential (GWP) of different greenhouse gases by mass (including feedback effects, carbon dioxide = 1)[11]

Gas	Lifetime (years)	GWP over 20 years	GWP over 100 years
Methane	12	86	34
HFC-134a	13	3790	1550
CFC-11	45	7020	5350
Nitrous oxide	121	268	298

Analysis of air bubbles in the ice of the Antarctic suggested that carbon dioxide levels were rather lower during ice ages than at other times.[12] The importance of potential feedback loops became recognised. New calculations in the late 1960s and early 1970s taking account of the effects of convection suggested that a doubling of global carbon dioxide concentrations would lead to a temperature rise of around 2°C, about one-third of the difference between the previous ice age and the present day. In 1979 the US National Research Council stated: 'When it is assumed that the carbon dioxide content of the atmosphere is doubled and statistical thermal equilibrium is achieved, the more realistic of the modelling efforts predict a global surface warming of between 2°C and 3.5°C, with greater increases at high latitudes.'[13]

It is important to recognise then that much of the early science of climate change was 'bottom-up', deriving from calculations ultimately based on the behaviour of the carbon dioxide (and other) molecules, rather than 'top-down', dependent on measurement of average global temperature and how it appears to vary at different levels of carbon dioxide concentration. Consistent measurement of emissions, atmospheric concentration and temperature trends at the global level emerged only in the middle years of the twentieth century.

Global carbon dioxide concentrations have been measured at the Mauna Loa Observatory in Hawaii since 1958. Ice bubble observations suggest that pre-industrial levels of carbon dioxide varied between 120 ppm (parts per million) and 280 ppm, the latter figure prevailing in the mid eighteenth century. Levels have subsequently grown inexorably by between 1 ppm and 2 ppm per year to cross the 400 ppm mark in 2013[14] – see figure 9.2.

Annual global emissions of carbon dioxide stood at around ten million tonnes (Mt) in the mid-1750s. By the time that Arrhenius was making his calculations this had risen to around 1,400 Mt. In 1929, just before the Great Depression, it had reached 4,200 Mt. The figure in 1950 was 6,000 Mt, in 1960 was 9,500 Mt, quadrupling again by 2013 – see figure 9.3.

However, when it comes to identifying a link between these concentrations and emissions and the average global temperature things become more complex. Given the effectively undisputed science that holds that the natural greenhouse effect (in the pre-industrial era) is or was responsible for several tens of degrees of warmth, it is very difficult to believe that increasing the concentration of greenhouse gases would have no overall warming effect. At least the burden of proof must surely lie with those who deny such a link. In fact relatively few of those who are described or derided as 'climate change sceptics' do deny it. Nonetheless, there are some who, such as Charles Anderson and Ole

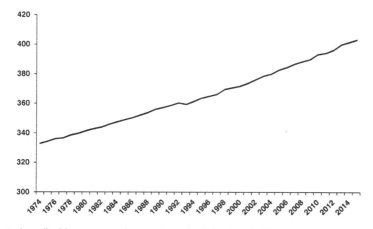

Figure 9.2 Carbon dioxide concentrations at the end of May (ppm), 1974–2015

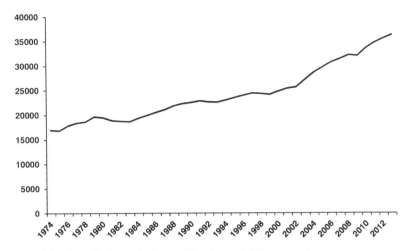

Figure 9.3 Global carbon dioxide emissions (million tonnes), 1974–2013[15]

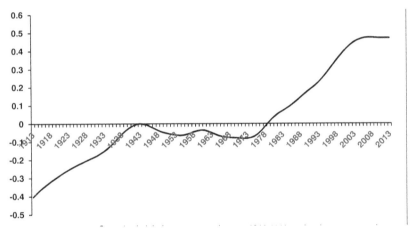

Figure 9.4 Smoothed global temperature changes, 1913–2013 against long-term trend (1943 as base year)[18]

Humlum, who argue that increasing carbon dioxide concentrations have no effect as the atmosphere is already entirely opaque to infrared at the frequencies that carbon dioxide absorbs the radiation[16] – although of course that argument would not apply to other greenhouse gases which absorb other wavelengths of infrared.

But climate is determined by many factors, not just the levels of greenhouse gases in the atmosphere. These include long-term ocean temperature cycles and currents, and changes in the amount of energy being released by the sun. Computer models of the climate have improved in their predictive ability over recent years but are still a long way from being entirely reliable.

The picture is further obscured by the difficulty in determining temperature trends over time. Measurement stations that may have been in unpopulated areas in say the 1950s may have become urbanised and suffer from heat pollution by 2000. If measurement equipment is moved to a different location, it can be difficult to be sure that any observed temperature variations are not caused by the shift in location rather than a real change over time.

With these caveats in mind, best estimates suggest that the global average temperature was around 0.9°C warmer in 2013 than it had been a century earlier – or in the words of Ivar Giaever, who in 1973 shared the Nobel Prize in Physics for his work with two colleagues on superconductors, it rose from around 280 K to 280.8 K. (In 2011 Giaever resigned from the American Physical Society over its use of the word 'incontrovertible' when discussing the evidence for anthropogenic climate change, asking the piquant question: 'In the APS it is ok to discuss whether the mass of the proton changes over time and how a multi-universe behaves but the evidence of global warming is incontrovertible?'[17]). However, the increase had not been uniform in the way that had been the case for global emissions and concentrations of carbon dioxide. Smoothing the data to reveal trends by taking account of random variability year-on-year suggests that between 1900 and 1940 there had been a significant increase, followed by a slight cooling to 1975, another significant increase to around 2005 and a plateau from 2005 to 2013. In fact the raw data suggest that the plateau may have begun in 1998 – or to give it its most extreme interpretation, of the 75 years from 1938 onwards the climate was warming in only 25 of these years (see figure 9.4).

This graph has been used by some who are sceptical about the extent of anthropogenic climate change to argue that carbon dioxide concentrations do not determine global temperatures, or at least that the climate models are at present unreliable and perhaps simplistic. The cooling between 1940 and 1975, for example, was accompanied by a 250 per cent growth in carbon dioxide emissions.

Evidence does suggest that historically there is a close relationship between temperature and atmospheric carbon dioxide concentrations. However, prima facie this might appear to be something of a two-edged sword for supporters of the anthropogenic climate change theory. Clearly it cannot have been emissions of carbon dioxide from human activity that were responsible for the variation of carbon dioxide concentrations in the atmosphere in prehistory – it might seem more plausible that the temperature was the independent variable and carbon dioxide concentrations vary with temperature, not the other way round. (For example, at higher temperatures there is more churn in the oceans, resulting both in releases of dissolved carbon dioxide and the bringing of carboniferous sediment containing sea shells to the surface where they could release some of their carbon dioxide.) Perhaps most likely, temperature and carbon dioxide levels reinforce each other, creating vicious (or virtuous) circles.

The mid-century global cooling also resulted in a cooling of fears about rising temperatures. Alongside continued scientific interest in increasing temperature grew a focus on 'global cooling'. As one review put it in 1975:

The central fact is that after three-quarters of a century of extraordinarily mild conditions, the earth's climate seems to be cooling down ... The Earth's average temperature during the great Ice Ages was only about seven degrees lower than during its warmest eras and the present decline has taken the planet about a sixth of the way towards the Ice Age average... [Some researchers] regard the cooling as a reversion to the 'little ice age' conditions that brought bitter winters to much of Europe and northern America between 1600 and 1900, years when the Thames used to freeze so solidly that Londoners roasted oxen on the ice and when iceboats sailed the Hudson River almost as far south as New York City.[19]

This was not the first time that such fears had been expressed: in the 1890s, for example, major media were warning of disastrous climate change as temperatures fell, with the *New York Times* carrying headlines such as 'Geologists think the world may be frozen up again', saying Canada could be 'wiped out' and lower crop yields would mean 'billions will die'.[20]

The global cooling scare was soon replaced by a resumption of fears about warming. A seminal moment came in 1988 when NASA scientist James Hansen published a major paper on the issue[21] and was credited with persuading the US Congress that the matter was real and of great importance.[22] (Hansen subsequently became one of Big Green's 'despised heroes', or perhaps 'part-time authorities', as they tried to negotiate the path between citing him as an unchallengeable source on climate change while rubbishing his ability and morality for advocating nuclear power.)

As it happened, Hansen's prediction proved wide of the mark – he predicted that the temperature would be some 1.9°C higher in the mid-2010s than it actually proved to be, as indicated in figure 9.5, despite underestimating the rate at which carbon dioxide would be released over the period.

Nonetheless, in a relatively short period of time climate change became a major international issue. A European Commission Directive in 1990 sought to ensure that emissions of carbon dioxide from the European Union as a whole would be no higher in 2000 than in 1990. The 'Rio Earth Summit' in 1992 saw the adoption of the UNFCCC – alongside conventions on biological diversity and on fighting desertification – which entered into force in March 1994. The ultimate objective of the Rio Convention was to stabilise greenhouse gas concentrations at a level that would prevent dangerous anthropogenic interference with the climate system. It states that such a level should be achieved within a time frame sufficient to allow ecosystems to adapt naturally to climate change, to ensure that food production is not threatened and to enable economic development to proceed in a sustainable manner. By 2013 it had 196 signatories, or 'Parties', including all UN Member States.[23] However, signing the Rio Convention and the subsequent Kyoto Protocol did not commit a country to meetings its targets – for this 'ratification' was necessary.

It is perhaps worth noting the extremely modest ambition of the Rio Convention when judged against what the science was suggesting would be necessary. Over 86 per cent of the world's traded primary energy use of 12,930 mtoe in 2014 came from oil, coal and gas, with less than 14 per cent coming from low-carbon sources. (Non-traded energy, adding perhaps an extra 10 per cent to the total, involves sources such as burning wood and using animal waste.) To stabilise the atmospheric concentration of greenhouse gases at double the pre-industrial level, i.e. 560 ppm, by 2050 it was calculated that some 50 to 70 per cent of annual global energy use of perhaps between 20,000 mtoe and 30,000 mtoe would have to be produced from very low carbon sources (unless carbon could be captured and stored efficiently for the very long term). This would translate into an 80 per cent reduction in emissions from the developed world. It was widely understood that even a doubling of the pre-industrial level of greenhouse gas concentration would be associated with a significant disturbance of the global climate.

The stated objective at Rio was that developed 'Annex 1' countries should emit no more carbon dioxide in 2000 than in 1990. Even this modest goal was not met by most countries while some of those which did, notably the UK and Germany, did so owing to fortunate by-products of other policies which had nothing to do with climate change, i.e. the 'dash for gas' and the closure of inefficient East German industry following German reunification.

The Fifth CoP to the UNFCCC was held in Kyoto in Japan in 1997 and led to a schedule of binding emission targets which became known as the Kyoto Protocol. Under this programme developed countries as a whole were expected to reduce their emissions of a basket of six greenhouse gases by a total of 5.2 per cent between the 'base year' of 1990 and the 'compliance period' of 2008–12, the particular target for each country or group of countries being negotiated with a view to their levels of economic development, current emissions, industrial profile, temperature and geography, etc.

The Kyoto Protocol was to come into force 90 days after the date on which no less than 55 Parties had ratified it, incorporating Annex 1 Parties accounting for at least 55 per cent of the total greenhouse gas emissions from Annex 1 countries in 1990. That the Protocol actually entered into force on 16 February 2005 demonstrates the challenge of bringing together an alliance representing even barely half of the world's greenhouse gas emissions. It was only when Russia acceded in 2004 (realising it would make a very large amount of money out of selling permits it did not need owing to the collapse of the communist economy) that the process could come into force. USA never ratified the Protocol while Canada withdrew from it in 2012.

In fact the Annex 1 countries' greenhouse gas emissions fell by 9.3 per cent between 1990 and 2011, including a 6.8 per cent fall in carbon dioxide emissions. This fall was much aided by the collapse of communism and reunification of Germany in the early 1990s and by the recession of the late 2000s and early 2010s.[24] Indeed, between 1994 and 2008 emissions were rising, as illustrated in figure 9.6.

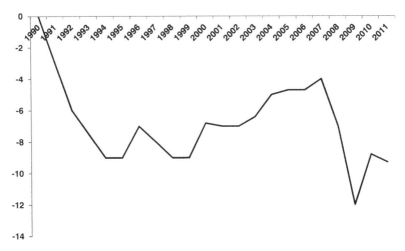

Figure 9.6 Percentage reduction in greenhouse gas emissions from Kyoto Annex 1 countries

This 'success' meant that global carbon dioxide emissions rose by 'only' 60 per cent during that period – the rise in greenhouse gas emissions being mitigated somewhat by falls in the other gases in the basket – because developing countries had no targets to meet.

The lack of ambition, in absolute terms, behind the Kyoto Protocol is perhaps best illustrated by the argument over the 'Kyoto mechanisms' – or 'flexibility mechanisms'. There is a very respectable argument that says that it is inefficient to allocate a rigid emission reduction target to each country. If, for example, it costs £1 billion to reduce emissions by one million tonnes of carbon dioxide in country A and £2 billion in country B, it would make more sense for country B to spend say £1.5 billion on emission reduction in country A than to spend £2 billion within its own borders – the overall emission reduction would be considerably greater but the cost much less.

This position was not universally held at Kyoto. Indeed, the conflict between those countries and organisations that wanted maximum flexibility (geographical, technical and temporal) in achieving emission limitation and reduction and those that believed that the bulk of emission controls should be achieved domestically by the biggest polluters was perhaps the fundamental dispute at the negotiations. Generally speaking Big Green wanted the big polluters to suffer pain by reducing their own emissions, even if they were already relatively efficient in energy use terms and even if to do so would result in a lower rate of global emission reductions.

The debate at Kyoto was to a considerable extent dominated by attempts to find a compromise between these two positions. The flexibility mechanisms that form part of the Kyoto Protocol can be viewed in this light. These mechanisms are:

- bubbles;
- tradable emission permits;

- joint implementation (JI);
- the Clean Development Mechanism (CDM).

The first three applied only to Annex 1 countries, i.e. those countries with emission limitation or reduction commitments.

'Bubbles' allowed groupings of developed nations, such as the European Union, to pool their emission reduction targets and distribute necessary measures internally. Thus within the overall EU limit Germany and the UK had relatively tough reduction targets while less economically developed members such as Greece and Portugal were to be allowed substantial emission increases (though in the event economic crisis allowed these countries to meet their Kyoto targets very comfortably). Tradable emission permits allowed developed countries with high compliance costs to buy permits from other developed countries with lower costs, the price of the permit being somewhere below the difference in abatement costs between the two countries involved. JI referred to projects funded completely or partially by one developed country in the territory of another, with credits for reducing emissions shared between participants.

The Clean Development Mechanism, although it resembled JI in some respects, was fundamentally different from the other Kyoto mechanisms and was by far the most controversial. The CDM would involve companies from Annex 1 countries investing in emission reduction schemes in developing countries. The host countries by definition did not have any emission limitation or reduction targets so any emission reduction credits that were generated in the course of a CDM project were to be used to offset emission control targets in developed countries and would be additional to the credits or permits in circulation within those countries. In most cases the host country would not benefit directly from carbon reductions associated with a CDM project, having no emission reduction or limitation targets of its own. It was allowed, however, that certified emission reduction credits could be shared with the host country, allowing the host to raise funds by selling them to agencies in developed countries. Other incentives could also be made available to the host country, notably financial aid for development.

There was great fear in some quarters that the CDM would simply result in 'emission target inflation', in effect creating a new source of emission permits from outside the total Annex 1 country cap emission reduction target of 5.2 per cent. There was obvious potential for gaming: for example, a scheme to construct a new factory, which would be more energy efficient than existing ones, might be cancelled only to re-emerge a few months later and a few metres to one side as a brand new scheme, thereby becoming eligible for emission reduction credits.

Ironically the country at the extreme pro end of the flexibility spectrum, the US, won the day as other countries feared this would be the price for the Americans ratifying the protocol: the mechanisms were duly included. The US then refused to ratify anyway. It is notable, however, that the Kyoto targets were so easy to reach that the flexibility mechanisms, particularly the CDM, were hardly used. The EU bubble did survive but the European Union carbon trading scheme saw its permit price collapse. By 2005 only 22 JI projects had been registered and the total savings claimed by 2012 under the mechanism were of the order of just 300 million tonnes of carbon dioxide. Between 2001 and 2012 CDM Carbon Emission Reduction (CER) credits amounting to one billion tonnes of carbon dioxide were issued.[25] Even if these all represented real reductions in emissions from those which would have occurred otherwise (a big if), this has to be judged against a period during which global carbon dioxide emissions alone were of the order of 367 billion tonnes. The price of CERs fell from around $20 per tonne in mid-2008 to 32¢ per tonne in 2012. The developed world simply did not need the CDM, or indeed much else, to fulfil its Kyoto commitments.

9.3 THE INTERGOVERNMENTAL PANEL ON CLIMATE CHANGE

The Intergovernmental Panel on Climate Change (IPCC), the leading international body for the assessment of climate change, was established by the UN Environment Programme (UNEP) and the World Meteorological Organisation (WMO) in 1988 to 'provide the world with a clear scientific view on the current state of knowledge in climate change and its potential environmental and socio-economic impacts.' The IPCC describes itself as a scientific body, under the auspices of the UN, which reviews and assesses the most recent scientific, technical and socio-economic information produced worldwide relevant to the understanding of climate change. It does not, however, conduct any research nor does it monitor climate related data or parameters. It is an intergovernmental body open to all member countries of the UN and WMO, boasting a membership of 195 countries by the mid-2010s. In 2007 the IPCC shared the Nobel Peace Prize with the former US vice-president, Al Gore.

Contribution to the work of the IPCC is made on a voluntary basis. Again in its own words: 'Review is an essential part of the IPCC process, to ensure an objective and complete assessment of current information. IPCC aims to reflect a range of views and expertise. The work of the organisation is policy-relevant and yet policy-neutral, never policy-prescriptive.'

The IPCC's structure includes a number of bodies.

- The IPCC Panel has overall control of the organisation's structure, processes and work programme and typically meets in plenary session once every year.
- The Bureau, elected by the Panel, consists of 30 members including individuals holding chairing responsibilities of the Working Groups and Task Force.
- The three Working Groups each have two co-chairs, one from a developed country and one from a developing country, and a technical support unit:

  Working Group I assesses scientific aspects of the climate system and climate change;

Working Group II assesses the vulnerability of socio-economic and natural systems to climate change, consequences and adaptation options;

Working Group III assesses options for limiting greenhouse gas emissions and otherwise mitigating climate change.

There is also a Task Force on National Greenhouse Gas Inventories.

Perhaps the most important activity of the IPCC is the publication of its Assessment Reports (ARs) updating the 'consensus' views of those contributing to the process. The first of these was issued in 1990 (with a supplementary report in 1992); the second in 1995; the third in 2001; the fourth in 2007; and the fifth in 2014.

The First Assessment Report in effect set the stage, stating that human activity was significantly increasing concentrations of carbon dioxide in the atmosphere and leading to increasing temperatures. The projected rate of warming was of the order of $0.3°C$ per decade, similar to natural variation, and it would be some time before climate science (and especially climate models) could confirm or refute this view.

Perhaps the Second AR, in 1995, was the most influential. It asserted that 'the balance of evidence suggests' that human activity was increasing the planet's temperature and that it would be a serious problem. The report in effect argued that the growing human population, standing at 5.7 billion at that point, and increases in energy demand were now damaging the balance between incoming and outgoing solar energy.

The 1995 report formed the basis for the negotiations and subsequent treaty at the Fifth CoP to the UNFCCC at Kyoto in 1997. However, although the US administration under President Bill Clinton and Vice-President Al Gore did sign the treaty it did not ratify the Kyoto Protocol, seriously undermining its credibility.

By the time of the Fourth AR the IPCC was arguing that it was now 'very likely' (rather than just 'likely') that human activity was responsible for at least some of the warming which had been observed. Among its conclusions were:

- warming of the climate system was unequivocal, evident from observations of increases in global average air and ocean temperatures, widespread melting of snow and ice and rising global average sea level;

- most of the global average warming over the previous 50 years was 'very likely' (greater than 90 per cent probability) due to human activity;

- the impacts of climate change would increase due to increased frequency and intensity of some extreme weather events;

- anthropogenic warming and sea level rise would continue for centuries even if GHG emissions were to be reduced sufficiently for GHG concentrations to stabilise, due to the time scales associated with climate processes and feedbacks;

- some planned adaptation to climate change was occurring but more extensive adaptation would be required to reduce vulnerability to climate change;

- unmitigated climate change would, in the long term, be likely to exceed the capacity of natural, managed and human systems to adapt;
- many impacts [of climate change] could be reduced, delayed or avoided by mitigation;
- the amount of carbon in the atmosphere was increasing at an even faster rate than previously;
- observed temperature increases were being mitigated to an extent by a blanket of other pollutants which were reflecting some sunlight away and so reducing the rate of warming but this was likely to be a temporary situation;
- alternative explanations for some of the warming, such as sunspot activity or heat pollution, were not credible;
- frozen areas were thawing, heavy rainfall was becoming more common, cold weather including frosts had become less frequent while hot weather, including heatwaves, had become more frequent.[26]

In effect AR5, in 2014, expressed the same findings but with an even greater degree of confidence.[27] Statements included:

- 'Warming of the climate system is unequivocal and since the 1950s, many of the observed changes are unprecedented over decades to millennia.'
- 'Atmospheric concentrations of carbon dioxide, methane and nitrous oxide have increased to levels unprecedented in at least the last 800,000 years.'
- 'Human influence on the climate system is clear. It is extremely likely (95 to 100 per cent probability) that human influence was the dominant cause of global warming between 1951 and 2010.'
- 'Increasing magnitudes of warming increase the likelihood of severe, pervasive and irreversible impacts.'
- 'Without new policies to mitigate climate change, projections suggest an increase in global mean temperature in 2100 of 3.7°C to 4.8°C relative to pre-industrial levels.'
- 'The current trajectory of global greenhouse gas emissions is not consistent with limiting global warming to below 1.5°C to 2°C relative to pre-industrial levels.'

IPCC work has received strong support from a number of academic and other institutions and groupings. In 2009 a collection of learned societies from Brazil, Canada, China, France, Germany, India, Italy, Japan, Mexico, Russia, South Africa, the UK and the US endorsed AR4 and called for urgent action.[28] Endorsement also came from the Australian Academy of Science (not a country always regarded as being in the forefront of climate change activism), which said:

The role of greenhouse gases in the atmosphere is qualitatively well understood. It is known that increasing the atmospheric concentration of the principal anthropogenic greenhouse gas, carbon dioxide, leads to higher mean global surface temperatures. It is known that carbon dioxide has increased very substantially during the last century, to the highest levels seen in the last 800,000 years and that this increase is primarily of anthropogenic origin. It is also beyond serious question that some carbon dioxide from human activities remains in the atmosphere for a very long time, as is the message that unless greenhouse gas emissions are reduced, an upward trend in global temperature will continue.[29]

For the most vocal climate change protagonists even the IPCC is too conservative. In the view of one commentator:

The process by which the IPCC conducts its deliberations – scientists and national government representatives quibbling at enormous length over wording and interpretation – is Byzantine at best and makes the group's achievements all the more impressive. But it sacrifices up-to-the-minute assessment of data in favor of lowest-common-denominator conclusions that are essentially beyond argument. That's a reasonable method but one result is that the 'shocking' conclusions of the new report in fact lag behind the most recent findings of climate science by several years.[30]

Lord (Nicholas) Stern, author of an earlier influential (but not unchallenged) report on climate change for the UK government[31], argued that AR5 understated the threat because it omitted factors such as the melting of frozen Arctic soil, expected to contribute significantly to global warming but 'left out of calculations because the size of their impact remains unclear.'[32]

9.4 THE ENDS JUSTIFIES THE MEANS?

The main criticism of the IPCC has come from sceptics who challenge both its conclusions and its processes and procedures. The feeling among many who are unconvinced by the IPCC is that its crypto-religious status, making any attempt to offer serious challenge appear an act of heresy, has allowed it to get away with a much lower level of rigorous scrutiny than would be the case if, say, Big Green did not approve of the messages it was promulgating.

Perhaps the first cracks in the apparent infallibility of the IPCC concerned a widely publicised statement in the 2007 Working Group II report, *Impacts, adaptation and vulnerability*:

Glaciers in the Himalaya are receding faster than in any other part of the world and, if the present rate continues, the likelihood of them disappearing by the year 2035 and perhaps sooner is very high if the Earth keeps warming at the current rate. Its total area will likely shrink from the present 500,000 to 100,000 km² by the year 2035.

The IPCC subsequently acknowledged that the date was incorrect, expressing regret for 'the poor application of well-established IPCC procedures in this instance'. The date of 2035 had been uncritically taken from a report by a Big Green organisation, the World Wide Fund for Nature (WWF), which by accident or design had misquoted an original source.

A reasonable person might take a view that given the vast amount of data the IPCC deals with it is entirely foreseeable and forgivable that the occasional detail goes astray. However, Robert Watson, who chaired the IPCC from 1997 to 2002, observed that the mistakes all appeared to have gone in the direction of making it seem like climate change was more serious by overstating the impact. 'The IPCC needs to look at this trend in the errors and ask why it happened.' Watson also expressed the view that challenge by sceptics should always be acknowledged and welcomed. 'The IPCC's job is to weigh up the evidence. If it can't be dismissed, it should be included in the report. Point out it's in the minority and, if you can't say why it's wrong, just say it's a different view.'[33]

Watson's concerns were in part prompted not just by errors in the IPCC's science but also by growing criticism of its structure and culture. This was compounded by the so-called 'Climategate' scandal of November 2009 (a second round following two years later), in which secret emails from researchers at the Climatic Research Unit of the University of East Anglia, one of the UK's principal research institutes into climate change (presumably because it would be the first to feel the effects of sea level rise) became public. Many of the emails consisted of discussions among participants in the IPCC report writing process and focused on how to sidestep inconvenient contradictions or uncertainties in the evidence.

Among the comments in the hacked emails were:

- 'The fact is that we can't account for the lack of warming at the moment and it is a travesty that we can't.' (Kevin Trenberth)
- 'I've just completed Mike [Mann]'s *Nature* trick of adding in the real temps to each series for the last 20 years (i.e. from 1981 onwards) and from 1961 for Keith [Briffa]'s to hide the [temperature] decline.' (Phil Jones)
- 'I can't overstate the HUGE amount of political interest in the project as a message that the Government can give on climate change to help them tell their story. They want the story to be a very strong one and don't want to be made to look foolish.' (Civil servant to Phil Jones)

- 'I've been told that IPCC is above national Freedom of Information Acts. One way to cover yourself and all those working in AR5 would be to delete all emails at the end of the process.' (Phil Jones)

- 'Any work we have done in the past is done on the back of the research grants we get – and has to be well hidden. I've discussed this with the main funder (US Department of Energy) in the past and they are happy about not releasing the original station data.' (Phil Jones)

- 'The trick may be to decide on the main message and use that to guide what's included and what is left out.' (Jonathan Overpeck, IPCC Lead Author)

- 'Mike [Mann], the figure you sent is very deceptive ... there have been a number of dishonest presentations of model results by individual authors and by IPCC.' (Tom Wigley, UEA)

- 'I can't see either of these papers [peer-reviewed studies questioning whether changing land use effects had been properly accounted for in the temperature record] being in the next IPCC report. Kevin [Trenberth] and I will keep them out somehow – even if we have to redefine what the peer-review literature is!' (Phil Jones)

Despite the climate sceptics' frenzied response to the emails, a careful reading does not strongly suggest that there was a serious attempt to misrepresent the essential messages coming from the data and analyses, so much as to stifle alternative views. A *Nature* editorial concluded that a fair reading of the emails revealed nothing to support the denialists' conspiracy theories.[34]

However, the effect on public confidence in climate science was not insignificant. Journalist George Monbiot, one of the most passionate proponents of the climate change hypothesis but one who is perhaps more driven by the science than by pure gut feeling, said:

It is true that climate change deniers have made wild claims which the material can't possibly support (the end of global warming, the death of climate science). But it is also true that the emails are very damaging. ... No one has been as badly let down by the revelations in these emails as those of us who have championed the science. We should be the first to demand that it is unimpeachable, not the last.[35]

Another adherent to the climate mainstream, Hans von Storch, said that the East Anglian researchers had violated a fundamental principle of science by refusing to share their data with other researchers, accusing them of 'playing science as a power game'.[36]

In 2010 the IPCC commissioned a report from the InterAcademy Council (IAC)[37] which had been created in 2000 by the world's science academies to mobilise the best scientists and engineers worldwide to 'provide high quality advice to international bodies such as the United Nations and the World Bank as well as to other institutions'. While the report concluded that the IPCC assessment process had been successful overall, it was critical of the organisation's governance and management, and in particular the absence of an elected Executive Committee and Executive Director; the dominance of the Lead Authors over the Review Editors; weaknesses in the process for responding to reviewer comments; inconsistencies across Working Groups on how to communicate uncertainty; the lack of justification for levels of certainty and probability being ascribed to various conclusions; and the lack of general transparency, for example the absence of any published criteria for selecting key participants in the assessment process.

As part of the process the IAC issued a questionnaire to participants in the meeting to finalise the wording of the Summary for Decision-makers.[38] Anonymous comments included:

- 'In my experience the summary for policy makers tends to be more of a political process than one of scientific précis.'

- 'This is a pure political process.'

- 'Often we see in the discussion that scientific merit gives way to political priorities.'

- 'This is an awful procedure and should be changed. It has far too much politics and the final version has little relation to the one suggested by the scientists.'

- 'It was a bitter process. At the end when time was out and everybody was tired something was changed and passed in a hurry and carelessly.'

Comments about other parts of the process, from past IPCC Lead Authors, revealed a similar picture.

- 'There are far too many politically correct appointments, so that developing country scientists are appointed who have insufficient scientific competence to do anything useful. This is reasonable if it is regarded as a learning experience but in my chapter in AR4 we had half of the Lead Authors who were not competent.'

- 'It is clearly noticeable that the [author nomination] process occasionally brings authors with poor knowledge or poor motivation into Lead Author positions.'

- 'The most important problem of the IPCC is the nomination and selection of authors and Bureau Members. Some experts are included or excluded because of their political allegiance rather than their academic quality. Sometimes, the 'right' authors are put in key positions with generous government grants to support their IPCC work, while the 'wrong' authors are sidelined to draft irrelevant chapters and sections without any support.'

- 'The team members from the developing countries (including myself) were made to feel welcome and accepted as part of the team. In reality we were out of our intellectual depth as meaningful contributors to the process.'

In a paper for the Global Warming Policy Foundation Professor Ross McKitrick from Canada, who had served as an Expert Reviewer for Working Group I's contribution to AR4, went further than the IAC, applying a more challenging approach reminiscent of the way Big Green attacks a body of which it does not approve.[39] McKitrick pointed out that IAC was hardly an independent body itself, its previous report having been a study promoting renewable energy co-authored by a committee that included the then Chair of the IPCC, Rajendra Pachauri, and two IPCC Lead Authors.

McKitrick claimed that criticism had been growing over whether the IPCC's assessment reports were as objective and comprehensive as they ought to be, citing three particular concerns.

- While the IPCC had long had critics, their number was growing and their ranks included new members who had in the past been advocates on its side.
- The IPCC played a very influential role in the world and it was imperative that its operations were unimpeachable but the oversight mechanisms of the IPCC did not appear to be adequate to assure this.
- There was a wide misunderstanding about the IPCC assessment process, such that it was often considered more formal and rigorous than it actually was.

McKitrick identified four characteristics of the practical workings of the IPCC:

- an opaque process for selecting Lead Authors (which was in the gift of the Bureau) – some Lead Authors had been criticised by others as being overly dominated by political considerations, many being employed by or serving as advisors to environmental activist organisations;
- the absence of any binding requirement for incorporating the full range of views, the language about procedures having to be 'comprehensive' being vague and inadequate;
- intellectual conflicts of interest, with Lead Authors regularly reviewing their own work and that of their critics;
- loopholes in the peer review sequence, with Lead Authors having the opportunity to overrule reviewers or wait until the close of expert review before rewriting the text.

McKitrick cited some specific cases of what he regarded as malpractice and counterexamples of the IPCC's claim to 'choose people based on their track record of publications and the research they have done'.

- The selection of Sven Teske, a climate campaigner for Greenpeace, as Lead Author for a report on renewable energy (SRREN), following which a non-peer-reviewed Greenpeace report co-authored by Teske and two lobbyists for renewable energy subsidies (Josche Muth from the European Renewable Energy Council and Steve Sawyer from the Global Wind Energy Council) was used as the main authority for the central claims in the report, as highlighted in the press release announcing its publication, despite being just one – the outlier, of course – of 164 scenarios.

- Sari Kovats who was selected to serve as an IPCC Contributing Author in 1994 when she was 25 years old, had no Ph.D. and no academic publications and was just starting a job as a research assistant at the London School of Hygiene and Tropical Medicine – she began a part-time Ph.D. programme in 2001, at which time she was promoted to a term as an IPCC Lead Author.

- The extraordinary dominance of advisors to campaign group WWF – 28 out of the 44 chapters in AR4 were authored by teams that included at least one member of WWF's (unpaid so presumably ideologically driven) advisory panel, including all 20 chapters of the Working Group II report and six of the 11 chapters of the Working Group I report; WWF campaign advisors served as Coordinating Lead Authors (CLA) for 15 of the AR4 chapters and in three cases both CLAs were WWF advisors, in one chapter eight authors being WWF advisors.

- The deletion of caveats about the statistical significance of apparent changes in temperature using simple models and their replacement with statements that in effect made the opposite claims or at least heavily diluted the original – e.g. replacing 'linear trend statistical significances are likely to be overestimated' with saying that they 'could be' overestimated.

The predominance of WWF in such matters is of particular concern. In its 2009 G8 Climate Scorecards, WWF perpetrated one of the most blatant pieces of climate 'science' fraud yet seen. In a sense one can sympathise – calculations based on actual greenhouse gas emissions from energy use annoyingly kept placing countries such as France and Sweden, with their large nuclear programmes, in the highest places in terms of reducing greenhouse gas emissions, clearly an unhelpful message from WWF's religious standpoint. WWF's response was an object in lesson in Big Green science. 'WWF does not consider nuclear power to be a viable policy option … To reflect this, a policy approach that favours the use of nuclear power was assessed in the following way: indicators for the "current status" were adjusted, by assuming that electricity from nuclear energy was produced with gas, the most carbon efficient fossil fuel.' Sure enough, by redefining 'very high volumes of carbon dioxide emissions' to include 'very low levels of radioactive releases' in such an arbitrary way, WWF was able to give the false impression that nuclear power was not a particularly effective way of reducing greenhouse gas emissions. Needless to say this sleight of hand was not mentioned in the publicity material surrounding the document, which, as ideologically required, succeeded in placing Germany, the doyen of simplistic 'environmentalism', as the top of the table.[40] One is bound to conclude that when this is the level of intellectual dissemblance which lies at the heart of WWF's values, no document in the climate change field in which they have been heavily involved can be trusted in any sense.

Again one wonders how Big Green would have reacted if say employees of or advisors to the oil industry had been given similar hegemony over these reports. Documents leaked to the media in 2013 appeared to show attempts at political interference in the production on AR5 and that several governments that had reviewed the draft objected to how the issue of the 'pause' in warming from 1998 onwards was tackled. In a leaked draft of the report, dating from June 2013, the IPCC had said that the rate of warming between 1998 and 2012 was about half the average rate since 1951. The draft said that the reduction in warming over the period 1998 to 2012, compared to 1951 to 2012, was in roughly equal measure down to natural variations in the climate and factors such as volcanic eruptions and the downward phase of the current solar cycle. But the documents, according to Associated Press, showed Germany calling for the reference to the slowdown to be deleted while the US urged scientists to include as its leading hypothesis that the reduction in warming was linked to more heat being transferred to the deep ocean. Both countries' governments had policies which stated their belief in man-made climate change. Belgium objected to using 1998 as a starting year for any statistics because it claimed it was a particularly warm year.[41]

Many people are quite comfortable with the idea that when a field of exploration is developing there will be times when contradictory data or alternative theories to explain one detail or another may emerge, without undermining the broad thrust of understanding. Damage to the role of science in driving policy comes far more from any impression that things are being 'sexed up', to use a phrase which came into particular vogue when inaccurate claims about weapons of mass destruction in Iraq were portrayed as being reliable, for political reasons. Climate scepticism has proved remarkably resilient in the face of the onslaught of climate science. One reason may be that too many exaggerated claims were made too early as to the confidence that could be placed in the results of that science. Far from scaring people and governments into action, this in turn may have contributed to the slow pace of policymaking.

9.5 PROFESSIONALISING CLIMATE SCEPTICISM

Concerns such as these and others over the objectivity of the IPCC led to the formation of a counterweight, the Nongovernmental International Panel on Climate Change (NIPCC), which emerged from an initiative by Fred Singer, a moderate climate sceptic. Singer is a trained atmospheric physicist though, like Big Green heroes such as Paul Ehrlich, he has something of a record of being on what history suggests is the wrong side of many scientific issues such as the link between CFCs and ozone depletion and the association between smoking and cancer. As a pointer to his position, in 2012 Singer challenged sceptics who claim that rising carbon dioxide levels do not cause temperatures to rise, that the concentration of carbon dioxide in the atmosphere is too small to have an effect and that natural variations in carbon dioxide levels dwarf human contributions.[42] He would undoubtedly be included in the 97 per cent of scientists who 'believe in anthropogenic climate change' by John Cook's criteria as described in the following section.

The NIPCC operates under the auspices of the Center for the Study of Carbon Dioxide and Global Change, a registered charity. It describes itself as: 'an international panel of nongovernment scientists and scholars who have come together to understand the causes and consequences of climate change. Because we are not predisposed to believe climate change is caused by human greenhouse gas emissions, we are able to look at evidence the Intergovernmental Panel on Climate Change (IPCC) ignores. Because we do not work for any governments, we are not biased towards the assumption that greater government activity is necessary.' It produced its first report in 2008, *Nature, not human activity, rules the climate*, followed in 2009 by *Climate change reconsidered: the 2009 Report of the Nongovernmental International Panel on Climate Change (NIPCC)*, with a foreword by Frederick Seitz, a celebrated physicist who inter alia had been president of the US National Academy of Sciences from 1962 to 1969 and recipient of the National Medal of Science and NASA's Distinguished Public Service Award. Even by IPCC standards the NIPCC is a modest affair, its first report listing 35 contributors and reviewers from 14 countries (although it should be noted that the IPCC itself only has around ten full-time staff plus its technical advisors). However, it also included the names of 31,478 American scientists (the 'Oregon petition') who signed a petition saying: 'there is no convincing scientific evidence that human release of carbon dioxide, methane or other greenhouse gases is causing or will, in the foreseeable future, cause catastrophic heating of the Earth's atmosphere and disruption of the Earth's climate.' In 2013 the NIPCC published *Climate change reconsidered II: physical science*, and in 2014 *Climate change reconsidered II: biological impacts*.

In its critique of the Summary for Policymakers associated with AR5, NIPCC listed 11 statements which it says represent a 'retreat from more alarmist positions struck in earlier ARs or in related research literature.'[43] These include:

- recognition of the 'pause' in warming since 1998;
- acceptance that temperatures in the Mediaeval Climate Anomaly between AD 950 and AD 1250 were at times as warm as those of the late twentieth century;
- identifying an increase of some 1.2 to 1.8 per cent per decade in the extent of Antarctic sea ice between 1979 and 2012;
- the admission that solar factors may be significant in determining the climate, something which had previously been dismissed as negligible;
- implicit acknowledgement that there was less certainty about projected

temperature changes than had been thought in 2007 and a reduction in the lower expected limit to levels consistent with the predictions of several 'climate sceptics'.

To be fair to the IPCC it is entirely to be expected that the early hypotheses in any complex scientific field should subsequently be revised or even discarded as new data become available. Inevitably as time goes by some areas of concern will be found to be less threatening than first appeared while other worries will grow in salience. It is a misunderstanding of 'science' to describe a theory as an 'error' simply because the emergence of more evidence subsequently suggests that a different interpretation is needed. It would have been much more suspicious if none of its estimates were revised down (alongside others being revised up) as a result of the appearance of better data or techniques. A body determined to take every opportunity to paint the issue in the most pessimistic light, as claimed or implied by the sceptical movement, might perhaps be more likely to gloss over such anomalies rather than to 'correct' them.

However, the NIPCC may have more of a point when it notes that revising the overestimates down received relatively little publicity either from the IPCC or from those commenting on AR5, while the counterexamples where things looked worse than they had in 2007 were trumpeted loudly.

The NIPCC also cited 13 examples of statements it claimed were misleading – like the '97 per cent consensus in favour of anthropogenic climate change' discussed in the following section, they may not all be technically untrue but may nonetheless give a false picture of reality. They include:

- 'Warming of the climate system is unequivocal and since the 1950s many of the observed changes are unprecedented.' (NIPCC notes that the increase after 1950 was akin to that between 1910 and 1940, for example.)
- 'It is virtually certain that the upper ocean has warmed from 1971 to 2010.' (NIPCC says that the claimed ocean temperature rise of 0.15°C is actually less than the accuracy limit of the instrumentation available until 2003; between the deployment of the Argo buoy network in 2003 and 2012 there was no statistically significant trend.)
- 'Emissions of carbon monoxide are virtually certain to have induced a positive radiative forcing.' (Carbon monoxide has a very short atmospheric residency, measured in weeks rather than years; the IPCC itself says elsewhere that it is difficult to quantify or project the effects of carbon monoxide and other species not generally regarded to be greenhouse gases.)
- 'The reduced trend in radiative forcing between 1998 and 2012 is primarily due to volcanic eruptions.' (There were no significant volcanic eruptions in this period.)
- 'The net radiative feedback due to all cloud types combined is likely positive. Uncertainty in the sign and magnitude of the cloud feedback is due primarily

to continuing uncertainty in the impact of warming on low cloud.' (The two sentences are mutually contradictory.)

- 'There is very high confidence that the maximum global mean sea level during the last interglacial period was at least five metres higher than present' (The implication that warmer temperatures in the current era could lead to a similarly large increase in sea level is unjustified – sea level was indeed higher during the Eemian interglacial around 125,000 years ago than it was during the current Holocene because orbital eccentricity then was much greater, with solar irradiation levels of some $230\,\mathrm{Wm^{-2}}$ compared to $90\,\mathrm{Wm^{-2}}$ at present.)

Summarising, the NIPCC claims:

With the same set of peer-reviewed scientific papers available to them, the scientists of the IPCC and NIPCC have come to diametrically opposing conclusions. IPCC scientists remain alarmist about the threat of human-caused global warming, even while they admit observations increasingly invalidate their model-based predictions. They are reluctant to acknowledge past errors and new research that challenge their hypothesis of human-caused dangerous climate change. In stark contrast, NIPCC scientists find no hard evidence for a dangerous human-caused warming. They find the null hypothesis – that observed changes in climate are due to natural causes only – cannot be rejected. NIPCC scientists remain open to new discoveries and further debate.

The aim of this treatment is not to take sides between the IPCC and NIPCC or indeed the pro- and anti-anthropogenic global warming (AGW) camps. It is to observe the contrast between the attitude of various commentators on climate change, from the political Left and Right, to the IPCC and NIPCC. For example, Greenpeace, while (legitimately) questioning the funding behind the NIPCC, attacks the organisation's integrity rather than challenging its claims, saying that:

- its purpose is not to give clarity on climate science, as the IPCC does, but to critique the IPCC;
- the scientists working for the NIPCC get paid; the IPCC scientists don't. (It might be more accurate to say that IPCC scientists are funded by outside organisations, including Big Green, while NIPCC scientists are not. In any case one might reflect that because someone is driven to take part in something without being paid does not necessarily mean that they are ipso facto more dispassionate and neutral – possibly quite the opposite);
- the NIPCC report only critiques papers published by deniers, whereas the IPCC critiques all papers, including those published by deniers.

Perhaps most hypocritically, Greenpeace, which has carried out an 'education' programme for schools, including visits from activists making statements about issues such as radiation which are flatly contradictory to mainstream scientific understanding as expressed by the United Nations, views it as scandalous that the NIPCC's supporting organisations should wish to see climate scepticism taught in US schools alongside the more 'orthodox' position.

Similar double standards have involved the treatment of another influential focus for climate change scepticism, the Global Warming Policy Foundation (GWPF), established in 2009 in the UK by Lord (Nigel) Lawson, widely respected former Chancellor of the Exchequer in the Thatcher years. The GWPF describes itself as unique, 'an all-party and non-party think tank and a registered educational charity which, while open-minded on the contested science of global warming, is deeply concerned about the costs and other implications of many of the policies currently being advocated.' It says it is funded mainly by voluntary donations from individuals and charitable trusts, not accepting gifts from energy companies. Its principles include:

- holding no official or shared view about the science of global warming, while recognising that the issue is not yet settled;
- a main focus on global warming policies and their economic and other implications;
- reliance on observational evidence and understanding the present rather than computer modelling or predicting the distant future.

From a neutral standpoint one might expect that a former finance minister might have some interesting ideas about the financial consequences of various courses of action designed to counter climate change – indeed, he might be rather better qualified to hold an opinion on such matters than a climate scientist. Nonetheless, when Lord Lawson was invited to take part in a discussion on BBC Radio with Sir Brian Hoskins, a respected climate change scientist and proponent from Imperial College, his presence provoked a large number of complaints from the climate change establishment. One complainant, Chit Chong, a former Green Party councillor, said that the encounter 'gave the impression that there's still debate about climate change. The BBC has the whole idea of balance wrong – in seeking balance they are creating imbalance and promoting untruths.'[44] Yet the Green Party rarely complains that time is given to someone who challenges the widespread understanding say of lack of observable health effects associated with low levels of radiation. Nothing creates scepticism and the policy paralysis that it engenders as efficiently as the impression that an alternative point of view, especially from someone as distinguished as Lord Lawson, is being suppressed.

9.6 CONSENSUS? WHAT DO YOU MEAN, CONSENSUS?

The peculiar nature of climate change dogma was well illustrated by a claim made in 2013 by a group of climate change activists associated with a website called *Skeptical Science*.[45] It was widely reported, including by such luminaries as the president of the United States and the UK Secretary of State for Energy and Climate Change, as revealing that '97 per cent of scientists agree climate change is real, man-made and dangerous' (in the words of an Obama tweet, or at least one put out in his name, not peer reviewed as far as can be established. Ed Davey, the UK minister at the time, was a little more measured, saying: 'in a recent analysis of 12,000 climate papers ... of the scientists who expressed a view 97 per cent said that climate change was happening and that it was a human-made activity.')

Luckily a leaked internal forum for staff members involved in the work revealed this exchange between Cook and another contributor, Ari Jokimäki, planning the publicity campaign before the results had been analysed.

COOK: 'It's essential that the public understands that there's a scientific consensus on AGW. So Jim Powell, Dana Nuccitelli [two of Cook's co-authors] and I have been working on something over the last few months that we hope will have a game changing impact on the public perception of consensus. Basically, we hope to establish that not only is there a consensus, there is a strengthening consensus. ...'

JOKIMÄKI: 'I have to say that I find this planning of huge marketing strategies somewhat strange when we don't even have our results in and the research subject is not that revolutionary either (just summarising existing research).'[46]

Cook also said: 'To achieve this goal [of persuading the public there was a vast consensus on climate change] we mustn't fall into the trap of spending too much time on analysis and too little time on promotion.'

Had this leak, and indeed the basic claim, been associated with a less untouchable orthodoxy, one suspects that someone would at least ask what such a statement – '97 per cent of scientists agree' – could possibly mean. How, for example, would one define a 'scientist' who qualified for inclusion in this figure? Does this refer say to the tens of millions of people with a science degree? Are they all scientists? Would anyone else qualify as a scientist? Or is the 97 per cent figure a kind of definition – a 'scientist' is someone with a 97 per cent likelihood of believing that climate change is real, man-made and dangerous? (The 'Oregon petition' for scientists and engineers who oppose climate change orthodoxy claims to have over 31,000 signatures – if this is reliable one would expect then that there was evidence of at least one million scientists who have expressed support for what seems to be the more orthodox position. Where is this evidence?)

More importantly, what exactly was Cook claiming? As noted previously, the basic science of climate change, involving the role of greenhouse gases in absorbing and reradiating infrared radiation, thereby warming the earth above what would otherwise be expected, is established, uncontroversial and certainly accepted by most if not all of those who are regarded as being in the 'sceptic' camp. It follows that there is relatively little doubt, at least among scientists, that human activity could in theory result in warming of the planet to some extent. Nor does a search of the literature reveal anyone who claims that the climate cannot change, in the sense of denying the existence of ice ages for example.

Cook and his fellow activists reviewed the abstracts of 11,944 climate papers and decided that 32.6 per cent supported anthropogenic climate change while only 0.7 per cent did not.

Leaving aside that 32.6 per cent is not 97.1 per cent (as Cook accepts), the paper includes three positions within the 'endorsing anthropogenic climate change' category, namely:

1. explicit endorsement with quantification (humans are the primary cause of climate change);
2. explicit endorsement without quantification (i.e. man-made climate change is happening but no view as the extent);
3. implicit endorsement (assumption that greenhouse gas emissions cause warming without explicitly stating humans are the cause).

To be counted as not supporting anthropogenic climate change a paper would have to express sentiments such as 'the extent of human-induced global warming is inconclusive'; 'a major proportion (undefined) to all of the warming of the twentieth century could plausibly result from natural causes'; or 'the human contribution to the carbon dioxide content in the atmosphere is negligible in comparison with other sources of carbon'.

However, Cook himself only identifies 65 of the papers as belonging in category 1, i.e. 0.6 or 1.7 per cent depending on how you look at it. Predictably Cook did not publish this figure but it emerged in a blog discussion of the results.[47] Another contribution to that discussion said: 'OK, so we've ruled out a definition of AGW being "any amount of human influence" or "more than 50 per cent human influence". We're basically going with AGW = "humans are causing global warming", i.e. no specific quantification, which is the only way we can do it considering the breadth of papers we're surveying.'

So it appears that Cook was actually saying that 97 per cent of scientists accept that the earth is warmer than the moon because of the insulating effect of greenhouse gases and therefore that increasing concentrations of greenhouse gases from whatever cause might be expected to have some effect on the climate, however small. This is hardly radical –

indeed, the vast majority of those who would regard themselves as climate sceptics in the sense of not 'endorsing' the devotion of vast resources to reducing greenhouse gas emissions would fit into the third of the above categories if not the second.

One example involved a paper by Nir Shaviv of the Hebrew University of Jerusalem. *Skeptical Science* itself had previously described Shaviv as a 'denier' but his work was classified as 'explicitly endorsing but not quantifying' global warming – in other words Category 2 above. Shaviv responded by saying that his work showed that climate sensitivity to greenhouse gas emissions is low and (revealingly), 'I couldn't write these things more explicitly in the paper because of the refereeing but you don't have to be a genius to reach these conclusions from the paper'.[48] (Shaviv seems to be suggesting that any comments which may arise from publishable data and arguments but which are challenging to the global warming dogma will not get past the refereeing process.)

So what Cook could have said, with equal accuracy and probably more meaning, is that a pitiful 0.6 per cent of the abstracts of papers looking at climate change take the view that human activity is the main cause of climate change, or 1.7 per cent of those papers which expressed a position on the issue.

One final caveat should be entered. Were it to be the case that there is a significant consensus about climate change, this does not of itself prove anything. Newtonian mechanics held enormous sway in the nineteenth century – then came the quantum and relativistic revolution and ideas had to be discarded or at least heavily qualified. One wonders whether, if the early twentieth century equivalents of Natalie Bennett or Lawrence Torcello had had their way, we would have had the fruits of the quantum revolution at all.

9.7 HOT, COLD OR LUKEWARM

Another example of an approach which was likely to cause raised eyebrows among neutral observers involved the 'pause' in global warming which was observed between 1998 and 2013. Various possible explanations come to mind. Perhaps the temperature increase has stopped and was nothing to do with greenhouse gas emissions, which have continued to rise rapidly through the period as did atmospheric concentrations of carbon dioxide. Maybe some other factor had come to the fore which obscured a continued greenhouse gas-driven increasing trend, such as a reduction in heat reaching the earth from the sun, it being known than the sun has its own 'cycles' of heat radiation: once this factor has worked its way through then warming would be expected to resume. Another proposal suggested that warming may have been continuing but in the period after 1998 was being manifested in different ways, for example warmer temperatures in the Arctic resulting in colder winters in Europe and the absorption of heat in depths of the eastern half of the Pacific, the main region which did not warm post-1998.[49] IPCC authors pointed out that the winter of 2009/10 was exceptionally cold in Europe but globally was one of the warmest, notably in North America, while in the winter of 2013/14 the eastern half of the US suffered through unusually cold and snowy conditions while there were record high temperatures in Alaska and along the US West Coast, and southern England faced widespread flooding.[50]

Whatever the truth, though, it is clear that the climate models which are sometimes portrayed as sources of a high degree of certainty about the future of the climate are in reality still at a relatively primitive stage of development. The explanations offered to show that the pause was consistent with an underlying warming trend were undermined somewhat by global warming advocates moving the target. For example, in 2009 Professor Phil Jones, in one of the 'Climategate' emails, said that a 'no upward trend' would have to continue for a total of 15 years before climate scientists should 'get worried'. (A cynic might observe that from a public point of view, rather than that of a research establishment in receipt of large grants, if warming had indeed ceased this would be a cause of less worry, not more.) By 2012 he was telling the media that 15 or 16 years was not a significant period: pauses of such length were to be expected.[51] Prima facie this appears a very unscientific attitude – one might expect a scientist to come out with a theory, test it against reality and, if reality fails to comply, abandon the theory rather than simply shifting the goalposts. Inevitably such an approach risks creating scepticism. Furthermore, it seems inconsistent to place such great reliance upon models of the climate when they predict significant warming while dismissing them as inadequate when they fail to predict or even reliably explain periods of cooling beyond the rather vague 'such things are expected'.

In the event, 2014 appeared to be the hottest year on record,[52] while July 2015 was the hottest recorded month since measurement began in 1880.[53] One sensed a sigh of relief that climate change looked real after all.

Another entertaining example surrounds the career of one Paul Ehrlich. In 2013 Ehrlich, much feted by the environmental movement as a siren voice on climate change, co-authored a report claiming to represent a 'consensus of global scientists' which blamed the warming climate for forthcoming disaster. 'By the time today's children reach middle age, it is extremely likely that Earth's life support systems, critical for human prosperity and existence, will be irretrievably damaged by the magnitude, global extent and combination of human-caused environmental stressors unless we take concrete, immediate actions to ensure a sustainable, high-quality future.'[54]

Ehrlich may have been representing a widespread view here but in 1968 he was explaining that the cooling climate, alongside other human activities, would lead to major food shortages in the US: 'in the 1970s ... hundreds of millions of people are going to starve to death'. His best estimate was that 65 million Americans would die of starvation between 1980 and 1989 and by 1999 the US population would have declined to 22.6 million. (The actual population figure was to be 273 million.) He added: 'If I were a gambler, I would take even money that England will not exist in the year 2000.'[55] In 1970 one of Ehrlich's Stanford colleagues, George Wald, predicted that '... civilization will end within 15 or 30 years unless immediate action is taken against problems facing mankind' while Senator Gaylord Nelson was saying in *Look Magazine* that by 1995 '... somewhere between 75 and 85 per cent of all the species of living animals will be extinct.'[56]

Ehrlich is in a sense consistent – humanity is going to destroy itself in around twenty years – though to maintain this position he has had to reverse completely his views on what will cause the catastrophe in dramatic not to say hilarious manner. The 'twenty-year' tipping point seems to be more in the nature of a horizon, always the same distance away wherever we find ourselves, than a location. What is interesting though is that he should remain a Big Green guru, for example being lavishly praised by Al Gore who himself has sought to occupy a position as a climate change crusader. Big Green is not always so forgiving of commentators whose words prove to be overtaken by events, for example still gleefully (mis)quoting Lewis Strauss's 1954 reference to (fusion) energy being 'too cheap to meter' as evidence that all nuclear proponents are fools. The damage which this inconsistency (a nicer word than hypocrisy) risks inflicting on the credibility of the science supporting climate change is significant.

The treatment afforded to those like Ehrlich by Big Green and the authorities stands in clear contrast to that handed out to those who point out anomalies in the climate change story. For example, data from insurers Munich Re suggest that there is little evidence that the costs of global weather-related disasters increased as a proportion of GDP between the Rio base year of 1990 and at least 2012: in fact they fell by about a quarter – see figure 9.7.

Among the conclusions that can be drawn from insurance data are:

- disasters associated with hurricanes, tornadoes, floods or droughts did not increase either in the US or globally in the period in question;

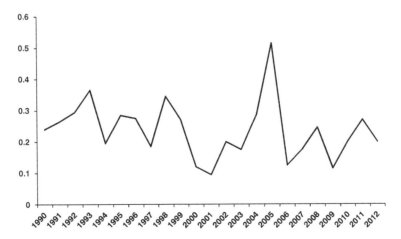

Figure 9.7 Losses associated with global weather disasters as proportion of GDP, 1990–2012[57]

- US hurricanes had not increased in frequency, intensity or damage caused since at least 1900, the same holding for tropical cyclones globally, at least since records began in 1970;

- floods did not increase in the US in frequency or intensity after at least 1950, flood losses as a percentage of US GDP dropping by about 75 per cent between 1940 and 2012;

- tornadoes had not increased in frequency, intensity or damage since 1950;

- droughts had for the most part become shorter and less frequent and covered a smaller portion of the US of the previous century.

As Warren Buffett, generally regarded the most successful investor of the twentieth century, said: 'The public has the impression that because there's been so much talk about climate that events of the last ten years from an insured standpoint and climate have been unusual. The answer is they haven't. I love apocalyptic predictions on climate change because they probably do affect rates.' [58]

Yet Roger Pielke, Jr, the author who brought these data to the attention of Congress, paid a heavy price despite having been a long-term advocate of stronger action to mitigate climate change and supporter of the IPCC. He became the subject of an investigation by Representative Raul Grijalva, ranking member of the House Committee on Natural Resources. Grijalva seemed to have no firmer grounds for this attack than that Pielke disagreed with the conclusions of one of President Obama's advisors.[59] This was despite Pielke's views on drought, for example, being consistent with those of the US Climate Change Science Programme, which said that: 'for the most part droughts become shorter, less frequent, and cover a smaller portion of the US over the last century, the main exception being the Southwest and parts of the interior of the West, where increased temperature has led to rising drought trends', and of the IPCC itself, which said: 'Globally there is medium confidence that since the 1950s some regions of the world have experienced a trend to more intense and longer droughts, in particular in southern Europe and West Africa, but in some regions droughts have become less frequent, less intense or shorter, for example in central North America and north-western Australia.' The communications director for the Democratic delegation on the Natural Resources Committee explained that such investigations were being launched into academics who had published widely, who had testified in Congress before and who seemed to have the most impact on policy in the scientific community – things that one might expect to be cause of admiration rather than suspicion in any field more driven by science than by politics. As one prominent academic institute put it: 'Grijalva's investigation is part fishing expedition, part innuendo campaign. It won't find nefarious funding of Pielke's research. But it will drag his good name and reputation through the mud, especially in an era where long debunked accusations take on a life of their own in the blogosphere. Long after Pielke's name is cleared accusations that his research is funded by the fossil fuels industry, and old links to the news stories that ran when Grijalva publicised the letters, will live on in cyberspace.'[60]

In 1980 one commentator made observations which would resonate, for example, both with those who were skilfully inventing the Linear No-Threshold model of radiation exposure in the 1950s and with those who seek a dispassionate assessment of science as a guide to action today.

> False bad news about population growth, natural resources and the environment is published widely in the face of contrary evidence. For example, the world supply of arable land has actually been increasing, the scarcity of natural resources including food and energy has been decreasing and basic measures of US environmental quality show positive trends. The aggregate data show no long-run negative effect of population growth upon standard of living. Models that embody forces omitted in the past, especially the influence of population size upon productivity increase, suggest a long-run positive effect of additional people.[61]

Whether driven by a desire to do good as one sees it or to attract the next round of funding, such an attitude towards the whole truth and nothing but the truth is only effective for as long as one's authority is unquestioned. That is no longer the case for Big Business or for Big Green.

9.8 SCIENCE TRUMPED BY POLITICS

Concern about climate change undoubtedly grew significantly throughout the 1990s and 2000s. There is evidence, however, that this might have peaked, as illustrated in figure 9.8.

In the US there is also some evidence that climate scepticism may have grown mildly through the 2010s, as shown in figure 9.9. Those saying that new reports of global warming were generally exaggerated stood at 42 per cent, though a majority (56 per cent) still believed the reportage was generally accurate or underestimated. There did seem to be a considerable decline (3 per cent in 2008 to 23 per cent in 2014) in those thinking that reporting was accurate, possibly a reflection of lower confidence that climate science was capable of delivering reliable predictions.

The politicisation of the issue is revealed by considering the attitudes of supporters of different parties. In 2014 half of Democrat voters believed that global warming news coverage underestimated the issue, with less than a fifth believing it was exaggerated. Among Republicans the figures were about one-sixth and two-thirds respectively (see figure 9.10).

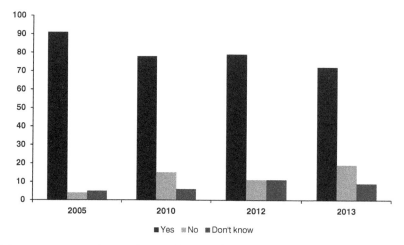

Figure 9.8 UK responses to the question: 'As far as you know, do you personally think that the world's climate is changing?'[62]

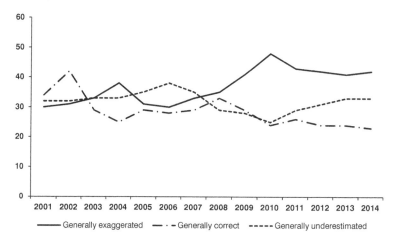

Figure 9.9 Percentage of US respondents stating that in news reportage global warming is generally exaggerated, underestimated or correct

Partisanship was also a major factor in believing where the scientific debate lay. Overall some 60 per cent of Americans believed that most scientists thought global warming was happening (the figure peaked at 65 per cent in 2008), with only 8 per cent believing that most scientists though it was not. However, only 42 per cent of Republicans said that most scientists believed global warming was happening, compared with 82 per cent of Democrats. 43 per cent of Republicans believed most scientists weren't sure about global warming, a position held by only 14 per cent of Democrats.[63]

It is not immediately obvious that one's political philosophy should be a relevant factor when it comes to evaluating scientific evidence and hypotheses. A more likely explanation is that respondents are choosing their sources of information and their interpretation of

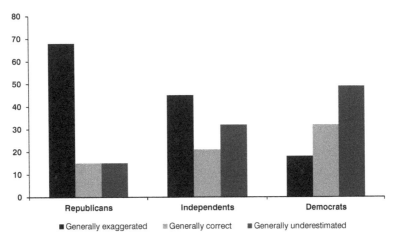

Figure 9.10 Percentage of US voters believing new coverage of global warming is exaggerated, correct or underestimated, 2014

those sources to fit in with pre-existing beliefs and biases. If Big Green's approach to the credibility of a source, an argument or a piece of evidence seems dominated by a Nietzschean consideration of whether it is useful in promoting the 'correct' underlying ideology rather than whether it fits reality, ultimately some people will notice this and be repulsed by it. This process is doubtless also at work when Big Business or the political establishment take a similar approach to the 'truth'.

What would seem to be clear is that the concept of true stable cross-party agreement on climate change, at least in the US, is a highly unlikely one. As Gallup put it when reporting these results: 'It is unlikely the two parties can agree on any road map when they can't even agree on whether or where they need to go.'

9.9 THE PRISONERS' DILEMMA RECAST

The reasons for the modest decline in concern about climate change are undoubtedly complex. As anxiety over the state of the economy grew, for example, following the world recession of the late 2000s it may be that people no longer needed climate change to justify their inherent level of unhappiness, as discussed in chapter 11. The arrogance of the keepers of the climate flame may also be a factor, even if they may happen to be broadly right about the science. Examination of the history of the climate change debate leaves one with a strong impression that some powerful groupings, notably the research establishment and Big Green, are so sure that disastrous climate change is a reality (or so sure that they should receive research grants) that they can justify all kinds of dubious activity on the grounds that 'the ends justify the means'. Green 'crusaders' – an interesting term of approval in the field as the mediaeval crusades were not characterised by a cool evidence-based approach and respect for opposing views that one might hope would underpin complex scientific issues – seem to seek to recreate the atmosphere of the Inquisition. Far from holding science and industry to account on behalf of the environment, as they seek to portray themselves, Big Green too often appears to be the staunchest defender of and apologist for bad practice – as long as the conclusion reached is the one it likes.

For there to be any justification at all for this there really would need to be no question at all about the underlying science. Yet this is not in the gift of science – all scientific theories are by their very nature provisional and subject to change. One way of telling a scientist from a politician is to compare the amount of time they put into proving themselves wrong to the amount they invest in proving themselves right. The systematic attempts to eliminate and demonise the sceptic as 'heretic' did not work with Galileo and it is unlikely to work now, while sincere belief in one's cause is not of itself a guarantor of a moral or beneficial stance (one didn't get the impression that Hitler or Stalin were fooling about). The outcome of trying to suppress debate is often to boost the unfashionable side of the case (anything unexpected – 'man bites dog' – becomes newsworthy). Modern communication technology makes it extremely difficult to keep suppression of alternative views secret, as Climategate showed.

Climate change is a different kind of pollution issue to many, perhaps to any, others. In most cases of environmental damage, it is relatively easy to demonstrate a link – both geographical and temporal – between the source of the pollution and the damage which it causes. It is also usually quite feasible to calculate within acceptable limits the costs of mitigating and the costs of not mitigating the problem. Indeed, most pollutants – particulates from burning coal, acid emissions, lead in petrol, radioactive releases, detergents in rivers, CFCs and their effect on the ozone layer – have a kind of 'industrial life cycle'. A new activity results in a build-up of the pollutant. Over time detrimental consequences of the pollutant are observed and in due course evidence can be found to reveal a cause-and-effect relationship. After a bit of argument estimates are made as to how releases of the pollutant may be mitigated and at what cost, often against the background of the 'polluter pays' principle. Either because of legislative action, public concerns or the development of new, more efficient technologies (which tend to have lower emissions) which either reduce the level of release or sidestep it entirely by using different approaches, the rate at which the pollutant is created begins to fall.

In the case of climate change the link between source and consequence is much more abstruse. It does not really matter where in the world the greenhouse gas is released, the ultimate outcome will be same. Furthermore, much of the damage will occur some decades into the future. The costs of mitigation are subject to enormous uncertainties, the costs of not mitigating even more so. The costs of unchecked climate change are likely to be disproportionately distributed between cold and hot countries or regions, countries close to sea level and those at higher altitudes, and so on. They may also not be simply proportionate to the amount of greenhouse gases emitted if there are indeed 'tipping points' at which a relatively small increase in atmospheric concentration could result in a very large effect. As a result it is much more challenging to come up with an international regulatory and institutional framework to reduce emissions by the enormous amounts required.

Climate change can be viewed as the biggest prisoners' dilemma of our times. In the prisoners' dilemma, two partners in crime face accusations for which there is strong evidence of a minor crime (attracting a one-year prison sentence) but insufficient evidence to secure conviction on a more serious crime. If neither conspirator 'shops' the other they will each get one year in prison on the lesser charge; if they both give up their partner in crime they each get five years. The police, keeping the prisoners in separate rooms of course, give each of them an offer – if you shop your accomplice we'll drop the other charge but he'll get ten years; if you don't shop him and he shops you then obviously you get the ten years and he goes free. This creates an intriguing position. The best outcome for both of them is that neither betrays their friend – they each get one year instead of five years. However (assuming of course that there would be no other repercussions for so doing), it is always in the interests of each of them to inform on the other. If suspect B does not turn Queen's evidence then suspect A reduces his term from one year to none by doing so; if suspect B does turn on his sidekick then suspect A reduces his term from ten down to five years by doing the same.

It seems clear, despite the somewhat uncritical way that climate science has been treated by Big Green and the history of exaggerated doomsaying, that it would be in the whole world's interest to treat climate change seriously. On many projections the global cost of not doing so would significantly outweigh the global cost of doing so, possibly by a great amount, though the margin of error in both of these quantities is enormous so this cannot be demonstrated beyond doubt at present. However, except in the vanishingly unlikely case that cutting greenhouse gases would be economic in the near future in the narrow sense of reducing costs even if carbon effects are ignored, it is unlikely ever to be in any country's interests to take expensive measures to reduce its own greenhouse gas emissions. If the rest of the world is reducing emissions any single country (with maybe one or two exceptions, e.g. China and the US) can 'freeload' and reap the benefit of lower costs to its industries by following cheaper, dirtier technology. By contrast, if the rest of the world is not reducing its emissions any country which did so would simply commit industrial suicide while its carbon-producing industries were 'exported', quite possibly to areas of the world with laxer environmental standards thereby actually resulting in an increase in global emissions of carbon and other pollutants.

There may be some grounds for optimism. CoP 21, held in Paris at the end of 2015, did end with an agreement among all 196 parties attending, the essence of which was to limit global warming to 2°C and to 'pursue efforts to' limit the increase to 1.5°C. Zero net anthropogenic greenhouse gas emissions is to be reached during the second half of the 21st century. This was certainly an impressive outcome. However, like Kyoto the agreement would onlybecome legally binding if joined by at least 55 countries which together represent at least 55 percent of global greenhouse emissions, requiring ratification from each of those countries' legislatures. To put the goal into perspective, at a time when large numbers of new coal-fuelled power stations were being opened in India, China, Japan, Germany and many other countries – in the first nine months of 2015 China alone approved plans to build more than 150[64] – to reach the 1.5°C goal so important to low-lying countries in particular, no new (unmitigated) fossil-fuelled facilities at all could be opened and every existing one had to replaced with a nuclear or renewable alternative when it reached the end of its economic life. If the world continued to build new fossil capacity – it was inconceivable that it would not – then a large amount would have to be retired well before the end of its life. The agreement that 'liability and compensation as a result of loss and damage will be explicitly excluded' and the failure to mention international aviation or shipping emissions looked ominous, as did the provision that any country could withdraw from the process without consequences (as Canada did from Kyoto in 2012).

So, while describing Paris as 'a huge disappointment', in the words of the CEO of Country Risk Solutions[65], may be an exaggeration, it remains the case that it may not prove possible to do anything serious about climate change from a logical point of view. However, if any progress is to be made it will require all sides to reject the temptation to spin the evidence for their own purposes while pursuing other goals which are incompatible with efficiently addressing the problem – closing nuclear stations early or exporting coal for use in foreign power stations being obvious examples. In the mid-2010s there was very little evidence that such a revolution was imminent.

ENDNOTES

1 Wolf M, 'Why the world faces climate chaos', *Financial Times*, 14 May 2013.

2 http://www.iea.org/newsroomandevents/news/2015/march/global-energy-related-emissions-of-carbon-dioxide-stalled-in-2014.html, IEA (2015), 'Global energy-related emissions of carbon dioxide stalled in 2014'.

3 http://www.carbontracker.org/wp-content/uploads/2014/09/Unburnable-Carbon-Full-rev2-1.pdf, Leaton J (2012), 'Unburnable carbon – are the world's financial markets carrying a carbon bubble?', Carbon Tracker Initiative.

4 As an example see http://www.timescolonist.com/opinion/letters/we-can-t-trust-radiation-consensus-1.813487, Kuzmyn N, 'We can't trust radiation consensus', *Times Colonist*, 5 February 2014.

5 http://theconversation.com/is-misinformation-about-the-climate-criminally-negligent-23111, Torcello L, 'Is misinformation about the climate criminally negligent?', *The Conversation*, 13 March 2014.

6 http://www.foxnews.com/science/2013/10/18/la-times-bans-letters-from-climate-skeptics/, 'LA Times bans letters from climate skeptics', *Fox News*, 18 October 2013.

7 http://www.bbc.co.uk/news/uk-politics-26187711, Hawkins R, 'Greens call for clear-out of climate change deniers', *BBC News website*, 14 February 2014.

8 http://wattsupwiththat.com/2012/01/08/the-moon-is-a-cold-mistress/, Eschenbach W, 'The Moon is a cold mistress', *Watts Up With That*, 8 January 2012.

9 http://www.wmconnolley.org.uk/sci/fourier_1827/fourier_1827.html, Connolley W (2001), 'Translation by W M Connolley of: Fourier (1827): 'Memoire sur les temperatures du globe terrestre et des espaces planetaires'.

10 http://science.widener.edu/svb/ftir/ir_co2.html, Bramer S (1996), 'Carbon dioxide', University of Widener, Chester, PA.

11 Myhre G *et al.* (2013), 'Climate change 2013: the physical science basis. contribution of Working Group I to the Fifth Assessment Report of the Intergovernmental Panel on Climate Change', IPCC: UN.

12 Etheridge D *et al.* (1996), 'Natural and anthropogenic changes in atmospheric CO2 over the last 1,000 years from air in Antarctic ice and firn', *J. Geophys. Res.*, **101** (D2) 4115–28.

13 http://www.nap.edu/openbook.php?record_id=12181&page=1, Ad hoc study group on carbon dioxide

and climate (1979), 'Carbon dioxide and climate – a scientific assessment', National Academy of Sciences: Washington DC.

14 ftp://aftp.cmdl.noaa.gov/products/trends/co2/co2_weekly_mlo.txt, NOAA ESRL Data.

15 http://cdiac.ornl.gov/ftp/ndp030/global.1751_2010.ems, Boden T et al. (2013), 'Global CO2 Emissions from Fossil-Fuel Burning,Cement Manufacture, and Gas Flaring: 1751-2010'.

16 http://www.principia-scientific.org/the-death-knell-for-the-co2-theory1.html, Kaiser K, 'The death knell for the CO2 theory', *Principia Scientific International*, 25 February 2013.

17 http://www.telegraph.co.uk/earth/environment/climatechange/8786565/War-of-words-over-global-warming-as-Nobel-laureate-resigns-in-protest.html, Sherwell P, 'War of words over global warming as Nobel laureate resigns in protest', *Daily Telegraph*, 25 September 2011.

18 http://www.cru.uea.ac.uk/documents/421974/1295957/Info+sheet+%231.pdf/c612fc7e-babb-463c-b5e3-124ac76680c5, Jones P (2014), 'Global temperature record', University of East Anglia.

19 http://denisdutton.com/newsweek_coolingworld.pdf, Gwynne P, 'The cooling world', *Newsweek*, 28 April 1975.

20 http://www.mrc.org/special-reports/fire-and-ice-0, Gainor D (2010), 'Fire and ice', Media Research Centre, Reston VA.

21 http://pubs.giss.nasa.gov/abs/ha02700w.html, Hansen J et al. (1988), 'Global climate changes as forecast by the Goddard Institute for Space Studies three dimensional model', *Journal of Geophysical Research: Atmosphere (1984-2012)*, **93**.

22 http://www.nytimes.com/1988/06/24/us/global-warming-has-begun-expert-tells-senate.html, Shabecoff P, 'Global warming has begun, expert tells Senate', *New York Times*, 24 June 1988.

23 http://unfccc.int/kyoto_protocol/status_of_ratification/items/2613.php, UNFCCC (2014), Status of ratification of the Kyoto Protocol.

24 http://unfccc.int/documentation/documents/advanced_search/items/6911.php?priref=600007639, UNFCCC (2013), 'National greenhouse gas inventory data for the period 1990-2011 – note by the secretariat'.

25 https://cdm.unfccc.int/CDMNews/issues/issues/I_P0QZOY6FWYYKFKOSAZ5GYH2250DRQK/viewnewsitem.html, UNFCCC (2012), 'Kyoto Protocol's CDM passes one billionth certified emission reduction milestone'.

26 https://www.ipcc.ch/publications_and_data/ar4/syr/en/mains1.html, IPCC (2007), 'Climate change 2007: synthesis report'.

27 https://www.ipcc.ch/report/ar5/index.shtml, IPCC (2013), 'Fifth Assessment Report'.

28 https://royalsociety.org/~/media/Royal_Society_Content/policy/publications/2009/7871.pdf, G8+5 Academies (2009), 'Joint statement: climate change and the transformation of energy technologies for a low carbon future'.

29 https://1-science.cdn.aspedia.net/sites/default/files/user-content/resources/file/climatechange2010_1.pdf, Allison I, Raupach M, *et al.* (2010), 'The science of climate change: questions and answers', Australian Academy of Science: Canberra.

30 http://www.nybooks.com/articles/archives/2007/mar/15/warning-on-warming/, McKibben B, 'Warning on warming', *New York Review of Books*, 15 March 2007.

31 Stern N (2007), *The Economics of Climate Change*, Cambridge University Press.

32 http://www.telegraph.co.uk/earth/environment/climatechange/10331898/Lord-Stern-IPCC-report-will-underestimate-climate-change.html, Collins N, 'Lord Stern: IPCC report will underestimate climate change', *Daily Telegraph*, 24 September 2013.

33 http://www.thetimes.co.uk/tto/environment/article2144989.ece, Webster B and Pagnamenta R, 'UN must investigate warming bias, says former climate chief', *The Times*, 15 February 2010.

34 http://www.nature.com/nature/journal/v462/n7273/full/462545a.html, 'Climatologists under pressure', *Nature*, **462**, 545, 3 December 2009.

35 http://www.guardian.co.uk/environment/georgemonbiot/2009/nov/25/monbiot-climate-leak-crisis-response, Monbiot G, Pretending the climate email leak isn't a crisis won't make it go away, *The Guardian*, 25 November 2009.

36 http://online.wsj.com/news/articles/SB125902685372961609?mg=reno64-wsj&url=http%3A%2F%2Fonline.wsj.com%2Farticle%2FSB125902685372961609.html, Johnson K and Naik G, 'Lawmakers probe climate claims', *The Wall Street Journal*, 24 November 2009.

37 http://reviewipcc.interacademycouncil.net/report.html, IAC (2010), Climate change assessments, review of the processes and procedures of the IPCC'.

38 http://reviewipcc.interacademycouncil.net/Comments.pdf, IAC (2010), 'Responses to the IAC questionnaire'.

39 http://www.thegwpf.org/images/stories/gwpf-reports/mckitrick-ipcc_reforms.pdf, McKitrick R (2011), 'What is wrong with the IPCC? Proposals for a radical reform', Global Warming Policy Foundation.

40 http://www.wwfblogs.org/climate/sites/default/files/WWFBinaryitem12911.pdf, WWF-Allianz (2009), *G8 Climate Scorecards*.

41 http://www.telegraph.co.uk/earth/environment/climatechange/10325562/Row-over-IPCC-report-as-nations-try-to-hide-lack-of-climate-change.html, Mendrick R, 'Row over IPCC report as nations try to hide lack of climate change', *Telegraph*, 21 September 2014.

42 http://www.independent.org/newsroom/article.asp?id=3263, Singer F, 'Climate deniers are giving us skeptics a bad name', *The Independent Institute*, 29 February 2012.

43 http://heartland.org/sites/default/files/critique_of_ipcc_spm.pdf , Idso C, Carter R, Singer F and Soon W (2013), 'Scientific critique of IPCC's 2013 Summary for Policymakers', NIPCC policy brief.

44 http://www.independent.co.uk/news/media/tv-radio/exclusive-today-programme-criticised-for-giving-platform-to-climate-sceptic-lord-lawson-9582970.html, 'Today Programme criticised for giving platform to climate sceptic Lord Lawson', *The Independent*, 3 July 2014.

45 http://iopscience.iop.org/article/10.1088/1748-9326/8/2/024024/pdf, Cook J. *et al.* (2013), 'Quantifying the consensus on anthropogenic global warming in the scientific literature', *Environmental Research Letters*. IOP Publishing.

46 http://www.populartechnology.net/2013/06/cooks-97-consensus-study-game-plan.html, 'Cook's 97% consensus study game plan revealed', *populartechnology.net*, 4 June 2013.

47 http://rankexploits.com/musings/2013/the-saga-continues/, 'The saga continues', *The Blackboard*

48 http://www.populartechnology.net/2013/05/97-study-falsely-classifies-scientists.html, '97% study falsely classifies scientists' papers, according to the scientists that published them', *populartechnology.net*, 21 May 2013.

49 http://onlinelibrary.wiley.com/doi/10.1002/grl.50382/abstract, Balmaseda M, Trenberth K and Källén E (2013), 'Distinctive climate signals in reanalysis of global ocean heat content', *Geophysical Research Letters*.

50 http://www.nature.com/nclimate/journal/v4/n10/full/nclimate2341.html, Trenberth K, Fasullo J, Branstator G and Phillips A (2014), 'Seasonal aspects of the recent pause in surface warming', *Nature Climate Change*, **4**, 911–16.

51 http://www.dailymail.co.uk/sciencetech/article-2217286/Global-warming-stopped-16-years-ago-reveals-Met-Office-report-quietly-released--chart-prove-it.html, Rose D, 'Global warming stopped 16 years ago, reveals Met Office report quietly released... and here is the chart to prove it', *Mail on Sunday*, 13 October 2012.

52 https://www.wmo.int/media/?q=content/warming-trend-continues-2014, World Meteorological Organisation (2015), 'Warming trend continues in 2014', United Nations.

53 https://www.ncdc.noaa.gov/sotc/global/201507, National Oceanic and Atmospheric Administration, 'Global analysis – July 2015'.

54 http://mahb.stanford.edu/consensus-statement-from-global-scientists/, MAHB (2013), Consensus statement from global scientists', University of Stanford.

55 http://archive.wired.com/wired/archive/5.02/ffsimon_pr.html, Regis E (2004), 'The doomslayer', *Wired*.

56 http://townhall.com/columnists/walterewilliams/2008/05/07/environmentalists_wild_predictions/page/full Williams W, 'Environmentalists' wild predictions', *Townhall.com*, 7 May 2008.

57 http://sciencepolicy.colorado.edu/admin/publication_files/2013.20.pdf, 'Statement of Dr Roger Pielke, Jr, to the Committee on Environment and Public Works of the US Senate', University of Colorado, 18 July 2013.

58 http://www.cnbc.com/id/101460458, Belvedere M, 'No climate change impact on insurance biz: Buffett', *CNBC*, 3 March 2014.

59 https://theclimatefix.wordpress.com/2015/02/25/i-am-under-investigation/, Pielke R, Jr, 'I am "Under Investigation" ', *The Climate Fix*, 25 February 2015.

60 http://thebreakthrough.org/index.php/voices/michael-shellenberger-and-ted-nordhaus/climate-incivility, Shellenberger M and Nordhaus T (2015), 'Climate McCarthyism is wrong whether Democratic or Republican', Breakthrough Institute, 26 February 2015.

61 http://www.eoearth.org/view/article/155723/, Simon J (1980), 'Resources, population, environment: an oversupply of false bad news', *Science*, **208** 1431–7.

62 http://www.ukerc.ac.uk/support/tiki-download_file.php?fileId=3371, Poortinga W, Pidgeon N, Capstick C and Aoyagi M (2013), Public attitudes to nuclear power and climate change in Britain two years after the Fukushima accident, UKERC.

63 http://www.gallup.com/poll/167960/americans-likely-say-global-warming-exaggerated.aspx?version=print, Dugan A, 'Americans most likely to say global warming is exaggerated', Gallup, 17 March 2014.

64 http://www.greenpeace.org/eastasia/publications/reports/climate-energy/climate-energy-2015/doubling-down/, Lyllyvirto L, Shen X and Lamini H (2015), 'Is China doubling down on its coal power bubble?', Greenpeace.

65 http://www.huffingtonpost.com/daniel-wagner/the-paris-climate-change-_b_8798614.html, Wagner D (2015), 'The Paris climate agreement is a huge disappointment', *Huffington Post*, December 14 2015.

10 | Politics and Energy: The Tripartite Relationship Among the Technical, Political and Public Realms in the Age of Inertia

10.1 Why politics matters

The often-cited energy 'trilemma', with economic, reliability and environmental requirements competing for attention, is actually a 'tetralemma', as political and social factors are also crucial (and indeed ultimately may well be the most powerful). The radically different responses to Fukushima in neighbouring countries shows politics to be a quasi-independent variable, though clearly to an extent affected by the other goals.

If something is not socially and politically sustainable then it is not sustainable, however attractive it might seem from an economic or environmental standpoint. The stance of the political establishment, locally, nationally and internationally, towards a major technical/scientific issue such as production of energy is enormously important. Nevertheless, the relationship between the public and political spheres in many countries has become rather fractious in recent years, while at least in most democracies political leaders are heavily and maybe even increasingly influenced by their perceptions of short-term public attitudes rather by than long-term scientific predictions.

The social/political dimension is more difficult to characterise than the horns of the traditional trilemma. It involves such disparate matters as public acceptability, the desire to help people who cannot afford their energy bills, sensitivity to local communities which have a relationship with a particular energy source and, of course, calculations of the electoral consequences of various policies in countries where this matters. It seems to have received rather less attention from the research community, though there is a growing literature devoted to decision-making and power structures in modern liberal economies.[1] However, its importance should not be underestimated. Radical changes in policy or regulatory requirements, or a perception that such changes may materialise at some point, can be extremely beneficial for or extremely damaging to the prospects of investment in any capital-intensive industry such as energy. In the worst case such fears may in effect act to prevent investment entirely. A lukewarm, or worse, political attitude towards the construction or operation of nuclear facilities or large hydropower dams, say, could increase the costs of electricity generation in a number of ways. There may be delays during construction or in achieving an initial operating licence, or interruptions in operation. Extra physical or operational security measures might be demanded, perhaps in response to a potential terrorist situation even if there is no direct evidence of a threat. Implementing such measures may be especially costly if they involve 'back-fitting' an existing design. The costs of site selection, evaluation and the licensing process itself can increase while those of transporting materials can escalate because of increased requirements for security against protest or the need to find new routes. The economic risk associated with uncertainty results in demands for higher rates of return on investment, an especially serious issue for technologies such as renewables or nuclear power which are highly capital-intensive even by energy standards.

There are many examples of energy policy being skewed to support a particularly pow-erful political lobby at the expense of supply security, low prices or the environment – the German Energiewende is only the latest example of such distortions (or perhaps contor-tions). And however much a government which has liberalised its power supply systems might argue publicly or dream privately that energy decisions are a matter for businesses, se-vere power outages or price rises will ultimately surely be blamed on the governing regime.

Although this book is examining some of the reasons why it is so difficult to take any kind of decision in the field of energy, energy is not unique in this respect, as the UK government's contortions over airport capacity in the south east of England over many years bears witness. No matter how urgent the need might become, doing anything about it seems to take forever, at least when one compares the rapidity of technological development during the early years of the Industrial Revolution, the late nineteenth century or the great wars of the twentieth century.

This chapter suggests that three phases can be identified in the tripartite relationship (or the three separate relationships) among the scientific/technical, political and public spheres over the course of the last century or so and to examine the implications for future decision-making.

In phase 1, the relationships are relatively deferential if not always harmonious. In ef-fect, politics (as the arena for decision-making), science/technology (as a source not only of technical input but also of legitimacy for those decisions) and the public (prepared to defer to both and to surrender some of their individual 'rights' as long as overall 'progress' is maintained) are able to work in considerable harmony. Decisions can be taken, often quickly and without a great deal of scrutiny, with the result that things get done but some-times badly. The predominating societal ethic is a utilitarian one: often there is a general feeling that there are real problems in society which need a solution, possibly a radical one. Science can even take on something of a religious aura, as discussed in chapter 11.

In phase 2, relationships become fraught as the three realms began to mistrust each other. There is a breakdown in the belief in 'progress' and the wide consensus that science and technology in society are 'good things'. Other social changes threaten the hegemony of the political/scientific establishment. People become more individualistic, traditional religion and its scientific successor is replaced by cults or alternative religions such as environmentalism, formal dress is replaced by informality especially among younger generations, popular music becomes more experimental, drug-taking becomes more common and young people are encouraged to 'tune in, turn on and drop out' by academic gurus such as Timothy Leary and Allen Ginsberg.[2] It is of course wrong to claim that all members of the population, or even more than a minority, ever directly take up such activities but a general decline in the awe with which royalty, politicians, clergymen and 'experts' are held becomes clear. The 'satire boom' of the early 1960s onwards, as epitomised in the UK by the TV show *That Was The Week That Was*, saw major political and business figures of the day lampooned in a way that would not have been thinkable a few years earlier. The rights of individuals became more important than had been the case when society was under more intense pressure, either from without (in wartime) or from within (in times of economic austerity or social unrest).

Phase 2 attitudes often emerge when society is going through a period of relative affluence, and facing few major threats. The relatively clear distinction between decisions that are the responsibility of government and decisions which are the domain of other agencies such as industry becomes blurred as many functions which had previously been regarded as social obligations are transferred into private hands, hands which are often mistrusted by a significant proportion of the population. The outcome is that decisions become much more thoroughly scrutinised – legislation is passed to require detailed consultation and impose heavier regulation, for example – leading to a state of affairs where decisions that do get taken may be better thought-out, but very few decisions actually get that far, no matter how urgent the need. The authority which had resided with scientists and politicians is gradually bestowed upon Big Green and others whose raison d'etre to a considerable extent is to stop things happening, driven by an ideology which is suspicious of material 'progress'.

In phase 3 – and this is highly speculative at this point in time – people gradually begin to realise that their new high priests, the enviro-fundamentalists, are no more worthy of unquestioning faith than the church, Parliament, the boardroom or the laboratory had been before them. Gradually it becomes clear that big decisions are needed from time to time. Inevitably these decisions will be detrimental to some sections of society but this alone can no longer be regarded as sufficient excuse for failing to act. Of course, it may be that it is simple wishful thinking to believe that such a phase is emerging or could ever do so, every bit as fanciful as an all-renewable world or major reductions in energy demand.

10.2 SCIENCE AND SOCIETY IN STEP – PHASE 1

In the post-Second World War years the scale of the challenge facing a world recovering from those terrible events led to broad recognition that bold decisions were needed in both technical and social spheres. Such decisions could be taken rapidly by the elected government of the day: the late 1940s onwards saw one of the biggest social/economic revolutions in British history through the nationalisation of major tranches of the UK economy, including in 1948 electricity supply. Many centralised power generation plants were built to serve growing demand, including the nine commercial Magnox nuclear power stations which began operating in the ten years between 1962 and 1971. There was an attitude among those in authority that members of the public, or pressure groups, who opposed a particular bit of scientific/technical 'progress' were either out of touch with the norms of society or would soon come round. The sense of shared purpose among people who had recently united to face an external enemy engendered a broadly utilitarian societal ethic, seeking 'the greatest good for the greatest number'.

This meshed well with many of the tenets which underlie statistical science, in which wider principles are generally of more interest than individual cases. If we follow the best available scientific advice in pursuing a dozen technologically innovative routes then it is likely that, in say one or two cases, events may subsequently show that the best available scientific interpretation underestimated the risks and some individuals might suffer. Nonetheless, a thoroughgoing utilitarian might argue that society would well be significantly better off by following all twelve opportunities rather than by abandoning all of them at an early stage and bending to the prejudices of those motivated by more ideological, mystical or obscure sources of belief. (It is hard to find a form of words that does not make utilitarianism seem rather cold and insensitive but utilitarians would argue that they are motivated purely by a desire to increase the total sum of human well-being – that is indeed the philosophical basis of utilitarianism – and that it is actually rather heartless to pursue the interests of those who shout loudest or are most photogenic at the expense of the unheard majority.) Jeremy Bentham, the founder of modern utilitarianism, proposed a 'felicific calculus' to determine mathematically the morality of various courses of action.[3] The highly visible success of science and technology in delivering longer, healthier, more leisurely lives for many people added to the enthusiasm. That it did not do so for everyone could be accommodated within the tenets of utilitarianism.

When society is relatively stressed and pursuing a utilitarian ethic, politicians often seem happy to delegate decision-making to the technical community. The Eisenhower regime in the US in the early days of civil nuclear technology, for example, took the view that nuclear science was by its nature too difficult for laypeople, including politicians, to understand so the responsibility not only for carrying out policy but also largely for forming it should lie with the Nuclear Regulatory Commission (NRC). It was perhaps inevitable then that a technocratic mode of decision-making became dominant, to the detriment of dialogue with and control by the normal democratic structures. The secrecy associated with military uses of nuclear materials doubtless exacerbated this.

Enthusiasm for science did not entirely evaporate as austerity came to an end. For example, at the Labour Party conference in 1963, Harold Wilson, leader of the opposition, famously declared that a New Britain would be 'forged in the white heat of [the scientific and technological] revolution'. Coming to power the next year, Wilson established a new Ministry of Technology which, in the words of its first minister, Tony Benn, would provide Britain with the role it was searching for since the demise of the Empire.[4]

However, the lack of serious challenge to the technocrats in charge of policy, either from alternative technical viewpoints or from different social values, had a darker side. In at least some cases, such as the development of the two Windscale nuclear piles, the first of which caught fire in 1957, it can be argued that government pressure forced, or perhaps permitted, scientists and engineers to take dangerous risks in the pursuance of urgent political demands. Such decisions were inevitably subject to certain systematic biases in favour of the technical mindset or financial vested interests, or to error as potential critics were drowned out by the 'experts'. The vast sums spent on nuclear research in the UK in the 1960s, resulting in the creation of a large number of apparently dead-end designs – at the end of which, in the view of many commentators, government chose the wrong one – do not now appear to have represented the best possible public policy. More recently, a significant contributor to the Chernobyl accident in 1986, in a society where challenge to the government and its institutions was even more limited, seems to have been overt and assumed demands on the operators from local politicians. The absence of a culture of challenge, either from society or from those working within the industry, did perhaps deliver some benefits from 'pushing the envelope' in terms of increased output. It also resulted in the plant being operated well outside its design parameters with fatal consequences.[5]

10.3 PHASE 2 – REACTION AND PARALYSIS

As science and technology lost something of their gloss, overt opponents of technology (or at least some technologies), generally treated as somewhere between a joke and a mild irritant in phase 1, became more vocal and influential. Environmentalism has long historical roots but it was perhaps in the 1970s that the modern anti-industrial Big Green movement began to develop seriously. Support for environmental goals led to the first Earth Day protests in 1970, establishing an annual event now organised in some 192 countries. Greenpeace, founded in 1971, advocated direct action, for example, to protect endangered species while other organisations such as Friends of the Earth placed more emphasis on public campaigning and research. However, while the different bodies varied in their styles, all developed some mix of direct protest – either physically interrupting activities which they opposed or organising campaigning marches, occupations of head offices and other events – and public proselytising. This included commissioning 'scientific' studies from friendly academics who could be relied upon to support their political stance, aping the behaviour of its opponents which it so heavily criticised.

One might think the decline of blind deference is a very good thing given the record of 'rushed' decisions in the post-war period. However, in phase 2 this ultimately becomes a rejection of informed advice – 'never trust an expert', as environmental 'expert' Amory Lovins once put it. Decision-making suffers, at least in the sense that 'difficult' decisions, defined as any decision which might upset someone affluent and articulate enough to complain, have to be avoided and the easiest route followed. In energy this may basically mean sticking with fossil fuels alongside perhaps a token smattering of renewables to keep Big Green relatively quiet.

To resolve an issue such as nuclear waste management in a long-term sense requires a firm decision, rather than the 'default' position of continued on-site storage (which might be a bit unpopular near the sites involved but where people are at least broadly used to the material being there and nobody is asked to take the whole lot). But if society is in a mode in which strong decision-making is generally unpopular then the temptation for politicians, however much they might prefer to show resolve, may well be to resort to perpetual consultation, research and review, in the hope that the problem will retreat from view until they are safely out of office. The wider the consultation the more the politicians can demonstrate that they are taking the issue seriously – and the less likely is resolution. As examined later, 'expert' groups dominated by individuals who won their place for reasons other than a sound and balanced understanding of the issues are set up to produce 'consensus' reports which are too vague to be of use in decision-making or dissolve in dispute between the technical and the social/political members, coming as they often do from different ethical standpoints.

Pursuing the nuclear example, although nuclear energy was not one of the earliest bugbears of the emerging Big Green organisations it grew in salience through the 1970s. The Sierra Club, founded in 1892, reversed its pro-nuclear energy stance in 1974[6] while the accidents at TMI and Chernobyl gave further impetus to these campaigns. The effect on nuclear development in the UK was stark (see figure 10.1). In the 27 years between 1953 and 1980 the UK ordered 18 nuclear power stations, including Calder Hall and Chapelcross which produced electricity as a 'by-product' of plutonium for the nuclear weapons programme, consisting of 40 reactors with a combined capacity in the order of 13.5 GW. One could also include two substantial prototypes, at Winfrith Heath and Dounreay of 100 MW and 250 MW output respectively, alongside various smaller machines. In the 35 years between 1980 and 2015 the UK ordered one nuclear power station with one reactor of capacity 1.2 GW.

Of course this striking contrast cannot be put down solely to the evaporation of public and political support for nuclear power during phase 2. The technical and economic performance of some of the AGR plants was disappointing and the discovery of large amounts of oil and gas in UK waters, coupled with the emergence of a new efficient technology in the Combined Cycle Gas Turbine (CCGT), were of great significance. But as the dismissive 1994 nuclear White Paper produced by the traditionally pro-nuclear

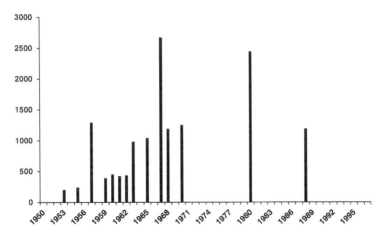

Figure 10.1 Date of commencement of construction of nuclear capacity in UK (MW)

Conservative government of the day showed, nuclear power in the UK had few friends by the mid-1990s.

10.4 POLITICS AND PEOPLE IN PHASE 2

During prolonged phases of prosperity, peace and comfort there may appear little need for radical political action to protect the fabric of our way of life, as there might be say at times of war, disease, famine or prolonged industrial unrest. Even for the UK generation approaching retirement age in the mid-2010s, memories of the last major round of power cuts in the UK, in the early to mid-1970s, are likely to be distant. For younger people even that memory is absent. It is perhaps natural that people should in effect take for granted the benefits of industry and science, such as secure power supplies and all that brings, to the extent that they may find it literally inconceivable to imagine life without them. And if there is a subconscious assumption that these benefits will always be available then it follows that people will focus more on the perceived downsides of the technologies in question rather than on making a balanced cost-benefit analysis. As a result there is less space for strong political decision-making in response to future threats which may be foreseen but nonetheless remain emotionally unimaginable.

As a reaction to the growing public scepticism about the role of science in decision-making and perhaps to the need to be seen to be 'doing something' without actually having to do anything, politicians towards the end of the twentieth century attempted to rebuild a relationship of trust with the public by downgrading technical expertise or even writing it out of the story. The panels which sprang up, charged with finding 'solutions' to matters such as radioactive waste management, the health effects of mobile phone masts, BSE (bovine spongiform encephalopathy or 'mad cow disease'), foot-and-mouth disease and so on, were increasingly populated by individuals with no technical knowledge of the topic in question and indeed often an antipathy towards such expertise, coupled with 'alternative experts' who would promote views not supported by mainstream responsible science. Woe betide any scientist who insisted on challenging a government policy merely because it was not supported by the science of the day.

During this phase the 'precautionary principle' added to the difficulties in making sound decisions which would deliver the greatest benefit to people at the time and in the future. It can be stated as follows: 'if an action or policy has a suspected risk of causing harm to the public or to the environment, in the absence of scientific consensus that the action or policy is not harmful, the burden of proof that it is *not* harmful falls on those taking an action.' This was often interpreted by opponents to a particular technology as meaning that development should be blocked until it could be proved that there is no problem, something which as discussed later is logically impossible. (Some commentators from within the scientific/industrial community have argued that, despite its theoretical shortcomings, the precautionary principle can be a valuable practical tool in designing regimes to manage emerging scientific techniques in ways that minimise potential downsides.[7] However, it needs careful handling if it is not to become 'we can't start this activity until we have had at least 20 years of experiencing it in action'.) In these circumstances science cannot fulfil its full potential to contribute to effective decision-making.

Politicians perhaps deserve some sympathy – when society is feeling comfortable it is often difficult to raise awareness that things cannot go on as they are at present, e.g. the world getting 86 per cent of traded energy from coal, oil and gas with the associated environmental and geopolitical implications. If a comfortable society rejects 'strong' political leadership aimed at problems that do not seem to be urgent, even those leaders who can see the growing cloud may not be listened to or re-elected if they speak out too loudly. Yet, as noted earlier, the timescales involved in investment in many industries, not least energy, can stretch to decades or even more. And when the serious problems re-emerge there will be plenty of people ready to criticise their political leaders for not having acted soon enough.

10.5 CHANGING SOCIETAL ETHICS

As noted already, a key difference between phase 1 and phase 2 societies, in this model, is in the underlying societal ethic. Utilitarianism is replaced by a more rights-based ethical atmosphere, variously referred to as Kantian, contractarian or deontological, though its roots stretch back to the concept of the 'social contract' to be found in the work of philosophers such as Hobbes, Locke and Rousseau. Some differences between these two moral theories, as they refer to technology in society, can be listed as per table 10.1.[8]

To take the matter of toxic waste management as an example, the switch from broad utilitarianism to a broad rights-based ethic effectively eliminated options such as the creation of international disposal sites in one or more of the less developed countries. Some would argue that this would have made sense on utilitarian grounds, e.g. allowing the choice of the most suitable geology and/or isolated areas away from centres of population, but it was generally recognised as being unfair in important respects. Similar objections were applied to 'dilute and disperse' approaches such as sea disposal, seen as carrying risks that material may contaminate third parties and indeed non-human life who may have no stake in the processes which created the waste. Other considerations, such as potential effects on neighbouring countries of waste creation and management and the desirability of long-term monitoring of wastes, especially radioactive wastes, in a retrievable condition, also become more important.

Utilitarianism itself has taken on a variety of forms but they all, in theory, offer the possibility of coming to a single 'best' answer, at least if one had access to sufficiently good information to chart the likely outcomes of each option with respect to human health and happiness. The rights-based approach, by contrast, inevitably throws up clashes between equally valid 'rights'. For example, local communities surely have some right over activities affecting their locality, and indeed in some countries have an absolute veto in law, but the country at large may also have rights and indeed obligations to deal with national issues in a way that is sustainable. Fracking, nuclear waste repositories, wind farms, tidal barrages, towers carrying high voltage power lines and many other issues in energy display this tension.

10.5 THE DECLINE OF DEFERENCE – ETHICS, EXPERTISE AND POLITICS

The demands that society places on its leaders vary considerably depending on whether the dominant ethic is utilitarian or rights-based. In the former case – typical, as suggested earlier, of societies under some kind of threat (war, economic decline, epidemics of disease, industrial strife, severe environmental degradation, etc.) – people tend to demand 'strong leadership'. This is seen most clearly during times of war. The rights of individuals tend to be subsumed in the needs of society at large – 'don't you know there's a war on?' – and big political gestures are often regarded as essential to maintain national morale.

By contrast, when society is feeling more comfortable and individual rights predominate, people seem to require leaders who are more consensual and do not take bold action. One can argue that Margaret Thatcher as UK Conservative prime minister was not notably different in her beliefs and style at the time of her political downfall in 1990 than she had been in 1982 at the time of the Falklands War or 1984 during the miners' strike. However, while her style was enormously successful, not least in electoral terms, during the years in which the UK faced serious strains following the industrial unrest of the 'winter of discontent' of 1978/9, it seemed to have become a liability by the late 1980s, a time of economic prosperity and the end of the Cold War: she was effectively

Table 10.1 The difference between utilitarian and rights-based theories

Utilitarian	Rights-based
Basic positions	
What is right is what promotes the best consequences for the most people.	All people have rights to live without having risks imposed on them, even if these risks also bring considerable benefits for other people, and to be protected from any risks that are so imposed.
The economic consequences of different safety standards and ethical positions are centrally important – resources spent in one way are not available to improve people's lives elsewhere.	Any risk is justified only with individual consent, compensation and equality of risk distribution.
Are risks ethically acceptable if they are at the same level as voluntarily chosen risks?	
Yes – the public should be consistent and accept risks that are lower than those associated with such activities as road travel. Risks have to be weighed against benefits.	No – it is not just a question of risk magnitude, but also of distribution and compensation. The risks associated with nuclear reactors or a radioactive waste repository, say, are perceived to be more catastrophic, less compensated and under less voluntary control than many other risks. Assuming a level of risk is ethically acceptable because it is 'normal' commits the naturalistic fallacy: we cannot derive 'ought to be' from 'is'.
'Natural (existing) standards' are used as the basis for determining safe level of other pollutants and should also be with radiation. Risk can be defined as a probability of some harm. Risks are quantitative and thus comparable.	Risk is also a function of qualitative components such as the degree of informed consent of individuals who may potentially be affected.
Uncertainty and ethical standards: should those who impose risk be presumed innocent until proven guilty, or the opposite?	
Policy should follow a rule of maximising average expected benefit. Worst cases are very rare; minimising the risk of their occurring or their consequences absorbs resources which would provide more benefits elsewhere.	Policy should follow a rule of avoiding the worst possible outcome. A small (close to zero) probability of catastrophe does not outweigh infinitely serious consequences.
Policy should follow the scientific norm of minimising 'false-positives' (i.e. concluding there was harm due to exposure when there was not). It should not be assumed that BSE-infected beef, GMOs, the MMR vaccine or radiation, say, have caused an injury unless all other possible causes can be ruled out. The burden of proof must be on the alleged 'victims' as it is too costly to do otherwise.	Policy should follow a norm of minimising false-negatives (concluding there was no harm when there was). Fairness and the right to equal treatment require those imposing a risk should have to prove (or demonstrate to a very high degree of likelihood) that the activity in question does not and will not cause injury.

Table 10.1 (continued)

Utilitarian	Rights-based
Equity issues	
The basic premise of equity among various groups is agreed but costs for the future can be discounted owing to the uncertainty involved, e.g. the likelihood of finding solutions to problems which today have no such apparent solutions.	Temporal and spatial considerations are not a morally relevant basis for discriminating in imposition of risks, so no discounting for geography or time should be allowed.
Poor communities that accept risks do so voluntarily – economic development benefits and wage differentials compensate for the risks to workers.	It is unfair disproportionately to burden these communities. Economic welfare does not justify such inequalities, as the human right to equal protection is inviolable.
Consent – is inequitable distribution of risk allowable, if parties consent?	
Economic development and risk-for-money trade-offs serve the greater good. Lower standards can be applied if there is informed consent. In any case no instances of consent are perfect. Sociological differences are relatively unimportant.	A stringent concept of consent should be applied. Consent must be informed and educated and those involved must have alternatives. (Generally those with highest levels of education and with alternatives tend not to consent to controversial industrial developments such as nuclear power plants; those who are least informed and have fewest alternatives are the ones who end up having such developments foisted on them.) Poor people should not be allowed to trade their basic health for jobs, whatever the level of compensation. How can future generations give consent?
Due process and compensation	
It is agreed that compensation is required to justify imposing higher risk.	Even where higher risk is potentially acceptable, i.e. where the consequences of this risk are not too high to make it unacceptable, compensation is required to justify imposing it, since lack of compensation harms victims by threatening their rights to due process. The acceptability of industrial risks (e.g. those associated with a large hydro dam, a fracking well or a nuclear power station) is a function of compensation.

Table 10.1 (continued)

Utilitarian	Rights-based
Some feel that if a facility is safe then risk imposers have little to lose by accepting full liability for accidents.	People are often concerned about small, uncompensated risks too. Industry and government liability is limited to less than costs of a worst-case accident, which violates the principle of the risk-creator providing compensation. If we cannot guarantee future generations will be compensated, we cannot justify imposing risk. Forms of compensation include financial funding, acceptance of liability and community control of health and safety monitoring.
Modes of political activity and attitudes towards institutions	
Traditional democracy is the best way to ensure that the interests of society are translated into appropriate policies.	Traditional elective democracy is often unable to protect the rights of individuals or small groups. 'Don't vote, it only encourages them – they are all the same anyway.'
Technical 'experts' have a vital role to play in ensuring that policies are devised which will deliver the population's requirements as expressed through the political system.	'Experts' should have no greater input to decision-making than those who will have to live directly with the consequences of those decisions. Deference to 'the great and the good' is an outmoded and entirely inappropriate attitude for a modern society.
'Direct action' is often simply a way of imposing the views of small minorities on society at large.	Local direct action is often justified as a way of protecting individual rights.

dismissed by her own MPs in 1990. By contrast, Tony Blair's more consensual electoral style as Labour prime minister between 1997 and 2007 worked well in the elections of 1997 and 2001. However, his support for President Bush over the Iraq invasion, in which it can be argued he showed similar resolve to that displayed by Mrs Thatcher in 1982, alienated considerable numbers of British voters and was to become the matter for which he was remembered long after his term of office finished. (It can also be argued that the nature of the direct risk to the UK, of which the Falklands were effectively part, was less clear in 2003 than in 1982. However, considerable effort had been expended, with some apparent short-term success, on erroneously persuading the British people that Iraq possessed weapons of mass destruction that threatened the UK.)

Whether a particular individual is a 'strong leader' or 'out of touch with ordinary people', or at the other extreme is 'consensual and democratic' or 'weak', may be as much a matter of the social and ethical environment in which they are operating as of their own particular actions. It was notable that David Cameron, when campaigning to become

Conservative Party leader in 2005, positioned himself as the 'heir to Blair' rather than as carrying the Thatcher flame forward.

10.6 A GROWING GAP BETWEEN PEOPLE AND POLITICIANS

A further feature of a society operating under a rights-based ethic seems to be reduced participation in traditional politics. If society seems to be enjoying a stable period of comfort and wealth then it may appear less vital to elect a government willing to take tough decisions, even if those decisions are necessary to preserve that comfort and wealth. A number of trends can be identified in many developed countries in the half-century from 1960, including:

- a fall in numbers voting in elections;
- a growing tendency to join single-issue pressure groups or political parties rather than the mainstream political parties;
- the resurgence of political satire, hardly new but more prevalent since the early 1960s;
- an increase in people asserting their individuality – the percentage of people agreeing that it is more important to 'fit in' than to be different from other people in the UK fell steadily from 49 per cent in 1980 to 27 per cent in 2002;
- growing informality in dress, social etiquette, etc., a reaction against or at least decline of pressure to conform with the traditional image 'authority' figures;
- a growth in the extent to which people trust friends, family and other individuals as a source of information over 'official channels'.

In the 2004 European Parliament elections in the UK, for example, on a turnout below 40 per cent, the two main parties, Conservative and Labour, between them collected fewer than 50 per cent of the total votes cast for the first time since the 1920s, while the secessionist UK Independence Party (UKIP) pushed the traditional recipients of protest votes, the Liberal Democrats, into fourth place.[9] By the 2014 European elections UKIP edged Labour and Conservatives out of first place. In the Scottish Parliament, and subsequently (and spectacularly) in the 2015 general election, the Labour Party saw itself being displaced in Scotland by the Scottish National Party (SNP), while the Green Party won its first Westminster MP in 2010. Membership of the main parties underwent steep decline, most notably the Conservative Party (which saw numbers fall from almost three million in 1951 to an estimated 134,000 in 2013) but also Labour (down from one million in the early 1950s to 190,000 in 2013) and the Liberals/Liberal Democrats, from almost a quarter of a million in 1960 to fewer than 50,000 in 2013 – see figure 10.2.

If the minor parties enter into government, alone or more likely in coalition, they often find the need for compromise on basic principles unmanageable, the SNP in

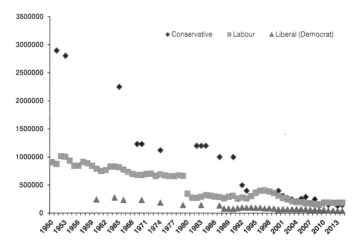

Figure 10.2 Estimated UK political party membership[10]

Scotland appearing a notable exception. Perhaps learning from the fate of the Liberal Democrats in the 2010–15 parliament (at the end of which their representation fell from 57 seats to eight) and the Green parties in Europe a decade or two earlier, the SNP, UKIP and the Greens in the UK all made it clear that they would duck the opportunity to join a formal governing coalition, should that opportunity be presented by the 2015 general election result.

Furthermore, political parties generally respond to a 'comfortable' electorate by moving towards what is perceived as the 'political centre' of the day. Reduced turnout may also reflect a widespread feeling that there is 'little difference' between the major parties (in contrast, for example, to the chasm between the policies of the UK Conservative and Labour parties throughout the 1980s, especially in 1983 – it is hard to recall Margaret Thatcher or Michael Foot, the left-wing Labour leader of the time, arguing about the 'centre ground').

Such observations have important implications regarding a government's scope for action in emergency situations. At times of comfort government may organise itself in such a way as to appear to be as consensual or non-interventionist as possible, e.g. by legislating to protect the rights of local communities to reject projects of national importance or by divesting itself of the mechanisms whereby it might intervene in a particular industry or business. This may leave the government in a weak position if the need should arise to take decisive action in response to a chronic emergency. The timescales associated with energy create particularly difficult challenges. If politicians are acting in consensual mode and are unwilling to take controversial decisions and stick to them then there is real danger that investors will not feel sufficiently confident to put their resources into heavily capital-intensive industries with very long investment horizons.

10.7 POLITICS AND SCIENCE OUT OF STEP

The third side of the triangle of public, politics and science/technology, i.e. the relationship between the political and technical spheres, also became more troubled as the 1970s and 1980s proceeded. (There is a sense of 'my enemy's enemy is my enemy' about the whole thing.) The relationship began to change from one of overt mutual support to one of coolness or, in some cases, antipathy.

For a while the 'establishment' could continue to take decisions despite growing public opposition – one suspects that it was largely unaware or dismissive of the changing societal ethic. In the energy field, for example, new nuclear reactors were ordered in countries such as the US during the 1970s and 1980s, though many were to be cancelled, as demand for electricity continued to grow. However, throughout the 1980s and 1990s (though it had been presaged by writers such as C.P. Snow a quarter of a century earlier[11]), the commonality of interest and the close interaction between scientific and political cultures began to disintegrate. In the UK in particular, the 1990s, with their sustained low oil and gas prices, exploitation of significant indigenous gas reserves and emergence of the new CCGT technology, cheaper and more environmentally acceptable than the predominant coal-fired stations of the previous decades, were a period of rare comfort in the energy field. A single policy, namely the switch from coal to gas for electricity production, was delivering relatively cheap electricity, secure supplies based on the UK's North Sea gas reserves plus a lot of new generating capacity and reductions in carbon emissions because CCGT emits less than half as much carbon dioxide per unit electricity generated as does coal-fired capacity. The public and political spheres could see no reason for any radical change in policy or any potentially unpopular decisions, despite warnings from some economists and researchers that the situation was not sustainable. Unwieldy and multi-stage licensing processes and increasing requirements for formal consultation were imposed for a range of different decisions, perhaps on the assumption that because the energy problem had now been 'solved' there would be no need for difficult decisions in the future.

It can be argued that the energy industries made it easier for governments to duck issues by fighting among themselves rather than by encouraging a holistic approach to impending challenges in the energy and environmental fields. Internal bickering – 'renewables are too expensive', 'fossil fuels are too dirty', 'nuclear power is too risky', 'fusion will never work', etc. – makes it more difficult for governments to show political leadership with respect to the energy sector (or, depending on one's viewpoint, makes it easier for governments to abdicate responsibility)[12]. Interestingly, the phase 2 attitude is no more accepting of open scientific debate than is that of phase 1, as demonstrated by the dismissal and even persecution of those who challenge Big Green orthodoxy on climate change, for example. It is just that the 'acceptable' dogmas become anti-industrial rather than technological.

During this period 'Big Science', as represented perhaps most spectacularly by the space programme, fell out of favour and governments, with a few notable exceptions (especially in Japan – see figure 10.3), drastically reduced expenditure on 'blue skies' energy research and development, as shown in table 10.2.

As time went by representatives of the political wing of Big Green were elected to local authorities, national parliaments and eventually national governments. After 1990 Green Party representatives at various times entered coalition governments or held the balance of power in France, Italy, Germany, Belgium, Finland and Sweden. From this position they could effectively undermine the scientific basis of decision-making, including the (incorrect) assumption that scientists could be regarded as founts of unbiased 'truths', and replace it with the pronouncements of activists and pressure groups who claimed (with dubious justification) to reflect the 'real' feelings of society better than traditional elected politicians or unelected advisors from industrial and commercial interest groups.

Table 10.2 Expenditure by IEA countries on energy R & D (2006 US$ millions)[13]

Year	1975	1980	1985	1990	1995	2000	2005
Conservation	333	955	725	510	1240	1497	1075
Fossil fuels	587	2564	1510	1793	1050	612	1007
Renewables	208	1914	843	563	809	773	113
Nuclear fission	4808	6794	6575	4199	3616	3406	3168
Nuclear fusion	597	1221	1470	1055	1120	893	715
Other	1030	1586	1063	1274	1648	1889	2508
Total	**7563**	**15034**	**12186**	**9394**	**9483**	**9070**	**9586**
Total excluding Japan	**6055**	**11596**	**8448**	**5842**	**5811**	**5349**	**5681**

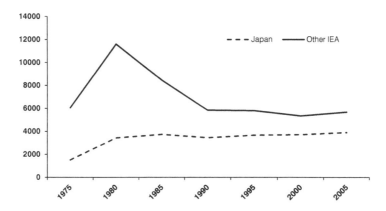

Figure 10.3 Expenditure on energy research by Japan and other IEA countries (2006 US$ millions)

In this venture the march of the anti-industrial (and to an extent anti-capitalist) movement was aided by the behaviour of some scientists, notably those working for major corporations, who seemed to have been prepared to allow their views to be used or covered up in such a way as to promote the financial interests of their sponsors. Research into the health effects of smoking tobacco became a cause célèbre, both because of the delay in the scientific community's taking up the cause and because of the failure of some scientists to make their findings public.[14] Science was seen to have 'failed' to lead to good decision-making, often because it had been engaged in the support of financial vested interests. The environmental disasters outlined in Rachel Carson's landmark book *Silent Spring* in 1962 and the chemical dumping scandal at Love Canal, New York, in the late 1970s were particularly influential, alongside nuclear accidents at Three Mile Island and Chernobyl and such events as the Bhopal chemical disaster in 1984.

Big Green is not made up of pathological liars. It is perfectly happy to cite orthodox scientific understanding over issues such as climate change if that science supports its ideology. But in general the Big Green priesthood displays a Nietzschean attitude to 'truth' – what is true is what is useful. Nietzsche claims that untruths can be just as useful as truths and makes a suggestion that the will not to be deceived could itself be seen as a will hostile to life, given that life itself is a matter of deception and error.[15] These thoughts lie deep in some strands of Big Green philosophy. This is to be contrasted with the approach of the many genuine environmentalists who, honourably, do publicly change their views as new evidence becomes available and challenge Big Green's focus on very small risks when doing so dilutes action against major threats such as water shortages or climate change.

However, the schism between science and politics was not caused simply by a few manipulative Greens. The two establishments perhaps never really understood each other in the first place. Despite occasional claims to the contrary in the mainstream or social media, neither 'politicians' nor 'scientists' are generally ill-intentioned or particularly divorced from 'ordinary people', at least when they set out. Doubtless with a few exceptions, both, in their own way, generally seek to apply some concept of 'truth' to the development of policy, with the belief that embedding this truth will lead to an improvement in human conditions. This is not to say that individuals within either culture are incapable of bias or of bending the detail to 'clarify' the bigger picture, as was seen during the Climategate scandal discussed in chapter 9, or even, in a small number of cases, of corruption. But however clearly it may appear to the contrary at times, politicians and scientists are first and foremost people. Indeed, it is unlikely that any single individual acts in 'scientific' or 'political' mode all of the time.

Nonetheless, there are important differences in the basic mindsets of the politician and the scientist which can lead to mutual misunderstanding and contribute in a major way to the paralysis in decision-making. The true scientist, again contrary to the popular view, inhabits a region of unremitting uncertainty. The phrase 'scientific fact' is near to an oxymoron. Far from living in a world of 'facts', scientists ultimately live in a world of intuition which is a form of guesswork, however well grounded that guesswork may be. At least in theory, and very largely in practice as well, science is an iterative process with no end point. Many scientists do effectively come close to pursuing the idealised Baconian model to a greater or lesser extent. They collect observations, either from fieldwork or as the result of experimentation. They then create hypotheses in an attempt to explain these observations – once more, in contrast to the popular stereotype, the formation of hypotheses often requires considerable imagination. The hypotheses lead to further predictions and the scientists design further experiments or carry out further field studies to test these predictions. If a hypothesis stands up – i.e. if its predictions are supported by observations – then that hypothesis will not be discarded: but this is not to say it is 'proven' in the sense that a mathematical or logical proposition can be proven. It can never reach that status – all that can be said of the very best hypotheses is that they accurately accord with or predict a wide range of real observations. Strictly speaking hypotheses do not even offer 'explanations', at least in the simple way that the pioneers of the enlightenment expected and hoped they would – they are better regarded as 'descriptions' or 'models', the answer to 'how' questions rather than 'why' questions. Of course, if a hypothesis creates predictions that are not borne out by experience then eventually the hypothesis will have to be abandoned or modified and new observations made to determine whether any new hypothesis that may emerge is better.

One important consequence of this – one that causes vast amounts of annoyance among the public and politicians alike – is that no one, not even a scientist, can 'prove a negative'. A single validated negative observation is enough to disprove a hypothesis, while a thousand positive ones will not prove it. Suppose a European of the sixteenth century were asked to describe a swan. One attribute that could have been mentioned was that swans are not black. As far as we know, all swans in Europe at that time were white so this was a reasonable 'hypothesis'. When the Dutch landed in Australia in the seventeenth century, however, they discovered a black swan. One single example of a black swan was enough to disprove the hypothesis that no swans are black. The hypothesis that no swans are black need not be abandoned completely – it might be modified to 'no swans indigenous to Europe are black', for example, and in such a form it may still have considerable usefulness – but that also would remain a hypothesis, not a fact.

This is of vital importance when it comes to considering complex scientific issues such as those found in the field of the environment or health. Ideally, when testing a hypothesis one would design an experiment where all other factors are kept constant except for the one which is under study. Typically one group of subjects – the 'experimental group' – would be subjected to exposure to, say, the alleged health risk in question, while a group of otherwise identical individuals – the 'control group' – would not be subjected to exposure. Any statistically significant difference in outcome for the two groups could then reasonably be ascribed to the different exposure of the alleged health risk in question.

Even if it were practical, it is clearly unethical to expose groups of human individuals to potentially serious health risks. Scientists are, therefore, often forced to observe large numbers of individuals in society and make their best attempt to control for other factors – this is known in medical science as 'epidemiology'. However, the complex nature of human life means that one can never be entirely sure that all relevant factors have been controlled for during the observations. To take a relatively straightforward example, cancer rates in towns such as Bournemouth in the UK tend to be well above the national average. This cannot be taken as proving that something in the natural environment in Bournemouth causes cancer. It could be, for example, that there is a famous cancer hospital in Bournemouth and a number of patients live nearby on a temporary basis. Or it could be that as a seaside town, Bournemouth attracts residents who have retired and cancer rates are higher in older people than in the population at large. A scientist would seek to compare cancer rates in Bournemouth with those of people of a similar age nationally in locations with similar access to cancer hospitals but even this raises the possibility that some other common factor may have been missed. Correlation is not the same as causation.

Many issues – environmental or health risks associated with fracking; BSE and its development in humans; the possible link between the MMR vaccination and autism; the health effects of low-level radiation, both natural and man-made (e.g. the excess of leukaemia near some nuclear establishments such as Sellafield in the UK); the consequences of erecting mobile phone masts or electricity pylons near people's homes; the effects of lead in petrol; the use of genetically modified organisms in agriculture – generally involve making sense of small numbers of data, usually gleaned from an uncontrolled environment. Unless a clear link can be identified the best that science can hope to offer in these circumstances is to say that current understanding suggests that the pathogen in question is not a significant threat to health but that further data might lead to a reappraisal of this position. Clearly the more data that are collected the more confidence there will be that the 'best guess' is a useful guide to action but there will never be absolute certainty.

A second consequence is the impression that a scientist who holds a hypothesis which is later proven to be wrong or inadequate is ipso facto a fool or a knave. In fact the duty of the scientist is to try to prove themselves wrong by designing experiments or studies that might invalidate a current or emerging hypothesis: the way to tell a 'scientist' from a 'politician' is to look at the ratio of the time they spend trying to prove themselves wrong against the effort they expend on defending their theories or dogmas even against a growing welter of contrary evidence. Newtonian mechanics is now known to be 'wrong' in the sense that it does not apply in all circumstances, especially to systems at the atomic level. The various atomic theories of a procession of great minds from Dalton to Bohr have similarly been superseded. But nobody would claim that these giant figures, who contributed so much towards the development of our understanding of matter and without whose work more recent theories could not have been created, were either stupid or dishonest.

The political mindset is almost the diametric opposite. Politicians often talk and sometimes appear to think in terms of certainties. An answer hedged in ifs and buts is regarded as of little use in public debate – the public and the mass media are often adept at finding questions that seem to require a 'yes' or 'no' reply and then accusing their victim of refusing to give such a response – but that is generally the only kind of answer the scientists have to give. The difference can perhaps be caricatured in this fashion.

- The scientist has a reasonable idea of what has happened in the past but can only make tentative and uncertain predictions about the future.
- The politician has absolute certainty about what will happen in the future (lots of nice things if you vote for their party, lots of nasty things if you vote for the other lot) but will argue until the bars close about what happened ten years ago and whose fault it was.

Dr Vincent Cable, who worked for Shell before entering Parliament in 1997 and rising to the rank of business secretary in the 2010 Conservative/Liberal Democrat coalition government before losing his seat in 2015, characterises the difference between the political and scientific modes of operation in terms of five factors:[16]

- speed – in the political world decisions often have to be taken very quickly;
- superficiality – politicians must cover a great deal of ground, especially if they are not in a ministerial position and therefore not supported by considerable civil service resources;
- spin – in politics, perception is often the only 'reality' which counts and there is always a danger that politicians will start to 'believe their own propaganda', thereby mistaking the art of presenting figures in the most favourable light for the situation actually being rosy;
- secrecy – scientists seek to make all available facts and reasoning known so that others can comment, while in government proper public review of decisions is often impeded by restrictions on the flow of information;
- scientific ignorance – many who enter politics do not have a scientific background, perhaps because the mindsets required for the two fields of activity are so different.

The later years of the twentieth century were a time of growing suspicion as these cultural differences became a source of friction. Scientists were becoming frustrated that politicians were requiring 'definitive' (and of course 'right') answers to simplistic questions of the kind that science cannot answer and accusing them of only being interested in the next election, while politicians were becoming increasingly annoyed at the difficulty of getting a simple 'straight' answer which could subsequently be relied upon and the scientists' inability to recognise the constraints of decision-making in a democratic context. It has been argued, for example, that scientists during the BSE affair of the 1980s and 1990s in the UK were both deliberately and inadvertently utilised to provide spurious scientific legitimation for policy decisions which government officials believed ministers, other government departments, the meat industry and the general public might not otherwise accept.[17]

The political mindset finds it especially difficult to admit to error, still more to apologise, and therefore, to change an attitude merely because there is a wealth of evidence showing that a particular policy does not or would not 'work', i.e. deliver on its stated aims. The open discussion of alternative possibilities, which is the very stuff of healthy scientific discourse, is regarded with increasing horror, at least within individual political parties where any sign of straying from the party line (certainly in public) is often regarded as an act of the most heinous treachery. An example of the breakdown of the relationship between politicians and academia emerged in the debate about the introduction of identity cards in the UK in mid-2005. The London School of Economics (LSE) published a 300-page report which, inter alia, questioned some of the government's assumptions on the costs of the scheme. According to the director of the LSE: 'Home Office officials demanded to see advance copies. Before they had been provided, the Home Secretary condemned the cost estimates as "mad". When it was published he described the analysis as a "fabrication" and one of the project mentors as "highly partisan". It is unfortunate that the government chose to adopt a bullying approach to critics whose prime motivation was to devise a scheme which might work, at an acceptable cost.'[18] In 2009 Professor David Nutt, the UK's chief drugs 'tsar' – maybe the word, which has become universal, is intentionally evocative of the unfortunate end of the Romanovs – was sacked for pointing to the evidence that cannabis, ecstasy and LSD were less dangerous than alcohol or cigarettes. The same fate befell the European Union's chief scientific advisor, Professor Ann Glover, in 2014, her dismissal being widely ascribed to her moderate pro-, or rather not vehemently anti-GMO views which made her unacceptable to Big Green. A letter in July 2014 from seven organisations including Greenpeace complained about Glover's 'opinions in the debate on the use of genetically modified organisms in agriculture'[19] and argued that the post should be scrapped. Despite voluminous support for Glover, including a letter from 40 scientific institutions with over 750 individuals stating: 'we cannot stress strongly enough our objection to any attempt to undermine the integrity and independence of scientific advice received at the highest level of the European Commission', Big Green got its way.

In practice the boundary between 'the scientist' and 'the politician' (if not 'science' and 'politics') may not be so clear. Some commentators, notably Thomas Kuhn[20], have argued that in practice science proceeds in ways highly reminiscent of the political process. Certain theories become 'accepted wisdom' and are associated with major scientific figures of the day. When evidence becomes available to challenge this accepted wisdom, often the first response of the scientific establishment is to reject this evidence and even to attack the integrity or competence of those who have discovered it. If, after a period of warfare between different factions, such a weight of evidence becomes available that the accepted wisdom is no longer tenable, a 'paradigm shift' may occur whereby a new view becomes accepted very rapidly. Those who until then have been seen as rebels now become the new establishment until such a time as their own theories begin to be challenged and a further period of warfare breaks out. Einstein, for example, vehemently rejected a key aspect of quantum mechanics, saying 'God does not play dice with the universe', until acceptance became unavoidable. However, influential though Kuhn's work has been it does not seek to overturn the view that 'good' science, through its ultimate recourse to experiment and observation, does proceed in a different, and ultimately more objective and realistic, way to other avenues of discovery.

Unsurprisingly, two such mindsets do not always fit together well. Though made in a satirical context, the following observation captures the tension well.

> There are times in a politician's life when he is obliged to take the wrong decision. Wrong economically, wrong industrially, wrong by any standards – except one. It is a curious fact that something which is wrong from every other point of view can be right politically. And something which is right politically does not simply mean that it's the way to get the votes – which it is – but also, if a policy gets the votes, then it can be argued that that policy is what the people want. And, in a democracy, how can a thing be wrong if it is what the people will vote for?[21]

The opposite can also be the case. When in 1988 the UK health minister, Edwina Currie, said that: 'Most of the egg production in this country, sadly, is now affected with salmonella', she was forced to resign, not because of the substance of her claim which was fundamentally accurate[22] but because of the political damage done by the unguarded way in which it was made.

The scientist equivocates not to avoid responsibility but because when dealing with precisely the public issues which are of most interest to politicians and the public, the available data often do not allow anything approaching a clear answer. But politicians and the media, so often coming from non-scientific backgrounds, believe science is about 'facts', not least because that is the way it seems to be taught in many British schools (one of the strongest reasons for including the teaching of a questioning attitude alongside the teaching of 'facts'). Although some observations may be reliable and replicable this does not mean that hypotheses based upon them can be regarded with equal confidence. Neither phase 1 deference nor phase 2 rejection represents a properly functioning relationship between science, politics and society.

10.8 CRITICISMS OF THE MODEL

Something as complex as human society does not conform to simple trends or explanations. Among the challenges that can be offered to any quasi-linear 'three-phase' model are the following.

The first is methodological. Any attempt to separate out one or two factors as 'determinants' of how society develops and how, and what, decisions are taken is inevitably simplistic. A great deal of literature has been generated by considering whether society is primarily determined by economics, technological, scientific or cultural factors. One of the frustrations of social science is that it is not generally possible to rerun a society over a period of say 20 years, changing just one variable, and thereby discover how that variable affects outcomes. Inescapably there is a great deal of impressionism rather than science in such an approach.

Secondly, even in societies which by the previous analysis are solidly in 'phase 1', the protection of individual rights is never entirely abandoned, notably through the judicial system. In the UK, in a case brought against the two Home Secretaries, Sir John Anderson and his successor Herbert Morrison, in 1941 by a German national known as Robert Liversidge, who was detained as an enemy alien, Lord (James) Atkin of Aberdovey said:

> In this country amid the clash of arms, the laws are not silent. They may be changed but they speak the same language in war as in peace. It has always been one of the pillars of freedom, one of the principles of liberty for which on recent authority we are now fighting, that the judges are no respecters of persons and stand between the subject and any attempted encroachments on his liberty by the executive, alert to see that any coercive action is justified in law.

This was a dissenting opinion, the other four Law Lords finding for Anderson, and most people seemed comfortable and even keen to see individual rights largely subsumed into the greater good, but even in wartime some recognition is given to the rights of conscientious objectors.[23] At any time some individuals will have attitudes characteristic of each of the three 'phases'. If one says that the switch from phase 1 to phase 2 in the UK began say in the mid-1960s this does not preclude some individuals or groups having 'got there early' or others never adopting the new fashion – for example, the governmental post of chief scientific advisor was created in 1964, an apparent recognition of the value of scientific advice. And as for society, so for the individual. History may tentatively suggest that a particular government can be regarded as broadly utilitarian or contractarian over a wide range of issues and that people will accept this – from the Second World War until at least 2025 in the UK only once (the Conservative government of Edward Heath, in office from 1970 to 1974) was a government led by either Conservative or

Labour party not re-elected at least once before losing office. However, this does not necessarily mean that 'the people' broadly hold to the same ethic with respect to issues which appear pressing and issues which appear unproblematic: very few people seem to hold a consistent utilitarian or contractarian ethic on all possible political issues.

Third, any attempt to analyse society in broadly Hegelian or Marxist dialectical terms – a thesis, an antithesis, a synthesis – risks implying that this is both an irreversible process and that we are observing something that could be described as 'progress'. In fact the economic decline of the 1970s, for example, driven largely by the huge increase in the oil price in 1973, presaged a more subdued decade in which the strong political leadership (or out-of-touch, even uncaring, authoritarianism, depending on one's standpoint) of figures such as Margaret Thatcher and Ronald Reagan could flourish in a way that might have been more difficult a decade earlier (or later). In some senses this resembled a step back to a 'phase 1' relationship between government and those governed, though there is no convincing evidence that science and technology regained its position as the religion of the people in the 1980s. Maybe, as philosopher John Gray has it, there is never any 'progress' in society akin to scientific progress, merely a never-ending cycle of changing fashion.[24] If so then even if the three phases can be identified they may exist more in a cyclical interrelationship than a linear one. For example, it is patently untrue that the 'decline in deference' is a recent phenomenon mapping comfortably onto a 'phase 2' relationship between science and society, before which politicians were universally respected. Comments like 'under every stone lurks a politician'[25] can be found throughout history: political satire has had many golden ages, such as that of Horace and Juvenal in ancient Rome from the first century BC to the second century AD and that of Pope, Swift and Hogarth in the eighteenth century.

Similarly it would of course be a mistake to imply that concerns about technology, either in general or in specific cases, only emerged when society as a whole had 'progressed' into phase 2. In 1272 England, for example, King Edward I banned the burning of 'sea coal' in London because of problems caused by smoke. Coal drove the Industrial Revolution and Liberals, such as the historian Macaulay, tended to argue that the well-being of the common people was not a matter of 'rose-covered cottages', in the phrase of Romantic poet Robert Southey, but of 'steam power and independence'. But by the middle of the nineteenth century writers such as Dickens[26] were severely critical of the effects of coal burning and industry on the quality of life and on human relationships. Early proto-Green thinkers in the US included Thomas Jefferson, Ralph Waldo Emerson and Henry David Thoreau while in 1864 George Perkins Marsh published *Man and Nature* in which he anticipated many concepts of modern ecology. By the second half of the nineteenth century the term 'conservationism' was being used to describe the management of vulnerable natural resources such as woodland and fisheries and also the prevention of pollution. Organised environmentalism can perhaps be traced back to the time of President Theodore Roosevelt, Gifford Pinchot and John Muir, who was the founder of the Sierra Club. The conservation movement urged the creation of state and national parks and forests and wildlife refuges. The term 'environmentalism' seems to have been coined in the 1920s.

By the 1930s environmentalism was an integral part of at least some political philosophies. A fairly typical exposition from 1934 reads: 'Separating humanity from nature, from the whole of life, leads to humankind's own destruction. Only through a reintegration of humanity into the whole of nature can our people be made stronger. That is the fundamental point of the biological tasks of our age. Life as a whole, rather than humankind alone, is now the focus of thought.' The author, Ernst Lehmann, captain of the airship Hindenburg, which had been ordered by Goebbels and was widely used for Nazi propaganda purposes until it famously caught fire in 1937, went on to say: 'This striving towards connectedness with the totality of life, with nature itself, a nature into which we are born, this is the deepest meaning and the true essence of National Socialist thought.'[27] Big Green continued to gain momentum after the Second World War: in the US conservationists were able to block a number of projects such as the proposed Hualapai (or Bridge Canyon) Dam that would have stood 230 metres high on the Colorado River in Arizona and flooded an area stretching some 100 miles upstream, an early example of environmentalists opposing a renewable energy scheme because of local damage.

Further, all sources of energy seem to have faced periodic changes in the way they are viewed by the public and by political decision-makers, whether they have emerged in times of stress or comfort. A 'new' fuel often emerges to high expectations, even excitement. A rise and fall in enthusiasm can be seen with regard to oil, nuclear power and renewable forms of energy such as onshore wind or hydropower. Even this observation should be tempered by a recognition that no energy sources have ever been met with a uniformly warm initial welcome, nor has any subsequently lost the entire support of the whole population. It is, however, probably true to say that the passion of the founding fathers (and in many cases, notably nuclear power, founding mothers as well) made it hard for the early pioneers to recognise that not everybody would share that devotion; and that the tendency to overstate advantages ultimately backfires and leads to a backlash among the intelligent public. Opposition to the military applications of nuclear technology began to be observed almost immediately after the explosions at Hiroshima and Nagasaki in 1945. The Atomic Scientists of Chicago was formed in 1945 to sound warnings against believing all of the official pro-nuclear pronouncements. The *Bulletin of Atomic Scientists* was first published in December 1945; in an early edition Albert Einstein wrote: 'The unleashed power of the atom has changed everything save our modes of thinking and thus we drift towards unparalleled catastrophe.' The forerunner of the Campaign for Nuclear Disarmament (CND), the UK pressure group opposing nuclear weapons, was formed in 1957.

Perhaps most seriously the whole argument can become circular – if a society does not show one element of what is 'expected' for say phase 2 there is a temptation to suggest that this is because it is not actually in phase 2 at all. So the previously described analysis seems to imply that when things get more difficult, not only should voter turnout increase but people should return to the mainstream political parties with all the compromises this involves, since it is recognised that tough decisions are needed and that such decisions will not please everyone – single issue approaches are no longer adequate.

Recent evidence is at best very mixed. As discussed later, voter turnout did increase in the US and UK from its low point in the late 1990s and early 2000s. However, in the UK there has been a long-term trend towards a reduction in the proportion of voters supporting one or other of the two main parties (Conservative and Labour), as shown in figure 10.4: although the 2015 general election bucked this trend despite the plethora of parties on offer the 'bounce' from 2010 was little more than two percentage points and insufficient to restore the figure to that which had been seen as recently as 2005.

It is very difficult to claim that society was remarkably more stressed during the 1960s, say, than it was around the time of the social and economic unrest of the late 1970s which brought the radical Thatcher government to power in 1979. In the 2010 general election, in the midst of global economic crisis, Conservative and Labour together collected fewer than two-thirds of all votes cast for the first time since 1918 (when Labour was in the process of replacing the Liberals as the main alternative to the Conservatives) while the 'other' parties outside the largest two along with the Liberal Democrats polled almost one eighth of the votes. It was the fragmentation of the minor parties' vote – notably the collapse of the Liberal Democrats – coupled with the first-past-the-post electoral system that delivered the overall majority to the Conservatives in 2015, not the increase in the Conservative share of the vote from 36.1 to 36.9 per cent. Indeed, Labour's share grew by a greater amount – from 29.0 to 30.4 per cent – but its number of seats fell significantly, largely owing to its disastrous performance against the SNP in Scotland. On these figures any attempt to draw a parallel between the position of the major parties and the level

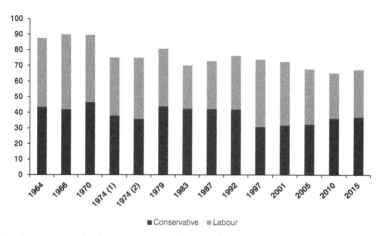

Figure 10.4 Percentage of voters supporting Labour and Conservative in general elections, 1964–2015

of social stress would be little more than a resort to heavy special pleading. In any case it is impossible even to consider a question such as 'Would people make the personal compromise necessary to choose one or other of the main parties for their own (and society's) "greater good" or had the habit of focusing on a single issue or small range of issues become too deeply ingrained?' if there are no objective criteria to judge the degree of stress a society were actually facing.

Perhaps safer ground for the broad theory is offered by the remarkable increase in the number of individuals joining political parties, or at least those outside government, around the time of the 2015 general election, as shown in Table 10.3. (Conservative numbers seemed to hold steady at around 150,000 but are not widely published.) For whatever reason, growing numbers of people appeared to be seeing political parties as relevant once more.

10.9 POLICY IN PHASE 2: CASE STUDIES

In idle moments it can be entertaining to speculate as to what effect the decision-making paralysis of the recent rights-based era might have had on a range of well-established technologies which emerged half a century ago or more. If air travel had been invented more recently would the potential for catastrophic disaster (a large aircraft suffering failure over a densely populated city) and the inevitable uncertainty over what level of safety could be achieved have prevented the whole industry from 'taking off'? The piping of explosive gas under and into our homes, coupled as it is with the occasional fatal disaster, would seem on the face of it to be worthy of rather more concern than some of the scares that have grabbed the public attention over the last two decades.

Table 10.3 Membership of UK political parties, 2013–15[28]

Party	Membership December 2013	Membership 2015
Labour	190,000	352,000 (August)[29]
Liberal Democrat	43,500	61,000 (May)
UKIP	32,000	42,000 (January)
Green	13,800	61,000 (June)
SNP	25,000	110,000 (June)

In one striking example of changing attitudes, in 1997 an Anglo-Irish company, Omega Pacific, bought a 400 km² former Ministry of Defence base at Trecwn in Wales. The company subsequently announced it wanted to use part of the site to store nuclear material – fuel rod cladding and isotopes used in medicine – claiming it would be quite safe in the labyrinth of tunnels up to 5km deep that once formed the largest underground munitions depot in Europe. The tunnels had been built during the Second World War and at one time held 4,000 tonnes of high explosives. Local campaigners, who had seemed quite sanguine about the tunnels storing high explosives, were up in arms about the proposals to store materials that by any estimation could not be regarded as being anything like as hazardous. Eventually the site was approved for a biomass plant which Friends of the Earth claimed would be inefficient and the Campaign for the Protection of Rural Wales said would create local pollution.[30]

Similarly, the very considerable dangers associated with smoking (including passive effects) and alcohol, one suspects, would have resulted in their being declared illegal if they were discovered today, like other drugs which are banned today but whose effects do not seem to be as severe or well established. If aspirin had been invented yesterday would it have been refused a licence because of its side effects, even though it has proven to be a highly successful and effective drug?[31] And as for a rail network making its way through virgin British countryside …

Be this as it may, recent years have been punctuated by a series of issues in which the best available scientific information has been rejected by the public and subsequently by government when it comes to formulating suitable policy. While each example has its own unique features, the similarities among the paths followed by controversies in widely varying scientific fields are striking. Typically they involve issues in which an activity is alleged to have an effect on a small number of people, in circumstances where no clear causal relationship can be determined, and exhibits a series of common themes, at least in 'phase 2' countries such as the UK of the 1990s and early 2000s.

- Problems arise where the activity in question is alleged to have severe effects on a very small number of individuals but where epidemiology, the study of the distribution of the disease among various populations, is unable to determine whether the activity and the alleged outcome are actually linked.

- Public concern about the issue in question emerges rapidly and unexpectedly – there seems no reliable way to predict when an issue will blow up in the public mind and the political and scientific establishments are often caught unaware.

- Initial attempts to dismiss these concerns may lead to growing dissatisfaction with the government's response among a relatively small number of highly vocal activists.

- The activists usually point to a small number of named alleged victims to counter the more statistical approach of the scientific establishment, serving to make the scientists appear more interested with numbers than real people. Statements such as: 'They're dealing with statistics – I'm dealing with people whose history I know'[32] became increasingly common. There is often complaint about a 'cover-up', Big Green demanding that government tell people 'the truth', by which they mean unequivocal admission that the effect is real even if the evidence strongly suggests otherwise.

- The media focus on the individuals who claim to have been affected by the activity in question and these individuals are often given at least the same degree of airtime and prominence as large-scale studies which imply a very small risk, if any. Big Green, so keen to express outrage if say a commentator sceptical about climate change is given any airtime at all, rarely if ever objects to the focus on individual cases of apparent harm, even if there is no credible science to suggest the case is linked to the claimed 'cause' in any way.

- Politicians ask scientists for black-and-white answers to questions such as 'is it [whatever the 'it' might be] safe?'

- Scientists are unable to answer unequivocally, first because of different conceptions of the word 'safe' – for the scientist nothing is entirely risk-free so a yes–no answer would be meaningless – and secondly because data are often limited, so preventing firm conclusions about the likelihood of harm.

- Governments therefore set up 'expert' committees to give credibility to the claims that the activity is harmless, sometimes at the same time as wide-ranging 'interim' restrictions are announced on the activity in question, presumably in the hope that this would allay public fears.

- The technical expertise of the committee is generally diluted, both through its membership and through its terms of reference, by the perceived benefit of including lay members and maverick 'academics' representing various pressure groups.

- The committee, when it reports, says 'we can find no reason to believe there is a significant problem here based on current evidence'. However, both in order to square the different political interests on the committee and because science cannot prove a negative, the report goes on to say 'but' the possibility of new evidence of harm emerging cannot be ruled out. In any case the activists often have an interest in rubbishing the report.

- The media pick up the 'but' – 'boffins fail to rule out major health threats from ...'

- Politicians are left with the choice of supporting the activity despite the equivocation of the expert committee or of continued vacillation and, in effect, erection of further barriers against the activity (perhaps invoking the 'precautionary principle'), whatever its benefits might be.

- In either case, the public is left with a feeling that something is wrong that was not detected at the start of the process. This serves to exacerbate mistrust in politicians and scientists alike and give extra impetus to the next such issue to appear on the public stage.

- The economics of the activity are adversely affected by the uncertainty over regulation and governmental action.

Though there are many to choose from, perhaps four case studies will serve to illustrate aspects of classic 'phase 2' quasi-scientific panics and the power that Big Green can wield.

Brent Spar

The use of fossil fuels has large negative impacts. However, Big Green is not above exaggerating these where it serves its purpose so to do, even at the cost of entirely avoidable environmental damage. A seminal example involved the Brent Spar in the mid-1990s.

Brent Spar was a North Sea oil loading buoy operated in the Brent oilfield, some 120 miles north-east of the Shetland Isles, by Shell UK. By 1991 it had been rendered redundant by the completion of a major oil pipeline. Shell spent some four years and commissioned 30 studies to determine how to deal with the structure, the overwhelming consensus being that sinking the rig in a deep North Sea channel would be the most environmentally responsible course of action as well as being the safest from the point of view of the workforce.

A consultancy, Fisheries Resources Services, was engaged to find a suitable site. Having identified three possibilities it carried out a comprehensive range of studies, including seabed visualisation using a remotely operated underwater vehicle; sediment sampling; and sampling of the different species living on the seabed. The North Feni Ridge, some 150 miles off the west coast of Scotland, was chosen. Modelling suggested that the rig would simple decay away slowly without significant adverse effects – indeed, as had happened during its working life the structure could provide a source of both shelter and nutritional iron to local sea life.

There was a problem, however, at least from the point of view of Big Green. This would also have proved the most economic option. In 1995 Greenpeace demanded that Shell surrender to its demands to dismantle the Brent Spar on land. When Shell held firm, Greenpeace launched what one commentator described as a three-week 'holy war'. Shell petrol stations suffered boycotts, personal attacks on staff and even an arson attack in Germany, though of course Greenpeace accepted no responsibility for the latter actions. Greenpeace issued estimates of the toxic dangers of sinking the rig, claiming it contained over 5,000 tonnes of oil. It was later to admit that the figure was dramatically overstated, the true figure being around 100 tonnes, saying its personnel 'feel extremely comfortable about the stance Greenpeace took against dumping this structure at sea' and claiming that the matter of the amount of oil aboard the platform was a minor matter[33] (in which case one might wonder why they had publicised it so widely in the first place). However, even accepting Greenpeace's assurances that the statement was a result of incompetence rather than deliberate falsehood, it had its effect: the public and government officials turned against Shell's plan and it became effectively impossible for the least detrimental environmental action to be pursued.

Shell convened a stakeholder group to consider alternatives to deep sea disposal, of which the present author was a member. Two things stick in the memory: the universal level of frustration among participants, bordering occasionally on anger, that Shell was prepared to cause unnecessary environmental damage rather than stand up to Big Green bullying (though of course there was recognition that Shell was in a difficult position not of its own making); and the indifference of Greenpeace to the outcome, to the extent of not sending a representative to the discussions. Eventually parts of the structure were cut up for use on the extension of the Stavanger Port Authority's quay at Mekjarvik in Norway, to accept roll-on roll-off ferries, while a significant volume of it was sent to landfill.

Based on off-the-record comments from Greenpeace officials, it appears that the organisation was well aware that the initial plans were probably the most environmentally beneficial in the short run.[34] The campaign of misinformation was driven by a desire to increase Shell's costs and therefore in the longer run make oil extraction less profitable – though given that the Brent Spar approach was in effect not just a redundant plant but a redundant technology in the age of international and intercontinental pipelines it is hard to see how this outcome could arise. There was also a view that increasing the overall level of environmental damage was a perfectly acceptable price to pay to maintain a quasi-religious dogma that North Sea disposal was a cardinal sin – rendering practical evidence-based decision-making irrelevant and begging the question as to why Greenpeace engaged in any argument, for example, about the amount of oil on the structure. In the phase 2 society Big Green, as one of the new religions of the time, can sail through the public revelation of such subterfuge with no more than the mildest of embarrassment.

Mobile phones

By the end of 2014 the number of mobile devices had overtaken the world population with an estimated 7.2 billion Global System for Mobile communications (GSM) and other mobile phone subscriptions across more than 200 countries.[35] The growth through the 2000s was extraordinary – from one billion customers in 2004 through two billion in 2006 and three billion in 2008. There are many other telecommunications and related systems in use, all of which result in exposure of the population to radiofrequency (RF) fields.

The extensive use of mobile phones suggests that users do not in general judge them to present a significant health hazard, instead welcoming the technology and bringing it into their everyday lives. Nevertheless, since their introduction there have been concerns about the possible impact of mobile phone technologies on health.

In 1999 the UK government set up the Independent Expert Group on Mobile Phones (IEGMP) to review the situation. Its report, *Mobile Phones and Health* (the Stewart Report), was published in May 2000.[36] It stated: 'The balance of evidence to date suggests that exposures to RF radiation below NRPB (National Radiological Protection Board) and ICNIRP (International Commission on Non-Ionising Radiation Protection) guidelines do not cause adverse health effects to the general population.'

As noted, however, such reports always have to offer a standard caveat, given that science is not capable of proving negatives. The Stewart Report caveat read:

There is now scientific evidence which suggests that there may be biological effects occurring at exposures below these guidelines. We conclude therefore that it is not possible at present to say that exposure to RF radiation, even at levels below national guidelines, is totally without potential adverse health effects and that the gaps in knowledge are sufficient to justify a precautionary approach. We recommend that a precautionary approach to the use of mobile phone technologies be adopted until much more detailed and scientifically robust information on any health effects becomes available.

It is not surprising that some studies have seemed to find a link between mobile phone use and one problem or another – for example, a Swedish study suggesting an increase in the risk of acoustic neuroma in people with more than ten years' use of mobile phones.[37] In scientific literature generally, for an effect to be deemed 'statistically significant' it has to have a lower likelihood of having occurred randomly than 5 per cent. It follows that one in twenty studies in any field would be expected to yield 'significant' results by simple chance. The vast bulk of subsequent studies from around the world have supported the main thrust of the Stewart Report's general conclusions. For example, the 2012 report of the UK government's Mobile Telecommunications and Health Research Programme (set up in response to the Stewart Report and which had spent some £14 million on research in the subsequent decade) concluded: '[A large case-control study headed by Professor Anthony Swerdlow of the Institute of Cancer Research] found no association between regular use of a mobile phone and the risk of leukaemia. There was also no evidence of a trend of increasing risk with the time since the mobile phone was first used, total years of use, cumulative number of calls or cumulative hours of use.'[38] A selection of other comments from international regulators and research bodies reads as follows.[39]

- Canada, 2014 (Royal Society of Canada): 'Available studies suggest that the basic restrictions recommended in the Canadian Safety Code do provide adequate protection against known adverse health effects across the radiofrequency range.'
- UK, 2014 (Biological Effects Policy Advisory Group (BEPAG), UK Institution of Engineering and Technology): 'BEPAG has concluded in this report that the balance of scientific evidence to date does not indicate that harmful effects occur in humans due to low-level exposure to EMFs. Our examination of the peer-reviewed literature published in the last two years has not justified a change in the overall conclusions published in our previous report (2012).'
- International Agency for Research on Cancer (IARC), 2014: 'Time trends in glioma incidence based on Nordic countries and the US exclude any large increase in incidence attributable to mobile phone use, albeit with reference to a relatively short time from initiation of exposure. No association was observed between mobile phone use and other cancers.'

- Nordic radiation safety authorities, 2013: 'The overall data published in the scientific literature to date do not show adverse health effects from exposure of radiofrequency electromagnetic fields below the guidelines or limits adopted in the Nordic countries.'

- Netherlands (Health Council of the Netherlands), 2013: 'The final conclusion from this systematic analysis is then: there is no clear and consistent evidence for an increased risk for tumours in the brain and other regions in the head in association with up to approximately 13 years' use of a mobile telephone but such risk can also not be excluded. It is not possible to pronounce upon longer term use.'

- Sweden (Scientific Council of Swedish Radiation Safety Authority (SSM)), 2013: 'Since 2011 numerous epidemiological studies on mobile phone use and risk of brain tumours and other tumours of the head have been published. The collective of these studies, together with national cancer incidence statistics from different countries, is not convincing in linking mobile phone use to the occurrence of glioma or other tumours of the head region among adults. Recent research on exposure from transmitters does not indicate health risks for the general public related to exposure to radiofrequency electromagnetic fields from base stations for mobile telephony, radio and TV transmitters or wireless local data networks at home or in schools.'

The World Health Organisation's summary is: 'A large number of studies have been performed over the last two decades to assess whether mobile phones pose a potential health risk. To date, no adverse health effects have been established as being caused by mobile phone use.'[40]

As already discussed, this is about as far as science can ever take us – no evidence of there being a problem but we must always bear in mind that data of that nature may emerge in the future so we need to be a bit careful. However, public concerns and opposition in many countries have been directed not towards mobile phones themselves but towards phone masts or base stations, despite the fact that the radiation exposure from these is much less than from direct use of a phone. (Exposures in proximity to 'picocells' have been found to be no more than a few per cent of the maximum given in safety guidelines for the public.) Major public protests have developed in response to particular applications to install phone masts, especially if they are near schools, hospitals or homes. Although in England and Wales the government issued an order whereby concerns over health could not be taken into account by local planning committees, many schemes were rejected on other grounds such as visual intrusion and impedance of footpaths which were clearly being used as proxies for health concerns or were withdrawn by the proposers, leaving some areas even in large cities without reliable network coverage.

One study of a local campaign against a mobile phone mast application brings out the importance of quasi-anecdotal evidence, impression and the media's response to it.

Although the news media were an important source of information, the campaigners primarily referred to anecdotal evidence to support their health concerns surrounding mobile phones. Tumours and headaches were the health problems most frequently cited. These worries reflect the medical conditions found in the group and not just media reports. One protestor feared that the additional background radiation from the proposed mast would increase her susceptibility to migraine attacks, a point she made in her evidence to the planning inquiry. She was convinced that mobile phones were a major cause of her migraines and used the model of flickering lights causing epileptic fits as an analogy. Another campaigner also suffered from migraines and was a member of the Migraine Action Association. They mentioned that the association's newsletter sometimes contained articles about the potential influence of mobile phones on migraines. There were also claims that pupils and teachers at a local senior school suffered disproportionately from headaches. This was blamed on a mobile phone mast which was situated on top of a teaching block at the centre of the school grounds. Two members of the group had been diagnosed with brain tumours and they both felt that mobile phone technology 'irritated' their tumours. The local newspaper highlighted these worries by reporting that one person had died from a tumour and four others in the area had similar medical conditions. As well as reference to individuals, cancer clusters associated with Menwith Hill Station, a UK Ministry of Defence satellite communications centre, and other mobile phone masts were mentioned.[41]

The phone mast controversy is another example of government and industry action serving not only as a response to public fears but also as a significant force in shaping those concerns. 'Almost by definition, what is a risk issue is itself determined by the extent and character of government reaction. There is also a more particular sense in which official risk responses potentially animate and cohere diffuse anxieties.'[42] Far from heading off potential accusations of complacency through a proactive strategy to 'keep ahead of public anxiety', as was claimed by the then health minister, Tessa Jowell MP, in 1999, the UK government's precautionary action in establishing IEGMP may actually have stimulated risk concerns, which increased after the Stewart Report was published. It seems that even balanced public information on negligible risks tends to increase anxiety on the assumption that 'there must be something to worry about if the government is taking action'.

Although the Stewart Report acknowledged that 'the balance of evidence does not suggest that mobile phone technologies put the health of the general population ... at risk', nevertheless the study called for a £7 million programme of further research and for leaflets to be included in future purchases of mobile phones warning of the possible risks. According to one commentator: 'In its rush to be open about communicating risk to the public, the government has simply forgotten that there was no risk to communicate.'[43] These leaflets would advise taking note of the specific absorption rate (SAR) of phones, which measures their heating effect, despite all sides to this argument accepting that such heating is not the problem. This suggests that recording anything that was easy to measure became the key goal, irrespective of the fact that it did not relate to the still-to-be-demonstrated 'non-thermal' effects.[44]

So the Stewart inquiry, like others, made significant concessions to the need to incorporate perceived public concerns and even prejudice, whether or not these were based in the science. The recommendations in the report required future research to take account of non-peer-reviewed and anecdotal evidence. The inquiry itself may have served to magnify such concerns by extending its remit beyond a review of the latest scientific knowledge on mobile electromagnetism to the non-scientific terrain of concerns pertaining to the siting of masts or base stations.

MMR vaccine

The MMR vaccine is a combined vaccine for immunisation against measles, mumps and rubella. It is usually administered to children around the age of one year, with a booster dose before starting school. It is widely used around the world – between its introduction in the 1970s and the mid-2010s over 500 million doses had been used in over sixty countries. It was adopted in the UK in 1988. Those receiving the vaccine may experience temporary side effects such as a rash or a mild fever, though in around 1 in 100,000 cases there may be a more severe allergic reaction.

At least from a utilitarian point of view inoculation was one of the success stories of twentieth century medicine. Measles infection rates were once so high that it is now assumed that anyone over the age of 65 years will have had measles at one time or another; by the mid-2010s rates among people under 30 in countries with vaccination programmes were below 1 per cent. Globally the disease still accounted for over 145,000 deaths in 2013, more than half of them in Africa, but vaccination programmes had reduced this figure by some 75 per cent since the turn of the millennium.[45] Mumps and rubella (German measles) also used to be very common. In rare cases mumps can cause sterility among males while the major risk in rubella is that if a pregnant woman is infected it can lead to significant congenital defects in her baby. All three diseases are highly contagious.

The MMR vaccine was designed to be a single-shot vaccine that protects against all three diseases. Significant improvements in reducing the incidence of the diseases can be attributed to widespread use of MMR.

Very few medicines are completely free of side effects or potential risks. A utilitarian society might take the view that if such a medicine brings considerable benefits to large numbers of people then a certain rate of problems should be accepted by those submitting themselves for treatment: neither disapproval nor penalties should be applied to the drug company unless clear dishonesty or negligence in carrying out trials could be proved. An extreme Kantian ethic may dictate that no case of detriment can be justified by benefits to other individuals, no matter how many individuals there may be or how much they may be benefited. Even if it were to result in the banning of the medicine or a huge increase in costs putting it out of the reach of many potential beneficiaries, pharmaceutical companies should be punished for any adverse effects that could be proven.

Controversy about vaccination began in the 1980s when a number of lawsuits were brought in the US against manufacturers of vaccines, alleging that they had caused a variety of physical and mental disorders in children. While these were inconclusive they did lead to a massive jump in the costs of the MMR vaccine as pharmaceutical companies sought to cover potential liabilities and some companies stopped selling it. Rates of take-up had started to fall in the UK when, in 1998, a paper was published in *The Lancet* by Dr Andrew Wakefield[46] which posited a connection between gastrointestinal symptoms and developmental disorders, most notably autism, in twelve children who had been given the MMR vaccine. At a press conference and in a video press release before the paper was published, Wakefield said that he thought single vaccines should be used until the MMR triple vaccine could be 'ruled out' as a cause of the children's problems. Parents of eight of the twelve children studied were said to have blamed the MMR vaccine.

Wakefield's publicity bid sparked a major health scare in the UK. The 'cause' was taken up by a variety of 'anti-establishment' figures and publications, notably the satirical magazine *Private Eye*, always on the lookout (often justifiably) for a cover-up.

The government and the NHS stressed extensive epidemiological evidence that failed to show any connection between MMR and developmental disorders. However, coming soon after another major controversy surrounding the potential link between BSE in cattle and the fatal Creutzfeldt–Jakob disease in humans, these assertions were disbelieved by some parents and other interested parties. The take-up of MMR dropped from 88 per cent in 1998 to 80 per cent in 2004. In some parts of London it was said to be as low as 60 per cent, well below the rate needed to maintain 'herd immunity' and avoid a measles epidemic: the World Health Organisation recommends that vaccination rates should be at least 95 per cent.

There were calls for the introduction of single vaccines. These were rejected by the government on several grounds – that the uptake rates of three separate vaccines would be lower than for a single dose of the triple vaccine; that single vaccines would be more expensive; but, most importantly, that the evidence strongly opposed the view that the MMR vaccine was a more serious health risk than single vaccination. A string of studies published mid-decade strongly implied that the government's position was correct in scientific terms.

- October 2004 – a review financed by the European Union assessed the evidence given in 120 studies and concluded that while the vaccine was associated with some positive and negative side effects, it was 'unlikely' that there was a connection between MMR and autism.[47]

- January 2005 – intensive research in a single county in Minnesota concluded that there was no link between MMR and autism and that an eightfold rise in the reporting of autism was due to an increased awareness of the disorder, a growth in services and changing definitions.[48]

- March 2005 – a study of over 30,000 children born in one district of Yokohama concluded that the rate of autism in children doubled after Japan abandoned the use of the MMR vaccine in April 1993. The authors' conclusion was: 'The significance of this finding is that MMR vaccination is most unlikely to be a main cause of Autistic Spectrum Disorders, that it cannot explain the rise over time in the incidence of ASD and that withdrawal of MMR in countries where it is still being used cannot be expected to lead to a reduction in the incidence of ASD.'[49]

Despite the evidence that any risks associated with MMR were at worst very small, the dispute followed a familiar course. Airtime given to researchers who could offer scientifically-based rebuttal of Wakefield's claims was often matched by time given to parents of offspring with ASD to express their views, sincerely held no doubt but not necessarily based on firm evidence, that MMR caused their children's problems. Understandably the latter tended to be treated more sympathetically by interviewers and programme-makers.

The common-sense view, perhaps, is that if problems emerge soon after an MMR jab then that jab is to blame. The scientific position, however, might be different. If over 90 per cent of children received the MMR vaccine, as was the case pre-1998, then by sheer chance it would inevitably be the case that some youngsters who sadly develop symptoms of autism would begin to show those symptoms soon after receiving their jab.

Epidemiology, while a blunt tool in some ways, can at least offer some pointers as to whether children are more likely to develop autistic spectrum disorders soon after being vaccinated. Furthermore, if it were indeed the measles component of the MMR vaccine that was associated with autism as Wakefield claimed then it is not clear to the non-specialist why a separate measles jab should not cause precisely the same problems, or at least why a similar scare should not develop.

These arguments rarely entered the public debate. The affair followed the usual pattern. 'Official' sources of information were disbelieved. (A particular political aspect of the dispute came when the prime minister, Tony Blair, quite reasonably one might feel, refused to confirm or deny whether his youngest child, Leo, had received the MMR vaccine, despite offering strong support for it in public. The then Chancellor of the Exchequer, Gordon Brown, was to act in the same way in 2004.) If some of those observing the issue were unconsciously looking for ways of justifying their inherent anxiety (as suggested in chapter 11), this may have made them inclined to believe even entirely unsupported claims by people who 'look right' and have no apparent vested interest in the issue. Meanwhile the effects of falling vaccination rates were predictable. By 2003 the number of measles cases in the UK had tripled to over 400, including at least three deaths.

There were those who swam against the tide. In an admirable peak-time TV programme broadcast in August 2005, *Should I worry about ... jabs?*, journalist Richard Hammond reviewed the issue with respect to his own 18-month-old daughter. He was unable to find any scientific support for the supposed link between MMR and autism. However, one mother said: 'I don't feel happy about the MMR jab and nothing will persuade me otherwise', while a survey suggested that only 1 per cent of the population trusted the government on health matters.

Nevertheless, the scientific picture and the background of the perpetrators of the scare story were revealed over time. In February 2004 it emerged that at the time that Wakefield had published his report he was being paid £55,000 to help lawyers seeking evidence of a link between autism and the MMR vaccine. This had not been revealed to either *The Lancet* or Wakefield's co-researchers. In 2008 the results of a six-year University of Columbia study into the issue, under the leadership of renowned immunologist Ian Lipkin, showed 'beyond reasonable doubt' that there was no detectable link between MMR vaccination and autism.[50]

The Lancet subsequently said that it should never have published Wakefield's study because of his 'fatal conflict of interest' and some of his co-researchers also strongly criticised the lack of disclosure[51] – ten of the 13 authors of the original *Lancet* paper formally retracted the claim of having found a possible link between MMR and autism. By 2010 the General Medical Council was declaring Wakefield to have been irresponsible;[52] in 2011 the *British Medical Journal* said the claim was fraudulent.[53]

To its great credit, in 2010 *Private Eye* published an editorial about its own role under the pseudonym of its medical reporter, 'M.D.'.[54] It acknowledged the dangers literally to human life which underlie the apparent 'morality' of phase 2 thinking and concludes that science must be given more credence and conspiracy theories less. It noted some of the key features of the kind of ruses that underlie many Big Green scares.

- Greater weight is given to a small number (perhaps just one) of studies which suggest a correlation than to a large number which find no case to answer.
- Conspiratorial behaviour among those who fail to publish data in support of the scare is assumed while huge vested interests – financial and political – among those who are pressing the case are ignored.
- There are often some within the scientific community willing to jump on a bandwagon, whether in the hope of receiving research funding or for other motives.

Private Eye got it wrong in its coverage of MMR. It gave undue prominence to unproven theories based on a small number of uncontrolled observations and paid far less attention to the weight of evidence from large comparative studies that failed to find any association between MMR and autism. And while the *Eye* cited potential conflicts of interest in many of the key supporters of MMR, it failed to point out any conflict of interest on the part of Andrew Wakefield. The 1998 *Lancet* paper that started the scare has now been removed from the medical literature on ethical grounds and Wakefield, its leading author, may soon be removed from the medical register. Clearly what precious research money there is should now be used to test more credible hypotheses for the causes of autism …

Overall the *Eye* sided with parents who were suing the vaccine manufacturers and its coverage was consequently one-sided. The parents were supported not just by Wakefield but by 27 expert witnesses who submitted files to support their case and an equal number behind the scenes … When the *Eye* joined the controversy in 2001 there were plenty of specialists willing to support the possibility that Wakefield might be right … [However], small trials and observations can eventually lead to proof but only if their findings can be replicated in larger trials.

Sadly few of the other cheerleaders for the scare saw fit to apologise and explain how they would learn from the issue. The *Daily Mail* had produced such headlines as 'MMR killed my daughter'; 'MMR fears gain support'; 'New evidence shows MMR link to autism'; 'MMR safe? Baloney. This is one scandal that's getting worse'; 'Scientists fear MMR link to autism'; and 'Why I wouldn't give my baby the MMR jab'.[55] By 2013 it was pursuing the scare by another route when it proclaimed that the MMR boycott: 'has left two million children at risk of measles: doctors say outbreak in Wales could spread to London'. In 2013 the publication reported, poker-faced, that 'the scare was based on a study that has since been discredited',[56] without mentioning that the scare was also based on the sensationalist reporting that accompanied it. A similar Orwellian *1984*-style volte-face afflicted other newspapers such as the *Daily Express, The Sun* and the *Daily Telegraph*.

One unusual feature, however, was that the British government, despite considerable pressure and criticism, stuck to its original policy by refusing to offer single vaccines. This was not a stance without political risk. Although a poll by the company ICM showed that 73 per cent of parents believed the vaccine was safe,[57] objections continued, perhaps in part because of a perception that the government's stance was 'heavy-handed and paternalistic', a typical way that firm decision-making is viewed from the starting point of a rights-based ethic.

In due course confidence in the MMR vaccine returned despite a rump of conspiracy theorists trying to keep the scandal going. Nonetheless, it was not until 2012 that levels of inoculation returned to those which had been seen before Wakefield's paper, reaching a record 92.7 per cent in 2013, as illustrated in figure 10.5, though still below the World Health Organisation's recommended figure of at least 95 per cent.

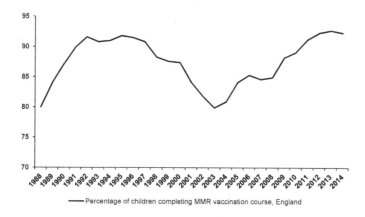

Figure 10.5 MMR vaccination uptake rates in England, 1988–2014[58]

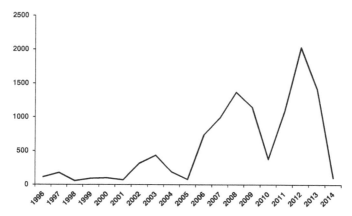

Figure 10.6 Confirmed cases of measles in England, 1996–2014[59]

The government's position was refreshingly responsible. To have surrendered to the calls for single vaccines in the face of evidence that MMR was, in all likelihood, 'safe' would have been to open the door to subsequent claims that the measles vaccine itself was dangerous while making continued use of MMR very difficult. Nonetheless, recovery from the episode will be slow, there being a lag between changes in vaccination rates and their effect becoming visible in the number of cases. Diagnosed measles cases, which stood below a hundred per year at the time of the scare, crossed the 2,000 mark in 2012 before falling back dramatically as immunity reasserted itself – see figure 10.6.

However much those purveying a warped sense of what is for the benefit of society may seek to portray themselves as on the side of the angels, the evidence is clear that the inevitable barrier, or at least interruption, that bad science creates for good decision-making can have serious consequences for some of the most vulnerable in society.

In the energy field the debates about low-level radiation and more recently about fracking have largely followed a similar pattern. The high initial costs and the consequent very long timescales involved in energy investment make such uncertainty particularly damaging to the attractiveness of new projects. For example, in 2014 the UK government's chief scientist, Mark Walport, commented in his Annual Report:

> History presents plenty of examples of innovation trajectories that later proved to be problematic – for instance involving asbestos, benzene, thalidomide, dioxins, lead in petrol, tobacco, many pesticides, mercury, chlorine and endocrine-disrupting compounds as well as CFCs, high-sulphur fuels and fossil fuels in general. In all these and many other cases, delayed recognition of adverse effects incurred not only serious environmental or health impacts but massive expense and reductions in competitiveness for firms and economies persisting in the wrong path.[60]

Disappointingly but perhaps inevitably some sections of the media chose to interpret this almost philosophical statement as: 'Fracking could be as damaging as thalidomide, tobacco and asbestos, government's chief scientific advisor warns in a new report'.[61] The title of the report, *Innovation: managing risk not avoiding it*, does make an appearance later in the article and gives a much more accurate impression of the point Walport was making, as does this passage:

> Innovation is essential for economic growth, health, well-being, security and resilience ... Many of the greatest periods of economic growth in the past have been driven by innovation. The need to innovate is a fundamental requirement for social and economic progress. Innovative economies are more competitive, respond better to change, see higher returns on investment and create increased living standards ... The innovators of the eighteenth and nineteenth centuries in Great Britain were motivated by and working against a backdrop of widespread poverty, high child-mortality and extremes of crime and squalor ... Innovation is not an unalloyed good – almost all innovations can cause both benefit and harm. Because of this, discussion of innovation has become almost inseparable from discussion of risk. Paradoxically, this discussion has become more prominent precisely because the innovations of previous generations have made our lives much safer and free of risk. People living in advanced economies have become more risk averse compared to previous generations. A common denominator of innovation in every generation is that it solves problems, creates wealth and new employment, while at the same time potentially disrupting the status quo of existing wealth and employment, and creating new problems and challenges.

None of this sensible and sensitive narrative made the pages of *The Independent*.

CERRIE on the cake

An example of attempted political interference in responsible science surrounded the short life of CERRIE, the Committee Examining Radiation Risks of Internal Emitters. This body was established in 2001 at the decree of the then environment minister, Michael Meacher. Despite his elevated position in the second rank of government Meacher had some strange ideas – just after he was sacked from the post in June 2003 and during the economic collapse of nuclear generating company British Energy he explained that because British Energy provided 23 per cent of the UK's 'electrical energy' over the previous year and the UK had a 'capacity margin' of 23 per cent then British Energy could be closed down without affecting power supplies.

Meacher, trained as a social administrator and a recognised expert on the treatment of elderly people in mental hospitals, was a long-time anti-nuclear campaigner who was barely constrained by the normal conventions of collective responsibility in government. In 1987, as opposition spokesman on health, he said children living near nuclear installations 'have literally grown up in killing fields'.[62] Meacher had succeeded in identifying two people with some way-out views on the dangers of radiation: Chris Busby (in many ways one of the 'pantomime dames' of the field, described by the one nuclear industry member of the committee as 'an aspiring politician who happens to have scientific qualifications' – he was the Green Party's spokesperson on science and technology and had stood for election to the European Parliament) and his friend Richard Bramhall, who together ran the Low Level Radiation Campaign (LLRC) which specialised in self-publishing Busby's own work to sidestep the inconvenience of getting it peer reviewed in the mainstream scientific press. Busby had produced a number of studies which created considerable local fear and concern, for example in north Wales, although all had been debunked by cancer research charities and other reputable commentators, notably the government's Committee on Medical Aspects of Radiation in the Environment (COMARE). Indeed, the LLRC claimed ownership for the idea of CERRIE and implied that it was set up by Meacher specifically to publicise their views.

> During the LLRC campaign to oppose transposition of the nuclear waste charter into UK law, LLRC developed a dialogue with the UK Environment Minister, Michael Meacher. Following meetings with Chris Busby, Richard Bramhall and Molly Scott Cato [another aspiring Green politician who was elected to the European Parliament in 2014], the Minister agreed to set up, jointly with the Department of Health, a new government committee whose remit would be to examine disagreements about the safety of the health risk models which presently underpin the regulation of discharges of man-made radiation to the environment.[63]

In classic 'phase 2' fashion CERRIE was set up with a blatant political aim – to endorse the views of those who would furnish support for the political views of the minister in charge

– rather than to determine the truth. It was chaired by widely respected scientist Dudley Goodhead. As well as Busby and Bramhall it also included members from Greenpeace (Pete Roche) and Friends of the Earth (Philip Day), three academics (Professor Jack Simmons, Professor Eric Wright and Professor Sarah Darby), three participants from the government regulator, then the National Radiological Protection Board, NRPB (Dr Colin Muirhead, Dr John Harrison and Dr Roger Cox) and Dr Richard Wakeford from British Nuclear Fuels plc (BNFL). The secretariat of three included two noted anti-nuclear campaigners, Ian Fairlie and Paul Dorfman. This was clearly not a committee created to err on the side of underestimating the effects of radiation.

As was perhaps unsurprising CERRIE produced a report in 2003 which was more pessimistic about the health effects of radiation than scientific consensus suggested. However, one unexpected bonus of the process was that it did at last expose Busby's work to peer review, under which scrutiny it effectively fell apart. Eventually Busby and Bramhall were reduced to producing their own 'minority' report in an attempt to salvage some of their reputation. All other members of the committee, including the anti-nuclear campaigners, signed up to the main report, while expressing areas of disagreement, and in a letter to *The Guardian*[64] made clear their full confidence in the report and in Goodhead's chairmanship of the committee.

Busby and Bramhall's vituperation was largely directed at the other anti-nuclear members of CERRIE, notably Fairlie, whom they bizarrely painted as being in the pocket of the supposed vested interests who were applying simple arithmetic to their work and finding it literally didn't add up. Indeed, there is often a puerile circularity in the attacks of Big Green and its fellow-travellers on those who heretically challenge them. In a variation on the 'no true Scotsman' fallacy, anyone who disagrees with Big Green ex cathedra pronouncements by definition is, in Busby's charming words, 'BIW – bastard, idiot or wimp'. (Characteristically Busby refers to his 'BIW theory' as if it were the outcome of a perceptive piece of research.) It follows then that their arguments can be dismissed as worthless.

As Wakeford notes: 'Interestingly, the row which surrounded this disagreement served to divert attention away from a report which on first reading seemed to offer some support to the views of the moderate anti-nuclear movement.'

Meacher, by then out of office, wrote a supportive foreword to the minority report, clearly betraying his disappointment that the mainstream anti-nuclear members and the neutral academics had, for reasons he did not elaborate on, chosen to freeze out the view that he knew, in his heart of hearts, was the 'right' one.

> Unfortunately it seems that the procedures which prevailed in the Committee, while they have allowed discussion of a wide range of topics, have produced a Final Report which does not accommodate a full and fair representation of all views. More seriously, from the point of view of taking this debate forward, the Report fails to explain the reasons for the continuing disagreements. This applies, in some cases,

to what look like quite basic issues. Take, for example, the question of whether there was or was not a significant increase in infant leukaemia across Europe after the Chernobyl disaster. Why does the Final Report present only one side? This is very worrying for it is hard to conjecture that, if the leukaemia peak was real, anything other than the radiation from Chernobyl could have caused it.

(Actually the main report devoted seven pages to considering the peer-reviewed literature concerning this matter.)

Note the language – 'all views' have to be represented even if they are so clearly refuted by the evidence that leading anti-nuclear campaigners distance themselves from them. Any report has to provide 'both sides' even if one side has been completely debunked. The murderer and the victim must get an equal hearing. (Oddly Meacher did not seem to have the same tolerance for those who argued against the orthodox view of climate change.) Further, note that for this politician, and perhaps for others, correlation is sufficient proof of causation – if an increase could have been found then only Chernobyl would need be considered as its cause and the goal would have been achieved.

10.10 DIFFERENT COUNTRIES, DIFFERENT POINTS ON THE JOURNEY

This book has focused on the UK and to a lesser extent other developed countries, where generally speaking the economies tend to be growing at a modest rate (a few per cent per year at most and vulnerable to periodic recessions during which growth evaporates) and energy demand more slowly still owing to ongoing improvements in energy efficiency (or reductions in energy intensity). It may even fall as heavy manufacturing industry is exported to areas with less rigorous regulation. From an energy standpoint the focus in such countries tends to be on replacement for existing plants. Looking more widely, when such countries are going through a 'comfortable phase' there is often a resistance to new technologies – if life is going 'well enough' then why take the (perceived) risks associated with mobile phone masts, low-level radiation or even vaccination? Even among neighbouring countries there are 'national characteristics' which appear to affect the result; for example, France was remaining relatively centralised and utilitarian in its approach while the UK was becoming more fragmented and rights-based. The French government showed great enthusiasm for the next stage of nuclear fusion research (ITER, the International Thermonuclear Experimental Reactor) through the 1990s and early 2000s despite its high costs (estimated at over €15 billion) and long timescale. By hosting the project France would reap some local economic benefits, though at a considerable cost to its exchequer, but the widespread public support seemed to be based on more than this, including elements of genuine excitement over the technology and a feeling that France should be at the forefront of technological development.[65] (That said, French environmental groups, notably Mediane, opposed the project, quoting 2002 Nobel Prize winner Matatoshi Koshiba who claimed that ITER did not meet 'a certain number of conditions, namely safety and economic costs' and invoked the precautionary principle to argue in effect that we should not start to carry out research into ITER safety until we know the results of that research.)

It is of course a mistake to view the developing and newly industrialised world as a homogeneous group of nations. One can find at least as many differences within this group as one can in the developed world. Nonetheless, one can speculate that certain characteristics of many developing or newly industrialised economies have systematic effects on the relationships between science/technology, politics and the people.

- The imperative for economic growth tends to result in most elements of society viewing industrialisation, science and technology as major benefits to their way of life.
- Because there are more severe challenges in everyday life, people's need to feel anxious, even if it is as great as that of people in developed countries, is more likely to be fulfilled by matters such as hunger, poverty, the need for education and health care, and so on.
- Levels of literacy may be lower in some cases than in developed countries and the penetration of the Internet and social media less widespread (although the mobile phone is becoming so), with the result that members of the public are less able to follow and participate in major debates of the day.
- The opportunities for well-paid employment associated with many scientific and industrial activities tend to be more valued in poorer countries. In the developed world many countries have seen a decline in their manufacturing sector (to the benefit of service industries) while much heavy manufacturing has been relocated to less affluent countries.
- The desire for many developing and newly industrialised nations to demonstrate that they are as advanced as developed countries has led to a similar attitude towards science and technology, especially at the high-tech end, as was seen in what are now 'developed' countries in the post-war years.

As a result it may be possible to regard the broad attitude of governments and people in some developing countries in the 2000s and 2010s as similar to that which pertained in what is now the developed world in the post-war period – in effect 'phase 1' societies. The investment in nuclear energy in China, India and South Korea, for example, is reminiscent of the early support given to the industry in the US and western Europe. It would seem possible, then, that politicians in these countries, as their economies develop, may find themselves facing similar pressures to those facing their counterparts in the developed world from the 1960s to 2000s.

Japan is a particularly interesting example. Of course it is a huge economy – the third largest in the world – yet in some ways until recently it has perhaps been more akin to a 'phase 1' society in that levels of deference appear very deeply ingrained in the Japanese culture and customs. Despite several cases of incompetence and dishonesty among industry and government, notably the Tokyo Electric Power Company (TEPCO) in the 2000s, the Japanese people in general do not seem to have developed the same degree of scepticism about the 'establishment' as was seen among their European and North American counterparts. It is a matter of interesting speculation as to whether the Fukushima accident will push Japanese society into a more questioning, less accepting mode. If so, how quickly and effectively the authorities accept such a change and enter into a free and mature relationship with critics may determine how long Japan's 'phase 2' lasts.

10.11 TIMESCALES

Even if this analysis has some truth behind it, does it really matter? Society could sail along at times of prosperity with politicians keeping a steady hand on the tiller, tinkering with perceived problems but making no dramatic gestures of the kind seen in 1945 or 1979. If and when major threats emerge to this complacency, either incumbent politicians or (more likely) a new radical generation can be elected to exercise 'strong leadership' and take the country in a new direction.

It is here, however, that the mismatch between the timescales on which the political and investment cycles work becomes particularly serious. A new nuclear power station, gas pipeline or LNG import facility, large wind farm or tidal barrage project cannot be started and completed overnight or even before the next election – it is a long and often tortuous process to get the necessary planning permissions, arrange finance, create a supply chain, build the plant in question and get it working. For example, the utilities funding Olkiluoto-3 in Finland made a final decision to apply to build the plant in 1998, after some years of feasibility and financial studies. The application was lodged in 2000, parliamentary approval was granted via a decision-in-principle in 2002 by a vote of 107 to 92, construction began in 2004 and commercial operation was still a distant dream by the middle of the 2010s.

It is possible to identify at least four sets of timescales, in which display considerable mutual tension.

Crises within the energy field can develop rapidly – and disappear just as quickly. Within little more than a year of the Great East Japan earthquake and tsunami in 2011 and subsequent Fukushima nuclear accident Japan had taken out of service all of its nuclear power stations, which had been providing 30 per cent of the country's electricity. Droughts in countries and areas dependent on hydropower have led to crises in countries such as China, Brazil, New Zealand and Norway. Trade union activity can interrupt supplies of coal. The price of oil has on a number of occasions (1973, 1979, mid-2000s) increased very rapidly, on others falling just as quickly – for example, halving from over $110 per barrel to below $50 per barrel in six months from the middle of 2014. Power shortages in California in 2000/1 appeared quite suddenly; the blockade of transportation fuel refineries caused by protests in much of Europe in September 2000 took barely a week to bring the economy to a virtual standstill; frequent disputes between Russia and Ukraine have led to gas pipelines being shut down, for example in mid-2014 as part of the ongoing dispute over Russia annexing Ukraine-held Crimea.

The political cycle is relatively short – typically four or five years between general elections, of which in the modern political world at least two years can be regarded as the 'election campaign'. There are other political cycles, e.g. the US 'midterms' and in the UK those involving local government, which tend not to be synchronised with presidential or parliamentary elections but which can have profound implications for implementing national policy in a particular location. Even when one party or coalition seems well ensconced, a change of power is always possible or a new party may emerge. All politicians must always have at least one eye on the electoral implications of their policies.

The investment cycle might be of the order of a decade or two. Its length is to a considerable degree dependent on the particular market structure. In competitive power markets investors generally prefer a 'quick' return in order to minimise economic risk – which inevitably tends to increase as timescales lengthen.

Finally, the timescales involved in making major changes to the energy industry can be much longer, frequently being measured in decades at least. There is often a long period between the emergence of a new concept or technology and its availability for commercial exploitation. In some industrialised countries a decision taken today to build a nuclear power station might be followed by up to a decade of planning and regulatory activity, a construction phase lasting five or six years and sixty years of operation. The installation of a new gas pipeline or major hydropower or tidal facility might have similar time horizons.

One common outcome of a failure to recognise these different timescales and their implications is confusion between short-term cyclical crises and long-term problems requiring long-term solutions. In the early 2000s, for example, a combination of a major increase in the oil price by the standards of the late 1990s (though not by those of a decade later), petrol protests which brought a number of European countries to a virtual standstill within a week in September 2000 and significant power cuts in California and Victoria in Australia led to the UK government launching a major energy review immediately after the 2001 general election. Concerns about the implications of the attacks of 11 September 2001 added to a sense of foreboding. However, by the time the report was ready the short-term crisis had resolved itself – in this case the oil price had fallen, attempts to repeat the petrol blockades had failed and secure supplies had returned in California. The report which emerges from such exercises, then, tends to find that there is no need for panic or decisive action because the problem has gone away.

For decades loud cries of 'wolf' have emanated from many in the Big Green movement about the imminent depletion of this or that resource ('peak oil' being a particularly risible example); the imminent collapse of civilisation and widespread starvation such as that predicted by Paul Ehrlich in the 1970s; the imminent bankruptcy of large utilities trailed by Amory Lovins in the 1980s; and the imminent rapid heating of the atmosphere foretold by James Hansen in the late 1980s. Such vested, or at least overpessimistic, interests tend to exaggerate the implications in their call for immediate and radical action and hence ultimately play into the hands of the 'do-nothing' (or 'nothing need be done') tendency. When no wolf makes itself visible Big Green simply shifts its doom date forward by a couple of decades, without seemingly recognising the damage such intellectual laxity does to the case for taking action against what may prove to be a genuine long-term threat. For just as it is a mistake to interpret the onset of a short-term problem as the 'beginning of the end', so it is wrong to believe that a short-term easing of the situation means the long-term issues have been resolved.

The conclusion is difficult to escape – politicians must find a way of pursuing policies that will head off the long-term problems, because by the time these problems are established to everyone's satisfaction it may be too late to do anything very effective about them. This could well mean they have to act as 'strong leaders' and force through controversial projects at a time when society, still feeling comfortable and not perceiving the precipice they are approaching, is not inclined to accept such leadership.

To take nuclear power as an example, the age profile of worldwide nuclear power capacity as illustrated in figure 10.7 is a striking one – over 75 per cent of global nuclear capacity was 25 years old or more by 2015. If the average life of a nuclear station is taken to be 50 years – perhaps optimistic for the older plants but possible (or maybe even pessimistic) for newer ones – then by 2030 some 21 per cent of the 2015 capacity (79,000 MW) will come offline, followed in 2040 by a further 52 per cent (198,000 MW). Even if all projects described as 'under construction' as of June 2015 were to be completed this would add only a further 65,500 MW, giving a total of around 168,000 MW in

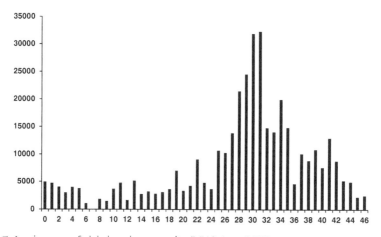

Figure 10.7 Age in years of global nuclear capacity (MW), June 2015

operation in 2040 (assuming some life extension). This would be a mere 44 per cent of installed nuclear capacity as of mid-2015.

The 'lost' 56 per cent of this capacity will have to be replaced. To replace it with more nuclear capacity will be a serious challenge because the period between ordering a new plant and when it starts to generate electricity tends now to be at least decade and possibly much more, at least in the developed world. The world achieved rapid nuclear growth in the 1970s and 1980s but this was despite phase 2 thinking, which delayed the completion of many plants that were ordered a decade or more earlier, not because of it. China was maintaining the plant-building tradition in the mid-2010s but again not with phase 2 thinking.

10.12 Towards a phase 3?

There seems to have been an underlying assumption in much discussion about new methods of encouraging lay 'participation' in decision-making that the move towards a rights-based ethic represents an irreversible change in society. Certainly it would appear that one factor which has contributed to increased participation at a local or single-issue level, the growth of communication through the Internet and modern social media networks, represents a permanent transformation. Political paralysis in the face of instant criticism for saying anything unpopular would seem to be embedded. Perhaps the breakdown of the extended family, also unlikely to be reversed given say the changing nature of employment, and the rising cost of housing in cities forcing children away from the family home is another example, as in many cultures the extended family is important in instilling appropriate values of respect and deference towards the elder members of that family.

This is not necessarily to say, however, that society at large has forever abandoned the utilitarian ethic in favour of individual rights. It is perfectly feasible that new threats to the stability of society – for example, prolonged power outages, terrorist attacks or severe environmental damage – may lead to a reversal of the trend and a return to calls for 'strong leadership', though one suspects that the element of almost 'blind faith' (or at least short-sighted acceptance) may not return as before.

It was noticeable, for example, that the 2004 presidential election in the US drew the largest voter turnout since 1968 (see figure 10.8) and the return of a candidate, George W. Bush, who was perceived by many Americans to have taken a strong stand against global terrorism after 9/11. The turnout rate continued to increase in 2008 and 2012 to levels rather higher than those of much of the interwar period but not touching those of the late nineteenth century. (The low turnout of 52.6 per cent in the 1932 election, in the depths of the Great Depression, may be a counterexample to the thesis that voting rates go up in times of hardship though turnout did rise to almost 60 per cent over the next two elections as war clouds began to gather.)

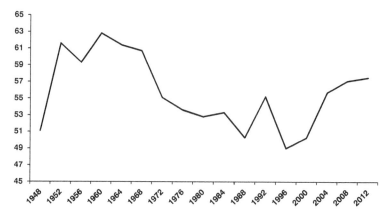

Figure 10.8 Percentage turnout in US presidential elections

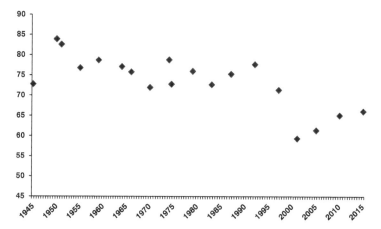

Figure 10.9 Percentage turnout in UK parliamentary elections

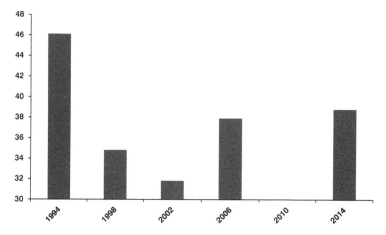

Figure 10.10 Percentage turnout in London Borough elections – excluding 2010

A similar effect can be seen in the UK. Turnout reached a post-war peak in the elections of 1950 and 1951 of well over 80 per cent, hovered about the 75 per cent mark between 1955 and 1992, and then fell back to below 60 per cent in 2001. By 2015 it had rebounded to over 66 per cent – see figure 10.9. The same was seen in elections for the London Boroughs (ignoring 2010 which was distorted by being held on the same day as a general election) – see figure 10.10.

Once again it is going far too far to put the increase in turnout purely, or even in major part, down to a growing sense of unease in the UK population over the threat of terrorism and then the economic recession. One can argue, for example, that the result of the 2001 election, after the Labour landslide of 1997, was so much of a foregone conclusion that many electors did not feel inspired to vote, whereas by 2010 the outcome looked much closer (and indeed resulted in a 'hung' parliament in which no party had an overall majority for the first time since February 1974, when the turnout was the third highest since the Second World War) while 2015 confounded most predictions by delivering an overall Conservative majority for the first time since 1992. Nor can one claim that President Bush or Prime Minister Brown entirely escaped the attention of the satirists – quite the opposite. But the sense that things needed to be done to address the economic crisis, and in the UK the re-emergence of some policy differences between the two major parties, may well have been factors behind the increasing turnout.

It is too cynical to accuse all politicians of being solely interested in winning the next election, even during phase 2. Alongside a genuine desire to do good, many leaders as their term of office proceeds turn their thoughts to their legacy. It seems highly unlikely that any politician would like to be remembered as someone who 'fiddled while Rome burned' with regard, for example, to failing to update the electricity infrastructure. In the past politicians have been particularly keen to be associated with exciting, large-scale technological developments which may have very long lifetimes indeed – air travel, the Channel Tunnel, nuclear energy, etc. – all of which required considerable government support of one kind or another to get started and whose main implications have only been seen long after the terms of office of their early political supporters. While it undoubtedly became more difficult to maintain a long-term strategic focus when society was going through a complacent phase, things might start to look different if phase 3 attitudes of 'constructive scepticism' take root and people start to accept once again that difficult decisions are necessary. By the middle years of the 2000s the energy question had once again started to look challenging as it had in the 1970s. In his foreword to a White Paper on nuclear energy in January 2008, the prime minister, Gordon Brown, said: 'More than ever before, nuclear power has a key role to play as part of the UK's energy mix. I am confident that nuclear power can and will make a real contribution to meeting our commitments to limit damaging climate change.' Admittedly it took nearly eight years for a plausible funding mechanism to emerge for Hinkley Point C, involving a considerable contribution from the Chinese state, despite a huge amount of activity 'behind the scenes'. However, at least the rhetoric had been couched in terms of a quasi-decision and despite predictable Big Green bleating there was little public objection, even after the Fukushima accident.

Indeed, the general aura of infallibility which surrounded Big Green in the 2000s showed signs of dulling by the mid-2010s as more people began to recognise that they too had an agenda and a vested interest in at best unbalanced 'information' and at worse simple dishonesty. It became perfectly respectable to criticise the more extreme outposts of environmentalism where before this would have been regarded as heretical. The sense of 'the ends justifying the means' – that Big Green 'owned' the environment and had a right to damage it or promote damage to it if and when it saw fit – started to fray round the edges. In 2014 twenty Greenpeace activists from seven countries, mainly from Germany (where Big Green was already under fire for promoting closure of nuclear plants, inevitably to the benefit of brown coal), chose to trespass on the sacred Nazca Lines in Peru in order to install a lurid yellow advertisement for renewable energy. The location, home of large representations of animals such as hummingbirds and monkeys, covers some 450 km², the lines etched into the fragile environment some 1,500 years ago still being deeply revered by the Peruvian people. The United Nations Educational, Scientific and Cultural Organisation (UNESCO) has referred to the treasure as 'one of the most impenetrable enigmas of archaeology'. Visitors, from ministers and presidents down (or up), need to have special permission to visit the site and to wear special footwear to mitigate the risk to the fragile environment. Greenpeace seems to have presumed that none of these restrictions would apply to them owing to the importance of their message. The invaders' footprints can now clearly be seen by the ancient lines or geoglyphs. The deputy culture minister, Luis Jaime Castillo, said: 'It's a true slap in the face at everything Peruvians consider sacred.'

The initial response from Greenpeace was: 'The surprise to us was that this resulted in some kind of moral offence.'[66] Presumably it also came as a surprise to them that this pronouncement did little to stem the growing tide of international condemnation. Castillo said: 'This has been done without any respect for our laws. It was done in the middle of the night. They went ahead and stepped on our hummingbird and looking at the pictures we can see there's very severe damage.' Greenpeace retorted that its activists had 'taken every care we could to try and avoid damage' and that their actions demonstrated the organisation's 'respect for people around the world and their diverse cultural legacies.'

Despite its obvious stunned amazement that anyone would dare criticise their impeccable behaviour, Greenpeace sought to minimise the public relations debacle by issuing a 'non-apology apology' which would have graced any politician wriggling to escape from a scandal without actually admitting to wrongdoing. It started with a justification and went on to express regret for any offence caused but, very carefully, not for the actions which caused it. 'The peaceful protest by Greenpeace in the area of the

Nazca lines was to demonstrate the impacts of climate change ... Without reservation Greenpeace apologises to the people of Peru for the offence caused ... We are deeply sorry for this.' And, masterfully, 'We fully understand that this looks bad ... we came across as careless and crass.' In other words the real problem here was the way Greenpeace looked and the way it came across. Finally, Greenpeace said it would 'cooperate fully with any investigation' – though not, it turned out, if it meant letting the Peruvian authorities have the names of the vandals so that appropriate action could be considered, still less anyone being sacked or resigning in order to take responsibility.[67] It took eight days for Greenpeace to produce a video accepting responsibility, though still seeming to imply that the widely circulated photographs of lasting damage and of activists in ordinary training shoes installing the advertisement were no evidence of harm or disrespect. Greenpeace's whole response seemed modelled on that of TEPCO after Fukushima – keep it quiet, play it down – with a similar outcome in terms of public anger and disillusionment.

In essence this stunt showed no less respect for truth and the environment than had the campaign against Brent Spar two decades earlier, though of course there the target was a big multinational, not a relatively poor and defenceless ancient community. However, by the mid-2010s Greenpeace could no longer get away with such arrogance. Several tens of thousands of critical tweets and Facebook postings lasted for well over a week. Self-righteousness could not protect Big Green any more.

While it is too early to be confident that a 'third phase' is becoming established, by the mid-2010s the decline in the role of and respect for science, or at least the adulation for those for whom science is only valued when it gives results commensurate with their personal dogma, may have been slowing or perhaps even reversing.

The synthesis which may emerge from the dialectic between phases 1 and 2 might look something like this. It becomes accepted that the appropriate use of scientific research will improve the quality of decision-making within areas involving an interface between technology and society. This is not to say that science will always deliver the 'right' answer nor that scientists should be given carte blanche in making the final decision. Elected politicians (and the legal system) rightly bring wider social issues to the table, not least by offering appropriate protection of the rights of individuals. One can envisage an iterative process whereby society, through the political process supported by research in social science, oversees scientific developments and communicates society's requirements while the scientific/technical community creates innovative processes and 'reports back' to the political establishment as to what is technically and economically feasible.[68] One manifestation of phase 3, it can be argued, is the growing split between evidence-based environmentalism and the dogma of Big Green, illustrated particularly in the debate on nuclear power. As early as 2001 US commentators were reflecting on the apparent conversion to the nuclear cause among prominent American environmentalists as John Kerry (unsuccessful Democratic presidential candidate in 2004 and subsequently Secretary of State in the second Obama administration from

2013) and President Clinton's environmental advisor Jerry Mahlman.[69] Over the next decade or so many environmentalists reappraised the potential role of nuclear power and publicly changed their position. Perhaps the most notable was James Hansen, credited with establishing climate change on the international political agenda in 1988. In 2013 he and three fellow climate and energy scientists, Ken Caldeira, Kerry Emanuel and Tom Wigley, released an open letter saying: 'With the planet warming and carbon dioxide emissions rising faster than ever, we cannot afford to turn away from any technology that has the potential to displace a large fraction of our carbon emissions. The time has come for a fresh approach to nuclear power in the twenty-first century.'[70] Other prominent pro-nuclear environmentalists in the US and Canada included Patrick Moore, a founder of Greenpeace in 1971, and Michael Shellenberger (president of environmental think tank the Breakthrough Institute).

By the mid-2010s the list in the UK included Stephen Tindale (former executive director of Greenpeace UK), Chris Goodall (a former Green Party parliamentary candidate) and (Lady) Bryony Worthington (former Friends of the Earth climate campaigner). Climate campaigner Mark Lynas, who announced his change of stance towards nuclear power in 2012, illuminated the way of thinking of Big Green when he wrote (in respect of his previous stances against GMO food which he also abandoned): 'In 2008 I was still penning screeds in *The Guardian* attacking the science of GM – even though I had done no academic research on the topic and had a pretty limited personal understanding. I don't think I'd ever read a peer-reviewed paper on biotechnology or plant science.' He was later to tweet: 'When I started climate campaigning 15 years ago I expected Exxon to be the main opponent. Turns out it's the Greens.' In March 2011 *Guardian* journalist and activist George Monbiot said: 'As a result of the disaster at Fukushima, I am no longer nuclear-neutral. I now support the technology.'[71] The following month he added: 'Over the last fortnight I've made a deeply troubling discovery. The anti-nuclear movement to which I once belonged has misled the world about the impacts of radiation on human health. The claims we have made are ungrounded in science, unsupportable when challenged, and wildly wrong. We have done other people, and ourselves, a terrible disservice.'

As Kuhn would predict, phase 2 thinking will kick back, as does any outdated ideology as the point of its displacement approaches. However, such conversions present a particular difficulty for Big Green. It is hard, though by no means impossible, for it to attack the personal integrity of individuals who have taken part in environmental campaigns against Big Business for many years. Further, it often appeals to the authority of these highly qualified environmentalists when they are saying the 'right' thing; more difficult while simultaneously seeking to deprive them of credibility when they are 'getting it wrong'. Fortunately Nietzsche comes to the rescue. Truth doesn't really matter, it is what is useful that is to be celebrated, so Big Green simply cites these individuals as authoritative sources when their message is palatable and denounces them as unreliable when it is not. Anti-nuclear propaganda group No2NuclearPower has even admiringly quoted Lord (Nigel) Lawson, former Conservative Chancellor of the Exchequer and a leading light in the climate sceptic movement, challenging the costs of the Hinkley Point C nuclear deal:[72] Lord Lawson is not usually a recipient of such praise from the Big Green quarter. The yardstick as to whether someone is a credible authority is simply whether they agree with Big Green on any particular issue.

This stance would look increasingly silly if there were indeed a developing 'phase 3' which viewed Big Green itself with scepticism and started to put more weight on properly conducted and 'unbiased' science, insofar as such a thing is possible, even when it produces the 'wrong' answer from the point of view of Big Green prejudice.

10.13 Is politics up to the challenge? Recreating a proper role for science

Properly used, science has a vital role to play in decision-making. The scientific method, while not offering sure and certain knowledge, especially when dealing with relatively uncommon potential health threats, is likely to provide advice which is closer to reality, and therefore more useful, than that emerging from religion, gossip, ideology or celebrity singers, sportspeople or chefs. This is not to denigrate the importance of non-scientific sources of 'knowledge' when it comes to developing our personal values and sense of right and wrong. Nevertheless, these sources are less reliable than the output of properly conducted science when it comes to interpreting the behaviour of the physical universe.

In this vital sense the analogy between science and religion has its limitations. As long as a suitable attitude is taken to uncertainty this must lead to better decision-making. Even Hume, who takes the extreme sceptical position that something having always happened in the past cannot be taken as proof that it will always happen in the future, argues that only a fool would live life on that basis.

> Should it be asked me whether ... I be really one of those sceptics who hold that everything is uncertain, I should reply that neither I nor any other person was ever sincerely and constantly of that opinion. I dine, I play backgammon, I converse and am merry with my friends and when after three or four hours of amusement I would return to these speculations, they appear so cold and strange and ridiculous that I cannot find in my heart to enter into them any further. Thus the sceptic still continues to reason and believe though he asserts he cannot defend his reason by reason.[73]

Such reflections are particularly important when decisions taken or ducked today will have implications long, long after the end of the term of office of the politicians taking them (or indeed the voters who elect them where that is the national custom). Even when politicians do have an eye on the long term it is not always clear how that perspective can be integrated with the short-term strains of the electoral timetable. The financing of much energy infrastructure – renewables, oil and gas pipelines, nuclear energy and so on – is highly vulnerable to any suspicion that politicians in the future may change the rules in a capricious and unpredictable way. As discussed in chapter 8, the initial investment costs of nuclear energy and renewables represent a higher proportion of total costs than is the case with CCGT. A prospect of making a fair rate of return on the project therefore requires a stable business environment for a rather longer period of time. If, say, it is perceived that a change in political control might bring with it more stringent regulations (or even a formal phase-out policy), or a major change in the subsidy or planning regime, then the economic risk associated with investment in low-carbon electricity generation may become very high and possibly unmanageable.

The challenge for politicians and the 'consumers' of their decisions, then, is twofold. First, how to reintegrate science into decision-making without making the mistakes of the past in which some scientists were given almost a free hand over policy development. This would entail creating institutions and incentives to build interdisciplinary scientific expertise, with a view to establishing reliable bases of 'facts', explaining different standpoints and outlining the advantages and disadvantages of various courses of action. Secondly, how to take strong and possibly unpopular decisions before the impending crises of, say, energy shortages and climate change become unmanageable (quite possibly after the politician in question has left office, thereby risking taking the short-term pain while receiving little of the long-term gain). Unless these challenges are overcome complex technologies such as nuclear energy, even if they have a useful potential role to play, are likely to be excluded on irrelevant or erroneous grounds. (This is not, of course, to argue that there are no justifiable grounds for opposing any particular technology.)

There are certainly political risks associated with taking firm action over controversial issues, especially when society is not yet ready to acknowledge the need for such action. But so too are there political risks in ducking difficult questions – the risk of being seen to be weak during or after one's term of office if and when society shifts back towards a more utilitarian ethic and starts demanding that 'something should be done' or 'something ought to have been done by now'. In this position politics once again becomes not merely a matter of getting through the next election with the minimum of fuss and controversy; some decisions are inevitably longer-term and the success or otherwise of politicians in dealing with them will cast long shadows.

What political support would be required to promote long-term thinking in energy and electricity? Part of the commitment would be a firm set of statements in favour of nuclear energy or renewables.

In the early to mid-2010s this support was forthcoming in both cases. All three major parties in Parliament supported nuclear new build and measures to support renewables, although there was more variation among the minor parties, with UKIP being pro-nuclear and anti-renewables while the SNP was the opposite, for example. While it may be broadly the case that the left of politics prefers renewables to nuclear while the right prefers nuclear to renewables this is not a hard and fast rule by any means – many of the UK's nuclear power stations, for example, were ordered by Labour governments, including most of the AGR programme.

The UK experience would seem to suggest, albeit very tentatively, that support for nuclear power tends to grow at times of societal stress and decline when society becomes more comfortable, while the opposite may be true of renewables. The early 2010s saw a coalescing of the political establishment around support for nuclear new build, the Liberal Democrats abandoning their in-principle opposition to nuclear power in 2012, while the right wing of the Conservative Party and UKIP, which emerged spectacularly through the European Parliament elections in 2014 followed by by-election successes in previously Conservative parliamentary seats in Westminster, began seriously to challenge the decade-old consensus around the desirability of renewables, especially onshore wind. A 2014 poll financed by the wind power industry suggested that some 78 per cent of Conservative MPs opposed development of onshore wind, the figure among Labour and Liberal Democrat MPs being 20 per cent and 9 per cent respectively, against only 45 per cent of the polled public in rural areas who would oppose wind farms in their area.[74] (That said, the weighted average opinion of MPs, at 47 per cent opposition, was remarkably close to the rural public figure.) However, there was little evidence of growing opposition to the concept of renewables among the public at this time.

(A caveat should be entered here. Energy still tends to be something of a back-of-one's-mind issue for many people in the UK, though this was less the case in the period of relatively high power bills in mid 2010s than it had been a decade previously. As such the interpretation of opinion polls must be undertaken carefully as the public response is particularly influenced by how the question is asked and what has been in the news recently. In the 1980s the nuclear industry would tend to ask questions such as 'Do you want to freeze to death next winter' and a 'no' response was recorded as support for nuclear power. The Green movement would prefer questions like 'Do you want to see your husband/wife/partner and children blown into atomic dust', which elicited a typical 'no' response of about 74 per cent and was duly reported as opposition to nuclear power.)

Perhaps the public face of the two technologies is different – nuclear power generally having been portrayed as a hard-headed practical solution to real problems of supply security, while renewables have been sold more in terms of a 'soft', maybe even Romantic, image, as shown by the repeated use of the term 'natural'. If so there may be a 'natural' tendency for those in utilitarian mode to support nuclear power (perhaps as the 'lesser of evils' rather than as something positive in its own right) and be sceptical as to the efficiency of renewables, while those who are more swayed by a consideration of rights might be emotionally attracted to renewables and viscerally opposed to nuclear power. (Once again these are very broad statements and do not imply that there is neither an emotional case in favour of nuclear power nor a data-driven argument for renewables.)

What is interesting is that for a surprising number of people it appears that their order of preference for fuel sources puts fossil fuels in the middle – those who feel instinctively attracted to nuclear power at least sometimes prefer fossil fuels to renewables while Big Green has demonstrated its liking for coal and gas over nuclear power in its response to the German Energiewende and the Japanese nuclear shutdown post-Fukushima.

A clear statement of support, especially if it carries a degree of cross-party consensus, is likely to have beneficial effects on the market's perception of the economic risks associated with funding new build, with corresponding reductions in the cost of capital (i.e. required rate of return) – up to a point.

However, political consensus cannot be banked. One of the features of a rights-based societal ethic is that consensus becomes highly prized in its own right – that way fewer individuals are upset even if the policy is ineffectual. After all, the background that creates the rights-based ethic is one where few major 'decisions' seem necessary: everything is apparently going fine. Centre parties do relatively well in such an atmosphere by arguing in effect that whether the Left or the Right emerges as the largest grouping, they will dilute their policies and drag politics to the centre. The UK Liberal Democrat pitch in the 2015 election – 'Labour can't be trusted with your money and the Tories can't be trusted to build a fair society' (or perhaps 'whichever you elect we'll dilute') – was clearly of this nature. It was, however, singularly unsuccessful.

Barely a month after that election Jonathan Reynolds, the Labour Party's shadow climate change minister, wrote to the new Secretary of State saying: 'I am asking you today to admit the [Hinkley Point C nuclear new build] project will not proceed and inform parliament what your alternative energy strategy will be.'[75] The election of an avowedly anti-nuclear Labour leader, Jeremy Corbyn, a few months later emphasises the point.

Another feature of consensus is that it becomes very difficult to challenge and the accepted wisdom becomes immune from proper scrutiny. There are political parallels to Kuhn's scientific paradigm shifts – a few mavericks challenge the consensus and are at first largely ridiculed or ignored until the inherent problems with the status quo become too great to be ignored, at which point something akin to a revolution of thought occurs. The breaking of the political consensus of the 1970s by the march of the neoliberalism of the 1980s, notably in the UK and US, is an obvious example, as is the post-war socialism of the Attlee government in Britain.

The (over)reaction of the Merkel government in Germany in the wake of the Fukushima accident is a case in point. Within a matter of weeks profitable nuclear power plants with years of good efficient operation in front of them were either closed down or given a maximum of eleven years on death row. Although the companies involved sued the German federal and state governments for several billions of euros, clearly the fear that such capricious responses could undermine future investment in the same way acts to deter investors from coming forward or leads them to demand a rather higher rate of return on that investment to compensate for such a risk (which may come to the same thing in policy terms if nuclear becomes priced out of the market). Ironically Germany also offers an example with regard to renewables. The defeat of the SPD in the North Rhine–Westphalia elections in May 2005 was accompanied by falls in the share price of companies involved in wind power which feared an early end to their generous subsidies should the right-of-centre CDU replace the SPD/Green coalition in a general election. (In the event, in the subsequent federal election the CDU bloc did not win an overall majority and ended up leading a 'grand coalition' with the SPD.)

Perhaps nuclear power could flourish in a stable market where such uncertainties were minimal or, less satisfactorily but more practically, subject to contractual guarantees such that proper compensation would be paid if the rules were to change. If they were to receive equal treatment with renewables, for example being immune from losing market share when the renewables were overproducing by being included in 'must-take' provisions or having such provisions removed from the marketplace altogether, and if they were not asked to contribute to the inevitably much higher grid costs associated with renewables, then at least businesses would be able to judge nuclear investment in terms of normal business risk associated with whether the plants could be built to time and cost and whether they would work effectively once built. (Whether renewables could thrive in such an environment is questionable.) Large power companies might well hedge their bets against very high gas prices by including some new nuclear in their portfolio, though this could not be guaranteed.

The alternative might be for government unequivocally to resume responsibility for the power mix, though not necessarily for operating power stations, and include a suitable proportion of nuclear power in that mix.[76] The UK Energy Act of 2013 was certainly a significant move in this direction, in effect a degree of reregulation to include, for example, more or less fixed future electricity prices for low-carbon technology, some considerable way from what a normal 'market' would look like. President Bush had moved in a similar direction in 2005 when he announced steps to reduce risk and uncertainty in the licensing process for new build, including a call for the Department of Energy to provide risk insurance to mitigate the additional cost of unforeseen delays.[77]

But the experience of the previous 20 years suggested that government could bring itself neither to leave the market alone for decades to find its own solution nor to resume formal responsibility for power outcomes. Indeed, having constructed a market in the 1990s to stimulate new build (almost exclusively CCGT), in the 2000s it rewrote the market rules to ensure that nobody silly enough to have put serious money into electricity in the previous decade would be allowed to make a reasonable return, or any return at all. All of the attempts to reregulate sufficiently to attract capital for low-carbon investment hit the same problem – what guarantees could one government give that a successor, perhaps in two or three parliaments' time, would not once again take advantage of a return to healthy capacity margins by removing the financial support mechanisms which had stimulated that investment in the first place?

10.14 THE ADVANTAGES OF THE GIRDING UP OF LOINS

By the mid-2010s it seemed at least possible that society was heading towards a period of greater stress in the field of energy and the environment than it had faced in the 1990s and early 2000s. It also seemed possible that it might respond by once more demanding of its politicians a sense of real direction and leadership. If so then some of barriers to taking decisions, rooted in a fear of being seen too dictatorial and scaring the horses, might be expected to dissolve. This could then free politicians who were so inclined to take difficult and potentially unpopular decisions to safeguard long-term prosperity, the well-being of the environment, social stability and so on. The payoff may come after their term of office, though history may well be kind to their vision.

In such a venture science can be a strong ally but only if politicians take the lead in reinstating technical expertise and research into the front line of the decision-making process. Society's obsession with very small risks should be challenged, not indulged and reinforced. The idea that difficult issues, such as the siting of a fracking well or a radioactive waste repository which may lead to conflicts between national and local priorities, can be consulted away, as long as enough people serve on enough committees, is naïve and may be diminishing, though the sacking of the European Commission's chief scientist in 2014 effectively for holding the 'wrong' views on GMO is not encouraging.

The technical community has its own part to play. It must of course be transparent in its dealings with government and also with academia: there must be no suspicion of scientists being leant on to hide or alter their findings if they cause short-term embarrassment or loss for the commercial company involved as is often the case during phase 1 (and is one of the causes of phase 2). But at the same time, as discussed in more depth in chapter 11, it is important that risks are not overestimated to satisfy some real or perceived political agenda. It is a myth, or at least an exaggeration, that the media are always looking for a scare story. During the Fukushima accident, for example, most of the mainstream British media showed exemplary balance in their reporting, giving very little airtime to the pantomime dames of the extremist wing of Big Green (or of the nuclear industry for that matter).

Evidence suggests that in a small number of countries (notably those in Scandinavia) the relationship between the political establishment and the people who elect it is mature enough to allow long-term decision-making with popular support, or at least acceptance. This often involves a programme of thoroughgoing consultation, coupled with rights of veto among local communities. However, this does not necessarily mean that a similar approach would result in implementable decisions in countries with different political cultures and histories. In some such countries the choice in practice may lie between:

- a strong political stance which can deliver projects of national importance and manage the tensions and disagreements which will inevitably arise; or

- interminable delay as the locus of decision-making (or rather its avoidance) moves further away from elected politicians and their civil servants and is dissipated among larger and larger numbers of 'stakeholders' bolstered up by a more and more impenetrable mire of regulation and legislation.

Many of the topics which have been discussed in this chapter, not least the future of energy, seem too important to be left floating in deliberative limbo. Good policy in these fields needs good science and a will to act on it. Neither phase 1 nor phase 2 could deliver both of these requirements.

ENDNOTES

1 See e.g. http://press.ecpr.eu/documents/content/9781907301131.pdf, ed. Torfing J and Triantafillou P (2009), *Interactive policymaking, metagovernance and democracy,* ECPR Press.

2 Beckman R (1983), *The downwave: surviving the second depression*, Milestone Publications, Portsmouth, UK.

3 http://www.utilitarianism.com/jeremy-bentham/index.html, Bentham J (1781), *Introduction to the principles of morals and legislation*, Batoche Books, Kitchener, Canada.

4 http://bufvc.ac.uk/wp-content/media/2009/06/the_white_heat_of_tech.pdf, National Film Theatre (2009), 'Projecting a modern Britain: the white heat of technology'.

5 Grimston M (1997), 'Chernobyl and Bhopal ten years on – comparisons and contrasts' in Lewins J and Becker M (eds), *Advances in nuclear science and technology*, **24**, Plenum: New York.

6 Sierra Club (1974), 'Sierra Club policies – nuclear power'.

7 http://www.iddri.org/Publications/Collections/Idees-pour-le-debat/id_0413_henry&henry.pdf , Henry C and Henry M (2004), 'L'essence du principe de précaution: la science incertaine mais néanmoins fiable' ('The nature of the precautionary principle – dubious science but nonetheless reliable'), Iddri Seminar no. 11, Paris.

8 Shrader-Frechette K, 'Nuclear power', in Chadwick R (ed.), *Encyclopaedia of Applied Ethics*, **3**, 343–51, Academic Press, San Diego, 1998.

9 http://www.parliament.uk/briefing-papers/RP04-50/european-parliament-elections-2004, House of Commons Library (2004), 'European Parliament elections 2004, Research Paper 04/50'.

10 http://researchbriefings.files.parliament.uk/documents/SN05125/SN05125.pdf, Keen R (2015), 'Membership of UK political parties', parliament.uk.

11 Snow C P (1959), *The two cultures and the scientific revolution*, Cambridge University Press.

12 President's Council of Advisors on Science and Technology (1997), 'Report to the President on federal energy research and development for the challenges of the twenty-first century', Washington, DC.

13 http://www.world-nuclear.org/info/Economic-Aspects/Energy-Subsidies-and-External-Costs/, World Nuclear Association (May 2015), 'Energy subsidies and external costs'.

14 Brandt A (2007). *The cigarette century: the rise, fall and deadly persistence of the product that defined America*, New York: Basic Books.

15 Nietzsche F (1882–87) translated Kaufmann W (1974), *Die fröhliche Wissenschaft*, Vintage Books.

16 http://www.odi.org.uk/RAPID/Meetings/Evidence/Presentation_3/Cable.html, Cable V (2003), 'Does evidence matter?', The political context, Overseas Development Institute.

17 http://spp.oxfordjournals.org/content/28/2/99.abstract, Millstone E and van Zwanenberg P (2001), 'Politics of expert advice: lessons from the early history of the BSE saga', *Science and Public Policy*, **28** (2).

18 http://www.timesonline.co.uk/newspaper/0,,2719-1677135,00.html, Davies H, 'LSE report on ID cards cost' (letter), *The Times*, 2 July 2005.

19 http://corporateeurope.org/sites/default/files/attachments/ngo_letter_on_chief_scientific_adviser_-_final.pdf, Greenpeace *et al.* (2014), Letter to President-elect of the European Commission.

20 http://www.marxists.org/reference/subject/philosophy/works/us/kuhn.htm, Kuhn T (1962), *The structure of scientific revolutions*, University of Chicago Press.

21 Lynn J and Jay A (1982), *Yes, Minister Volume 2*, BBC Books, London.

22 http://www.telegraph.co.uk/news/uknews/1366276/Currie-right-on-salmonella.html, Millward D, 'Currie "was right" on salmonella', *Daily Telegraph*, 26 December 2001.

23 See Rozenberg J, 'When law collides with foreign affairs', *Telegraph*, 2 July 2002.

24 Gray J (2004), *Heresies: against progress and other illusions*, Granta Books.

25   http://classics.mit.edu/Aristophanes/thesmoph.pl.txt , Aristophanes (441 BC), 'The Thesmophoriazusae'.

26   Dickens C (1854), *Hard times*, Wordsworth Classics.

27   Lehmann E (1934), 'The biological will: methods and goals of biological research work in the new Reich', Munich.

28   http://researchbriefings.files.parliament.uk/documents/SN05125/SN05125.pdf, Keen R, 'Membership of UK political parties', House of Commons Library Briefing Paper No. SN05125, 11 August 2015.

29   http://www.ibtimes.co.uk/labour-membership-hits-342000-jeremy-corbyn-attracts-left-wing-surge-support-1519841, Silvera I, 'Jeremy Corbyn election attracts 40,000 new Labour Party members in five days', *International Business Times*, 15 September 2015.

30   http://www.walesonline.co.uk/news/wales-news/green-light-given-80m-biomass-8912120, Morgan S, '£80m biomass plant approved at former site of Europe's biggest arms depot', *Wales Online*, 24 March 2015.

31   http://www.independent.co.uk/voices/commentators/dr-hamish-meldrum-there-is-no-such-thing-as-a-100-per-cent-safe-drug-307422.html, Meldrum H, 'There is no such thing as a 100 per cent safe drug', *The Independent*, 22 August 2005.

32   http://www.theguardian.com/lifeandstyle/2005/oct/08/weekend.johnharris, Allis-Smith J, quoted in Harris J, 'Blast from the past', *The Guardian*, 8 October 2005.

33   http://www.independent.co.uk/news/greenpeace-comes-clean-on-brent-spar-error-1599641.html, Schoon N, 'Greenpeace comes clean on Brent Spar error', *The Independent*, 6 September 1995.

34   http://www.jonentine.com/ethical_corporation/2002_6_Stakeholder_Dialogue.htm, Entine J, 'Stakeholder dialogue can be dangerous to your health', *Ethical Corporation Magazine*, June 2002.

35   http://www.independent.co.uk/life-style/gadgets-and-tech/news/there-are-officially-more-mobile-devices-than-people-in-the-world-9780518.html, Boren Z, 'There are officially more mobile devices than people in the world', *The Independent*, 7 October 2014.

36   http://webarchive.nationalarchives.gov.uk/20101011032547/http://www.iegmp.org.uk/report/text.htm, 'Mobile Phones and Health', May 2000.

37   http://www.ncbi.nlm.nih.gov/pubmed/15475713, Lönn S, Ahlbom A, Hall P and Feychting M (2004), 'Mobile phone use and the risk of acoustic neuroma', *Epidemiology* 15 (6) 653–9.

38   http://www.mthr.org.uk/documents/MTHRreport2012.pdf, Mobile Telecommunications and Health Research Programme Management Committee (2013), 'Report 2012', Public Health England.

39   http://phonesar.info/expert-reviews.cfm, SAR (2014), 'Expert reviews'

40   http://www.who.int/mediacentre/factsheets/fs193/en/, WHO (2014), 'Electromagnetic fields and public health: mobile phones, Fact sheet No.193'.

41   https://www.academia.edu/834934/Mobile_phone_masts_protesting_the_scientific_evidence, Drake F (2006), 'Mobile phone masts: protesting the scientific evidence', *Public Understanding of Science*, 15, 387–410.

42   https://www.academia.edu/279373/Comparing_National_Responses_to_Perceived_Health_Risks_From_Mobile_Phone_Masts, Burgess A (2002), 'Comparing national responses to perceived risks from mobile phone masts', *Health, Risk and Society*, 4 (2).

43   http://www.spiked-online.com/Printable/0000000053FA.htm, Kaplinsky J, 'Mobile moans', Spiked Science, 2000.

44   www.durodie.net/pdf/TheDemoralizationofScience.pdf, Durodié W (2002), 'The demoralisation of science', Conference 'Demoralization: morality, authority and power', Cardiff University, UK.

45   http://www.who.int/mediacentre/factsheets/fs286/en/, WHO (2015), 'Measles: fact sheet no. 286'.

46   Wakefield A *et al.* (1998), 'Ileal-lymphoid-nodular hyperplasia, non-specific colitis, and pervasive developmental disorder in children', *The Lancet*, 351.

47   http://www.ncbi.nlm.nih.gov/entrez/query.fcgi?cmd=Retrieve&db=PubMed&list_uids=12922131&dopt=Abstract, Jefferson T, Price D, Demicheli V and Bianco E (2004), 'Unintended events following immunisation with MMR: a systematic review', *Vaccine*, 21 (25–26) 3954–60.

48   http://archpedi.ama-assn.org/cgi/content/abstract/159/1/37, Barbaresi W, Katusic S, Colligan R, Weaver A and Jacobsen S (2005), 'The incidence of autism in Olmsted County, Minnesota, 1976–97 – results from a population-based study', *Archives of Pediatric and Adolescent Medicine*, 159 (1) 37–44.

49   http://www.blackwell-synergy.com/links/doi/10.1111%2Fj.1469-7610.2005.01425.x?cookieSet=1, Honda H, Shimizu Y and Rutter M (2005), 'No effect of MMR withdrawal on the incidence of autism: a total population study', *Journal of Child Psychology and Psychiatry and Allied Disciplines*, 46 (6) 572–9.

50   http://www.plosone.org/article/info:doi%2F10.1371%2Fjournal.pone.0003140, Hornig M *et al.* (2008), 'Lack of association between measles virus vaccine and autism with enteropathy: a case-control study'.

51   http://www.staffnurse.com/nursing-news-articles/mmr-autism-link-study-476.html, Staffnurse.com, MMR-autism link study, 23 February 2004.

52   http://www.theguardian.com/society/2010/jan/28/andrew-wakefield-mmr-vaccine, Boseley S, 'Andrew Wakefield found irresponsible by GMC over MMR vaccine scare', *The Guardian*, 28 January 2010.

53 http://www.bmj.com/content/342/bmj.c7452, Godlee F (2011), Wakefield's article linking MMR vaccine and autism was fraudulent, *BMJ*, **342**.

54 http://www.private-eye.co.uk/sections.php?section_link=columnists&article=196&, 'M.D.', 'Second Opinion: M.D. peer reviews the Eye's MMR coverage', *Private Eye*, 16 February 2010.

55 http://www.theguardian.com/society/2013/apr/25/mmr-scare-analysis, Greenslade R, 'The story behind the MMR scare', *The Guardian*, 25 April 2013.

56 http://www.dailymail.co.uk/news/article-2311432/MMR-boycott-left-2m-children-risk-measles-Doctors-say-outbreak-Wales-spread-London.html, *Daily Mail*, 19 April 2013.

57 BBC Radio Four (2002), 'MMR vaccine – should we be given the choice?'

58 http://www.hscic.gov.uk/catalogue/PUB18472/nhs-immu-stat-eng-2014-15-rep.pdf, Health & Social Care Information Centre (2015), 'NHS immunisation statistics, England 2014-15', Government Statistical Service.

59 http://webarchive.nationalarchives.gov.uk/20140505192926/http://www.hpa.org.uk/web/HPAweb&HPAwebStandard/HPAweb_C/1195733833790, Public Health England (2014), 'Confirmed cases of Measles, Mumps and Rubella 1996-2013'.

60 https://www.gov.uk/government/uploads/system/uploads/attachment_data/file/377945/14-1190a-innovation-managing-risk-report.pdf, Walport M and Craig C (2014), 'Innovation – managing risk, not avoiding it: Annual Report of the Government Chief Scientific Adviser 2014', Government Office for Science, London.

61 http://www.independent.co.uk/news/uk/home-news/fracking-could-be-as-damaging-as-thalidomide-tobacco-and-asbestos-governments-chief-scientific-adviser-warns-in-new-report-9891931.html. Smith L, 'Fracking could be as damaging as thalidomide, tobacco and asbestos, government's Chief Scientific Adviser warns in new report', *The Independent*, 28 November 2014.

62 Cited in http://iopscience.iop.org/0952-4746/24/4/E02/pdf/0952-4746_24_4_E02.pdf, Wakeford R (2004), 'Reflections on CERRIE', *Journal of Radiological Protection*, **24** (4).

63 http://www.llrc.org/, LLRC (June 2003), 'Committee Examining Radiation Risk from Internal Emitters (CERRIE) – a new model of scientific advice gathering for Government: the progress and the problems', *Radioactive Times*, **5** (1).

64 http://www.theguardian.com/environment/2004/oct/28/energy.guardianletters, 'Radiation experts fall out', *The Guardian*, 28 October 2004.

65 http://www.dw-world.de/dw/article/0,1564,1631650,00.html, 'France wins nuclear fusion plant', *DW-World.de*, 28 June 2005.

66 http://www.theguardian.com/environment/2014/dec/10/peru-press-charges-greenpeace-nazca-lines-stunt, Collyns D, 'Greenpeace apologises to people of Peru over Nazca lines stunt', *The Guardian*, 11 December 2014.

67 http://www.bbc.co.uk/news/science-environment-30422994, McGrath M, 'Greenpeace sorry for Nazca lines stunt in Peru', *BBC News website*, 11 December 2014.

68 http://www.psi.org.uk/docs/2003/esrc-energy-grimston-nuclear.doc, Grimston M (2003), 'Nuclear power', in report of PSI/ESRC conference, *Projects and policies for step changes in the energy system – developing an agenda for social science research*, ESRC.

69 Morano M (2001), 'Greens going nuclear', Science and Environmental Policy Project: Arlington, Virginia.

70 http://www.cnn.com/2013/11/03/world/nuclear-energy-climate-change-scientists-letter/index.html, 'Top climate change scientists' letter to policy influencers', *CNN*, 3 November 2013.

71 http://www.monbiot.com/2011/03/21/going-critical/, Monbiot G, 'Going critical', *monbiot.com*, March 21 2011

72 http://www.no2nuclearpower.org.uk/nuclearnews/NuClearNewsNo56.pdf, Roche P, No2NuclearPower, *NuClear News No. 56*, November 2013.

73 Hume D (1739), *A treatise of human nature*.

74 http://comres.co.uk/wp-content/themes/comres/poll/REG_Windpower_Onshore_Wind_MPs_Survey_Summer_2014.pdf, ComRes (2014), 'REG windpower onshore wind MPs survey'.

75 http://www.politics.ie/forum/environment/157492-how-safe-european-nuclear-power-98.html, 'French reactor problems cast doubt on UK nuclear power plant', *Politics.ie forum*.

76 http://www.ippr.org/assets/media/publications/pdf/when-the-levy-breaks_Jun2015.pdf?noredirect=1, Garman J and Aldridge J (2015), 'When the levy breaks: energy bills, green levies and a fairer low-carbon transition', IPPR: London.

77 US Embassy London (2005), 'White House outlines Bush energy proposals'.

11 | PUBLIC PERCEPTIONS AND THE STRANGE CASE OF RADIATION

Experience shows that social and political factors can be just as influential as economic, security of supply or environmental considerations when it comes to designing and implementing policy in the energy field. This chapter will focus on the relationship between radiation and the public, not just as an issue in its own right but as an illustration of how misunderstanding relationships of this kind by industry and governments can severely damage the prospects of taking well-reasoned decisions.

Decision-making in the nuclear power field has been severely, if not fatally, hampered and delayed by perceptions of radiation. Concerns about the effects of radiation on human health, especially in the context of a major plant accident or a potential leak from a radioactive waste repository, have underpinned decisions to prevent plants being built, or prevent their operation once they have been built, or being closed early once they have been operated. These concerns have had their effect both directly, through government and regulatory responses, and indirectly, by pushing up the costs and economic risks of nuclear power until it is effectively unaffordable, certainly within anything that resembles a competitive market in electricity generation and supply.

Between 1978 and 2001, for example, some 25 nuclear reactors with a combined capacity of some 16 GW, and one MOx fuel production plant, were closed or halted in advanced stages of construction for non-economic reasons in six OECD countries (Austria, Germany, Italy, Spain, Sweden and the US), some as a direct result of referenda.[1] Italy closed all its plants; countries such as Germany, Sweden and the Netherlands adopted formal phase-out policies by law; Switzerland imposed a ten-year moratorium on new construction in 1990[2] and Belgium took a policy decision to phase out nuclear power. A number of countries without operating nuclear power plants, such as Australia, Austria, Denmark, Greece, Ireland, Norway and Poland, put in place legal or policy obstacles to any developing nuclear power, thereby reducing the potential export opportunities for companies which had built nuclear stations in the previous decades. In one notorious case the Shoreham nuclear station at Long Island in New York was closed in 1989 before commercial operation began because it was refused an operating licence on the grounds that it could not comply with evacuation requirements that were introduced after construction had started. The Long Island Lighting Company was effectively bankrupted by the affair.[3] Similarly, the Zwentendorf plant in Austria was refused its operating licence after a referendum in 1978.

After a relative thawing of political attitudes to nuclear power through the 2000s which saw the phase-outs and moratoria in a number of countries replaced by proposals for a new build, the Fukushima accident in March 2011 prompted further antipathy, though the response varied widely from country to country. As noted in Chapter 7, over the course of the following year or so Japan took its entire nuclear fleet offline. Although formally only six reactors were closed permanently it was clear that many of the others would never resume operation. Germany forced the closure of its oldest seven nuclear plants and readopted its phase-out policy with a 2022 deadline. Switzerland's minister for energy and the environment, Doris Leuthard, announced that all plans for new nuclear plants would be suspended 'until we know for certain if our own safety regulations are enough in light of these new developments';[4] Belgium also reintroduced a phase-out policy. Even China, which in early March 2011 had announced proposals for some 90 GW of new nuclear capacity to be online by 2020, said that it would scale back these plans, premier Wen Jiabao announcing a halt on approvals for new nuclear plants and the vice chairman of the China Electricity Council confirming that the nuclear programme would be slowed down over the subsequent decade,[5] although new starts in construction projects did recommence later.

Yet the evidence strongly suggests that nuclear power is among the safest of the major energy technologies; it has among the lowest greenhouse gas emissions; and it is the only technology which has demonstrated that it can be built sufficiently quickly to make a significant dent in the hegemony which fossil fuels have enjoyed since the start of the industrial revolution.

So this chapter will ultimately address a puzzle – why is the safest source of energy we have yet come up with regarded by a significant number of people as being the most dangerous? And an associated question – how do we make sure that the next major release of radioactive material, which like the previous ones will have very few direct health effects, does not turn into a human disaster of the scale of Fukushima? Although it is extremely unlikely that anyone will die from radiation exposure, at least among those who were off-site at the time of the accident, there have been major detrimental effects on the quality of life and indeed deaths among those affected, caused purely by the fear of radiation that had built up beforehand and the countermeasures that the authorities took afterwards. Various attempts have been made by researchers and the nuclear industry to explain this apparent discrepancy, involving various interpretations of the public's relationship with radiation and indeed with science as a whole.

11.1 A BRIEF HISTORY OF THE BRIEF HISTORY OF NUCLEAR SCIENCE

The terms 'atomic' and 'nuclear' energy have been used in a largely interchangeable way since the technology emerged in the 1940s. Current understanding characterises 'atoms' as made up of an extremely dense 'nucleus' which contains almost all of the mass of the atom in question, surrounded by a cloud of 'electrons'. The nucleus consists of two types of subatomic particle which have almost exactly the same mass (given a unit '1' in the descriptions of nuclei). The proton has an electrical charge of +1; the neutron is uncharged. In the weird subatomic world the electrons which surround the nucleus can be viewed either as particles, with a tiny mass (less than one 1,800th of that of a proton or neutron) and an electrical charge of -1; or they can be viewed as 'waves' or as a 'cloud'. A neutral atom, as found when an element is in its pure form, will contain the same number of protons as electrons, i.e. it will be electrically neutral. Chemical reactions, such as burning, involve the interaction of the electrons of various atoms to form more stable states, usually (though not always) giving out energy as they do so. However, electrons, strange and fascinating though they are, are of little direct relevance to the nuclear physicist. Similarly protons and neutrons are now themselves believed to be made up of 'quarks' but for the purposes of this discussion can be regarded as fundamental particles in their own right.

Our current understanding evolved over many centuries. The ancient Greeks were fascinated by many aspects of the physical world, including of course the nature of matter itself. The earliest ideas about matter and atoms were developed by Greek philosophers between 450 BC and 380 BC. Democritus (c.430–360 BC) argued that matter was 'discontinuous', i.e. if we started chopping up a piece of say graphite we would eventually come to a point where we could not chop it up further while it continued to be graphite. Democritus used the word 'atomos' or 'indivisible' to describe these ultimate building blocks of matter. By contrast, Aristotle (384–322 BC) suggested that matter was 'continuous', i.e. that if you chopped a piece of graphite in two, then cut it in two again and so on, you could in principle continue to do so forever and still have 'graphite' at the end.

More recent commentators such as Bertrand Russell have argued that there was little evidence available on which to choose between these theories – that only luck put Democritus on the right side and Aristotle in the wrong.[6] In the first century BC, Lucretius, in *De Rerum Natura*, did offer an attractive logical argument for the atomic theory. He observed that any material, even hard rock, is subject to irreversible decay. Further, when things were mixed together, say flour and water to make bread, it was difficult if not impossible to return the mixture to its original unmixed state. However, throughout nature there were examples of new 'unmixed' substances, such as the water and cereal from which flour can be made, being created. So the real mystery was why had everything in the world not simply decayed away, as one might expect if matter could continually be eroded away to nothing. Lucretius argued that the properties of materials must derive from something internal which, although too small to be perceived by human senses, was unchanging and never decayed, retaining its precise nature forever and being available to be recast into fresh 'unmixed' material. Lucretius also argued that there must be quite big gaps, or 'voids', between these 'atoms' (or molecules as we might say today), in order to explain how gases and liquids can change shape and metals can be beaten into different shapes, without their basic properties being changed.

Such arguments from logic notwithstanding, for centuries there were no real experimental techniques available to test either of these hypotheses and the continuous theory commanded considerable support right up to the early nineteenth century.

The flourishing of chemistry – defined in 1661 by the great British scientist Robert Boyle as 'the subject of the material principles of mixed bodies [what we might today call 'chemical compounds']' – from the mid seventeenth century led to the discovery of a number of important natural laws.

Boyle, a supporter of the atomic approach to matter, had offered a definition of the word 'element' in the chemical sense as 'any substance that cannot be decomposed into a simpler substance' (thereby seeking to replace the ancient Greek concept, which had survived for more than 2,000 years, that the world is made up of combinations of just four elements: earth, wind, fire and water). So for Boyle, although there is a 'unit' of carbon dioxide which cannot be broken up without the substance remaining as carbon dioxide, it can be broken up into oxygen and carbon. However, the resulting particles of oxygen and carbon, in Boyle's world view, cannot be further broken down. So carbon dioxide is a 'compound' and its smallest unit is a 'molecule', while carbon and oxygen are 'elements' and their smallest units are 'atoms'.

In 1789 Frenchman Antoine Lavoisier, 'the father of modern chemistry' (and a victim of Madame Guillotine five years later), formulated the Law of Conservation of Matter (or Mass) which states that matter cannot be created or destroyed – in other words the mass of the products of any physical or chemical change or reaction must be exactly the same as the mass of the starting materials. Around the turn of the nineteenth century another great early chemist, Joseph Proust, published the 'Law of Definite Proportions' which said that in any particular chemical compound, say carbon dioxide, the ratio of the mass of carbon to the mass of oxygen making up the compound (in this case 3:7) will always be the same no matter how the compound is made. In 1804 the great Mancunian physicist John Dalton, after whom the school of nuclear science at the University of Manchester is named, added the 'Law of Multiple Proportions': 'if two elements react to form more than one compound, the amount of element B that combines with a fixed mass of element A in the various compounds will be related by a simple ratio of small numbers'. So carbon reacts with oxygen to form two different compounds, carbon monoxide and carbon dioxide. In the first 1g of carbon combines with 1.33g of oxygen; in the second 1g of carbon combines with 2.67g of oxygen, i.e. the ratio of the mass of oxygen combining with 1g of carbon in the two compounds is 1:2. The law holds even in more complicated compounds, e.g. in methane 1g of carbon combines with 0.222g of hydrogen; in propane 1g of carbon combines with 0.250g of hydrogen, a ratio of exactly 8:9.

Dalton used these laws as the basis of a new atomic theory which had the following tenets.

1. Elements are made of extremely small particles called atoms.

2. Atoms of a given element are identical in size, mass and other properties; atoms of different elements differ in size, mass and other properties.

3. Atoms cannot be subdivided, created or destroyed.

4. Atoms of different elements combine in simple whole-number ratios to form chemical compounds.

5. In chemical reactions atoms are combined, separated or rearranged.

In 1869 the Russian Dmitri Mendeleev organised the 60 or so known elements at the time into a table based on their 'atomic weight' (the weight of an atom, on a scale on which the hydrogen atom has a weight of 1). He noticed that there was a regular pattern, or 'periodicity', among the elements. As he put it: 'I began to look about and write down the elements with their atomic weights and typical properties, analogous elements and like atomic weights on separate cards and this soon convinced me that the properties of elements are in periodic dependence upon their atomic weights.' So, for example, lithium, sodium, potassium, rubidium and caesium are all light metals which react vigorously with water and air and which form a chloride in which one atom of chlorine combines with one atom of the element. The elements next up in the table of weights – beryllium (the next element up from lithium), magnesium (the next up from sodium), calcium (next to potassium etc.), strontium and barium – are similar to each other but different from the lithium family, being denser, somewhat less reactive and forming a chloride in which two atoms of chlorine combine with one atom of the element. Yet another group – fluorine, chlorine, bromine, iodine – are non-metals which form compounds with say calcium in which one atom of calcium combines with two of the element. As well as helping Mendeleev to identify elements where the published atomic weight was wrong, the table was particularly impressive in that he noted that some elements seemed to be 'missing'. He predicted not only their existence but also their general properties. When scandium, gallium, technetium and germanium were eventually discovered they were found to be very like Mendeleev's postulated elements ekaboron, ekaaluminium, ekamanganese and ekasilicon. A hypothesis which makes predictions which are then subsequently proven by experimentation or investigation is by its nature more impressive than one which seeks to work backwards from a set of observations but which either makes predictions that cannot be tested or makes predictions which turn out to be wrong.

Not so simple

The atomic theory jostled with continuous theories of matter until the early twentieth century. However, it also came under attack from the other direction as evidence began to emerge that atoms themselves might not be the smallest units in nature. The atom was (and is) still regarded as the basic unit of a particular element but it came to be thought that atoms themselves were made up of smaller entities or 'subatomic particles'.

As early as 1838 Michael Faraday (1791–1867), 'the father of electronics', had noticed that if air was pumped out of a glass tube and an electric current passed through it, an arc of light would be seen stretching from the cathode (the negative electrical pole) to the anode (positive pole), except for a small dark space directly in front of the cathode itself (the 'cathode dark space' or 'Faraday dark space').[7] The outcome of the experiment underwent two changes as vacuum pumps became more efficient over the subsequent decades. The cathode dark space lengthened towards the anode until, in a very efficient vacuum, the glow observed by Faraday had practically disappeared. However, the glass behind the anode itself started to glow and any object between the cathode and the back wall, such as the anode itself, left a shadow in this glow. In 1869 a German physicist, Johann Hittorf, realised that there must be an invisible ray of something, either particles or an energy beam, being emitted by the cathode. The arc of light seen in Faraday's time had been caused by this beam striking air molecules, causing them to glow. In 1876 Hittorf's compatriot Eugen Goldstein named the phenomenon 'cathode rays'.

Two theories emerged – that the rays were charged atoms or that they were a new kind of electromagnetic radiation (i.e. pure energy, like light but more intense). In 1897 in Cambridge, J.J. Thomson exposed cathode rays to a magnetic field at right angles to their direction of travel and found that they were bent off course. This showed that they were electrically charged and so, in the prevailing models of the day, must be particles, not energy waves. But a variety of experiments, including measuring the amount of deflection for a magnetic fields of a certain strength, led him to calculate that the particles were much lighter than the lightest atom (that of hydrogen), by a factor of about 1800. Furthermore, it did not matter how the particles were created – they behaved the same way. Clearly this was a fundamental particle which was much smaller than an atom. Thomson called it the electron and it was soon realised that it was electrons which carry the current in electrical circuits.

The discovery of radiation and the collapse of old assumptions

In Paris in 1896, physicist Henri Becquerel left a photographic plate in a drawer under some pitchblende, a mineral containing the elements radium and uranium, with a metal Maltese cross between them. A few days later he developed the plate and found it had darkened, except for a negative shadow of the cross. He realised that the pitchblende was emitting some invisible ray with sufficient energy to affect the photographic plate but which could not penetrate the cross. Becquerel called these rays 'radiation' and his doctoral student, Marie Curie, named the process 'radioactivity'. (Like Thomson, Bequerel received a Nobel Prize in Physics for his troubles, sharing it with Curie, who was later to win a Nobel Prize in Chemistry in her own right, and her husband Pierre, 'father of Irene Curie' who would herself become a Nobel laureate, as discussed later.) Madame Curie discovered that other elements alongside uranium, notably thorium and radium, also produced 'Becquerel rays'. In 1899 New Zealand-born Ernest Rutherford, 'the father of nuclear physics', showed that 'spontaneous radiation' consisted of three types, characterised by their penetrating ability. Alpha radiation, later revealed to be the same as helium nuclei consisting of two protons and two neutrons, is absorbed by a few centimetres of air, a sheet of paper or the dead outer layer of our skin; beta radiation consists of electrons and can be stopped by a few millimetres of aluminium; gamma radiation (which is 'pure' energy similar to light or ultraviolet radiation but more energetic) can pass through many metres of concrete and many centimetres of lead.

The radioactivity of any radioactive substance dies away over time. A certain quantity of a particular radioactive material containing say a trillion (1,000,000,000,000) atoms can undergo a trillion radioactive 'decays' (i.e. emit a trillion alpha or beta particles – gamma is a bit more complicated as it is not a particle as such). Whether a particular atom decays at a particular moment is a matter of chance and cannot be accurately predicted but statistically speaking the amount of radiation being emitted is proportional to the amount of the substance present. So if it takes say ten minutes for half a trillion of the atoms of the substance in question to decay and become non-radioactive then it would take another ten minutes for half of what is left (i.e. a further 250 billion) to go (leaving one-quarter of the original trillion atoms); after another ten minutes we would be down to 125 billion atoms (one-eighth of the original) and so on. A particular radioactive material is said to have a constant 'half-life' sometimes given the symbol $t_{\frac{1}{2}}$; this concept was discovered by Rutherford in 1907.

Each different radioactive material has a different half-life. So the most common version, or 'isotope' (see later), of the element uranium, uranium-238, has a half-life of around 4.5 billion years. Coincidentally that is more or less the same as the lifetime of the earth, so half of the uranium-238 which was incorporated into the earth when it was formed is still here. The artificially-produced isotope lithium-8 has a half-life of around 0.8 seconds (there are plenty of isotopes with shorter half-lives than that). Sodium-24 has a half-life of around 15 hours (see figure 11.1), while the important 'fission product' (discussed later) caesium-137 has a half-life of 30 years.

Paradoxically, elements with very short or very long half-lives are of relatively little concern if they are found in the environment. If an element has an extremely short half-life then although it is intensively radioactive, to all intents and purposes it has all decayed away within a short period of time before it has had a chance to reach the local population. (Technically speaking no radioactive material ever decays away completely but after eight half-lives, i.e. around six seconds in the case of lithium-8, its radioactivity drops to below a thousandth of the original.) If by contrast it has an extremely long half-life then it is persists for a long time but gives out its radiation extremely slowly so is not very hazardous (radiologically speaking – heavy metals like uranium may of course also be chemical poisons). It is those with intermediate half-lives, perhaps a few days to a few years, which represent the most difficult-to-manage potential danger if they are released into the environment – radioactive enough to be

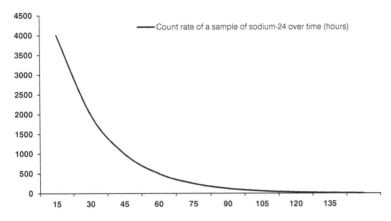

Figure 11.1 Illustration of half-life of sodium-24 (count rate per second)

a significant health risk in the wrong circumstances but not so radioactive that they exhaust themselves very rapidly.

Marie Curie noted that as radium gave off radiation it also produced quite a lot of heat. This was a puzzle, as the Law of Conservation of Energy stated that energy could neither be created nor destroyed, only converted between various kinds, e.g. heat energy, kinetic energy, chemical energy, potential energy. In 1900 Max Planck, 'the father of quantum physics', published an equation which in effect said that energy, like matter, was not continuous but came in discrete 'packets' or photons. Einstein's famous equation linking matter and energy ($E = mc^2$) appeared in 1905. Clearly something was going on which was outside the prevailing understanding of the day. From the time of Newton to the early twentieth century there was an assumption that the tiny (submicroscopic) world was simply a smaller version of the world which we perceive through our senses. The quantum revolution which accompanied and underpinned the development of modern theories of the atom destroyed this comforting notion. We can never hope to 'conceive' of the nature of the subatomic world using the senses with which our evolution has equipped us. Greater knowledge has led to a greater realisation that the subatomic world is much stranger than we thought and the best we can hope for are mathematical models which describe or predict it to a higher or lower degree, but can never 'explain' it. Quantum theory not only upset the old certainties about the nature of matter: it brought into question the very concept of 'certainty' itself. This was in time to become a serious challenge to confidence in the infallibility of the scientific project which had held sway since the Enlightenment.

11.2 FURTHER DEVELOPMENTS IN ATOMIC THEORY AND THE EMERGENCE OF NUCLEAR FISSION

Of course, as is always the case in scientific or any other revolutions, not everybody embraced the new thinking at once. Einstein himself was famously unwilling to accept one conclusion of his work on relativity – that the universe was expanding – and introduced a random constant into his work to 'correct' for it. He also rejected the introduction of statistical probability and uncertainty into what looks to the naked eye to be real and solid matter – 'God does not play dice'. 'Classical' or semi-classical descriptions of the atom continued to emerge and in some cases to offer reasonable descriptions of some phenomena at the atomic level, if not actually 'explaining' them. In 1926, for example, the Austrian Erwin Schrödinger proposed an effectively classical equation to describe how 'quantum states' (the positions of particles which are allowed by quantum physics) change over time, in an attempt to fit the uncomfortable observations that underpinned quantum theory into the classical world view. In 1935 Schrödinger's imaginary cat would become one of the most famous animals alive (or dead, or neither depending on the outcome of his 'experiment'). He used a thought experiment in order to demonstrate that what was known as the 'Copenhagen interpretation' of quantum mechanics, which seemed to have the bizarre conclusion that a cat could exist in a state somewhere between alive and dead until the mechanical system (involving a subatomic particle) which threatened its life was observed, must be nonsense. In fact Schrödinger was confirming that the quantum world is just weird and cannot be conceived of by a brain which has evolved to deal with the macrosystems of everyday life, where a literally unimaginable numbers of events with various probabilities add up to something which looks reliable and predictable.

In 1913 Niels Bohr, the great Danish physicist and 'the father of the theory of atomic structure', proposed that the atom was made up of a small dense nucleus surrounded by electrons which exist in 'orbits' or 'shells' round this nucleus. Electrons can only jump from one orbit to another; they cannot exist in between them. This version took into account some of the findings of quantum physics, i.e. that the electrons can only have specified energy levels, but still envisaged the atom as essentially 'like' larger arrangements such as the solar system with its central sun and orbiting planets.

Bohr was building on work supervised by Rutherford and carried out by his associates Hans Geiger (of the radiation detection counter fame) and Ernest Marsden between 1908 and 1913. In this work, alpha particles, which as noted before are quite massive in atomic terms, being equivalent to the nuclei of helium atoms, were fired at gold foil. If, as had previously been assumed, atoms were more or less uniform in their internal structure then one would expect these huge atomic fragments to crash through the foil pretty much unaffected. In fact some of the alpha particles were deflected well off their path by as much as 150 degrees (in other words nearly coming straight back at the source). Rutherford memorably reported: 'It was quite the most incredible event that has ever happened to me in my life. It was almost as incredible as if you fired a 15-inch shell at a piece of tissue paper and it came back and hit you. On consideration, I realised that this scattering backwards must be the result of a single collision, and when I made calculations I saw that it was impossible to get anything of that order of magnitude unless you took a system in which the greater part of the mass of the atom was concentrated in a minute nucleus. It was then that I had the idea of an atom with a minute massive centre, carrying a charge.' If an atom is typically around 10^{-10} metres across, its nucleus is only around 10^{-14} metres across (i.e. one 10,000th of the width of the atom or one trillionth of its volume). The phenomenon that Rutherford observed occurs because of huge electrostatic repulsion between the physically tiny (even by atomic standards) positively charged alpha particle (+2 units) and the physically tiny positively charged gold nucleus (+79).

Up to this point elements in the Periodic Table had still been arranged by reference to their atomic weight. However, in 1913 radiochemist Frederick Soddy noticed that a variety of atoms of different weights deriving from studying various radioactive elements as they 'decay' (i.e. undergo radioactive processes, changing the element to which they belong in the process) could not be separated chemically and indeed behaved identically in chemical reactions. These included, for example, 'mesothorium' (weight 228), radium (weight 226) and 'thorium-X' (224). This implied that they all belonged to the same element. At the same time J.J. Thomson was finding evidence for different versions of non-radioactive elements such as neon which could also have different weights. But if this were true then one assumption behind the work of Dalton, Mendeleev and other early champions of modern atomic theory was wrong. The atoms of a particular element need not be identical to each other. The word 'isotope' – literally the 'same place' in the Periodic Table – is believed to have been coined in 1913 by Soddy's cousin, Dr Margaret Todd, to refer to atoms of the same element which have different atomic masses or weights.

Presumably the number and arrangement of their electrons was the same in these different isotopes of a particular element because electrons were known to be responsible for chemical behaviour. As matter must on balance be electrically neutral, this suggested that all atoms of a particular element must have an equal positive charge in their nucleus but a different charge to atoms of other elements. This in turn suggested that there might be more to matter than the two particles then postulated, i.e. the negatively charged electron and its positively charged and much heavier counterpart, the proton. The development of the mass spectrograph in 1919 by Francis Aston, based on earlier work by J.J. Thompson, allowed this to be confirmed experimentally.

In 1920 Rutherford had suggested that inside the nucleus of the atom there were protons and 'neutral pairs' made up of a proton and electron very closely bound together. In 1932 James Chadwick showed that there was in fact a third 'fundamental particle', the 'neutron' – produced, for example, when alpha radiation is fired at beryllium atoms – which was effectively the same mass as the proton but had no electrical charge. The atoms of a particular element must have the same number of protons but may have different numbers of neutrons. Natural uranium, say, has an 'atomic number' of 92 (i.e. its nuclei all contain 92 protons) but in nature consists almost entirely of two isotopes, uranium-238 (with 92 protons and 146 neutrons, described as 'mass number' 238), making up some 99.3 per cent of the total, and uranium-235 (with 92 protons but 143 neutrons) which accounts for the other 0.7 per cent. Soddy's mesothorium, radium and thorium-X were all isotopes of radium (atomic number 88) with 140, 138 and 136 neutrons respectively. This is also true of smaller atoms – chlorine (atomic number 17), for example, is made up in nature of around three atoms of chlorine-35 (18 neutrons) to one of chlorine-37 (20 neutrons), giving an apparently puzzling atomic mass of 35.5.

In general, though with many exceptions, especially among artificially produced isotopes, large atoms are more likely to be radioactive. So we do not know of any 'stable', i.e. non-radioactive, isotopes of elements with an atomic number of more than lead's 82. Two smaller elements, technetium, number 43, and promethium, 61, also do not have any stable isotopes – there does not seem to be a straightforward answer as to why. For smaller atoms there is a 'stability' range, i.e. a ratio of protons to neutrons where most isotopes are stable. Very small atoms tend to be stable when they have the same number of neutrons as protons or a few more. Carbon (atomic number 6) has stable isotopes with six and seven neutrons; carbon-14 (six protons and eight neutrons) is radioactive, a feature which makes it valuable in determining the date of various artefacts. For larger atoms the stable range usually contains rather more neutrons than protons. Barium, atomic number 56, has stable isotopes with between 76 and 82 neutrons; lead, atomic number 82, has four stable isotopes with between 122 and 126 neutrons.

If an atom has too many neutrons to lie in this stability range it may undergo 'beta decay', i.e. eject a beta particle, or electron, and increase its atomic number by one while leaving its mass (number) unchanged. In effect a neutron falls apart to make a proton and an electron, reminiscent of Rutherford's 'neutral pairs' breaking up although we do not now believe that is what is happening. So carbon-14, with six protons and eight neutrons, undergoes beta decay to make nitrogen-14 with seven protons and seven neutrons. Alternatively a larger atom such as uranium-238 may undergo 'alpha decay', shedding two protons and two neutrons to reduce its mass number by four and also reduce its atomic number by two and creating thorium-234, which is also radioactive. The end product of the U-238 'decay series' is lead-206.

One implication of the extremely dense nucleus is that it is hard to break it up or to change it in other ways by bombarding it with charged particles. Electrons are too small and in any case are prevented by the laws of quantum mechanics from getting too close to the nucleus. Protons or alpha particles are themselves positively charged so they have to possess vast amounts of energy to overcome the electrical repulsion as they approach the target nucleus.

Neutrons, however, are a different matter. They are massive enough to cause a potential change when absorbed by a nucleus but, being uncharged themselves, they are not repelled by the positive charge of the nucleus. This opens up significant possibilities which were explored vigorously in the decade after the neutron's discovery and led directly to the use of fission as a source of energy, both constructive and destructive.

Active nuclear transmutation – the dream of the alchemist comes true

The protoscientific 'discipline' of alchemy had roots going back to ancient Egypt, although the word itself seems to have first been used in the fourteenth century. Among its defining objectives were the creation of the 'philosopher's stone', the development of an elixir of life and the ability to transmute one element into another, notably base metal into gold.

The process of replacing alchemy with the science of chemistry can be traced back at least as far as Boyle. The final goal of transmuting the elements was to be realised in the mid twentieth century. One element could indeed be turned into another, though the creation of artificial gold remained elusive. In 1934 Irene Joliot-Curie (daughter of Marie and Pierre) and her husband Frederic Joliot found that bombarding nuclei with protons at very high energy could sometimes cause them to undergo transmutation into artificial radionuclides, i.e. new nuclei not found in nature that would then undergo radioactive decay. The discovery of artificial radiation earned Joliot and Joliot-Curie the Nobel Prize and also confirmed a fourth 'fundamental particle', a positively charged version of the electron known as the 'positron', which had been identified in cosmic radiation in 1932 and was now found when artificial isotopes such as magnesium-23 underwent radioactive decay.

The next year the great Italian-American Enrico Fermi found that a much greater variety of artificial radionuclides could be formed when neutrons were used instead of protons. Fermi discovered that when the vast majority of substances are bombarded with neutrons, either the neutron bounces off or the substances absorb them, to create heavier atoms. Sometimes a proton is ejected in the neutron's place. However, uranium-235 did something very strange – when it was bombarded with neutrons it would produce lighter atoms as well. At the end of 1938, Otto Hahn and Fritz Strassmann in Berlin confirmed that the lighter elements created when uranium was so bombarded included barium and others which were about half the mass of uranium.

The new phenomenon was furnished with an explanation the following year by two more German pioneers, Lise Meitner and her nephew Otto Frisch. The neutron causes the uranium atom to undergo 'nuclear fission', to break into two not quite equal parts. Hann and Strassmann then showed that the fission also releases two or three more neutrons which could, in principle, cause fission reactions in other uranium atoms, thereby potentially creating an accelerating, or self-sustaining, 'chain' reaction. This was soon confirmed experimentally by Joliot and his associates in Paris and by Leo Szilard working with Fermi in New York. The mass of the fission products plus the free neutrons produced is very slightly less than that of the original uranium plus the neutron and that difference is expressed as a large burst of energy. This was an early first experimental confirmation of Einstein's famous equation $E = mc^2$, published over thirty years previously.

In 1932, before the discovery of fission, Rutherford had dismissed the idea of 'nuclear energy'. 'We might in these processes [Cockcroft and Geiger's experiments bombarding lithium with protons to produce alpha particles and energy – after the Second World War, John Cockcroft would become the first director of the atomic energy research establishment at Harwell] obtain very much more energy than the proton supplied but on the average we could not expect to obtain energy in this way. It was a very poor and inefficient way of producing energy and anyone who looked for a source of power in the transformation of the atoms was talking moonshine. But the subject was scientifically interesting because it gave insight into the atoms.' Szilard responded the following year by conceiving of, and patenting, the idea of a chain reaction. In 1938 he and Fermi observed that bombardment of uranium with neutrons resulted in an increase in the number of neutrons present. As Szilard later said: 'We turned the switch and saw the flashes. We watched them for a little while and then we switched everything off and went home. That night, there was very little doubt in my mind that the world was headed for grief.'

Other basics of nuclear science emerged very rapidly in those years just before the outbreak of war among the leading countries in the field, Germany, US, France and the UK. Bohr found that slowing the neutrons down after they had been released in a fission reaction would reduce the likelihood that they would simply bounce off the next uranium atom they encountered and would instead be absorbed to cause the next fission. Szilard and Fermi proposed using a 'moderator' to slow down the emitted neutrons. The Germans identified graphite as an excellent candidate. However, in Germany graphite was manufactured using boron carbide rods. Boron is a powerful absorber of neutrons and the traces of boron in the graphite were sufficient to prevent a chain reaction developing by 'mopping up' too many neutrons. Szilard had the US manufacturers create boron-free graphite which proved highly effective. The early story of nuclear technology is full of such stories – an apparently random side issue, in this case how graphite was manufactured, arguably changing the whole course of history. By 1942 the German authorities had decide that an atomic bomb was impractical but this was unknown to the Allies. Assumptions that the German Uranverein project was continuing was a major incentive for the wartime development of the atomic bomb by the US and Britain.

The greater tendency of uranium-235 to undergo fission when compared to uranium-238 gave rise to the idea that 'enriching' uranium by increasing the proportion of uranium-235 would result in a material which would more readily undergo fission. As the two isotopes are chemically identical (by definition) they could not be separated by normal chemical means. So it was necessary to exploit the very slight difference in their masses by creating uranium hexafluoride gas (UF_6) and forcing it through a porous membrane. As the uranium-235 hexafluoride is very slightly lighter than its uranium-238 counterpart it passes through the membrane very slightly more quickly so by repeating the process many times the proportion of uranium-235 in the mixture can be increased. It is a very expensive, energy-intensive and slow process but was the only alternative available at the time it was developed in 1940.

One further important consideration was introduced in 1939 by Francis Perrin, 'father of the French atomic bomb', who introduced the concept of the 'critical mass' of uranium required to produce a self-sustaining release of energy. Below this mass too many neutrons are lost from the surface of the uranium to allow a self-sustaining reaction to develop. Perrin's group in Paris demonstrated that a sustainable fission reaction could be produced in a mixture of uranium and water (the water acting as a moderator) as long as neutrons were introduced into the system from some external source. They also showed that some substances which absorb neutrons could be used to slow down the process so it did not run away with itself.

Another important figure was Rudolf Peierls. He was German-born and had studied under Werner Heisenberg (of the 'uncertainty principle' fame, one of the key anti-common-sense tenets of the new physics) but he was studying at Cambridge when Hitler came to power in his homeland. From April 1939 he presided over the UK nuclear energy project, which was initially directed towards military applications. (Because of their origins he and another German national living in the UK, Otto Frisch, were banned from working on radar because it was regarded as being too sensitive.) Peierls calculated that the amount of highly enriched uranium needed to make a bomb was of the order of 1 kg, rather than the several tonnes which had previously been assumed.

Conceiving the atomic bomb

The development of the atomic bomb is outside of the scope of this book and has been well documented elsewhere, notably in Margaret Gowing's seminal *Independence and Deterrence*.[8] As noted above, during the war it was established that a chain reaction could be sustained with 'slow' (moderated) neutrons in a mixture of uranium oxide and heavy water (water consisting of oxygen and the second isotope of hydrogen, twice as heavy as normal hydrogen and known as deuterium). It was also shown that when uranium-235 and uranium-238 absorb slow neutrons, uranium-235 has a much greater tendency to undergo fission while uranium-238 absorbs the neutron and undergoes successive beta decays to form a new element, mass number 239 and atomic number 94. Bretscher and Feather, in Cambridge, argued that element 94 should be readily fissionable by slow and fast neutrons and had the added advantages that it was chemically different to uranium and therefore could easily be separated from it. This was confirmed in independent work by a team led by Edwin McMillan and Glenn Seaborg in the US in 1940. Seaborg named the element 'plutonium' in 1941. (In 1997 element 106 was to be named after Seaborg himself, the only time an element has been named after a living person. It was the tenth element that Seaborg had discovered or co-discovered.)

11.3 THE PASSING OF THE DOG COLLAR — RELIGIOUS RELIGION, SCIENTIFIC RELIGION AND ENVIRONMENTAL RELIGION

Before looking in more detail at how these discoveries played out in the public sphere it is worth looking at how the scientific and technological enterprise itself came to be viewed in the period since the emergence of modern science, a theme touched on in chapter 10. Traditional religion underwent something of a decline as the Enlightenment and subsequent Industrial Revolution took hold of national philosophy. However, this decline did not remove the apparent need for society to imbue some group of individuals with magical powers, authority or even 'infallibility'. Maybe the first cultural enterprise to move into the gap left by the decline of traditional religion was science.

The transition from religion to science was a very long and at most partial affair – organised religion still had considerable influence in the 2010s. Sometimes the relationship has been oppositional, sometimes more harmonious. In 1543 Copernicus had publicised the theory of heliocentrism – that the sun, not the earth was at the centre of the solar system. When Galileo announced in 1610 his findings that Venus underwent a cycle of phases rather like the moon and that there appeared to be satellites circling Jupiter he set in train a sequence of events which saw the Catholic Church, through the Inquisition, declare heliocentrism as heretical. Galileo himself spent his last decade under house arrest to prevent his continuing promotion of ideas which challenged the place of man in the universe and indirectly the place of the Church in society. One suspects such treatment would only be meted out to climate change sceptics today, at least if Big Green had its way.

But the relationship was not always so fractious. In the later seventeenth century it became a common, perhaps even natural, notion that humanity could reach salvation through science. Many of the great thinkers of that time took it for granted that science was central to their sacred endeavour. Nature was God's creation and to study it was simply one of the many ways to celebrate His glory. Such celebration was still understood to be the proper destiny of the soul, the meaning of human life. A stream of thinkers from Fontenelle through Locke, Voltaire and Kant to Franklin and Paine argued that humanity was, or could be, on a journey from barbarism to ultimate civilisation and that science, or 'natural philosophy', could provide the path. In the late eighteenth century, for example, Nicolas de Caritat, Marquis de Condorcet, introduced his most famous work saying that it would seek:

to show by appeal to reason and fact that nature has set no term to the perfection of human faculties; that the perfectibility of man is truly indefinite; and that the progress of this perfectibility, from now onwards independent of any power that might wish to halt it, has no other limit than the duration of the globe upon which nature has cast us. This progress will doubtless vary in speed but it will never be reversed as long as the earth occupies its present place in the system of the universe and as long as the general laws of this system produce neither a general cataclysm nor such changes as will deprive the human race of its present faculties and its present resources.[9]

The Collins English Dictionary offers the following definitions of the word 'religion':

- belief in, worship of, or obedience to a supernatural power or powers considered to be divine or to have control of human destiny;
- any formal or institutionalised expression of such belief;
- the attitude and feeling of one who believes in a transcendent controlling power or powers;
- something of overwhelming importance to a person;
- the practice of sacred ritual observances, sacred rites and ceremonies.

To Karl Marx, religion was 'the sigh of the oppressed creature, the heart of a heartless world and the soul of soulless conditions. It is the opium of the people'. In his view religion was the painkiller that made life bearable: people turned to religion to feel good and to anaesthetise the pain in their lives. However, it could only be a temporary fix and one that was often harmful, just like drugs. It clouded the judgment of the working class by giving them a false sense of happiness, thereby making it less likely that they would challenge the ruling bourgeoisie.

To Freud, religion was an illusion deriving its strength from the fact that it fell in with our instinctual desires. It represented an attempt to get control over the sensory world in which we are placed by means of the 'wish-world' which we have developed inside us as a result of biological and psychological necessities. Among his many statements on religion are:

- 'A religion, even if it calls itself a religion of love, must be hard and unloving to those who do not belong to it.'
- 'It is humiliating to discover how a large number of people living today, who cannot but see that this religion is not tenable, nevertheless try to defend it piece by piece in a series of pitiful rearguard actions.'
- 'The different religions have never overlooked the part played by the sense of guilt in civilization. What is more, they come forward with a claim ... to save mankind from this sense of guilt, which they call sin.'

Bertrand Russell: 'Religion is based ... primarily and mainly upon fear. It is partly the terror of the unknown, and partly ... the wish to feel that you have a kind of elder brother who will stand by you in all your troubles and disputes. Fear is the basis of the whole thing – fear of the mysterious, fear of defeat, fear of death.'

Russell, writing between the wars, was among the many thinkers of the time who believed that the coming of science would relieve people of the need for religion.

In this world we can now begin a little to understand things, and a little to master them by the help of science, which has forced its way step by step against the

Christian religion, against the Churches and against the opposition of all the old precepts. Science can help us to get over this craven fear in which mankind has lived for so many generations. Science can teach us, and I think our own hearts can teach us, no longer to look round for imaginary supports, no longer to invent allies in the sky, but rather to look to our own efforts here below to make this world a fit place to live in, instead of the sort of place that the Churches in all these centuries have made it.

The tone of Russell's writing, and of other advocates for the replacement of religion with science, seems to imply that science can remove from us the source of our fear and uncertainty. Others, however, recognised that science might simply become another way of dealing with the fear that our lives are ultimately meaningless –a replacement religion rather than a replacement 'for' religion. Some thirty years earlier Nietzsche, in expanding on his famous aphorism 'God is dead', had expressed a fear that science and technology would simply take control and 'be treated as the new religion, serving as a basis for retaining the same damaging psychological habit that the Christian religion developed'. For Einstein, science did not in any way obviate the need for a religious and spiritual connection to the universe. 'Science without religion is lame, religion without science is blind.' Also: 'A knowledge of the existence of something we cannot penetrate, of the manifestations of the profoundest reason and the most radiant beauty – it is this knowledge and this emotion that constitute the truly religious attitude; in this sense, and in this alone, I am a deeply religious man.'

There is no doubt that the practising of traditional religion did decline, and decline steeply, in many developed countries through the twentieth century. Baptisms into the Church of England, expressed per thousand births, fell by about three-quarters, especially after the Second World War, as shown in figure 11.2.

Clearly science is not identical to religion. While religion tends to defer to ancient authority and scholars who serve to interpret it but not to challenge it, science relies at least to an extent more on hypothesis, experiment and observation. In particular, the scientific revolution brought with it a concept of 'progress' which was much less prevalent in the philosophy of an unchanging God. For many philosophers from the time of Parmenides onwards there was a belief that reality was a unified whole and that change was impossible. Only the activities of our senses lead to the illusion of change. Society did change and progress in early periods, often in response to major events such as disease, war, religious conversion or invasion but also because of technological developments, say in agriculture, building or weaponry. However, the pace of change was such that a human lifetime would often pass without significant development of basic social, technological or economic circumstances. As novelist H.G. Wells says: 'Through the long generations, the ancestor of all of us fought and bred and perished, changing almost imperceptibly.' But as science emerged, each new discovery or theory

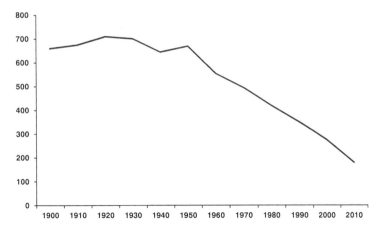

Figure 11.2 Baptisms per thousand births in UK, 1900–2010

being based on and improving upon the one before, a new sense of 'progress' emerged and began to be applied to society as well. However, science and religion are both human activities for which cultural and psychological drivers are important and which share certain matters of tone. So it can be argued that science, like religion, tends to categorise people as 'believers' or 'heretics' and treat them accordingly. The fervour with which UK Darwinist Professor Richard Dawkins attacks those who do not accept the tenets of atheistic scientism has an unmistakeable religious tone. Individuals are vilified or, in modern style, refused research grants if their views, however well founded in experiment or observation, are too far from the accepted orthodoxy, or dogma, of the day, as explored in more detail in chapter 9. Science has its priesthood: figures who seek to take over from traditional religious figures as the unchallengeable wielders of holy knowledge. High-ranking scientists in the energy field, perhaps particularly with respect to nuclear power in the post-war era, could exercise a mesmeric influence over the political and public establishment of the day to the extent that it became very difficult to criticise them. This scientific priesthood reserves the right to decide which ideas are 'scientific' and which are not.

Philosophically evolution is no more 'scientific' than creationism – certainly neither can be 'proved'. One suspects that Hume might argue that the scientific/technical world view is a self-consistent system with no external reference point to 'reality'. Consider a reasonably typical scientific statement: 'We know how oil came into being and we can tell with considerable accuracy how long the process took.' There are at least three distinct ways in which any individual actually does not 'know' these things but takes them as articles of faith.

First, it is of course in principle simply impossible to check whether what we 'know' about oil is correct, as it is impossible to go back in time to check. Secondly, even if we could it would not be possible for any individual to check the theory about oil being laid down, as we are not given sufficiently long lives or the sensory apparatus to allow us to do this. And thirdly, in what sense could any individual personally 'know' such a thing? No single person can have done the fundamental scientific research which led to this conclusion, spanning as it would the sciences of geology, physics, biology, chemistry and doubtless a dozen others.

In effect, then, a lot of what a particular scientist 'knows' they have read in a book or heard by word of mouth. The people who wrote the book in turn will only have carried out a tiny proportion of the direct underlying research if any: they will have based almost all of their beliefs on things they read in another book or someone else has told them. And so on.

In any case the scientific and spiritual world views may not be inconsistent. There could be a creating God who set up the laws of physics, mathematics and logic and then allowed the universe to unfold according to those laws.

So science, like traditional religion, is a self-consistent system without an external point of anchorage. It is as consistent to argue that God created the world, including creating a fossil record, to imply that the world is much older than it really is, perhaps to keep us interested or to test how honest we can be with ourselves about what is fact and what is faith, as it is to argue that the fossil record 'proves' that oil was laid down millions of years ago by a particular chemical–biological process. In fact, one could even argue that since creationism has fewer steps in it (just the one) than evolution, applying Occam's razor would imply that we should prefer the former. 'The truth about physical objects must be strange. It may be unattainable, but if any philosopher believes that he has attained it, the fact that what he offers as the truth is strange ought not to be made a ground of objection to his opinion.'[10] Traditional religion is certainly ineffable but at its heart the scientific approach is no less so.

The enthusiasm for science and the progress which it fuelled did not come through the two centuries following Condorcet's optimistic pronouncements intact. In 1807, for example, Wordsworth published his sonnet, 'The world is too much with us', in which he attacked the 'decadent material cynicism of the time' and the effect of industrialisation on mankind's relationship with 'nature':

> The world is too much with us; late and soon,
> Getting and spending, we lay waste our powers;
> Little we see in Nature that is ours;
> We have given our hearts away, a sordid boon!
> This Sea that bares her bosom to the moon,
> The winds that will be howling at all hours,
> And are up-gathered now like sleeping flowers,
> For this, for everything, we are out of tune;
> It moves us not. Great God! I'd rather be

A Pagan suckled in a creed outworn;
So might I, standing on this pleasant lea,
Have glimpses that would make me less forlorn;
Have sight of Proteus rising from the sea;
Or hear old Triton blow his wreathèd horn.

However, ongoing improvements in the quality of life as industrialisation proceeded served as a counterweight to the finer feelings of the (generally speaking) wealthy artistic classes. The discovery of radiation came at a time when many countries were fascinated by the prospect of many benefits that science and technology could bring into people's lives. The 'second industrial revolution', characterised by electricity, early mass communication (notably the telephone), the private motor car, corporate governance of major industries rather than individual owners, mass production (Fordism and Taylorism) and the growth of trade unions, had very different effects on people's lives from the first Industrial Revolution a century earlier, with its urban squalor and extreme unevenness in distribution of wealth.[11] Between 1900 and 1950 in the UK infant mortality fell from 140 per 1,000 births to 33; life expectancy grew from 45 years (men) and 49 years (women) to 65 years (men) and 70 years (women); and the country's GDP per capita (in 2014 monetary values) rose from £6,100 to £8,800.[12] The temptation to imbue science with the same awe and reverence as had once been bestowed upon the Church proved irresistible.

11.4 A LOVE–HATE RELATIONSHIP

Radiation seems to tap into some of our deepest 'race memories', such as the Eden myth (forbidden knowledge and being cast from paradise), the small boy playing with fire, 'the Sorcerer's Apprentice', as well as some more positive ones such as the alchemist's dream, mankind's triumphing over nature and taking dominion over the earth.[13] Further, writers such as H.G. Wells were postulating 'atomic bombs' some decades before fission was discovered.

Never before in the history of warfare had there been a continuing explosive; indeed, up to the middle of the twentieth century the only explosives known were combustibles whose explosiveness was due entirely to their instantaneousness; and these atomic bombs which science burst upon the world that night were strange even to the men who used them. Those used by the Allies were lumps of pure Carolinum, painted on the outside with unoxidised cydonator inducive enclosed hermetically in a case of membranium. A little celluloid stud between the handles by which the bomb was lifted was arranged so as to be easily torn off and admit air to the inducive, which at once became active and set up radioactivity in the outer layer of the Carolinum sphere. This liberated fresh inducive and so in a few minutes the whole bomb was a blazing continual explosion.

The catastrophe of the atomic bombs which shook men out of cities and businesses and economic relations shook them also out of their old established habits of thought, and out of the lightly held beliefs and prejudices that came down to them from the past. To borrow a word from the old-fashioned chemists, men were made nascent; they were released from old ties; for good or evil they were ready for new associations. In the map of nearly every country of the world three or four or more red circles, a score of miles in diameter, mark the position of the dying atomic bombs and the death areas that men have been forced to abandon around them. Within these areas perished museums, cathedrals, palaces, libraries, galleries of masterpieces and a vast accumulation of human achievement, whose charred remains lie buried, a legacy of curious material that only future generations may hope to examine.[14]

11.5 THE PUBLIC'S RELATIONSHIP WITH RADIATION IN THE EARLY DAYS (1895 TO 1930)

Radiation, in the form of X-rays, had been used to treat cancer almost since its discovery by Wilhelm Röntgen in 1895. Radium therapy for the autoimmune disease lupus began trials in 1901, with considerable success, to be followed by inserting radium chloride sources into the body to fight cancer. It was a small step to ascribe all kinds of physical benefits to the new phenomenon.

Radon soon developed a strong reputation.[15] The water at Hot Springs in Arkansas had been valued for their reputed health-giving properties for many years. Indeed, in 1832 Congress established the Arkansas Hot Springs as the first federal reservation, a forerunner of the national park system. In 1903 J.J. Thompson wrote a letter to the journal *Nature* in which he described his discovery of the presence of radioactivity in well water. Soon it was realised that many more of the world's most famous springs contained radioactive materials owing to 'radium emanation', now understood to be radon gas produced by the decay of radium deposits in the ground through which the springwaters flow.

Correlation often equals causation in the media and political mind and it was unsurprising that many assumed that the radioactivity of the springs underlay their health-giving properties. In 1910 the US surgeon general, George Torney, wrote: 'Relief may be reasonably expected at the Hot Springs in various forms of gout and rheumatism, neuralgia, metallic or malarial poisoning, chronic Bright's disease, gastric dyspepsia, chronic diarrhoea, chronic skin lesions, etc.' Dr C.G. Davis, in the *American Journal of Clinical Medicine*, added that: 'Radioactivity prevents insanity, rouses noble emotions, retards old age and creates a splendid youthful joyous life.' Professor Bertram Boltwood of Yale provided the scientific basis for these miraculous properties, explaining that the radioactivity was: 'carrying electrical energy into the depths of the body and there subjecting the juices, protoplasm and nuclei of the cells to an immediate bombardment by explosions of electrical atoms, stimulating cell activity, arousing all secretory and excretory organs, causing the system to throw off waste products and acting as an agent for the destruction of bacteria.' (The experimental evidence for these conclusions seems to have been lost.) Health spas and resorts enjoyed booming trades: many of them changed their names to include the word 'radioactive' or 'radium'. Nor was the benefit to be restricted to human health. One farmer reportedly wanted to add radium to his chicken feed so that the chickens would lay hard-boiled eggs; another argued that fertilising fields with radium would improve crop production and produce better tasting food.

An early problem to be identified was a practical one. Radon, being a 'noble gas' which does not interact much with other elements, does not remain dissolved in water for very long. So to get the health benefits of radon it was necessary to drink the water at the spa – it could not be bottled for remote usage, thereby depriving the poor and the infirm of these benefits.

By 1912 R.W. Thomas patented the Revigator, a device which could add radon to drinking water in the home. Sales reached several hundred thousand despite it selling for $29.50 in 1929 (over $400 in mid-2010s values). The Revigator involved a jar made of radium-bearing ore which produced radon which would dissolve in the water overnight – 'a perpetual health spring in the home'. Other devices like Thomas Cone's 'Radium Emanator' could be placed in water and were small enough to be carried in a suitcase, so granting the benefits of radon to those who travelled for a living. Radiation Emanation Bath Salts, containing 'Epsom salts – radium chloride, 1 microgramme', were described by the manufacturer, the Denver Radium Service, as being good for nervous disorders, insomnia, general debility, arthritis and rheumatism. The directions read: 'Empty contents in a quart of hot water. After a few moments add to regular bath solution. Remain in bath 45 minutes with cover over top of tub. Upon leaving bath relax in bed for one hour.' (The concept, though not radium's part in it, survives in the name of Radox bath salts, which were first manufactured in 1908 and were supposed to 'radiate oxygen'.)

The American Medical Association (AMA) was concerned that the public was being fleeced by frauds. Fearing that some of the devices might not deliver the dose of radiation they promised, from 1916 to 1929 it established guidelines by which its approval would only be granted if the apparatus generated more than 75 kBq (i.e. sufficient to create 75,000 particles of radiation per second) of radon per litre of water in a 24-hour period. The AMA therefore struck an important blow for ensuring that people received the irradiation they were paying for.

Some people came to assume that the direct ingestion of radium would be even more effective than letting it wash against the skin. From the early 1920s and right into the early 1930s it was possible to purchase radium-containing salves, beauty creams, toothpaste (radon was thought to fight dental decay and improve the digestion), ear plugs, chocolate bars, soap, suppositories and even contraceptives. For the sufferers of respiratory ailments there were the Radium Nose Cup and the Radium Respirator ('Radium: scientists found it, governments approved it, physicians recommended it, users endorse it, we guarantee it, SURELY IT'S GOOD.')

Out came the charlatans, of course. J. Bernard King, manufacturer of the Ray-Cura, a quilted pad that he said would emit radium emanation into the diseased portions of the body to kill the germs, came to grief when it was revealed that the pad contained ordinary soil rather than radium ore. The Degnen's Radioactive Solar Pad claimed to get its energy from the sun and had to be charged in sunlight for several minutes prior to use. However, the manufacturers of its highly priced competitor, the Radiendocrinator, appealed to science when they retorted that the idea of charging in the sun was 'the purest of nonsense; there is not a shred of truth known to modern science that substantiates such a theory.' The Radiendocrinator, by contrast, was made of 'refined radium' encased in 14-carat gold and shipped in an embossed velvet-lined leatherette case (at a price of $150, or $2,000 in 2014 prices). Giving one example as to how their Radiendocrinator might be used, the manufacturers advised men to: 'Wear the adaptor like any athletic strap. This puts the instrument under the scrotum as it should be. Wear at night. Radiate as directed.' Whether this was underpinned by better science than the 'pure nonsense' of the Radioactive Solar Pad is hard to judge at this distance.

11.6 CHANGING UNDERSTANDING

Almost from its discovery in 1898 some of the harmful biological effects of radium became apparent. The first case of 'radium dermatitis' was reported in 1900 by Becquerel, who carried a small ampoule of radium in his waistcoat pocket for six hours and reported that his skin became ulcerated. Marie Curie experimented with a tiny sample that she kept in contact with her skin for ten hours and noted that an ulcer appeared several days later.[16] (Mis)handling of radium has been blamed for her death from aplastic anaemia.

So amid the enthusiasm for all things radioactive, darker themes were emerging. As early as 1915, California Senator John Works questioned the efficacy of radium treatment for some cancers, which he said seemed to be made worse by the procedure.[17] In the same year the British Röntgen Society adopted a resolution to protect people from overexposure to X-rays, probably the world's first organised effort at radiation protection. However, it was the emergence of 'radium jaw' which prompted one of the first serious public challenges to the unquestioning acceptance of medium-level radiation as a life-giver.

Radium, when mixed with zinc and copper in paint, is one of the few radioactive materials which correspond to the public image of stuff that glows in the dark (and indeed is probably largely responsible for that misconception). From 1910 onwards paint containing radium was used for watches, nuclear panels, aircraft switches, clocks and instrument dials. A typical self-luminous watch using radium paint would contain around 1 microgramme of radium.[18]

In the mid-1920s a lawsuit was filed against the United States Radium Corporation by five dying 'Radium Girl' dial painters who had painted radium-based luminous paint on the dials of these watches and clocks. The dial painters routinely licked their brushes to give them a fine point, thereby ingesting radium. Their exposure to radium caused serious health effects which included sores, anaemia and bone cancer. Chemically radium is very similar to calcium (they belong to Group 2 of Mendeleev's Periodic Table) so it gets deposited in the bones, where its radioactivity degrades marrow and can cause mutation in bone cells.

During the litigation it was determined that the company's scientists and management had taken considerable precautions to protect themselves from the effects of radiation – suggesting that at least among the research community some were recognising that the radium health fad certainly had its limitations and may even be based on a myth – yet had not seen fit to protect their employees. Worse, for several years the company had attempted to cover up the effects and avoid liability by insisting that the Radium Girls were instead suffering from syphilis. This flagrant disregard for employee welfare (and personal reputation) had a significant impact on the formulation of occupational disease labour law.[19]

There were other examples. Radithor, a liquid, came in half-ounce bottles with each bottle guaranteed by the manufacturer to contain 75kBq of radium (the minimum level which the AMA demanded before it would recognise a product as offering health benefits). Eben Byers, well-known Pittsburgh industrialist, US amateur golf champion and man about town, was so convinced of the product's worth that he averaged three bottles a day until his death from radium poisoning in April 1932. His widely publicised demise, coming on top of deaths among the Radium Girls, helped cool the public's appetite for these radioactive cure-alls.

Manufacturers of the devices countered by cautioning against excessive doses of radium and recommending moderation. Nevertheless, radium's heyday as a medical miracle was coming to an end, to be replaced by an appreciation of its potentially lethal properties.

Despite legal restrictions, some radioactive products were manufactured into the 1960s. For example, radium contraceptives were advertised and sold in the early 1950s by a Denver company. The 1960s saw the production of the Gra-Maze Uranium Comforter in La Salle, Illinois: a quilted pad containing uranium ore that was meant to be placed on whatever part of the body was ailing. Unlike Ray-Cura, on which it was based, this device made no false claims, containing uranium as advertised. Production ceased in 1965 following a search and seizure operation by federal agents. About the same time a similar fate befell the operations of the Ionic Research Foundation in Winter Park,

Florida. The main product of this company was the Ionic Charger, a device intended to add radon to drinking water. The manufacturer explained that people: 'have been brainwashed by bureaucratic screaming about fallout. The truth of the famous spas has been lost sight of.' His literature claimed the product would have a sedative effect on the nervous system and that 'highly-strung' individuals 'become less irritable and lose their distressing tendency towards insomnia.' Other examples from the 1960s included the Lifestone Cigarette Holder, 10 cm long and containing a small quantity of radium. Inhaling the smoke over the radium would make the tobacco sweeter and milder, protect users from lung cancer and promise them beautiful faces and excellent health.

In 1985, 20,000 Endless Refrigerator/Freezer Deodorisers, made of green plastic mixed with thorium-containing monazite sand, were distributed in the US. Users were told to hang it in the refrigerator where the emitted radiation would purify the air by destroying odours. Japan saw the emergence of the NAC Plate, which looked like a playing card but which contained low-grade uranium ore on one side. The plate was to be slipped into a pack of cigarettes where the radiation 'denatures and reduces nicotine, tar, and harmful gas', helping the user to enjoy 'golden moments of watching (the) smoke rise slowly' so 'with your nerves relieved and refreshed you can get back to work.' But at least such devices were not designed directly to irradiate human tissue.

Claims for the benefits of radiation exposure have not entirely died out today. The Free Enterprise Radon Health Mine is the largest of six operating uranium 'health mines' in Boulder, Montana. Its website, listing 26 conditions which can be addressed by 'radon therapy' (including ankylosing spondylitis, diabetes, migraine, eczema, multiple sclerosis, gout, bronchitis, breast cancer and high blood pressure), describes radon as:

> one of mankind's oldest therapies ... the use of hot springs with high radon content dating back some 6,000 years ... Clinical, double-blind and randomised controlled studies in many countries report findings substantiating claims of pain and symptom relief, supporting the observations of benefit equal to that reported by visitors to the Free Enterprise Radon Health Mine ... Therapeutic applications of low dose radiation start with optimizing the body's natural hormones (endorphins, enkephalin, adrenaline, insulin, histamine) for pain control and stress relief ... DNA repairs and cell recovery are also obvious in the research findings. A key tumour suppression gene – the P53 – is activated and helper T-cells are enhanced by means of the decrease of suppressor cells. Overall, stopping or suppressing the progress of the disease process followed by recovery is to be expected. With further comprehensive research, optimum individual dose rates for preventive measure, health maintenance, post-operative recovery and healing of injuries will be found in the future.[20]

11.7 Radiation all around and a brief divergence into common sense

With so much natural radioactive decay going on around us, plus the radiation that can travel largely undisturbed in the vacuum of space, some of which penetrates the earth's atmosphere, radiation is a common feature of the natural world in which we live and have evolved. The biological effect of certain dose of radiation – strictly of the amount of energy absorbed by cells of the body from radiation passing into or through our bodies – is expressed in term of 'Sieverts' (Sv), named after Rolf Sievert, 'the father of Swedish medical physics', or its subunits the millisievert (mSv), or microsievert (μSv). In the UK we get an average annual radiation dose of around 2.6mSv. Most of it – about 84 per cent – comes from a variety of natural sources – see figure 11.3. More than half of the total comes from two isotopes of radon gas (sometimes called 'radon' and 'thoron') which can accumulate in living areas.

Dose varies significantly depending on geographical location, altitude and (to a lesser extent) lifestyle. The cracked granite geology of the south-west of England leads to much higher levels of radon accumulating in homes and therefore to a substantially higher radiation dose – see figure 11.4 (colour section).[21] The average radiation dose to which individuals in Cornwall are exposed is three times as great as the UK average (7.8 mSv per year against 2.6 mSv per year). Of course there are individual areas and dwellings in Cornwall which deliver much greater doses than the average and there are places elsewhere in the world where the dose can be higher still, even to levels well over 100 mSv per year.

Common sense is not necessarily an infallible guide to scientific truth – indeed, as discussed earlier, it is a positive barrier to grasping the implications of quantum mechanics. However, while what we might regard as medium to high doses of radiation

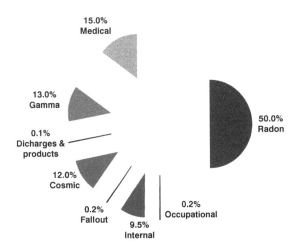

Figure 11.3 Average radiation doses from various sources in UK

would be expected to cause biological damage owing to the potentially disruptive energy that they deposit in cells, this is less clear when it comes to doses of the level that are prevalent in the normal environment.

The common sense question is this. Given the wide variations in background levels of radiation, why have we not evolved to be able to detect it through our senses? After all, a wide range of harmful environmental stimuli, such as bright sunlight or the smell of hydrogen sulphide, lead us to recoil to protect ourselves from harm. This is something which presumably has evolutionary advantages. If radiation represents a significant risk at dose rates of a few mSv per year then we might expect that those of our ancestors who could detect it might have a better chance of passing their genes on than those who couldn't. That we cannot suggests that radiation does not represent harm at these levels, probably because we have evolved mechanisms to repair any damage that might arise.

11.8 HOW DANGEROUS IS RADIATION – THE LINEAR NO-THRESHOLD (LNT) DOSE SPECULATION

Whatever the 'truth' might be, talking up the potential detriment associated with low levels of radiation obviously serves a useful purpose for those who for financial or other reasons would like to see nuclear power fail. As journalist and Green activist George Monbiot said at the height of the Fukushima crisis in 2011: 'Over the past fortnight I've made a deeply troubling discovery. The anti-nuclear movement to which I once belonged has misled the world about the impacts of radiation on human health. The claims we have made are ungrounded in science, unsupportable when challenged and wildly wrong. We have done other people, and ourselves, a terrible disservice.'[22] Sadly the authorities, both regulatory bodies and indeed the nuclear industry itself, have often shown themselves to be willing accomplices in fostering exaggerated fears of radiation, no matter how much damage these exaggerations may cause to the well-being of the public at large.

By the mid-1950s there was quite a lot of evidence about the health effects of relatively high doses of radiation gained from studies of various populations who received doses, for example, from medical sources (both administering and receiving the treatment) and also among the populations affected by the Hiroshima and Nagasaki atom bombs.

There are two sets of potential health effects associated with radiation exposure: 'non-stochastic' and 'stochastic'. Non-stochastic effects are those which will (almost) certainly occur within a short time of the individual involved receiving a sufficiently high dose of radiation. 'Acute radiation syndrome' (ARS) sets in if the individual receives a dose of more than around 1 Sv (1,000 mSv, i.e. 400 times the annual average dose) in a short period. It can result in three sets of symptoms: haematopoietic, gastrointestinal and neurological. Haematopoietic (blood) disorders, which can be detected after a dose of 1 Sv or even less, involve a drop in the number of blood cells, potentially causing conditions such as anaemia (from a shortage of red blood cells) and inability to fight infections (caused by a lack of white blood cells). Gastrointestinal problems, which typically appear after doses of 6 Sv or more, can involve nausea, vomiting, loss of appetite and stomach pains. Neurological effects may include dizziness or headaches and tend to emerge after very high doses of at least 10 Sv.

Below around 4 Sv or 5 Sv the sufferer will usually recover from ARS. Above this level exposure is usually accompanied by death, although there are individual differences – one of the survivors of the Goiânia incident (see below) appears to have received a dose of 7 Sv. Death itself often takes up to 28 days, a period through which the victim may appear to recover from initial symptoms before suddenly succumbing.

However, after the end of the Second World War deaths from ARS have been relatively few. There has been a total of around 120 such deaths from a total number of ARS cases of below 500. The single biggest category has been overdoses during radiotherapy, caused by mistakes in calibrating equipment, which accounts for perhaps 40 cases. Twenty-eight workers died of ARS as a result of the Chernobyl accident and a similar number owing to accidents on Soviet nuclear submarines. There have been four murders (including Mr Alexander Litvinenko in London in 2006) and five 'criticality' accidents in research or production facilities. The theft or loss of medical sources containing radioactive materials has resulted in around 20 deaths. One of the most widely reported of these, at Goiânia in Brazil in 1986, involved the theft of a radioactive source from an abandoned hospital. Two hundred and fifty people were found to have radioactive material on their skin, 20 were treated for ARS and four died.

Research and public attention has been focused far more on the potential late or delayed effects of radiation, especially cancer. (Despite occasional claims to the contrary the evidence for genetic effects of low-level radiation seems weak – for example, there is no reliable evidence for detectable genetic effects among children conceived by Hiroshima and Nagasaki survivors after the bombs were dropped.)[23] These effects are stochastic (or 'statistical') – if 1,000 people each receive a certain dose of radiation, say 100 mSv (a total of 100 'person-Sv'), around five excess cancers might be expected during the lifetime of those exposed. In other words, if 300 of them might be expected to die from cancer in normal circumstances, the figure should be 305 for the exposed population. However, it is impossible to say with any certainty which of the 1,000 people involved would get cancer, nor which of the 305 who did get cancer contracted the disease as a result of their radiation exposure. A further complication is that, unlike acute radiation effects which are seen very soon after exposure to radiation, stochastic effects can often lie 'latent' for some years before expressing themselves. The survivors of the Hiroshima and Nagasaki atomic weapons, for example, showed an increase in leukaemia which reached a peak around five years after the bombs were dropped and then fell away. Much later, after a gap of around 20 years or more, an increase in solid cancers was seen.

A single dose of 50 mSv, or a dose of 100 mSv spread over a year, are the smallest where it can be shown with statistical confidence by measuring the population that the risk of developing cancer over the following decades will increase as a result of the exposure. Above this level the risk seems proportional to the dose – in other words a population of 1,000 people who received a dose of 1,000 mSv (the threshold at which ARS starts to become apparent) would be expected to suffer from an excess of 50 cancers when compared to a similar population which did not receive such exposure.

In part the difficulty is differentiating between deaths caused by radiation and deaths from other causes in the population at large. By the normal laws of statistics, the number of deaths in a population varies from year to year, both because of variations say in the weather and from sheer chance.

The number of registered deaths in England and Wales over a decade up to 2013 varied between 514,250 (2004) and 484,367 (2011), as shown in figure 11.5.

The population grew from perhaps 53 million to 57 million over this same period, but as can be seen this change does not explain the pattern of the number of deaths, which declined between 2004 and 2011 and remained below 2004 levels in 2013.

A variety of factors may underpin this change – for example, a disease epidemic (or absence thereof), a change in healthcare or diet, or a change in the average age of the population. However, at least some of the variation is likely to be simply random, or at least impossible to tie to any particular environmental factor.

Let us now assume that radiation is just as dangerous, proportionally, at a dose of 1 mSv as it is at a dose of 100 mSv. This would imply that there should be an average of one twentieth of an excess case of cancer among 1,000 people exposed to this dose, or 50 cases per million people. Assume that through some catastrophe the whole population of England and Wales received an extra dose of 1 mSv every year throughout their lifetimes. In a population of 57 million this would, under this assumption, represent an extra 2,800 deaths each year (assuming no improvement in cancer treatment rates, of course). That figure would be entirely lost against the natural variation from year to year – what in the jargon of statistics is called the 'noise' in the system. Even if cancers alone were considered it would be impossible to tell just by looking at the figures whether there was an effect or not – the 'noise' would drown out the 'signal'. All we can say with any certainty is that if there is an effect of radiation at very low doses – positive or negative – it is too small to be measured.

As figure 11.6 illustrates, some substances may remain proportionally dangerous right down to a zero dose. Some do not seem to have any health effects below a certain level, or 'threshold'. Others become beneficial at very low doses (referred to as 'hormesis' and the basis, for example, of inoculation against diseases). One person taking 100 aspirins at one go would die. Three people taking 33 might suffer some health difficulties. One hundred people each taking one might do rather well out of it – indeed, some doctors advise some categories of patient, for example those suffering from heart disease, to take

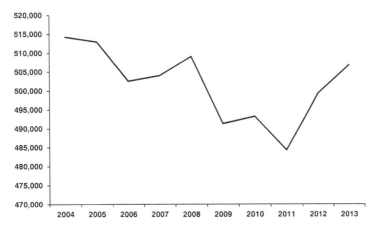

Figure 11.5 Annual deaths in England and Wales, 2004–13[24]

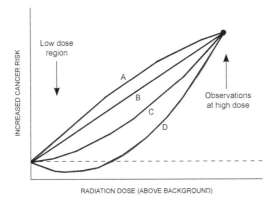

Figure 11.6 Possible relationships between radiation dose and risk of developing cancer at low doses

regular low doses. It is difficult to find evidence of any biological insults that do follow a Linear No-Threshold model. Claims to the effect have been made for the chemical benzene and for asbestos but these seem to be more in the way of working hypotheses to help design protection systems or to decide compensation claims than anything based on clear observations.

For some years the prevailing model, or speculation, used by international bodies, most notably the UN Scientific Committee on the Effects of Atomic Radiation (UNSCEAR), has been the 'Linear No-Threshold' (LNT) model, which does hold that radiation remains proportionally dangerous irrespective of dose rate – i.e. 1 mSv has the same risk of causing harm whether it is received over one minute or one year. The principal justification offered for the LNT is theoretical and very simple. A single particle of radiation hitting a single DNA molecule in a single cell nucleus of a human body can initiate a cancer. The probability of a cancer initiation is therefore proportional to the number of such hits, which is proportional to the number of particles of radiation, which is proportional to the dose. Thus the risk is linearly (i.e. proportionally) dependent on the dose. In fact, the philosophy of radiation protection actually claims that the authorities 'prudently assume for the purposes of protection that the detriment follows the LNT model'.[25] As we shall explore below, it is highly questionable whether assuming radiation is more dangerous than it probably is represents 'erring on the side of caution' when its overall consequences are taken into account. A variety of bodies alongside UNSCEAR, such as the US National Research Council and the US Environmental Protection Agency, have supported LNT, while others, including the French Academy of Sciences and National Academy of Medicine, and the US Nuclear Regulatory Commission have either rejected it or questioned its basis.

One of the main critics of the LNT speculation was the late Bernard Cohen. He argued:

> The problem with the very simple argument [supporting LNT] is that factors other than initiating events affect the cancer risk. Our bodies have biological defence mechanisms which prevent the vast majority of initiating events from developing into a fatal cancer … There is plenty of very direct and obvious evidence on this. For example, the number of initiating events is roughly proportional to the mass of the animal; more DNA targets means more hits. Thus, the simple theory predicts the cancer risk to be proportional to the mass of the animal. But experience indicates that the cancer risk in a given radiation field is roughly the same for a 30 g mouse as for a 70 kg man and there is no evidence that elephants are more susceptible than either.[26]

Citing animal studies, he went on to postulate that, as is the case with other kinds of inoculation, the immune response to radiation damage is stimulated by low-level radiation (LLR). He referred to a wide range of evidence, derived from radiation workers, atom bomb survivors, breast cancer patients, the Radium Girls and others with occupational exposure to radium ingestion. He noted lower overall cancer rates in the Rocky Mountain states of the US (where radiation levels are highest) than in New England, all of which he claimed suggested that below around 200mSv lifetime exposure, radiation exposure was protective against cancer.

In particular, with respect to a study carried out by the University of Pittsburgh, Cohen goes through a range of possible confounding factors, including average affluence, levels of smoking, degree of industrialisation, air quality and other lifestyle and environmental issues, and concludes:

> Since no other plausible explanation has been found after years of effort by myself and others, I conclude that the most plausible explanation for our discrepancy is

that the linear no-threshold theory fails, grossly over-estimating the cancer risk in the low dose, low dose rate region. There are no other data capable of testing the theory in that region. An easy answer to the credibility of this conclusion would be for someone to suggest a potential, not implausible, explanation based on some selected variables. Either they or I will then calculate what values of those variables are required to explain our discrepancy. We can then make a judgement on the plausibility of that explanation.

It does not appear that the challenge was taken up.

11.9 Vested interests at work

The aim of considering this as a scientific dispute here is not to come down on the side of LNT, hormesis, or any other theory. It is rather to examine how the (self-appointed) guardians of public well-being and environmental protection might have been expected to respond, based on their records elsewhere, and contrast this with their actual reaction. Several vested interests do very well out of a general public fear of low-level radiation and heavy expenditure to reduce doses of radiation, whether or not such expenditure would make a significant contribution to human health or could be better spent researching other diseases. As one beneficiary of spending on radiation and health research admits: 'I can think of no other area of toxicology (e.g. asbestos, lead, smoking) with so many studies.'[27]

The risk of agenda-driven conclusions in scientific research has always been an issue but it seems to have become more acute as the practice of science has changed from an activity characterised by committed amateurs motivated by a pure search for knowledge and largely supported by philanthropists, into a profession providing quite well-paid employment but requiring its participants to persuade government funding councils or private companies (directly or through educational foundations) to grant finance to their proposals programme rather than to those of their competitors. Prior to the Second World War support for research by the government of the US, for example, was largely focused on government missions and carried out by federal employees in federal establishments. The experience with weapons development during the war revealed the potential impact of the results of scientific research on national needs.[28] In 1950 Vannevar Bush, who had headed the Office of Scientific Research and Development during the war, wrote: 'The government should accept new responsibilities for promoting the flow of new scientific knowledge and the development of scientific talent in our youth. These responsibilities are the proper concern of the government for they vitally affect our health, our jobs and our national security.'[29]

In such an environment a 'winning' bid is more likely to be one that purports to address and then solve a 'problem', especially one with apparent public policy implications, rather than simply expand the sum of human knowledge and understanding. Scientists may well still be motivated by an urge to gain 'pure', accurate knowledge, not least because work which is demonstrably inaccurate will eventually lose favour with the funding authorities and even with politicians, though this can take a depressingly long time in some cases. But, inevitably, considerations of how to attract the next pay packet, or television appearance, will create pressure towards a particular type of result. Studies which absorb large amounts of resource to find no evidence of a supposed link between say a pollutant and a disease can sometimes be portrayed as a waste of money and are often not regarded as newsworthy – when in fact of course ruling out of various non-viable options is a vital component in coming to an understanding of any phenomenon.[30]

Study of letters exchanged by various participants around the time of a meeting held in 1956 by the Biological Effects of Atomic Radiation I (BEAR I) committee Genetics Panel of the US National Academy of Sciences (NAS) is illuminating.[31] BEAR I's remit was to provide decision-makers with the best scientific advice concerning radiation exposure on which to base their stance towards the atmospheric atomic weapons testing and potential genetic effects of the fallout. However, there were at least three sets of vested interests in play who would benefit from the establishment of a LNT model, with the associated powerful propaganda message that 'there is no safe dose of radiation', i.e.:

- those promoting a political campaign against nuclear weapons;
- those seeking research grants into radiation, genetics, etc.;
- those in the oil industry in particular with a desire to limit the development of nuclear power.

Many politically active members of the scientific community, well connected to members of BEAR I, opposed the further development of nuclear weapons and were developing public campaigns to support this cause (in much the same way, though for the opposite objective, as Eisenhower's use of his 'Atoms for Peace' speech in 1953, with its enthusiastic vision for nuclear power, to allay public concerns about nuclear weapons technology). It was clear that, whatever the truth might be, statements supporting LNT would offer considerable assistance to these activists by raising fears of alleged hazards associated with fallout from atmospheric atom bomb testing.

Perhaps less honourably, other members of BEAR I wanted to maximise their chances of winning funds for a prolonged programme of general genetic research and saw public fears, whether justified or not, as a considerable benefit. Contemporary correspondence strongly implies that members of the Genetics Panel saw their role in the BEAR I committee as involving advocacy of funding for radiation research. For example, Theodosius Dobzhansky wrote to Milislav Demerec saying:

I myself have a hard time keeping a straight face when there is talk about genetic deaths and the tremendous dangers of irradiation. I know that a number of very prominent geneticists and people, whose opinions you value highly, agree with me.

Demerec replied:

Let us be honest with ourselves – we are both interested in genetics research and for the sake of it we are willing to stretch a point when necessary. But let us not stretch it to breaking point! Overstatements are sometimes dangerous since they result in their opposites when they approach the levels of absurdity. The business of genetic effects of atomic energy has produced a public scare and a consequent interest in and recognition of [the] importance of genetics. This is to the good, since it will make some people read up on genetics who would not have done so otherwise and it may lead to the powers-that-be giving money for genetic research which they would not give otherwise.

By the 1950s the fossil fuel industry was one of the world's biggest and most profitable, with a vested interest to prevent low-cost competitors entering the marketplace. A 1987 biography of Warren Weaver, director for Natural Sciences for the Rockefeller Foundation from 1932 to 1959, described the beginnings of the NAS study of radiation in 1955. The Rockefeller Foundation inevitably had (and until very recently continued to have)[32] an interest in maintaining the position of oil and natural gas in energy supply systems. Those fuels had made John D. Rockefeller the world's first dollar billionaire (and in today's money terms probably the world's richest person ever) and created the resources for the Foundation to spend on supporting suitable scientific research. The Rockefeller Institute was one of the largest sources of funds for basic science in the US.

Press articles in 1954 suggested that the public was confused about the effects of radiation. At a Rockefeller Foundation board meeting, attendees asked Detlev Bronk, NAS president and a member of the Rockefeller Foundation board, if there was a way to produce some definitive answers. The NAS proposed forming six committees to investigate the issue and the Rockefeller Foundation agreed to provide the funds to produce the reports. Warren Weaver served as the chairman of the Genetics Committee for the first BEAR reports. Of the other members of the committee, at least four had received Rockefeller Foundation grants before 1956 and several continued receiving grants afterwards. In the *New York Times* of 13 June 1956 the committee's report was front page news. The article read: 'A committee of outstanding scientists reported today that atomic radiation, no matter how small the dose, harms not only the person receiving it but also all of his descendents [sic]. The six committees studied the radiation problem in the fields of genetics, pathology, meteorology, oceanography and fisheries, agriculture and food supplies and disposal and dispersal of radioactive wastes'. The Genetics Panel report received the most attention. Yet the primary data that were available to this committee was derived from experiments using X-rays on fruit flies, most of which were conducted by recipients of Foundation grants and members of the committee.

This reads as something of a grand conspiracy which may or may not be a realistic representation of all that went on. What is interesting is the response, or lack of response, from Big Green. Were a story of this nature to be concerned with say claims that climate change is an exaggeration or that renewables have serious side-effects Big Green would, one suspects, be falling over itself to attack the funding streams and other potential vested interests of the perpetrators.

More frustrating is the role that the regulators and the nuclear power industry have played in lending the Big Green claims credibility. After all, in a sense none of this would matter if the outcome were a system of radiological protection which genuinely reduced the harm being done to individuals. In reality, however, the outcome is precisely the opposite. These measures may have had some success in reducing the dose of radiation sustained by some individuals. It is a much more difficult task to argue that they have on balance improved their quality of life. As one group of researchers put it: 'LNT was a useful model half a century ago. But current radiation protection concepts should be based on facts and on concepts consistent with current scientific research and not on opinions. Preconceived concepts impede progress: in the case of LNT models they have resulted in substantial medical, economic and other societal harm.'[33]

A combination of the way the establishment responds to incidents involving radiation from nuclear power – not from other sources – and the way the issue is communicated by the 'nuclear family' itself has been a major contributor to a huge and unnecessary detriment to many individuals, not least at Fukushima. In common with Three Mile Island, Fukushima doesn't seem to have caused any deaths from radiation, though at least one worker received compensation under the plant's worker compensation scheme after developing cancer – causation was described as 'unclear'.[34] Even at Chernobyl the demonstrable death toll which resulted from radiation exposure was very small compared to the direct effects of events such as Bhopal or the Banqiao dam failure. What created the human misery in all of these events was the reaction – not so much the immediate precautionary evacuation but what followed and ironically what preceded.

11.10 THE IRRATIONALITY OF THE RATIONAL

As noted in chapter 3, in the initial hours and days after the Fukushima accident, progressively larger zones around the stricken plant were evacuated. This was an entirely sensible response while the accident progressed and until a reasonable assessment could be made of the likely doses and any other consequences that local people might face.

But as the weeks stretched into months and then years, tens of thousands of people were refused permission to return to their homes. The agriculture and fisheries of the area were severely undermined by a series of actions that one suspects were taken to be 'on the safe side' but which displayed little if any evidence of concern for the effects directly on the lives of the individuals and indirectly on their peace of mind. Though there are small areas of the world which have been evacuated, for example because of the consequences of lead mining,[35] they have received much less attention than the exclusion zone around Chernobyl or Fukushima, despite being considerably more dangerous. It follows, to the rational non-expert, that the levels of radiation throughout these exclusion zones must represent a higher risk than any other man-made threat on the planet.

These actions were entirely out of proportion with the radiological harm likely to befall the population even if one adopts the LNT model. The philosophy of the LNT, so valuable to those early vested interests and their successors – that 'there is no safe dose of radiation' – certainly put heavy pressure on the authorities to overreact. Even so, it seems hard to defend the sheer irrationality of the underlying communication philosophy both after and indeed before the accident.

11.11 ATTITUDE AND 'KNOWLEDGE'

The nuclear industry and its fellow-travellers have traditionally responded to what they perceive as the public's excessive fear of radiation in terms of a mixture of anger, despair and incomprehension. The public is 'irrationally' afraid of radiation and/or badly informed, deficits that must be addressed.

An educated and informed public is something devoutly to be wished. However, the often-heard plea among protagonists in complex issues – that they must 'educate the public' so that they will come to the 'right' conclusion (and therefore allow or even force politicians to take the 'right' decisions) – is unconvincing.

The relationship between knowledge and attitude is far from being a simple linear one. For example, in one study participants were asked to state their attitudes towards carbon capture and storage (CCS) associated with the production of electricity using gas. They were then given some information which was designed to be unbiased and then asked how their views had changed, if at all. While the attitudes of those who had been in favour or against beforehand did not change much, a clear majority of those who had not held a firm view became much less favourable towards the technology, as shown in figure 11.7.

One person's unbiased information is another's propaganda. The very concept of an 'independent' commentator is flawed – everyone has vested interests involving their own reputation, their sense of moral worth (or superiority), their religion or quasi-religion, their employer, their political party or any one of a myriad of other possibilities. It is interesting to note that in these scientific-societal disputes both 'sides'

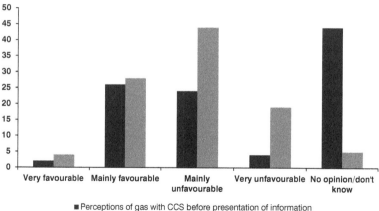

Figure 11.7 Participants' rating of gas with CCS before and after receiving information (%)[36]

are usually passionate in their desire for public education and are highly frustrated at the ill-informed attitude of the population at large. 'If only they understood the real issues they would stop worrying/worry a lot more about fracking/radiation/vaccination/ genetic modification/BSE/prions/phone masts ...' So in the UK the rethink caused by Fukushima should be 'exploited by the nuclear industry to educate the public, show how much safer things are and promote the developments.'[37] In the US executives talked in similar terms.[38] In India it was claimed that government hubris had resulted in there having been 'no attempts to educate the masses on nuclear power'?[39] Big Green made a similar plea, pledging to 'bust myths and nail the facts' about nuclear power as part of its public information campaign.[40]

From time to time attempts are made to 'bring the sides together', to find a set of 'facts' that can be accepted by all shades of opinion as a basis for discourse and policy-making. Such efforts to create authoritative 'referees' have rarely yielded much fruit – it is not necessarily in the interest of either side, and certainly cannot be in the interests of both, for a particular version of 'the truth', with all its inevitable inconvenient complexity, to become accepted. In any case the existence of such a body would not change the fact that the 'professional' supporters and opponents of a technology such as nuclear energy are often both extremely well informed. Furthermore, even when the facts are not seriously in dispute diametrically opposed conclusions can be drawn. One obvious example is the Three Mile Island accident in 1979, cited by opponents of nuclear energy as powerful evidence of the inherent dangers of the technology and by supporters as demonstrating that even in the event of a major accident, releases of radioactive materials were very limited and nobody was hurt, a tribute to the inherent safety of nuclear energy. There is actually not much evidence that pro-nuclear people know more 'facts' about radiation than anti-nuclear people. Some studies show a weak correlation which could just as easily be explained by an assumption that those who are naturally inclined towards nuclear power may be more likely to be motivated to find out something about it. Other studies fail to find a reliable correlation at all. Anti-nuclear campaigners often know a great deal about the issues but draw a different conclusion. In fact it seems that those who are uninterested in nuclear issues have a lower level of knowledge than either people who describe themselves as either pro- or anti-nuclear.

Even leaving aside who gets to decide what 'accurate' information is – not least what should be included and what should not – such an approach brings considerable dangers for those who see it as a way of winning hearts, minds and wallets.

Of course some very basic understanding of the issues is necessary to allow participation in any controversy. However, to demand a highly sophisticated level of public debate leads to a second difficulty. Each of us only has a certain amount of mental 'space' to devote to issues of complex scientific controversy. There may be individuals who can claim to be experts in each one of the dozen (hundred? thousand?) complex issues that might arise in the course of a normal day but if so they are extremely rare. Why should any individual without a direct personal or professional interest in the nuclear debate, say, choose to immerse themselves in this rather than in MMR, BSE, phone masts, genetically modified organisms, road building or any one of the other issues on offer?

Society simply cannot move forward on the basis that all, or perhaps even any, of its citizens must be fully informed and engaged in all such debates. Somehow a path has to be found which can command support among the vast majority who will not be well acquainted with the fine details of every issue. Of course, a more widespread understanding of what science is really about would help in all cases but would not alter the need for each of us to accept somebody else's appraisal of many complex issues.

In fact, what might the cost be of winning that battle for headspace? It is not fanciful to assume that for many people the only way of attracting this level of attention would be to give them a reason why the issue is important – and by far the most likely reason for coming to that conclusion would be because the issue in question represents a threat. The very fact that someone seems so keen to persuade me that what they are doing is perfectly safe is bound to lead me to wonder what is going on. One does not automatically buy the secondhand car from the dealer calling herself 'Honest Jane'.

Three common assumptions

Traditional approaches to risk perception seem to be based on three assumptions which are rarely questioned.

- People get worried because they see things to get worried about.
- Every time something is (publicly) made a bit safer people feel a bit less worried about it.
- Giving people 'accurate' information will make their perceptions more 'rational'.

In fact the truth may be closer to the following.

- We live our lives at a fairly constant level of comfort and discomfort – different from individual to individual but largely independent of the general quality of our lives at any particular time.
- Every time something is (publicly) made a little safer people think it must have been a bit more dangerous than 'they' were letting on before.
- Our perceptions are governed far more by whether we feel we can trust the person giving the message and whether the message is common sense, than by the 'facts' being presented.

Modern Western society is extraordinarily safe by any historical or geographical standards, as measured, for example, by life expectancy. 'In the developed world, life today is a much less risky business, day-to-day, than is the historical or geographical norm. The world has become a terrifyingly safe place. Clean, healthy, predictable, relatively crime-free, children don't die much, most people live to a ripe old age, journeys are no longer a dangerous adventure, epidemics don't sweep us away – life is dull, dull, dull.'[41]

Yet societal concerns about low levels of risk run at a high level, most notably in periods when there seem to be few external threats. 'How extraordinary! The richest, longest lived, best protected, most resourceful civilisation, with the highest degree of insight into its own technology, is on its way to becoming the most frightened.'[42]

It seems that many of us live at our own level of anxiety and unhappiness whatever the external environment. Schopenhauer reflected on this some two centuries ago.

All willing springs from lack, from deficiency, and thus from suffering. Fulfilment brings this to an end; yet for one wish that is fulfilled there remain at least ten that are denied. Further, desiring lasts a long time, demands and requests go on to infinity; fulfilment is short and meted out sparingly. But even the final satisfaction itself is only apparent; the wish fulfilled at once makes way for a new one. No attained object of willing can give a satisfaction that lasts and no longer declines; but it is always like the alms thrown to a beggar, which reprieves him today so that his misery may be prolonged till tomorrow. Therefore, so long as we are given up to the throng of desires with its constant hopes and fears we never obtain lasting happiness or peace.[43]

An alternative view, then, is that, as we each live our life at a fairly constant level of anxiety – perhaps conforming to a 'Law of Conservation of Misery' – and cast around our world for justifications or 'candidate risks' onto which to hang that anxiety, 'risk perception' may be more a matter of finding risks to justify our anxiety than actually being frightened by a particular risk. If a particular risk becomes unattractive as a 'hook' onto which we can hang our anxieties – say the war is won or the economy is improving – we do not cheer up, merely seek out another risk to replace the justification of our anxieties. Paradoxically, times of 'real' threats to safety and security, e.g. wartime, natural disaster, often result in a degree of societal comfort – low suicide rates and levels of industrial disputes, etc. – as a 'solution' can be imagined. Some people of the era used to claim that the years of the Second World War were among their happiest in at least some ways. In his classic study on suicide, Emile Durkheim, 'the father of sociology', noted that suicide rates seem to fall during times of war, which he ascribed to greater social cohesion.[44] Data from England and Wales seem to support the observation, with falls during the periods of the world wars of 1914–18 and 1939–45 (see figure 11.8). Industrial disputes also become less frequent.

By contrast, times of real safety, when life is going well and there is no external threat to bind us together, may lead us to seek more nebulous justifications for our anxiety – justifications that may have no clear 'solutions' and so may give a more disturbing quality to our inherent anxiety. One example is the often noted observation that fear of crime does not seem to be well correlated with actual crime levels and (anecdotally) may even be inversely related. For example, the fear of being a burglary victim did not change between 1984 and 2004 although the burglary rate doubled through that period.[46] Harrow Borough in London, which has the lowest crime rates in the capital, has reported that 'fear of crime remained high in Harrow, a disproportionate reaction when compared to the level of crime in the Borough which was low'.[47]

The question for risk managers, then, may be not whether their activities are causing needless concern among members of the public but is the stance they are taking on risk communication likely to increase or reduce the probability that their activity will be perceived unconsciously as a credible candidate for individuals to use as justification for their background anxiety levels.

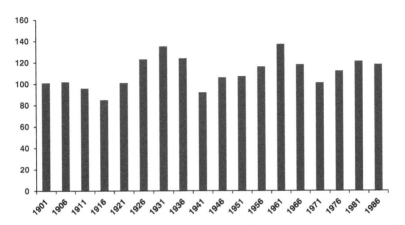

Figure 11.8 Suicide rates per million in England and Wales in five-year periods (starting years)[45]

It is an interesting if as yet unresearched question as to what might make an activity a good candidate in this respect. Possible features (alongside the Slovic factors as discussed later in this chapter) might include:

- nonsensical – the arguments simply do not stand up to common sense;
- high profile – if I am not constantly reminded about a risk I'm unlikely to get worried about it;
- relevance – I can see how this risk might impinge on me;
- few or no perceived benefits – I can get worked up about this risk without feeling guilty about the benefits it brings or having to reconcile myself to feeling an extra degree of discomfort ('cognitive dissonance');
- sense that the representatives of the companies or other bodies involved in the activities are in some sense 'not like us' – for example, overimpressed by the technology at the expense of apparent concern over the human effects;
- 'protesting too much' – spending an awful lot of time telling us not to worry.

Historically, the nuclear industry seems (inadvertently, of course) to have done a great deal to put itself in the frame as a credible societal risk.

The female 'problem'

For some years there has been a clear and persisting gender difference when it comes to attitudes to nuclear power. Women are consistently more sceptical than men. Whatever one's view of nuclear power this is an interesting observation as it is highly unlikely that nuclear power will actually be good for half of the population and much worse for the other half.

The general industry response has veered between a redoubling of the effort to furnish women with the facts so that they will change their mind and greater use of female executives and others to communicate to female audiences. Neither approach has been met with notable 'success' from the industry's point of view. (Conversely one can imagine Big Green's disappointment at the relative enthusiasm which men display towards the technology.)

It seems unlikely that there is any biological reason why men and women should take different views say on the technology of radioactive waste disposal (as they do). There is no evidence that female professional engineers have systematically different attitudes towards nuclear power than do their male colleagues, the same applying to men and women say in careers like the performing arts, whose practitioners tend to be broadly sceptical of nuclear power.

An alternative explanation may go along the following lines. There may be two opposing world views in place. One sees the world of technology as an inherently exciting place and tends to focus on the technical aspects of a particular industry –

perhaps a 'mechanistic' or 'artificial' attitude. The other sees the world more from a human angle, judging technologies not in terms of their inherent characteristics but through a sense of concern for their implications for future generations, the environment and similar matters – a 'societal' and 'naturalistic' attitude. In reality each individual exists somewhere on a continuum between the two extremes rather than falling into one simplistic category or the other. Nevertheless, since in many cultures boys are brought up to value the former – for example, in the kind of toys they are given to play with and the kind of careers they are encouraged to consider – while girls are socialised into the latter, these two approaches map reasonably well onto a male–female or even 'masculine–feminine' division.

The technocentric nature of the nuclear industry has tended to result in an emphasis of the machinery at the expense of its place in society. The message so promoted has at times delivered two impressions: that those who work in the nuclear industry are more interested in the technology and economic–technical issues than they are in the outcome for people; and that the argument for nuclear energy is basically a technical one, i.e. that it is less dangerous than some people think it is. While such arguments work well with others in society who share the mechanistic attitude, it can often backfire with respect to those of the 'natural' disposition, serving to reinforce the suspicion that the advocates of nuclear energy are fundamentally unaware of or uninterested in the potential consequences of their actions on the wider society and the environment.

If this is the case then simply using women to promote the mechanistic argument might be ineffective in developing a proper dialogue on the issues. It suggests that what may be lacking in the whole discussion of energy is a narrative, its place in our lives and the implications of severe shortages as a counterweight to the heavy focus on the downsides of each particular way of generating energy. Of course such a narrative may not favour any particular energy source but it may be an important element in fostering the recognition that however we do it, continued investment in energy sources is of vital importance to the health and well-being of the present and future generations, perhaps especially the most vulnerable members who would face the most severe problems if the energy on which welfare and health systems depend should be compromised.

Buried messages

For many people it is extremely difficult, given the activities and statements of those associated with nuclear safety, to believe that the direct health consequences associated with even quite large releases of radioactive material are actually quite limited. A fairly typical response to Fukushima, for example, was that of the US NRC commissioner, Dale Klein, who chaired TEPCO's Nuclear Reform Monitoring Committee set up after Fukushima: 'The committee's goal is to ensure that TEPCO develops practices and procedures so an accident like (Fukushima) will never happen again.'[48] Similarly, a Nuclear Energy Agency question and answer briefing asked: 'What is being done to ensure an accident like Fukushima can never happen again?'[49]

Use of such language is not entirely restricted to the nuclear industry. However, the response of BP to the enormous Deepwater Horizon oil spill in 2010, for example, seems to have been subtly different. One commentator bemoaned that they could not see 'even a hint [in its statement about the event] that BP is doing everything (or anything) within its power to learn from this horrible spill so that it is unlikely to ever happen again, and if it does, so they will be able to respond more quickly and effectively next time.'[50]

In speaking so much and in such a way about safety – implying that absolute safety is both possible and necessary – the nuclear industry seems to be saying something along these lines:

> We'd like you to believe that radiation and nuclear materials are not particularly dangerous and you are overreacting, but as you'll see from the way we treat them they are actually uniquely hazardous – not just as a matter of degree but as a matter of essence. No other industry drops packages from a great height onto a steel plate set in concrete or talks about burying its waste in mountains or several hundred metres underground. No other industry speaks about permanent or at least long-term exclusion of people from their homes in the case of an accident. No other industry talks about an accident 'never being allowed to happen again'. In other words, a single serious accident would be an extremely damaging event, a real catastrophe, in effect in a different category from any other kind of environmental disaster. But at the same time we will use our superiority and cleverness to try to persuade you that actually what we are doing is quite safe and you can trust it to us.

The public (or some of the public) – rationally – may be responding as follows:

> This just doesn't make sense. Either this stuff is hugely dangerous and a major accident would be a global catastrophe, as your actions and half of your rhetoric suggests – in which case, since I know accidents will always happen, e.g. through human error or because they haven't thought of everything, I should be worried. Or it is ok, as the other half of your rhetoric claims – in which case you are wasting vast amounts of resources that could be going to lower my power bills or increase your profits. Either way, I think this is a pretty good candidate for me to attach my anxiety to, especially as you nuclear folk don't seem to think the way I do or to understand how much I care for the people around me and for my community and environment.

Churchill once said something along the lines that that when confronted by someone with a point of view, first he was influenced by who was saying it, then by how they were saying it, and only then by what they were saying. People largely go on impressions not content – does the messenger seem to understand our concerns, is she or he 'one of us' or are they some rather distant clever dick who just talks about machines and statistics? (Again one suspects there are very good evolutionary reasons for developing a talent for telling if people with whom we interact are sincere and whether they know what they are talking about.) Three weeks after they have attended a talk on a topic outside of their normal field of interest, many people can remember whether it was good and whether they felt they could trust the speaker but may not recall much of what they actually heard. As the old joke goes: 'He didn't look like a conman ... If he'd looked like a conman he wouldn't have been a very good conman.'

Anecdotally, when Christopher Hinton, a founding board member of the UK Atomic Energy Authority, addressed local residents at Thurso Town Hall in 1955 over the establishment of nearby Dounreay for fast reactor research, he responded to a question asking why the site was chosen by saying that the technology was too dangerous to be placed close to population centres. This straight-talking approach, reportedly, did much to allay local suspicions and fears – people felt they were getting the truth even though that truth may not be particularly appetising.[51]

11.12 RATIONALITY AND CORRECTNESS

The nuclear industry has traditionally accused the public of being 'irrationally' afraid of radiation, though as noted before, this only seems to hold specifically with respect to radiation from nuclear power. As also suggested earlier, this seems unlikely.

It is perhaps first worth reminding ourselves of the difference between an irrational belief and an incorrect one. An irrational belief is one that has been formed in contradiction to (or at least without regard to) the available evidence – an incorrect belief may be perfectly rational but based on incorrect assumptions.

When it comes to the charge of irrationality, just because the public (or a considerable proportion of the public) may have a 'mistaken' view of the health effects of radiation, in the sense that their perceptions do not accord with the best available scientific evidence, this does not mean that these views are 'irrational', i.e. have formed without a cause. It is surely irrational to assume that people make their minds up randomly on such issues, especially when so many do so at the same time.

A more sophisticated version of the irrationality theme considers the work done on risk perception by Paul Slovic and his co-workers in Oregon in the late 1970s, who demonstrated that risks of the same absolute magnitude tended to be overestimated if they were regarded as new and unfamiliar, as outside the control of the individual, and as representing a small risk to a large number of individuals rather than a large risk to a few people.[52] Car travel, for example, is familiar, voluntary and identifies its victims clearly, while radiation from nuclear power is regarded as a 'new' and undetectable risk (though as already noted this should perhaps be taken as a reason to be less worried about it), one which is imposed on people rather than seeking their consent, and one which could affect huge numbers of people, albeit representing only a small risk to any particular individual.

While this is a valid analysis, it does not seem to address the observation that there are many cases in which the public seems stoically, if not obstinately, comfortable about radiation exposure – in air travel, medical contexts, high radon areas and so on. One might expect climate change to be in a similar position – we can't directly detect carbon dioxide or its consequences; it is certainly outside the control of any individual or even any country; and it has potentially catastrophic long-term effects on very large numbers of people. However, it does not seem to prompt the level of unease and distress that Slovic's work might lead us to expect. Perhaps some perceived risks are too great to serve the purpose as credible justification for our inherent uneasiness: if they were taken seriously they would engender much more anxiety than is 'required'.

Nor does a simple 'natural or man-made' distinction fully explain fear about radiation. There have been several examples of man-made radioactive material from medical and other sources being released into human environments, as shown in table 11.1.[53]

Table 11.1 Some radiation-related health incidents

Country	Year	Number exposed	Number exposed to high doses	Number of deaths
Mexico (Mexico City)	1962	?	5	4
China (Anhui Province)	1963	?	6	2
Algeria (Setif)	1978	22	5	1
Mexico (Juarez)	1983	≈ 4,000	5	0
Morocco (Casablanca)	1984	?	11	8
Brazil (Goiânia)	1987	249	50	4
Ukraine (Kramatorsk)	1980s	?	17	6
China (Xinhoa)	1992	≈ 90	12	3
US (Indiana)	1992	≈ 90	1	1
Thailand (Bangkok)	2000	?	10	3
UK (London)	2006	?	1	1
India (Mayapuri)	2010	?	8	1

These incidents, despite their severity, did not appear to cause significant or long-lasting radiophobia. When Alexander Litvinenko was murdered in London, for example, polonium-210 could be detected in taxis, restaurants and hotel rooms, yet rarely was there any focus on the radioactivity angle (nor any attendant radiophobia) – it was predominately a KGB story.[54] Even descriptions of the post-mortem as 'one of the most dangerous ever undertaken in the western world' and claims that 'many thousands of members of the public, including British residents and visitors from overseas, might have been at risk from radioactivity', presented at the Inquiry into his death in 2015, seemed not to result in public panic or even notable concern.[55]

In Budapest in late 2011 there was a brief public scare over the detection of iodine-131 in airborne samples, with fears that it might have come from the Paks nuclear plant or another further afield. After investigation, however, it was found that the material had been released over ten weeks from the Institute of Isotopes. At this there seems to have been a collective sigh of relief – it was not the 'dangerous' (nuclear power) type of radioactive stuff but the 'nice' kind, connected in some way with medicine.[56] Similarly, the oft-claimed observation that people are afraid of or concerned about radioactive waste is only partially true – there seems to be no fear (and no anti-nuclear campaign) concerning the production and storage of waste associated with medical (or indeed industrial) uses of radioactive materials.

It may be, then, that it is the 'public' (including the media) who are rational and the industry irrational in communication issues. Here are some examples of where it can be argued that the industry's messages may be having the opposite effect to the desired.

- 'Radioactive waste is not very dangerous – about as hazardous as petrol or paint-stripper – and we are going to bury it 800 metres underground if somebody will let us.'

The industry's irrational belief over many years has seemed to be that people will be reassured by this. Actually, the rational response for the public is more like, 'this is obviously the most dangerous stuff mankind has ever produced (we don't bury anything else 800 metres underground), so we should be scared. And what's more these jokers must think we are idiots if they expect us to believe it is not very dangerous at all, so we won't believe them ever again. Help!'

- 'Ships carrying spent fuel are specially reinforced, triple hulled and have gun turrets to repel terrorists, and if they sank it would not matter as water is a good shield against radioactive material and radiation.'

Again, the industry irrationally believes this will reassure people. In reality, people are very unwilling to believe that such vast expense and effort has been devoted to preventing something that would have no consequences if it did occur. Either the material is in need of very careful handling because it is so dangerous or it can be allowed to sink safely – but, rationally speaking, it cannot be both of these things.

- 'We have spent a fortune on a monitoring system that can pick up radioactivity many thousands of times below danger levels.'

Industry's irrational belief – people will be reassured by this. Public's rational response – this simply cannot be true. Either the industry (or someone working on its behalf) has wilfully wasted a vast amount of my money, so shouldn't be trusted, or it is lying about the dangers involved. Surely nobody would spend a fortune on detecting something that can do no harm.

- 'The industry only emits a few per cent of its allowed releases each year and carries out many tests beyond regulatory requirements.'

Industry's irrational belief – people will be reassured by this. Public's rational response – the regulators are making it easy on the industry by setting such lax limits, we obviously can't trust them to look after our interests. We should be pressurising the regulators to become tougher and do their job properly.

- 'Safety is the top priority.'

Industry's irrational belief – people will be reassured by this. Public's rational response – wow, this really is dangerous. And if safety really is more important than generating electricity or cost, for example, then why not just stop doing it? This is incomprehensible – what do these people really think?

Former BNFL board member Harold Bolter gave a good example.

We tried to counter the public's indignation about the closure of the beaches [near Sellafield in 1983] by showing them how much BNFL had spent on reducing routine discharges of radioactive materials into the sea and how small these discharges now were by comparison with what they had been at their peak. We did so by showing a very small box, representing the current discharges, set inside a very much larger box, representing the earlier discharges, with a suitable explanatory caption designed to show what a splendid job the company had done, with no expense spared. Unfortunately, the research we carried out showed that the public had taken entirely the wrong messages from the ads. Instead of being impressed by the reductions BNFL had achieved, they questioned why there were any discharges to the Irish Sea at all. They also came to the conclusion that the peak discharges must have been extremely dangerous for BNFL to spend so much money on reducing them.[57]

Who is to say that this is an 'irrational' response from the public? Indeed, one could argue the opposite – that anyone who did take the industry's claims at face value may actually be the irrational one.

11.13 OTHER INDUSTRIES' APPROACHES

The contrast with other industries is an interesting one. One observation is that most industries recognise they do not need widespread active public support to thrive – indifference is usually quite sufficient. The almost pathological need to soothe and allay all fears appears unique to the nuclear industry.

The aircraft industry does not take out full-page advertisements (similar to the way the former waste management agency Nirex did over radioactive waste in the 1980s, for example), explaining that if we get on one of their planes it is very unlikely the wings will fall off. The message is more frequently along the lines that if you get on one of their planes they will take you to Tahiti. Again Bolter makes the point.

Immediately after the war redundant aircraft were available for conversion into passenger-carrying planes and the burgeoning travel industry wanted to develop overseas tourism. But people were afraid of flying. The travel industry could have spent its advertising budget trying to convince the public that flying was safe – probably worrying more people than it reassured by doing so – and may even have considered doing so. Instead, it spent its money on showing pictures of the exotic holiday destinations which were being opened up, with their sun-kissed beaches, blue skies and cheap food and drink. It effectively deflected public attention towards these pleasures and away from the dangers. The travel industry's advertising does exactly the same today.

The oil industry, as noted earlier, also seems to have a rather different approach to its nuclear counterpart's. A major oil spill, by almost any realistic yardstick, causes considerably more environmental damage than a major nuclear accident. Yet oil appears to face a rather more manageable regulatory regime and little public clamour for it to be closed down. The differences between the relationship of the oil industry and its accidents and the equivalent relationship with regard to nuclear power may include:

- nuclear power very rarely has large accidents, the oil industry has them on a fairly regular basis, thereby allowing people to see that the environment does recover from such events (this is obviously not an argument for more nuclear accidents);
- the nuclear industry tends to super-comply with regulations, the oil industry is less anxious to go well beyond regulatory requirements (or at least to publicise doing so), with the result that the oil industry regulators do not face criticism that their regime is too lax and easy to meet;
- the oil industry does not speak about 'making sure such events never happen again', the nuclear industry regularly makes such statements;
- the oil industry does not claim that 'safety is the top priority'.

11.14 LESSONS FROM FUKUSHIMA

The technical lessons from the Fukushima accident may be modest. The world no longer builds 1970s nuclear technology any more than it builds wind generators to the designs of the early 1900s (in each case twenty or so years since the invention of the technology). If governments and regulators are up for it, there could be considerable lessons in how to minimise the human misery of major industrial events, especially those involving radiation. However, for this to be achieved the focus must move from a mechanistic 'reduce the level of radiation exposure at all costs' approach to 'do the most good and [crucially] the least harm' when designing and implementing responses to such incidents.

Five years after the Fukushima accident there was no credible evidence of radiation-related fatalities associated with the accident, even among those who stayed on site to mitigate the effects of the accident in the days and weeks after the tsunami, though as noted earlier at least one operator qualified for compensation having developed cancer of 'uncertain' causation. One survey found 57 cases of thyroid cancer among 296,000 children tested but there seemed to be little difference between rates in areas directly around the Fukushima plant evacuation zone and areas further away and researchers were unwilling to associate the cases with events at the Fukushima power plant.[58]

However, there had been clearly observed psychological illness, including a significant number of suicides.[59] Psychological effects also dominated the health consequences of Three Mile Island[60] and even Chernobyl (where there was demonstrable off-site health damage, notably several thousand thyroid cancer cases resulting in over ten deaths).[61] New York psychiatrist Evelyn Bromet reported that more than a decade after the accident, mothers of young children who were evacuated from the Chernobyl area had twice the rate of posttraumatic disorders found among the general population.

Levels of stress clearly affected the confidence of Japanese residents to return to their homes even outside the evacuation zone. A health questionnaire sent to Fukushima residents by Fukushima Medical University showed that about 15 per cent of 67,500 respondents indicated high levels of stress on the Kessler Psychological Distress Scale, against normal rates of about 3 per cent, while 21 per cent scored highly on a checklist used to screen for post-traumatic stress disorder. One resident, whose family left their home 18 miles to the north of the plant, said, 'If it really is safe, I want them to come back. But it's hard to know. Different people say different things, and that adds to my stress. I don't know whom to trust.'[62]

While cruel and opportunistic exaggerations of the potential health effects of radiation (anti-nuclear activist, Arnold Gundersen, saying: 'Fukushima is the biggest industrial catastrophe in the history of mankind') are undoubtedly partly to blame, one suspects that the inadvertent message that had come from the Japanese nuclear industry for many years also contributed. Once again, by making sweeping claims for safety, they not only left themselves open to credibility collapse when the accident did occur, they implied that such an accident might indeed be 'the biggest industrial catastrophe' at least in Japan's history.

The Japanese response to the accident served to reinforce this impression. Most damaging was the prolonged forcible exclusion of people from their homes, long after the accident itself had been brought under control. In addition the Japanese authorities set a level of activity in food at one-fifth of the internationally accepted 'safe' level (itself indefensibly low given the understanding of the health effects of radiation even accepting LNT), and decontaminated groundwater from the hills above the stricken reactors, which had not been anywhere near the plant itself though may have encountered some ground fallout at the height of the crisis, to levels below the international standards for drinking water before releasing the water into the sea.

One suspects that these were intended as confidence-building measures. If so the psychological outcome will have been disappointing. The underlying message appeared to be that international standards of safety cannot be trusted – a particularly unhelpful message when the Japanese government was making much of how it was seeking help from international experts to deal with the problem – and that if even drinking water would be too radioactive to be released into the sea then the damage done to the ocean during the month after the accident, with vastly greater releases of radioactivity into the sea, must be huge and probably irreversible. Once again there seems to have been little reflection on what the rational non-expert would make of all of this.

At the Japan Atomic Industry Forum Conference in April 2014 one of the speakers bemoaned the fact that people in Japan do not know that natural and man-made radiation are the same thing. (The audience laughed at such extraordinary ignorance.) It was obvious to speaker and audience alike that correcting this misconception would put the public mind at rest over the hazards associated with the exclusion zone and by implication smooth the way to Japan's idle nuclear plants reopening.

However, considerable parts of the 20 km exclusion zone round the plant exhibited very low levels of contamination, if any. The 2013 UNSCEAR report into the accident[63] showed that even after one year most of the exclusion zone to the south and south-west of the plant in particular lay outside the 1 mSv per year dose 'contour'. By contrast, areas of Japan with high background radiation, for example the radon-emitting health spa at Misasa, which like its counterparts in the US and elsewhere was long-feted for its health-bringing properties (and where no adverse health effects could be detected), remained free from pressure for evacuation. In such places the background dose comfortably exceeded the total dose (background plus contamination) of almost all of the exclusion zone.[64] On its website the MisasaOnsen Ryokan Cooperative proudly tells the PLJapanese people: 'Misasa hot springs contain one of the highest levels of radium in the world. Radon is a weak radioactive gas produced by the decay of radium. When inhaled into the body, radon improves the metabolism and boosts the immunity and our natural healing power. The fantastic illumination of the Japanese-style lodgings and this "radiation hormesis effect" are familiar to both tourists and people visiting for health treatment purposes.'[65] There are other areas of the world where background doses are dramatically higher than any found in Japan.[66]

So the intelligent non-specialist in Japan was faced with a conundrum. They knew that four years after the accident all but a few of the original evacuees were still prevented from – not advised against but prevented from – returning home. They knew, or could easily find out, that the authorities were saying that total radiation levels in some areas of the zone were lower than natural levels elsewhere in Japan and the world – indeed, nuclear enthusiasts were often making this point. The most rational conclusion to draw would be that natural and man-made radiation must be different. (To put it another way, practically speaking man-made and natural radiation are obviously dramatically different. A dose of the one would result in my being forcibly excluded from my home and having my entire life disrupted and destroyed; the same dose of the other would result in no such hell being visited upon me.)

If these intelligent non-specialists could be persuaded that this was not the case and that natural and man-made radiation are equivalent, then this would not free them for having to find a way of understanding the actions of the authorities. They would be left with two other possibilities: systematic lying by the authorities as to the levels of radiation in the zone; or a cruel and unnecessary destruction of the entire way of life of the exiled residents, by forcing them away from their homes for such a long period and for no significant health benefit. Indeed, if improved physical health were the aim it might have made much more sense to evacuate the population of Tokyo into the exclusion zone, thereby helping them to avoid the airborne pollution that kills over three million people every year in the world's major cities, than the opposite.

It is anyone's guess which they would choose – probably the one about the authorities lying about levels of contamination rather than the other, which is closest to the truth. But whichever it is, it is not immediately obvious that the overall effect would be an easing of public concerns. Far from being irrational, it would seem more likely that the public is employing its logical faculties in an attempt to find a credible rationalisation for what the authorities have actually done.

In this sense the focus on communication techniques and educating the public may be beside the point. If the actions themselves are exaggerated then no clever communication techniques will make the truth seem plausible.

In order to get the issue into a proper perspective and thence to improve society's capability of taking good decisions in a reasonable timeframe, the very concept of 'erring on the side of caution' needs to be reviewed. Whatever the motivation might have been it is impossible to argue that what was done to the population of Fukushima was ultimately in their best interests when all effects, not merely the highly unlikely risk of radiation damage at the doses the action was designed to avoid, are taken into account. One can argue that truly 'erring on the side of caution' would have involved allowing people to choose whether or not they stayed in the area once the initial uncertainty had passed, supporting them if they chose to do so but allowing them to continue their lives as normally as possible. Undoubtedly there would still be psychological stress associated with this course, with people having been trained for many decades to expect a nuclear accident to be more severe than it actually was, but it is hard to believe that the overall outcome would have been worse. The message concerning the true level of hazard would also presumably be beneficial for future events.

A 2012 MIT study questioned the concept of 'conservative' measures. One group of mice was exposed to low-level radiation for five weeks and the effects of this were compared with another group exposed to the same amount of radiation in one burst and a third control group exposed only to normal background levels.[67] The researchers observed that the low dose rate group showed no significant change in the levels of various kinds of DNA damage (base damage, double-strand breaks and induction of a DNA damage response) when compared to the control group. By contrast, the group receiving the dose in a single burst did show such responses. One of the researchers, Jacquelyn Yanch, said: 'You really want to call into question how conservative in your analysis of the radiation effect you want to be. Instead of being conservative, it makes more sense to look at a best estimate of how hazardous radiation really is.' It is only a short step further to ask whether the word 'conservative' is the best one to describe policies which impose such misery on the population for no detectable reduction in radiation harm.

Of course the level of radiophobia which has been nurtured by previous complacency and overreaction stretching over decades would not be conquered at once. But as Einstein said at the very beginnings of the atomic age, no problem can be solved from the same level of consciousness that created it. If the overreaction has indeed been a major cause of public disquiet then this disquiet will not be reversed by ever greater overreaction.

INES

One more lesson from Fukushima concerns the International Nuclear Events Scale (INES), established in 1989 following Chernobyl and described as 'a tool for promptly communicating to the public in consistent terms the safety significance of reported nuclear and radiological incidents and accidents, excluding naturally occurring phenomena such as radon, to be applied to any event associated with nuclear facilities, as well as the transport, storage and use of radioactive material and radiation sources.' Events would be rated on a seven-point scale, broadly analogous to the Richter scale for earthquakes (i.e. each level would be around ten times as serious as the level before). Levels 1 to 3 were 'incidents' and levels 4 to 7 'accidents' (see figure 11.9). On this scale the Windscale fire and TMI were both rated 5, and Chernobyl rated 7.

INES's primary purpose was to: 'facilitate communication and understanding between the technical community, the media and the public on the safety significance of events, aiming to keep the public, as well as nuclear authorities, accurately informed on the occurrence and potential consequences of reported events.' In comparing relatively

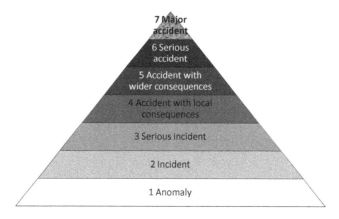

Figure 11.9 International Nuclear Events Scale (INES)

minor categories, up to 'serious incident' or 'accident with local consequences' it has proved useful. However, the idea that by reducing the complexities of a major nuclear event into a single number and that INES would help people to compare the severity of different events and thereby get a handle on what was going on, while appearing superficially attractive, proved impossible to deliver at Fukushima.

The higher levels are defined as follows.

- Level 5: limited release of radioactive material likely to require implementation of some planned countermeasures; several deaths from radiation; severe damage to the reactor core; release of large quantities of radioactive material within an installation with a high probability of significant public exposure.
- Level 6: significant release of radioactive material likely to require implementation of planned countermeasures.
- Level 7: major releases of radioactive material with widespread health and environmental effects requiring implementation of planned and extended countermeasures.

Chernobyl was clearly level 7 on all counts: an exclusion zone is still in place; there were long-lasting food restrictions; over 6,000 cases of thyroid cancer were identified by 2005 including around 15 deaths.

However, when it comes to Fukushima one could make a case for below level 5 – there were no deaths from radiation (the level 4 definition includes 'at least one death from radiation') and indeed even the workers on site during and after the crisis did not sustain sufficient exposure to succumb to acute radiation syndrome. Overall, people in Fukushima were expected on average to receive less than 10 mSv due to the accident over their whole lifetime, compared with the typical Japanese lifetime background dose of 170 mSv.

But one could also argue, on the above definitions, that the event was level 5, level 6 or level 7 (except that there were no widespread health effects from radiation exposure).

When it comes to judging whether the use of INES was 'prompt' and 'facilitated communication and understanding between the technical community, the media and the public on the safety significance of events', then the picture is not a flattering one.

Responsibility for declaring the INES level of an accident lies with the authorities within the country where the event is happening. Japan's Nuclear and Industrial Safety Agency (NISA) originally claimed the accident was at level 4 – deeply implausible very soon after the emergency started as in each of the initial four or five days it seemed that overnight (UK time) something serious had happened to make things look more serious than first thought. There were then two days when the overnight news, if not positive, was at least not getting any worse. At that point, on 18 March, NISA raised the level to 5 (again something that was heavily criticised by the French and Austrians, among others, who on 22 March said they regarded it as a level 7). Naturally the media asked what had happened overnight that the Japanese were not telling them about and which had caused NISA to raise the rating.

Similarly, after two weeks had elapsed during which the situation had been extremely serious – for example, carrying the fear that if one reactor took a turn for the worse it might be necessary to evacuate the whole site and abandon the struggle to control the other units – there was a further two weeks' wait before the news came through on 12 April that the Japanese had raised the accident to level 7.

In effect the authorities in Japan seemed to treat INES like an academic exercise, waiting until they had firm evidence of the state of affairs before making their declaration. Peer review is a powerful tool in normal circumstances when there is no particular rush for a piece of work to be published. Such a luxury is not available during an emergency, when people will be looking not only at the substance of an announcement but the context in which it has been offered.

Secondly, the final assessment according to the risk communicators was that Fukushima was 'like' or 'as serious as' Chernobyl. Once again this was not a helpful impression to give to anyone trying to understand the implications. As noted in chapter 3, Chernobyl resulted in both on-site deaths and off-site health consequences (including some deaths from thyroid cancer) while Fukushima was certainly not associated with any 'discernible' deaths from radiation and it is hard to believe that the dose involved will cause hidden health consequence either, at least off-site. Two hundred and fifty people suffered from ARS at Chernobyl, none at Fukushima. At least five times as much radioactivity was released at Chernobyl as at Fukushima and a much greater area was contaminated. That said, at Fukushima four reactors were in play, at Chernobyl only one.

This attempt to 'make information more accessible' (or 'dumb down', depending on one's viewpoint) appears to be another example of technical rationality working against common sense. How much effort was made to consider how the intelligent layperson would respond to such an approach? The nature of large nuclear accidents is surely that no two will be alike – the next one will not be because of a valve sticking open, a plant being run at a very low temperature or a tsunami – it will be something else entirely which we have not yet really considered. As such a single numerical scale is only likely to exacerbate communication challenges.

11.15 A HELPFUL RETHINK

The radiation and perception problem ultimately does not entirely lie with the LNT theory itself, though undoubtedly the 'no safe dose' argument is a powerful one for those who seek to exaggerate the potential hazards. If the levels of risk which emerge from LNT are treated in a manner commensurate with similar risks from other activities and, in particular, with the full risks associated with taking drastic action in response to modest challenges, then there is no reason why the beneficial uses of radioactive materials, including in power generation, should meet insuperable obstacles.

Such considerations have led to a reappraisal not so much of the science of low levels of radiation exposure but the way it is explained and used. In 2012 UNSCEAR published what amounted to a commentary on LNT as follows.

- Uncertainties at low doses are such that UNSCEAR does not recommend multiplying low doses by large numbers of individuals to produce estimates of radiation-induced health effects within a population exposed to additional doses at levels equivalent to or below natural background levels. The process which led to the oft-claimed '4,000 long-term deaths from Chernobyl' is an obvious example.

- If a population receives long-term exposure at radiation levels typical of the global average background levels (around 1 mSv to 13 mSv per year), health effects should not be attributed to this exposure unless two conditions are met: that spontaneous occurrence of that issue was low while the radiosensitivity of that issue were very high; and that the number of cases was high enough to overcome the inherent statistical uncertainties. So thyroid cancers in the population affected by Chernobyl could reliably be attributed to the accident because it is a rare condition which is known to be associated with iodine-131 which was released in substantial amounts during the accident. However, no other major radiation-related health effect could be.

After Fukushima some US researchers questioned whether the US Environmental Protection Agency 'action levels' for long-term evacuation following a nuclear incident were appropriate.[68] In 2013 the White House Office of Management and Budget completed a review of the Environmental Protection Agency's protective action guidance for radiological incidents. This recommended that clean-ups after nuclear plant accidents or 'dirty bombs' should not have to comply with a single set of public health guidelines, irrespective of the features of the incident in question, established during the 1980s by the EPA Superfund programme. Instead, a principle of 'optimisation' should be adopted which would allow for bespoke remediation standards for a given incident. The report argued that the Fukushima accident demonstrated that abandoning normal EPA standards would be beneficial in some cases, noting that contamination was detectable over an area the size of Connecticut. To remediate such an area to EPA Superfund criteria would be prohibitively expensive for a very modest radiological health benefit.[69]

The challenge ultimately is how to square the use of LNT as a guide to minimising the radiological risk associated with a major release of radiation with an approach which does not create much greater negative effects on the quality of life and peace of mind of those who may be affected. Fukushima surely demonstrates that at present that balance has not been found, that reducing very small exposures to radiation has taken disastrous precedence over the bigger picture. By the end of 2013 reports suggested that the number of deaths among the excluded population caused by the effects of the evacuation, both through the failure to receive proper medical care and stress-related cases including suicides, exceeded the 1,600 deaths which had accompanied the earthquake and tsunami, while no deaths could be attributed to radiation.[70] As one researcher said, the urgent need was: 'not just to prevent illnesses or injuries but even more so to improve the living conditions of the people.' Fukushima's owner TEPCO was subsequently held accountable for the suicide of Hamako Watanabe, who burned herself to death after falling into depression: the company paid almost $500,000 in compensation to her widower.[71]

While few would seek to defend TEPCO, it is nonetheless disturbing to reflect that this family's tragedy was not caused by radiation but by some combination of factors associated with the response to radiation. Tempting as it is to blame heartless anti-nuclear campaigners for creating entirely unnecessary fears, this does not capture the breadth of actions taken over many years by very many agents which have led to this situation.

UNSCEAR chose the phrase 'no discernible increase' to express the idea that currently available methods will not be able to demonstrate an increased incidence in the future disease statistics due to irradiation from Fukushima. It made clear that this phrase did not equate to absence of risk nor did it disregard the suffering associated with any such cases. It is a most useful statement. There are many challenges, not least the direct death rates associated with the use of fossil fuels for power generation and the forthcoming devastation expected to be caused by climate change, which are highly discernible and which judicious use of radiation, through nuclear power, could mitigate and indeed already has done so. A NASA study calculated that by 2013 global nuclear power had prevented some 1.84 million air pollution-related deaths and the equivalent of 64 billion tonnes of carbon dioxide emissions.[72]

A case could perhaps be made that nuclear power is not the best way to achieve these benefits. But the treatment of radiation, and the rational though disproportionate fears it has fostered, is a major barrier to timely and effective decision-making in this field. A radical reappraisal of the whole strategy, based on human rather than technical rationality, is well overdue.

Counterproductive attempts to allay public fears by treating radiation as more dangerous than it clearly is have served both to stoke up further fears and thereby exacerbate the stress-related health effects of a major accident, and to pile further costs onto nuclear energy, making it less able to contribute to global energy goals. Reversing this path will not be an easy step to take.

ENDNOTES

1 http://www.oecd-ilibrary.org/energy/nuclear-power-in-the-oecd-countries_9789264188525-en, IEA (2001), Nuclear power in the OECD.

2 http://www.swissinfo.ch/eng/Home/Archive/Voters_throw_out_nuclear_ban.html?cid=3316614, Bierling B, Voters throw out nuclear ban, *swissinfo.ch*, 18 May 2003.

3 Fagin D (1995), 'Lights out at Shoreham – anti-nuclear activism spurs the closing of a new $6 billion plant', *newsday.com*.

4 http://www.bbc.co.uk/news/world-europe-12822969, Foulkes I, 'Swiss search for strategy on nuclear', *BBC News website*, 23 March 2011.

5 http://www.ft.com/cms/s/0/eda6a298-4fdb-11e0-a37e-00144feab49a.html#axzz1LmKmp73f, Hook L, 'China suspends approval of nuclear plants', *Financial Times*, 16 March 2011.

6 Russell B (1945), *A history of Western philosophy*, Simon and Schuster.

7 Faraday M (1838), 'VIII Experimental researches in electricity – thirteenth series', *Philosophical Transactions of the Royal Society of London*, **128**,125–68.

8 Gowing M (1974), *Independence and deterrence – Britain and atomic energy* 1945–52, Palgrave Macmillian.

9 Condorcet (1795), 'Sketch for a historical picture of the progress of the human mind'.

10 https://www.andrew.cmu.edu/user/jksadegh/A%20Good%20Atheist%20Secularist%20Skeptical%20Book%20Collection/The%20Problems%20of%20Philosophy%20-%20Bertrand%20Russell%20-%20secure.pdf, Russell B (1912), *The problems of philosophy*.

11 See ed. Chant C (1989), *Science, technology and everyday life* 1870–1950, Open University Press.

12 http://www.parliament.uk/briefing-papers/RP99-111/a-century-of-change-trends-in-uk-statistics-since-1900, House of Common Library (1999), 'A century of change: trends in UK statistics since 1900', Research Paper 99/111, parliament.uk.

13 Weart S (1989), *Nuclear fear, a history of images*, Harvard University Press.

14 http://www.gutenberg.org/files/1059/1059-h/1059-h.htm, Wells H.G. (1914), *The world set free*.

15 Frame P (1989), 'Radioactive curative devices and spas', *Oak Ridger*, 5 November 1989.

16 Emsley J (2003), *Nature's building blocks: an A-Z guide to the elements*, Oxford University Press.

17 Works J (1915), 'The public health service … speech … in the Senate of the United States, January 5 and 6 1915', US Government Printing Office.

18 http://www.vintagewatchstraps.com/luminous.php, 'Vintage Watch Straps'.

19 http://www.rst2.edu/ties/radon/ramfordu/pdffiles/The%20Radium%20Girls.pdf, Kovarik W, 'The Radium Girls' in Neuzil M and Kovarik W (2002), *Mass media and environmental conflict*.

20 http://radonmine.com/why.php, *Free Enterprise Radon Health Mine*.

21 http://www.ukradon.org/information/ukmaps, Public Health England (2007), 'Radon areas in England and Wales'.

22 http://www.monbiot.com/2011/04/04/evidence-meltdown/, Monbiot G (2011), 'Evidence meltdown', published in *The Guardian*, 4 April 2011.

23 http://www.epidemiology.ch/history/PDF%20bg/Beebe%20and%20Hamilton%201975%20future%20research%20on%20atmoic%20bomb.pdf, Beebe G and Hamilton H (1975), 'Review of thirty years study of Hiroshima and Nagasaki atomic bomb survivors', *Journal of Radiation Research*.

24 http://www.ons.gov.uk/ons/rel/vsob1/mortality-statistics--deaths-registered-in-england-and-wales--series-dr-/index.html, ONS (2014), 'Mortality statistics: deaths registered in England and Wales (Series DR)'.

25 http://iopscience.iop.org/0952-4746/34/4/E13/article, Coates R (2014), 'Radiation protection – where are we after Fukushima?', *Journal of Radiological Protection*, **34** (4), IOP Publishing Ltd.

26 http://www.solarstorms.org/Threshold.html, Cohen B (1998), 'Validity of the Linear No-Threshold theory of radiation carcinogenesis at low doses', Uranium Institute

27 http://www.ianfairlie.org/news/childhood-leukemias-near-nuclear-power-stations-new-article/, Fairlie I (2014), 'Childhood leukaemias near nuclear power stations: new article'.

28 http://www.nsf.gov/nsb/documents/1997/nsb97186/nsb97186.htm, US National Science Board (1997), 'Government funding of scientific research: a working paper of the National Science Board'.

29 Bush V (1950), 'Science – the endless frontier', 40th Anniversary Edition, Washington, DC: National Science Foundation, 1990.

30 http://scholar.lib.vt.edu/theses/available/etd-09142007-000938/unrestricted/Dissertation.pdf, Selzer M (2007), 'The technological infrastructure of science', PhD thesis, Virginia Polytechnic Institute and State University.

31 http://www.ncbi.nlm.nih.gov/pubmed/24993953, Calabrese E (2014), 'The Genetics Panel of the NAS BEAR I Committee (1956): epistolary evidence suggests self-interest may have prompted an exaggeration of radiation risks that led to the adoption of the LNT cancer risk assessment model', *Arch. Toxicology*, **88** (9) 1631–4.

32 www.bbc.co.uk/news/world-us-canada-29310475?print=true, 'Rockefellers to switch investment to clean energy', *BBC News website*, 23 September 2014.

33 http://www.ncbi.nlm.nih.gov/pmc/articles/PMC2663584/, Tubiana M, Feinendegen L, Yang C and Kaminski J (2009), 'The Linear No-Threshold Relationship is inconsistent with radiation biologic and experimental data', *Radiology*, **251** (1) 13–22.

34 http://www.bbc.co.uk/news/world-asia-34579382, 'Japan to pay Fukushima worker cancer compensation', *BBC News website*, 20 October 2015.

35 http://catastrophemap.org/wordpress/?page_id=759, 'Toxic Apocalypse Hall of Fame', *catastrophemap.com*.

36 http://www.tyndall.ac.uk/publications/journal-article/2011/public-perceptions-ccs-emergent-themes-pan-european-focus-groups-a, Upham P and Roberts T (2011), 'Public perceptions of CCS: emergent themes in pan-European focus groups and implications for communications', *International Journal of Greenhouse Gas Control*, **5** 1359–67.

37 http://www.theengineer.co.uk/home/blog/energy-events-put-spotlight-on-transport-and-nuclear-industry/1008760.article, Ford J, 'Energy events put spotlight on transport and nuclear industry', *The Engineer*, 23 May 2011.

38 http://www.foorumnm.com/news.php?news_id=256119, Reuteman R, 'A case for the comeback of nuclear power – an executive talks about the need to educate the public', Planet-Profit Report, 28 March 2011.

39 http://indiatoday.intoday.in/story/nuclear-power-koodankulam-row-fukushima-nuclear-disaster/1/178531. html, Joshi M, 'No attempts to educate the masses on nuclear power', *India Today*. 20 March 2012.

40 http://www.foe.co.uk/campaigns/climate/issues/nuclear_index.html, Friends of the Earth (2012), 'Five need-to-know facts about nuclear power'.

41 http://www.ukriversguidebook.co.uk/forum/viewtopic.php?t=55947, Toynbee P (2000), *Radio Times*.

42 Wildavsky A (1979), 'No risk is the highest risk of all', *American Scientist*, January-February 1979.

43 Schopenhauer A (1819), tr. Payne E, *The world as will and representation* (Volume 1), Dover Books.

44 Durkheim E (1897), *Le suicide*, Paris.

45 http://www.ons.gov.uk/ons/taxonomy/index.html?nscl=Suicide+Rates, Office of National Statistics (2014), 'Suicide rates'.

46 http://homepage.ntlworld.com/gary.sturt/crime/fear%20of%20crime.htm, Sturt G (2006), 'Fear of crime'.

47 http://www2.harrow.gov.uk/mgAi.aspx?ID=48967, Harrow Borough Council (2009), 'Presentation by Borough Commander to Overview and Scrutiny Committee, 10 February 2009'.

48 http://www.naturalnews.com/037583_TEPCO_Fukushima_catastrophe.html, Huff E, 'TEPCO finally admits catastrophic Fukushima disaster was completely avoidable', *Natural news*, 18 October 2012.

49 http://www.oecd-nea.org/press/press-kits/fukushima-faq.html, OECD/NEA (2012), 'Fukushima FAQs'.

50 http://www.psychologytoday.com/blog/work-matters/201005/bp-why-cant-they-say-they-are-sorry-and-trying-make-sure-it-will-never-happ, Sutton R (2010), 'BP: why can't they say they are sorry and trying to make sure it will never happen again?', *Psychology Today*.

51 http://www.internet-promotions.co.uk/archives/dounreay/doun3.htm, Cashmore S (1998), 'Highland archives, Dounreay 1954 – the fear and the facts', UKAEA.

52 Slovic P, Fischhoff B and Lichtenstein S (1979), 'Rating the risks', *Environment*, **21**.

53 Nénot J C (1993), 'Les surexpositions accidentelles', CEA, Paris.

54 See for example: http://www.telegraph.co.uk/news/worldnews/europe/russia/6615872/Alexander-Litvinenko-A-very-Russian-poisoning.html, Volodarsky B, 'Alexander Litvinenko: a very Russian poisoning', *Daily Telegraph*, 2 December 2009.

55 http://www.telegraph.co.uk/news/uknews/law-and-order/11381789/The-assassination-of-Alexander-Litvinenko-20-things-about-his-death-we-have-learned-this-week.html, Tweedie N, 'The assassination of Alexander Litvinenko: 20 things about his death we have learned this week', *Daily Telegraph*, 31 January 2015.

56 http://www.reuters.com/article/2011/11/17/us-europe-radiation-iaea-idUSTRE7AG1F820111117, 'Hungary

isotope lab likely radioactive source: IAEA', *Reuters*, 17 November 2011.

57 Bolter H (1996), *Inside Sellafield*, Quartet Books.

58 http://www.japantoday.com/category/national/view/57-fukushima-children-suffering-from-thyroid-cancer, '57 Fukushima children suffering from thyroid cancer', *Japan Today*, 26 August 2014.

59 http://www. nippon.com/en/features/c00705/, 'Effects of the Fukushima Dai-ichi nuclear disaster', *Nippon.com*, 27 February 2012,

60 http://www.pddoc.com/tmi2/kemeny/index.htm, Kemeny J, et. al. (1979), 'Report of The President's Commission on the Accident at Three Mile Island'.

61 http://www.ncbi.nlm.nih.gov/pubmed/18049228, Bromet E and Havenaar J (2007), 'Psychological and perceived health effects of the Chernobyl disaster: a 20-year review', *Health Physics*, **93** (5) 516–21.

62 http://www.intelihealth.com/IH/ihtIH/E/333/8014/1481430.html, Foster M, 'Stress emerges as major health issue in Fukushima', Associated Press, *The Washington Times*, 8 March 2013.

63 http://www.unscear.org/docs/reports/2013/13-85418_Report_2013_Annex_A.pdf, UNSCEAR (2013), 'Levels and effects of radiation exposure due to the nuclear accident after the 2011 great east-Japan earthquake and tsunami', United Nations.

64 http://www.ingentaconnect.com/content/tandf/bher/2004/00000010/00000006/art00013, Tanooka H and Sobue T (2004), 'Cancer mortality studies in Misara, a radon hot spring in Japan: a summary up to 2003', *Human and Ecological Risk Assessment*, **10** (6) 1189–94.

65 http://spa-misasa.jp/eng/about/index.html, MisasaOnsen Ryokan Cooperative, 'Welcome to MisasaOnsen Hot Springs' (accessed 20 December 2015).

66 http://www.ncbi.nlm.nih.gov/pubmed/11769138, Ghiassi-nejad M et al. (2002), 'Very high background radiation areas of Ramsar, Iran: preliminary biological studies', *Health Physics*, **82** (1) 87–93.

67 http://www.ncbi.nlm.nih.gov/pmc/articles/PMC3440074/ , Olipitz W et al. (2012), 'Integrated molecular analysis indicates undetectable DNA damage in mice after continuous irradiation at ~400-fold natural background radiation', *Environmental Health Perspectives*, **120** (8) 1130–36.

68 National Council on Radiation Protection and Measurements (2013), 'Decision making for late-phase recovery from nuclear or radiological incidents'.

69 http://www.nti.org/gsn/article/white-house-advances-controversial-nuclear-incident-response-guide/, Guarino D, 'White House advances controversial nuclear incident response guide', *NTI Global Security Newswire*, 2 April 2013.

70 http://japandailypress.com/evacuation-related-deaths-now-more-than-quaketsunami-toll-in-fukushima-prefecture-1841150/, Torres I, 'Evacuation-related deaths now more than quake/tsunami toll in Fukushima Prefecture', *Japan Daily Press*, 18 December 2013.

71 http://www.therakyatpost.com/world/2014/08/26/fukushima-nuclear-plant-operator-found-liable-womans-suicide/, 'Fukushima nuclear plant operator found liable for woman's suicide', *Rakyat Post*, 26 August 2014.

72 http://pubs.acs.org/doi/abs/10.1021/es3051197?source=cen&, Kharecha P and Hansen J (2013), 'Prevented mortality and greenhouse gas emissions from historical and projected nuclear power', *Environ. Sci. Technol.*, **47** (9) 4889–95.

12 | Conclusions – unblock the arteries or wait for thrombosis?

12.1 A historical review – how did we get into this mess?

In the energy world at least, the period between the appearance of the earliest sources of electricity generation – coal, wind, solar, hydropower, wave – in the 1870s to 1900s and the first major exploitation of the newest large-scale source, nuclear power, in the 1960s and 1970s (coming to fruition in the 1980s) can be described as 'the age of the big decision'. There was enormous and rapid development and deployment of new technology – as noted in the previous chapters, electricity use grew at a vast rate in developed countries from the 1880s onwards. By the early years of the twentieth century there was widespread use of hydropower, wind power and solar power – at least until the decisive technical and economic edge offered by centralised electricity production in very large units and national power grids killed off wind, solar and wave in the 1930s – as well as a system which became increasingly dominated by coal and, subsequently and to a lesser extent, oil. Nuclear electricity emerged as a concept in the early 1950s; by the late 1980s it was providing about one-sixth of the world's electricity.

Throughout this period society became hugely richer in a material sense – life expectancy grew, infant mortality rates fell and buying power increased enormously. Despite fears over many centuries that the growing human numbers would spell imminent disaster, the world's much greater population on average lived longer and healthier lives than ever before. In developed countries almost everyone came to enjoy a standard of living beyond the wildest dreams of even the very richest in society before the Industrial Revolution. The spirit of invention, of 'progress', largely dominated public action and private thinking, although as is always the case there was a substantial number of critics of the prevailing ethos. Science and technology took on a quasi-religious aura of infallibility; critics were heretics needing no serious rebuttal. In the 1930s Rudolf Carnap was saying: 'When we say that scientific knowledge is unlimited, we mean "there is no question whose answer is in principle unattainable by science".' In the 1950s Jawaharlal Nehru said: 'It is science alone that can solve the problems of hunger and poverty, of insanitation and illiteracy, of superstition and deadening custom and tradition, of vast resources running to waste, of a rich country inhabited by starving people.'

12.2 A CHANGING 'ATMOSPHERE'

But this paradigm – decision-making protected by the progress imperative and thereby not subjected to proper scrutiny – did not come without cost. Over time at least three problems or potential problems become clear.

First, the 'establishment', be it political or scientific/technical, could not always be trusted either to be honest or to be accurate. One element of the dishonesty of the whole technocratic regime was a tendency to downplay the potential for inaccuracy. It is not within the gift of science to give definitive 'correct' answers to complex questions such as the future of the climate, the health effects of various substances or processes, or the extent of a particular natural resource. Although Lewis Strauss's 1954 speech is mainly remembered for an alleged rosy view of nuclear power, its scope was much wider than that. Even leaving aside that he was apparently referring to fusion rather than fission (he certainly made no reference to nuclear fission), his speech is actually a paean to science and technology as a venture as much as it is to specific instances. 'It is not too much to expect that our children will enjoy in their homes electrical energy too cheap to meter. It is not too much to expect that our children will know of great periodic regional famines in the world only as matters of history and will experience a lifespan far longer than ours, as disease yields and man comes to understand what causes him to age.' Sixty years later and these gifts had indeed largely turned out to be too much to expect.

Secondly, largely untrammelled technological deployment was having growing environmental consequences. The localised effects of human activity had been apparent for centuries but it became gradually accepted that the regional and global environment could not be used as an unlimited dumping ground for noxious substances. Rachel Carson's 1962 book *Silent Spring*, examining the effect of pesticides on bird populations and also obfuscation by the companies involved over what they knew about the phenomenon and its causes, was highly influential in changing attitudes.

Thirdly, even where science had been successful in delivering improvements in material standards of living it was clear that this in itself did not create societies which were happier, once a certain basic standard had been reached. In 1943 Abraham Maslow published his celebrated 'hierarchy of needs' theory, suggesting that our psychological needs can be portrayed as a pyramid (see figure 12.1), each becoming important when, but only when, all those lower down have been met.[1] The scientific/technological revolution had proved highly effective in helping growing numbers of people to fulfil the first and perhaps the second level. It was much less clear to some commentators that the kind of society which emerges from that revolution, governed by what Thomas Carlyle a century earlier had dubbed the 'cash nexus' whereby social values are reduced to economic gain, could deliver on the sense of love and belonging which is more characteristic of societies built on social hierarchies and 'organic connections'.

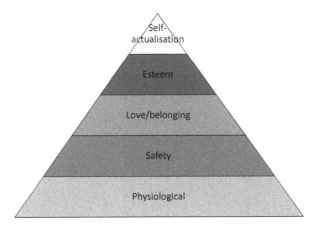

Figure 12.1 Maslow's 'Hierarchy of Needs'[2]

Whatever the merits of these various observations might be, the age of the big decision gradually gave way to a period where decision-making become a much more difficult business as faith in both political and technological establishments declined. Paradoxically, this shift in the way that science and technology were viewed by at least some people was made easier because technology had delivered those very considerable successes in terms of Maslow's first two levels of need. The effective conquest of hunger and famine in developed countries afforded to growing numbers of people the 'luxury' of shifting from a utilitarian ethic, in which difficult decisions can be taken if they benefit the greater number even at the expense of minority who may face an adverse outcome, to a rights-based one in which 'love' and 'belonging' took the front ground. As society went through a long period of enjoying the fruits of science and technology it lost sight of the fact that those fruits depended on the compromises and disadvantages that came with them. Environmental and social detriment could be limited, not least by further technological improvement, but they could never be eliminated. If a sense develops that the benefits of the technological society, not least secure electricity supplies, can literally be taken for granted (it having been so long since they were seriously under threat), then it follows that one is more likely to focus on the negative implications.

In the energy field this second phase coincided with and was perhaps enabled by a sense that the challenge had been met. The 'dash for gas' in the 1990s was certainly a late example of effective large-scale decision-making but one can argue it was hardly 'difficult' (what civil servants might describe as 'courageous' when trying to put their ministers off a particular course of action). It delivered secure electricity supplies relatively cheaply and with the collateral benefit of falling greenhouse gas emissions, at least if one ignored the effects of exporting heavy industry overseas thereby displacing greenhouse gas emissions geographically but not reducing them overall. The change of

UK government in 1997 made very little difference to the roll-out of liberalisation in electricity production. The 2000s saw very little investment in new power capacity, nor was there much of an argument for doing so given the excess capacity that was already in place by the turn of the millennium. The market rules were changed to make it very difficult for anyone to make a decent return on investment in the power industry, the independent power producers went bust, vertical integration and consolidation followed and the seeds were sown for a very precarious outcome.

By the mid-2010s this indulgence had led to a dreadful hangover. On the one hand government was left trying to persuade private companies to carry out public policy at private rates of return, in effect nationalising the risk while privatising the potential profit. On the other, as power prices inevitably rose on the back of tightening supply and generators started to enjoy capacity payments as an inducement to keep their existing plant open, the danger grew that companies would start to make a decent rate of return but still be too afraid to invest it in new capacity. The possible, maybe even likely, outcome had consumers facing both worsening security of supply and rising prices.

In such an atmosphere power generators may invest something in the power source which is 'easiest' in a market context, i.e. CCGT, though probably not enough to maintain healthy dispatchable plant margins, required in case several power stations should break down during a cold snap. Evidence from the UK showed that even with very generous subsidies available companies were unwilling to invest in nuclear or renewables, no new nuclear station having been ordered more than eight years after the then government had said it was necessary and a number of offshore wind farms having been cancelled. A similar paralysis affected gas production – the precipitate decline in North Sea production, coupled with the shortage of storage capacity that had accompanied the era in which the UK could in effect rely on its gasfields as proxy stores, did not serve to encourage the exploitation of what may have been very considerable reserves of shale gas. The UK government seemed powerless to act on its oft-stated policy objective of developing this resource in the face of local objections.[3]

Evidence from Japan after the closure of its nuclear stations in response to the Fukushima accident in 2011 suggests that security of supply will be maintained almost at any cost, at least in the immediate future. Greenhouse gas emissions targets were simply abandoned and the cost of fuel for electricity generation doubled, resulting in Japan running a trade deficit for the first time in over 30 years, but nonetheless there were no unusual power outages except for a few in the area directly affected by the earthquake and tsunami in the weeks following the events. However, although outages may be delayed by some time there does come a point at which decisions to invest in new capacity become unavoidable. By 2015 Japan had announced plans to build 43 new coal-fired powered stations with a combined capacity of over 20GW to replace older, less reliable plant which had been pressed into service.[4]

12.3 REAPING THE WHIRLWIND

So by the mid-2010s the need to take decisions on new plant had become urgent. Yet there seemed little prospect of the kind of investment needed, estimated in the hundreds of billion pounds, being provided in the timescale required. Fewer mistaken decisions were being taken, it was true, but this could be ascribed largely to the absence of decisions of any description. Burgeoning legal requirements for 'consultation' and a growing level of regulation and licensing may have reduced the likelihood of negative consequences of policy being overlooked but they also added hugely to rising costs and created serious delays in implementing necessary decisions. It is difficult to argue that this was a better outcome for the majority of people in society.

This book has looked at some of the underlying reasons for the decision paralysis which has converted a world which could, for example, bring huge amounts of nuclear power capacity online very quickly in the 1980s into one where no progress has been made in reducing the proportion of primary energy coming from fossil fuels since the mid-1990s, in a period during which the total amount of energy being used increased enormously. But can anything be done to counter these developments and create an atmosphere which combines the ability to take tough decisions with a proper degree of scrutiny and societal accountability?

12.4 IT'S THE ECONOMICS, STUPID – OR THE STUPID ECONOMICS

It may seem too obvious to state, but nonetheless … in a competitive marketplace in which electricity policy is determined by commercial entities, economic considerations, i.e. the financial interests of the players, will dominate the decision-making environment. One of the points of introducing competition into power supply was that command-and-control models tended to gold-plate security of supply, indulge in technology-centred flights of fancy and deliver higher costs than necessary. Market discipline brought down the operating costs of plant, stimulated innovation and reduced capacity margins, resulting in falling power costs and in many countries prices too.

But it is illogical in the extreme to introduce private companies into the industry and then complain that they behave like private companies. If society wishes to deliver on non-market goals in electricity generation, in terms of environmental protection and maintaining security of supply, the investors and operators will need to be paid to pursue them.

With most commodities the interests of producer and consumer largely coincide. However, as discussed in chapter 8, the combination of our inability to store large amounts of electricity with the variation in demand through the day and year and the severe consequence for the customer if demand is not fulfilled means that the interests of generator and user diverge significantly. Power cuts are no great tragedy for the power producer – indeed, in some circumstances they can be beneficial, as the California crisis of 2000/1 demonstrated. For the customer they can be devastating.

Perhaps, then, the challenge of finding a single set of policies that both delivers new build when required and promotes efficient operation of existing assets is an impossible one. Unless and until liberalisation has completed a full cycle, returning to a plant mix which is similar in both quantity and age to that which pertained at the start of full liberalisation in 1998/9, it will be impossible to judge this issue or to decide if liberalisation was 'successful'. In the mid-2010s the signs were not good.

There may, however, be a way of sidestepping the matter. If governments are relatively good at safeguarding a healthy capacity margin by ensuring that new plant is built in time and that a suitable mix of different power capacity is maintained, then perhaps that role should be returned to them. After all, as noted previously, the level of political 'engagement' (or interference) in the marketplace in the first quarter of a century of the liberalisation experiment to promote non-market outcomes was profound and persistent. If competitive markets are better at 'sweating assets', operating existing capacity efficiently and effectively, then maybe that role should remain with them. A system might then evolve whereby the government unequivocally retakes formal responsibility for the capacity mix and quantum – something which de facto it never really relinquished despite liberalisation – but once the plant is built it can be franchised out or sold on to private companies to operate within a competitive generating market. (Whether it would then make any sense to persist with the faltering supply market, at least for household consumers, is another matter – EMR already moved some way towards a 'single buyer' model to support the Contracts for Difference regime.) The most obvious political stumbling block to the command-and-control model working in the post-war period in the UK was perhaps the malign power of the mining unions, something which has long been consigned to the dustbin of history.

There are of course flaws in such an approach, as there are with any. But it would at least return us to the days when it was clear what a 'decision' would look like and who was ultimately responsible for taking it, something which in principle could lower one of the barriers to timely decision-making which has blighted the market for most of the twenty-first century.

12.5 MANAGEMENT OF SCIENTIFIC UNCERTAINTY

It is an annoying fact that 'science' cannot offer anything more than well-informed guesswork when it comes to major issues of uncertainty. This book has cited many examples of quite confident predictions from industry and Big Green alike which have not come to pass. If 'too cheap to meter' is worth a good laugh then so are the ridiculous prognostications of the peak oil brigade stretching back over a century, or the environmental doomsayers of the 1970s and 1980s such as Amory Lovins or Paul Ehrlich.

A scientist cannot be criticised if their well-reasoned hypothesis should subsequently prove wrong or inadequate to deal with new data. Certainly any scientist, or worse a party politician masquerading as a scientist, who claims that a scientific debate is over and that any 'heretic' should be banished is both missing the point and threatening the effectiveness of future decision-making.

But this inability to deliver immutable truth does not put science on the same footing as anecdote, religion or cult thinking. It is dangerous to regard science as basically no different from any other human-based source of opinion which is less reliant on objective evidence (and less amenable to potential 'falsification' in the Popperian sense). Just because every scientific viewpoint is provisional and subject to being overturned does not mean that every 'scientific' position is equally well supported by evidence and therefore equally likely to be an effective guide to policy.

Three broad approaches to managing this dilemma may present themselves. Natalie Bennett, leader of the UK Green Party in the mid-2010s, suggested that anyone with a different view from what she, rightly, perceived to be the majority opinion on climate change should be barred from acting as an advisor to government in any capacity.[5] Several similar statements to stifle dispute and investigation have been made from time to time, especially in relation to climate change from both 'sides', including at least one alleged attempt by a climate sceptic to suppress a film examining how some 'climate deniers' appear to have financial links to the fossil fuel industry.[6]

Attempting to silence debate in this way, both by blocking the access that ministers might have to alternative viewpoints and by threatening the livelihood of anyone with a different take on the issue however well argued and supported, may on the surface seem to make firm decision-making easier. However, there are at least two practical difficulties with such a course, alongside any attachment in principle that one might have to the concept of free speech and discourse. First, science proceeds by allowing alternative hypotheses to be offered and tested in an atmosphere of critical openness. Very often, of course, the alternative hypothesis fails but there have been many key examples where either individual elements of an accepted hypothesis have been overturned or even where an entire way of thinking proves wanting, the shift from Newtonian to quantum mechanics being perhaps the prime example in recent times. Science as an enterprise may sometimes need to be dragged kicking and screaming into a new paradigm as the forces of conservative thinking can be deeply entrenched but the weight of evidence generally wins out in the end.

The second difficulty is in deciding whose version of scientific 'truth' should be the accepted one and who gets to decide which voices should be silenced. Were Big Green to follow Bennett's approach on issues in which it swims against the scientific tide – the health effects (if any) of low-level radiation exposure, fracking or genetically modified organisms (GMOs) – it might find itself with very little to say. In practice Big Green will happily cite the worst-argued, worst-supported study with a poker face as long as its conclusions are sufficiently supportive of its dogmatic position on the issue in question. Whereas in the climate debate such dissenting studies are to be rejected out of hand, in other fields they are portrayed as evidence that there is no 'mainstream' view or that that view is wrong. 'Someone has produced a report I agree with' is taken as synonymous with 'I was right all along'; 'someone has produced a report I don't agree with' translates as 'there are dark forces at work which must be discredited'. Big Green follows a thoroughgoing Nietzschean approach to 'the truth' – whether something is true or not is not the question, the key thing is whether it is 'useful' in pursuing its political goals. Again, Big Green is not alone in this: Big Business, Big Green's soulmate in so many ways, also has something of a record of selectivity, exaggeration or underplaying what it deems to be 'scientific' in pursuit of other goals, though in its case those goals tend to be more financial than ideological. Policy, though, ends up following the vested interest of whichever of these undemocratic forces should have the upper hand at a particular time, not the best available evidence and interpretation of the day.

The second broad approach is to treat all viewpoints as equally valid, however well or badly they are supported by more or less 'objective' evidence. In effect this became the norm in the phase 2 society. In the media, anecdote, especially from those directly affected say by an accident but even vox pop from the random person in the street, was very frequently put up against valid scientific research and given like prominence and treatment. 'Expert' groups set up to look into an issue were increasingly dominated by pressure groups and 'stakeholders' with no real understanding of or expertise in the technical issues in question. This attitude has the merit of seeming to be very democratic and ensuring that all voices are heard – underlying the rights-based ethos is a sense that all views have equal validity. However, it neither makes the best use of the prevailing scientific opinion of the day – which must acknowledge the appropriate degree of uncertainty but is not open to all and every interpretation imaginable – nor acts as a spur for decisions to be taken in a timely fashion. Indeed, the very opposite tends to be the case. Interminable 'reviews' coupled with membership sometimes hand-picked not to come to a reasonable conclusion, if any conclusion at all, rarely serve to strengthen the hand of those who see the need to take firm and potentially unpopular action.

12.6 A MIDDLE WAY

So an alternative strategy, a 'third way' to coin a phrase, is needed. That is the easy statement. What form it should take is another matter. However, it should have the following features.

- Peer review must retake its central role in judging the validity of a scientific statement and the way it is communicated. The elevation of anecdote and views which are diametrically opposed to the majority of peer-reviewed studies to the same level as those studies, in some misguided attempt to be 'balanced', must be abandoned but so too must the idea that a view which just happens to be different should be suppressed – as long as it is successfully peer reviewed it should be taken as substantial and worthy of appropriate consideration.

- Publications should be subject to 'accreditation' by a government-appointed body as trustworthy, to avoid the risk of disreputable or incompetent sources self-publishing using their friends and fellow ideologues as their 'referees' and being taken more seriously than their work merits.

- Material which has not been through such accredited peer review should not be banned or rejected out of hand but where it offers a viewpoint seriously out of kilter with the bulk of peer-reviewed material its unrefereed status should always be stated clearly when it is being discussed.

- The media should be involved in developing this approach and encouraged to apply its principles when giving airtime or column inches to different views – i.e. making clear whether the participants in a debate have views and publications which have been through the accredited peer-review process or not, rather than either treating all views as equally valid or seeking entirely to exclude those whose views and work have not (yet) reached the standard of peer review required.

- Specialist committees appointed to consider a matter would be made up of a majority of peer-accredited participants; alternative voices could be heard but their specific expertise (or lack thereof) would be made clear.

So, in the infamous case of Lord (Nigel) Lawson 'against' Sir Brian Hoskins debating climate change and what to do about it on the BBC, as discussed chapter 9,[7] Hoskins would have been described as a peer-accredited source on climate change but not on economics while Lawson would have been given the opposite billing. (There may be a grey area: someone who has held the office of Chancellor of the Exchequer could clearly claim great expertise in terms of the economy without necessarily having published widely on it. Perhaps a category of 'accreditation through experience' should be considered.)

Quite how this might be administered is of course a challenge – ideally it would be democratically accountable but immune from political interference and systematic bias and it is not clear whether both of those goals can be met simultaneously. The closest body to being able to provide this oversight function in the UK might be the House of Lords (as long as it remains unelected) but there is unlikely to be an ideal solution. That of itself, however, should not be taken as a reason to perpetuate the current unsatisfactory situation.

It would be beneficial to complement such an approach with a programme of public education on the philosophy of science. This should not focus on the particular issues associated with a particular dispute – the detail required might well be beyond all but a handful of individuals and the very action of trying to engage with people may lead to suspicions, as discussed in chapter 11 – but as to what science is, what it can and cannot offer and how its findings can best be used in shaping public policy. It would do little harm for these concepts to be more deeply rooted in our political and media establishments as well.

12.7 THE INCONVENIENT BUT UNAVOIDABLE LAWS OF PHYSICS

Nobody could blame the big multinational companies which cream off enormous subsidies from hard-pressed power consumers for spinning their messages in such a way as to maximise the likelihood of further subsidies in the future. It's what multinational capitalism does and on balance technological advances have made a great contribution to human well-being despite the flaws of the institutions that make them.

However, there comes a point where the spin risks getting in the way of good decision-making by creating a false expectation and distorting the true effects of various policies.

A serious debate leading to better decision-making must address a number of myths. These are some of the most important.

- The 'new renewables' myth – that for some reason wind and solar have not been given a fair crack of the whip and their failure to flourish in the mid twentieth century was because of funding starvation and a conspiracy from Big Energy, not inherent characteristics.

This argument is important to Big Green because it is part of its case against subsidising nuclear power is based around a claim that nuclear is 'mature' but renewables are 'new'. In fact, as outlined in chapter 4, solar photovoltaics was invented in 1883 and electricity from wind in 1887. Wind power in particular received heavy research funding and deployment in the early twentieth century but fell out of favour when electricity grids brought reliable power to isolated areas and destroyed the case for localised production. By contrast hydropower grew rapidly and has continued to do so as in effect its electricity output can be 'stored' and released when required by raising or lowering water levels in the dam.

Unless it is accepted that the variable renewables failed nearly a century ago because of their own characteristics, not a lack of support, effective decision-making may be delayed waiting for a renewable future that can never come without a breakthrough in storage technology, which at present seems unlikely. Interestingly, many renewables companies, trade bodies and supporters seem to try to slip in an argument about onshore wind being 'mature' when trying to portray it as a proven and effective technology, only to back off rapidly when asked if they really do support the dropping of all subsidies to allegedly 'mature' technologies, not just ones from which they don't profit or which they don't like.

- The 'grid parity' myth – that the economic effects of renewables can be expressed in terms of the cost of installation divided by the units of electricity delivered and therefore that renewables are not much more expensive than the alternatives – coupled with the 'falling costs' myth, that the costs of deploying renewables fall as the quantity increases because of falling costs of the generating units.

In reality, and as discussed chapter 4, the costs that the variability of renewables imposes on the system as a whole are very considerable; for example, in terms of requirement for a much more extensive grid, the need to compensate dispatchable generators for loss of market, the costs of dealing with excess power when too much renewable electricity is being generated and the greater greenhouse gas emissions per unit of electricity generated caused by operating fossil capacity at low thermal efficiency. These effects become much larger as the penetration of renewables into generation systems grows, resulting in an upwards pressure on costs which may well outweigh any reductions in the cost of manufacturing the generation units themselves. The implications of renewables should be stated in terms of 'system costs' not 'levelised costs', as what is important is the final price that has to be paid by domestic and industrial consumers, not how the costs are allocated to different parts of the electricity supply system.[8] To fail to do so will result in much higher costs, or to put it another way whatever subsidy is available will deliver lower greenhouse gas emissions reductions than it would if used more effectively.

- The energy efficiency myth – that energy efficiency and reductions in energy use are the same thing and are proportional, i.e. a 5 per cent improvement in energy efficiency delivers a 5 per cent reduction in energy use.

As discussed in chapter 5, this is patently simplistic and almost always untrue. In the UK in the mid-2010s there was much concern about the relatively low levels of labour productivity. This was not because there was a hope that if productivity improved the levels of unemployment would rise, as it would take fewer people to create the goods or services in question. The argument was the precise opposite: improvements in labour productivity would improve the UK's international competitiveness, thereby creating more jobs. It is a credit to Big Green spin doctors that they have largely succeeded in keeping the parallel argument out of discussions on the energy productivity of the economy, but by the mid-2010s there were signs that even the IPCC and the European Union were accepting that 'rebound effects' had to be considered when making claims about reduced energy demand based on improving energy efficiency. Future projections of energy use should take these accounts fully into effect.

12.8 Normalising nuclear

Experience of the mid-1970s to the mid-1990s and subsequently suggests that nuclear power is the only major energy technology capable of being deployed fast enough to make a serious difference to global dependence on fossil fuels. Over that period the proportion of traded primary energy from non-fossil sources doubled from about 6 per cent to about 13 per cent where it has stubbornly remained ever since despite the attempts to bring renewables into the mix.

Yet nowhere does the change in the attitude towards taking big decisions show itself more clearly than in the nuclear field. It took less than five years to move from four flickering light bulbs powered by the world's first nuclear electricity, in late 1951, to connecting the world's first commercial-scale nuclear power station at Calder Hall in north-west England to the grid in mid-1956. Between January 2008, when the UK government made clear its desire for a nuclear revival and without any serious political dissent, and the middle of the subsequent decade no new plants were under construction in the UK, let alone had been brought into operation, though a possible funding arrangement had at last emerged. Costs of installation, in real terms, had grown by almost a factor of ten and European projects were taking a decade or more to bring to fruition, though this was not the case in the Far East which had become the centre of gravity for nuclear new build.

By the mid-2010s it was quite difficult to find anyone outside the vested interests of the renewables industry and Big Green, certainly in the UK, who made an environmental or resource case against nuclear power. Even Friends of the Earth said it was abandoning its opposition to nuclear power on those grounds. However, many commentators from the financial world were making a strong case that nuclear power was simply too expensive to build, certainly within the context of liberalised power markets. The 'strike price' reached for the proposed Hinkley Point C nuclear station, an index-linked £89.50 or £92.50 per MWh to run for 35 years plus loan guarantees at a time when wholesale power prices had remained below £50 per MWh for some time, illustrated the point. Yet even such inducements did not result in an early investment decision as project leaders EDF sought financial partners for the £16 billion scheme, described by one respected financial analyst, Peter Atherton, as 'the most expensive object ever built' (or rather not built),[9] warning that the deal could look 'economically insane'.

To stand any chance of regaining its position as the most attractive technological contributor in fighting climate change, nuclear power would need to rein back these costs and demonstrate that it could deliver on its projects, both in terms of time and cost. This in turn would require that nuclear power fundamentally be treated in the same way as other energy sources, for example when it came to dealing with its overall safety record.

As noted in chapter 3, there is nothing inherent in nuclear power that prevents it being built and operated economically, or at least much closer to the costs of fossil fuel alternatives than is the case in Europe today. It was done in the past and it is being done in Asia–Pacific today. The challenge is to create an industry in countries such as the UK that can rediscover that art.

As discussed in chapter 11, the nuclear industry's product-focused and technocratic approach has been a significant contributor to its failure to fulfil its enormous potential in fighting climate change and ensuring security of supply.

Nuclear power has an almost perfect safety record in the West, there never having been a properly supported example of anyone suffering off-site health consequences because of an accident at a nuclear power establishment in the European Union or the Americas. Yet the industry and its supporters seem obsessed with 'improving nuclear safety'.

Of course, any industry learns from mistakes and accidents as part of its normal progress and development, without making wild claims about 'making sure this never happens again'. Nuclear technology would continue to develop as more experience was gained without safety becoming such an obsession. While it might be a source of great theoretical interest to the scientists and engineers involved, it has two deeply unfortunate consequences. First, it adds hugely and unnecessarily to the capital costs of construction. Secondly, far from putting people's minds at rest as one suspects is intended, such talk and the action which accompanies it serves to reinforce the idea that there is something uniquely hazardous about nuclear power and that a major nuclear accident would be in a different category from any other industrial accident imaginable.

There is widespread acceptance and recognition that releases of radiation from two of the three large plant accidents, at Three Mile Island (US, 1979) and at Fukushima (Japan, 2011), are extremely unlikely to have caused discernible health consequences for anyone off site at the time. The third, Chernobyl (Ukraine, 1986), caused several thousand thyroid cancers with over ten fatalities and caused several tens of deaths on site from radiation exposure. However, in terms of the direct health effects all these incidents were middle-ranking industrial accidents of the kind that are not unusual globally, and certainly come nowhere near the death toll and other devastation associated with major hydropower dam failure, coal mining accidents or detriment to air quality, let alone the potential consequences of climate change. There are some who make exaggerated claims for the negative health effects of radiation exposure but one suspects they would fail the Natalie Bennett test of fitness to be heard if they held similar 'off-message' views about climate change.

Yet in their overall effects Chernobyl and Fukushima were indeed human disasters. Evacuation, exclusion and other measures which would not have been contemplated for similar levels of risk associated say with chemical spills caused devastation to the lives of those affected, with, for example, suicide and psychological illness reaching serious levels. This damage was caused not by what happened at the time of the accidents but what happened beforehand, in schooling people that radiation is more dangerous than it is, and afterwards by the measures taken to reduce radiation exposure rather than to reduce detriment.

If one wanted to design an accident preparedness and response system which had as its goal limiting damage to quality of life rather than simply reducing radiation exposure, what might one do?

- Abandon the so-called 'as low as reasonably achievable' principle, which reinforces the idea that very low levels of radiation exposure represent a risk which is in any way significant (if it exists at all).

- Never as a matter of policy pursue goals of reducing exposure below internationally accepted levels, implying that radiation is even more dangerous than current estimates suggest.

- Step by step 'normalise' radiation by treating it as we would other potential health hazards of the same severity – so, for example, if the centres of London, Tokyo, New York or Beijing are not subject to long-term evacuation despite the problem of air quality then neither should the area round any incident involving radioactive material which would represent less of a threat to life.

- Recognise that the damage to the quality of life caused by forcible exclusion is not a constant but is heavily influenced by the nature of the population and area in question, so abandon the concept of formula-driven forcible exclusion beyond a short period of compulsory evacuation while the emergency is brought under control, if necessary.

- Stop producing messages that cannot be true – e.g. that radioactive waste is not terribly dangerous when compared to other toxins that we produce but that it needs much more careful treatment; that artificial radiation is the same as natural radiation but people will be evacuated from areas with very low levels of artificial radiation although not from areas with much higher levels of natural radiation. This requires the industry and its regulators to stop behaving as though (man-made) radiation were more hazardous than it is.

The concept of 'erring on the side of caution' should be critically examined: in practice it has usually been used to mean taking extra measures (to reduce exposure to radiation) rather than taking fewer (to reduce destruction of a way of life and therefore general well-being).

Applying the same overall standards of safety and projected health detriment to nuclear power radiation would lead to major improvements in the value for money that society would gain from investment in energy. This would even more be the case if indeed liberalised markets in constructing generation were abandoned and nuclear (and renewable) projects could be built at public sector rates of return rather than those of the private sector. China has been building what are in effect 'Generation II' nuclear plants at a rate and cost which is reminiscent of the achievements in the West in the mid-1980s. If one takes the view, strongly supported by evidence, that the most dangerous nuclear power station is the one which doesn't get built, the practical alternatives for producing reliable power all being demonstrably much more dangerous, then it may follow that there is a case for banning 'irresponsibly safe' Generation III reactors and returning to the Generation II designs which were deployed so rapidly and successfully in the 1970s and 1980s. A utilitarian world view would almost certainly take this position.

12.9 THE CLIMATE CHANGE CONUNDRUM

Ongoing scientific debate on climate change should be not just tolerated but encouraged. In a sense Big Green is right in saying that the basic scientific debate is 'over'. There is very little, if any, dispute over the notion that carbon dioxide in the atmosphere acts as a greenhouse gas, resulting in a climate warmer than would otherwise be the case, nor that other gases including artificial ones, which did not exist in the pre-Industrial Revolution atmosphere, are exacerbating the effect. The theory goes back to the first half and much of the underlying science to the second half of the nineteenth century. Studies of two locations on the earth's surface have confirmed that increasing carbon dioxide concentrations of around 22 ppm caused increased radiative forcing of around 0.2 Wm^{-2} between 2000 and 2010.[10]

However, there is huge uncertainty about the implications of increasing greenhouse gas emissions and concentrations. The long 'pause' in atmospheric warming from the late 1990s to 2013 was not predicted by the models being promoted in the 1990s. Indeed, as noted in chapter 9, in the 75-year period from 1938 to 2013 the climate was warming during only 25. Some argue that the temperature will rise more slowly than the models imply (and they have the experience of the early 2000s on their side) while others argue that other factors masked the underlying temperature increase which will reassert itself in due course (and may have done so in 2014, a particularly hot year, and 2015, hotter still) or that warming is still happening but has shifted to the oceans rather than the atmosphere.

There is another layer of debate concerning the costs of mitigation against adaptation. Some researchers argue that the costs of reducing greenhouse gas emissions rapidly would be so great, especially for those in developing countries seeking to improve their lives by addressing Maslow's 'physiological' and 'safety' needs, that resources would be better directed to adapting to changes as they arise. Others believe that the imperative is reducing greenhouse gas emissions themselves so rapidly that the global concentration of carbon dioxide does not exceed around 550 ppm, double the level that pertained at the start of the Industrial Revolution: this is effect the pledge given at the Paris summit at the end of 2015. These are genuine issues of enormous importance. Any crude Big Green or Big Industry attempt to silent debate would almost certainly result not only in suboptimal understanding of the nature of the challenge but also risk losing public confidence in the honesty of the climate change proselytes or their opponents. The 'Climategate' leaked emails had precisely that effect. To take the US as an example, the proportion of respondents saying that they were worried about climate change fell from a high of almost 70 per cent in 1999 to just over 55 per cent in 2014.[11]

Dealing with climate change is going to be hugely challenging even in the best possible circumstances. As discussed in chapter 9, in the case of most pollutants it is possible to draw a fairly precise line of causation, both geographically and temporally, between the source of pollution and the damage which results. It is therefore generally possible to create credible estimates of the financial costs both of the pollution damage and of mitigation measures. By applying the 'polluter pays' principle it is at least feasible to introduce market measures or emission limits to address the problem.

This is not the case with climate change. It does not matter where the greenhouse gas is released; its effects will be effectively the same. Much of the damage will affect future generations rather than today's. Different countries might face radically different problems, ranging from complete destruction for low-lying islands through maybe even to potential benefits if permafrost melts and new lands become fertile. As noted above, the degree of scientific uncertainty associated both with the amount of damage that might be associated with different increases in atmospheric greenhouse gas emissions and with the costs that would be incurred to reduce them is considerable. Climate change is by its nature the great prisoners' dilemma of our time – it is (probably) in all countries' interests that everyone acts to reduce greenhouse gas emissions but never in any individual countries' interests to take action. If everyone else is reducing greenhouse gas emissions one country can 'freeload' on this and enjoy lower costs and therefore more economic success, while if everyone else is allowing their emissions to grow a single country would simply commit economic suicide, for no global benefit, by forcing its heavy industry to relocate to cheaper areas which might well have lower overall environmental standards. (If this is not the case and it is actually economic to reduce greenhouse gas emissions then of course there is no need either for government action or indeed for concern – reductions will happen simply through the laws of economics. This does not seem to be the case though, however much Big Green might wish to pretend it to be so.)

The situation is exacerbated by the political agendas of various players. With the exception of a small though growing number of 'true' environmentalists who are prepared to accept uncomfortable truths and their implications if mainstream refereed science reveals them, most players would quite like to reduce greenhouse gas emissions but not at any cost. Governments will accept growing greenhouse gas emissions if there is a perception of high costs or lost votes; electricity consumers will accept higher greenhouse gas emissions if it protects secure supplies; Big Green will accept higher greenhouse gas emissions as a price worth paying for closure of nuclear plants before the end of their technical or economic lifetimes. The planet has very few 'bad weather' friends in this respect.

Japan post-Fukushima suggests that if security of supply is threatened then greenhouse gas emissions will be abandoned very rapidly, with very little protest from the public or from Big Green. Maintaining secure supplies at reasonable cost would appear to be essential if we can properly value the longer-term more controversial challenges of global warming. This would imply a different attitude to the famous energy 'trilemma' – far from needing to 'balance' the needs of the climate 'against' those of secure supplies and the economy, the climate might only be protected if the other imperatives are properly served.

For over 20 years the world has fiddled with climate change policy and rhetoric while the planet has begun to burn, or at least simmer threateningly. While politicians, industry, Big Green and indeed the public all have higher priorities than addressing the one environmental issue which has the potential to threaten the human species, this poor record is likely to persist. In 2014 the world used some 20,000 TWh of electricity; estimates suggest this might reach 50,000 TWh by 2050. Even deployment of new nuclear power at the rate of the mid-1980s, when nearly 65 GW of capacity came online in 1984/5, would result in a nuclear capacity of approximately 1.1 TW in 2050, representing perhaps 8,500 TWh per year or 17 per cent of global electricity use, which is little more than the peak proportion reached in the mid-2000s (though nearly three times those levels in absolute terms). To achieve a similar energy output from renewables, even ignoring the profound problems caused by their variability, would need installation of something in the order of 80 GW of wind or over 200 GW of solar per year over the next three decades. And this is only considering electricity, not entire energy use which may be three times as much as will be provided by electricity alone. Without a single-minded focus and determination to take extremely tough decisions very quickly it seems next to impossible for us even to hold greenhouse gas emissions down to today's unsustainable levels. Perhaps an early climate shock may serve to focus minds but the picture does not look promising and will not do so even if all players put climate change first. And there seems no prospect even of that.

12.10 'TAKING THE POLITICS OUT OF ENERGY'

As (allegedly fictional) senior civil servant Sir Humphrey Appleby says in *Yes Minister*, in the field of controversial public policy it is often a good idea to get the difficult bit out of the way in the title. So a White Paper protecting government secrecy would be called 'Freedom of Information' and so on. As his boss, Sir Arnold Robinson, said, 'The less you intend to do about something, the more you have to keep talking about it – the Law of Inverse Relevance.'[12]

There are frequent calls to 'take the politics' out of energy (or health, or education …) as a way of removing unhelpful argument and implementing sensible policies with which everyone will be happy. This may or may not be desirable – it is entirely unfeasible, at least in a democracy.

Energy infrastructure investment involves timescales of several decades. It would seem bizarre, then, to leave its governance to a competitive market or even to a political establishment that can change its stance radically at or even between five-yearly parliamentary elections. The offered solution usually involves something like creating an 'independent' board for policymaking, accountable to but independent of the government of the day.

However, when it comes to energy as a whole where disputes are often a matter of different views of what would constitute a healthy society as they are of disagreements over 'facts', any attempt in effect to remove the powers of a newly elected government to deliver on its promises to the electorate would be impossible to implement. Furthermore, any policy that is pursued today needs to be able to be adapted to a wide range of possible futures, for example one in which climate change proves to be much less serious than is now thought and one in which it proves to be much more so. In practice then the assumption that a policy is set in stone, say because of parliamentary consensus at a particular point in time, could result in insufficient effort going into preparing to adapt the policy to changing circumstances and increasing knowledge.

One could even argue that the point at which there is cross-party agreement on an issue, as there was, for example, in the mid-2010s over nuclear new build and expanding renewables, is the point where alternative views stop being put forward and orthodoxy can become more important than evidence. Scrutiny and challenge are valuable in science and politics alike.

While society retains a rights-based ethic, and elects parties which reflect this, it is unlikely that the difficult decisions required in energy will be forthcoming. As noted previously, the most likely outcome would be the continuing growth in the use of gas, and perhaps coal in those countries such as Germany which promote its use, for electricity consumption, it being by far the 'easiest' to deliver in a competitive power market. Decisions which properly value security of supply and the environment may become possible when things have reached crisis point, but the Japanese experience post-Fukushima suggests that that day may be a long time coming. When it does arrive it may be very difficult, if not impossible, to implement large-scale alternative policies in time to make a real difference.

12.11 THE BIG BANG APPROACH

The above analysis seems to point towards a single question. If nothing can be done seriously to redress the balance and create an atmosphere in which utilitarian decision-making becomes possible once more until a great crisis befalls us all (probably in terms of secure power supplies but maybe also because of environmental damage); and if a great crisis will probably not occur until the problem has become so severe that no solution could be implemented perhaps for decades; then could anything be done to bring the scale of the devastation to the public eye earlier and so push society into a utilitarian mindset?

It seems a shame to spend a long time writing a book about energy only to finish it with a suggestion which is clearly insane. Of course, the instincts of every scientist and engineer, alongside most politicians and many in the Big Green movement, are to head off disaster for as long as possible except where it would suit their ideological agenda. But if the broad analysis of why decision-making has become so difficult is correct – if at its heart is a society which can't see the point of big tough actions because it can't envisage the future towards which we are inexorably heading – then the opposite approach might be needed.

Should developed countries plan programmes of rolling blackouts to give people a flavour of what is in store, or might be? Is there a danger that such an exercise, could it ever even be contemplated, would result in either being so realistic as to cause very severe hardship, including many deaths and huge economic damage, or so unrealistic as to make people feel it might not be so bad after all? And how could the government of the day ever hope to survive having deliberately put its people through such an ordeal?

Clearly mad. And as Sherlock Holmes tells us: 'How often have I said to you that when you have eliminated the impossible, whatever remains, however improbable, must be the truth?' But Holmes does not tell us what to do when all the options are impossible. The alternative would require some firm agreement among all sides that every effort must go into all possible technical approaches, including fracking and nuclear power (to provide dispatchable power output), electricity storage (which may significantly increase the practical scope for renewables and improve the economics of nuclear power) and carbon capture and storage (for the environment). There would need to be an honest appraisal both of why the variable renewables failed despite the enthusiasm with which they were pursued after their invention over a century ago and of the full system costs involved in managing their variability – phrases like 'new renewables' and 'grid parity' should be abandoned. Industry must accept that using the atmosphere as a free dumping ground for its carbon and other gaseous emissions cannot continue; the rest of us must accept the higher costs that will result. Attacks on the energy companies for excess profits while simultaneously paying them to keep plant open has to go. Either we accept that private or quasi-private entities operating in a marketplace will have to be allowed to make a reasonable rate of return on investment or we return to the government the responsibility for the plant mix, neither of which is a very palatable outcome but both probably better than the recent fudge. We have to be more mature about managing risk and abandon the fruitless and expensive quest for 'perfect safety', especially in the nuclear field, recognising that not only is it impossible, but that it also ends up creating fears rather than laying them to rest.

In short, everyone will have to give up on a considerable proportion of their deepest desires if we are to come close to providing a viable future. And just as a power system which depends on a vast number of small schemes all behaving in the 'right' way rather than the centrally planned system which succeeded it proved unsustainable, so it is extremely unlikely that so many players should make so many changes to their quasi-theological stances in so short a time. Twenty-five years of experience after the Rio Convention serves only to emphasise the point.

Planned and deliberate power outages with all the misery they would bring or everyone focusing on the real issues and accepting both abandonment of their own deepest-held ideologies and some of the views of their opponents. Which of these is the more deliverable? Ultimately – like much in the energy field – that is a matter not of scientific proof but of personal judgment.

ENDNOTES

1 http://psychclassics.yorku.ca/Maslow/motivation.htm, Maslow A (1943), 'A theory of human motivation', *Psychological Review*, **50** (4) 370–96.

2 http://psychclassics.yorku.ca/Maslow/motivation.htm, Maslow A (1943), 'A theory of human motivation'. *Psychological Review*, **50** (4) 370–96.

3   http://www.independent.co.uk/news/uk/politics/lancashire-councillors-decision-to-reject-fracking-a-serious-setback-for-shale-gas-in-uk-10353892.html, Bawden T, 'Lancashire councillors' decision to reject fracking a serious setback for shale gas in UK', *The Independent*, 30 June 2015.

4   http://www.bloomberg.com/news/articles/2015-04-09/japan-s-new-coal-plants-threaten-emission-cuts-group-says, Watanabe C, 'Japan's new coal plants threaten emission cuts, group says', *Bloomberg Business News*, 9 April 2015.

5   http://www.bbc.co.uk/news/uk-politics-26187711, Hawkins R, 'Greens call for clear-out of climate change deniers', *BBC News website*, 14 February 2014.

6   http://www.theguardian.com/environment/2015/mar/11/climate-sceptics-attempt-to-block-merchants-of-doubt-film, Goldenburg S, 'Climate sceptics attempt to block Merchants of Doubt film', *The Guardian*, 11 March 2015.

7   http://www.independent.co.uk/news/media/tv-radio/exclusive-today-programme-criticised-for-giving-platform-to-climate-sceptic-lord-lawson-9582970.html, 'Today Programme criticised for giving platform to climate sceptic Lord Lawson', *The Independent*, 3 July 2014.

8   https://www.pik-potsdam.de/members/Ueckerdt/system-lcoe-working-paper, Ueckerdt F *et al.* (2013), 'System LCOE: what are the costs of variable renewables?'

9   http://blogs.spectator.co.uk/coffeehouse/2014/12/why-is-britain-building-the-most-expensive-object-ever/, Clark R, 'Is the Hinkley C nuclear power station the most expensive object ever built in Britain?', *The Spectator*, 1 December 2014.

10 http://www.nature.com/nature/journal/vaop/ncurrent/full/nature14240.html, Feldman D *et al.* (2015), 'Observational determination of surface radiative forcing by CO2 from 2000 to 2010', *Nature*, **519** 339–43.

11 http://onlinelibrary.wiley.com/doi/10.1002/wcc.321/epdf, Capstick S *et al.* (2014), 'International trends in public perceptions of climate change over the last quarter century', *WIREs Climate Change*.

12 Lynn J and Jay A (1989), *The complete Yes Minister Vol. 1*, BBC: London.

INDEX